DA

Advances in

Marine Biology

Cumulative Subject Index Volumes 20–44

Volume 45

Advances in
Marine Biology

Cumulative Subject Index Volumes 20–44

Edited by

Alan J. Southward
Marine Biological Association
The Laboratory, Plymouth, UK

Paul A. Tyler
Department of Oceanography
University of Southampton, Southampton, UK

Craig M. Young
Oregon Institute of Marine Biology
Charleston, USA

Lee A. Fuiman
The University of Texas at Austin Marine Science Institute
Texas, USA

Volume 45

2003

ELSEVIER
ACADEMIC
PRESS

Amsterdam • Boston • Heidelberg • London • New York • Oxford • Paris
San Diego • San Francisco • Singapore • Sydney • Tokyo

ELSEVIER Ltd
The Boulevard, Langford Lane
Kidlington, Oxford OX5 1GB, UK

First edition 2003

Library of Congress Cataloging in Publication Data
A catalog record from the Library of Congress has been applied for.

British Library Cataloguing in Publication Data
A catalogue record from the British Library has been applied for.

Academic Press   Academic Press
An imprint of Elsevier Science   *An imprint of Elsevier Science*
84 Theobald's Road, London WC1X 8RR, UK   525 B Street, Suite 1900, San Diego, CA 92101-4495, USA
http://www.academicpress.com   http://www.academicpress.com

ISBN: 0-12-026145-6

⊗ The paper used in this publication meets the requirements of ANSI/NISO Z39.48-1992 (Permanence of Paper). Printed in Great Britain.

CONTENTS

CONTENTS OF VOLUMES 20–44

VOLUME 20

The Biology of the Clupeoid Fishes 3

J. H. S. Blaxter and J. R. Hunter

Competition between Fisheries and Seabird Communities 225

R. W. Furness

The Scallop Industry in Japan 309

R. F. Ventilla

VOLUME 21

Recent Developments in the Japanese Oyster Culture Industry 1

R. F. Ventilla

Marine Toxins and Venomous and Poisonous Marine Plants and Animals ... 59

F. E. Russell

VOLUME 22

Assessing the Effects of "Stress" on Reef Corals 1

B. E. Brown and L. S. Howard

Nutrition of Sea Anemones 65

M. Van-Praët

Effects of Environmental Stress on Marine Bivalve Molluscs 101

H. B. Akberali and E. R. Trueman

Growth in Barnacles 199

D. J. Crisp and E. Bourget

VOLUME 23

Population and Community Ecology of Seaweeds 1

A. R. O. Chapman

Aspects of the Biology of Frontal Systems 163

J. LE FÈVRE

The Ecology of Deep-Sea Hydrothermal Vent Communities 301

J. FREDERICK GRASSLE

VOLUME 24

The Biology of Pycnogonida 1

F. ARNAUD AND R. N. BAMBER

Siphonophore Biology 97

G. O. MACKIE, P. R. PUGH AND J. E. PURCELL

Relationships between the Herring, Clupea harengus L., and its Parasites .. 263

K. MACKENZIE

Comparative Physiology of Antarctic Fishes 321

J. A. MACDONALD, J. C. MONTGOMERY AND R. M. G. WELLS

Natural Variations in ^{15}N in the Marine Environment 389

N. J. P. OWENS

VOLUME 25

Predation on Eggs and Larvae of Marine Fishes and the Recruitment
Problem 1

K. M. BAILEY AND E. D. HOUDE

Recent Studies on Spawning, Embryonic Development,
and Hatching in the Cephalopoda 85

S. V. BOLETZKY

Parasitology of Marine Zooplankton 117

J. THÉODORIDÈS

Sandy-Beach Bivalves and Gastropods: A Comparison between
Donax serra and Bullia digitalis 179

A. C. BROWN, J. M. E. STENTON-DOZEY AND E. R. TRUEMAN

VOLUME 26

Biology of the Atlantic Halibut, *Hippoglossus hippoglossus* (L., 1758)...........   1

TORE HAUG

Egg Quality in Fishes   71

B. KJØRSVIK, A. MANGOR-JENSEN AND I. HOLMEFJORD

Diel Vertical Migrations of Marine Fishes: an Obligate
or Facultative Process?   115

J. D. NEILSON AND R. I. PERRY

The Development and Application of Analytical Methods
in Benthic Marine Infaunal Studies   169

BRENDA J. BURD, AMANDA NEMEC AND RALPH O. BRINKHURST

Plankton Production and Year-class Strength in Fish Populations:
an Update of the Match/Mismatch Hypothesis   249

D. H. CUSHING

VOLUME 27

The Biology of the Penaeidae   1–461

W. DALL, B. J. HILL, P. C. ROTHLISBERG AND D. J. SHARPLES

VOLUME 28

Field Investigations of the Early Life Stages of Marine Fish   1

M. R. HEATH

Biology of Living Brachiopods   175

M. A. JAMES, A. D. ANSELL, M. J. COLLINS, G. B. CURRY, L. S. PECK
and M. C. RHODES

The Burrowing Habit of Marine Gastropods   389

E. R. TRUEMAN AND A. C. BROWN

VOLUME 29

Turbulence, Phytoplankton Cell Size, and the Structure of Pelagic Food
Webs   1

T. KIØRBOE

Autotrophic and Heterotrophic Picoplankton in the Baltic Sea 73

J. KUPARINEN AND H. KUOSA

Spermatophores and Sperm Transfer in Marine Crustaceans 129

T. SUBRAMONIAM

The Bristol Channel Sole (*Solea solea* (L.)): A Fisheries Case Study 215

J. HORWOOD

VOLUME 30

Meiobenthos of the Deep Northeast Atlantic.. 1

M. VINCX, B. J. BETT, A. DINET, T. FERRERO, A. J. GOODAY, P. J. D. LAMBSHEAD,
O. PFANNKUCHE, T. SOLTWEDEL AND A. VANREUSEL

The Biology of Oniscid Isopoda of the Genus *Tylos* 89

A. C. BROWN, F. J. ODENDAAL

Social Aggregation in Pelagic Invertebrates.. 155

D. A. RITZ

An Appraisal of Condition Measures for Marine Fish Larvae 217

A. FERRON AND W. C. LEGGETT

The Biology of Seamounts 305

A. D. ROGERS

VOLUME 31

Hybridization in the Sea 1

J. P. A. GARDNER

Reproductive Biology of Marine Cladocerans 79

D. A. EGLOFF, P. W. FOFONOFF AND T. ONBÉ

The Role of Microscale Turbulence in the Feeding Ecology of Larval Fish ... 169

J. F. DOWER, T. J. MILLER AND W. C. LEGGETT

Adaptations of Reef Corals to Physical Environmental Stress 221

B. E. BROWN

Harmful or Exceptional Phytoplankton Blooms in the Marine Ecosystem...... 301

K. RICHARDSON

VOLUME 32

VOLUME 33

VOLUME 34

The Effects of Fishing on Marine Ecosystems   203

S. JENNINGS AND M. J. KAISER

A Biogeographical Perspective of the Deep-Sea Hydrothermal
Vent Fauna   355

V. TUNNICLIFFE, A. G. MCARTHUR AND D. MCHUGH

VOLUME 35

Population Genetics of Bathyal and Abyssal Organisms   3

S. S. CREASEY AND A. D. ROGERS

Growth Performance and Mortality in Aquatic Macrobenthic Invertebrates ...   153

T. BREY

VOLUME 36

The Biochemical Ecology of Marine Fishes   1–325

G. E. SHULMAN AND R. M. LOVE

VOLUME 37

The Assessment of Marine Pollution – Bioassays with Bivalve Embryos and
Larvae   3

E. HIS, R. BEIRAS AND M. N. L. SEAMAN

Population Structure and Dynamics of Walleye Pollock,
Theragra chalcogramma   179

K. M. BAILEY, T. J. QUINN II, P. BENTZEN AND W. S. GRANT

VOLUME 38

The Enhancement of Marine Fish Stocks   1

J. H. S. BLAXTER

The Biology of Pandalus   57

B. I. BERGSTRÖM

VOLUME 39

The "Exxon Valdez" Oil Spill in Alaska: Acute, Indirect and Chronic Effects on the Ecosystem 3

C. H. PETERSON

Reproduction and Development of Marine Peracaridans 107

W. S. JOHNSON, M. STEVENS AND L. WATLING

Remote Sensing of the Global Light-Fishing Fleet: An Analysis of Interactions with Oceanography, other Fisheries and Predators 261

P. G. RODHOUSE, C. D. ELVIDGE AND P. N. TRATHAN

VOLUME 40

The Parasite Fauna of the Atlantic Cod, *Gadus Morhua* L. 1

W. HEMMINGSEN AND K. MACKENZIE

Biology of Mangroves and Mangrove Ecosystems 83

K. KATHIRESAN AND B. L. BINGHAM

Structural, Histochemical and Functional Aspects of the Epidermis of Fishes 253

G. ZACCONE, B. G. KAPOOR, S. FASULO AND L. AINIS

VOLUME 41

Interactions between Phytoplankton and Trace Metals in the Ocean 3

MICHAEL WHITFIELD

The Sea Cucumber *Holothuria scabra* (Holothuroidea: Echinodermata): Its Biology and Exploitation as Beche-de-Mer 131

JEAN-FRANÇOIS HAMEL, CHANTAL CONAND, DAVID L. PAWSON AND ANNIE MERCIER

VOLUME 42

Protobranch Bivalves 1

JOHN D. ZARDUS

Shelled Opisthobranchs 69

PAULA M. MIKKELSEN

The Scaphopoda 139

PATRICK D. REYNOLDS

Pleurotomarioidean Gastropods.. 238

M. G. HARASEWYCH

VOLUME 43

Ecology and Biogeography of Marine Parasites 3

KLAUS ROHDE

Fecundity and Life-history Strategies in Marine Invertebrates....................... 87

EVA RAMIREZ LLODRA

Ecology of Southern Ocean Pack Ice... 173

ANDREW S. BRIERLEY AND DAVID N. THOMAS

Biological and Remote Sensing Perspectives of Pigmentation in
Coral Reef Organisms.. 279

JOHN D. HEDLEY AND PETER J. MUMBY

VOLUME 44

A Synthesis of Growth Rates in Marine Epipelagic Invertebrate
Zooplankton 1

A. G. HIRST, J. C. ROFF AND R. S. LAMPITT

Biology of Early Life Stages in Cephalopod Molluscs................................... 143

S. V. BOLETZKY

Movements of Marine Fish and Decapod Crustaceans: Process,
Theory and Application.. 205

S. J. PITTMAN AND C. A. MCALPINE

Culture of Harpacticoid Copepods: Potential as Live Feed for Rearing
Marine Fish 295

C. J. CUTTS

CUMULATIVE SUBJECT INDEXES
FOR VOLUMES 20-44

Abdomen (*Pandalus*) **38**: 63
Abiotic habitat factors in benthic
    studies **26**: 226, 227
Abiotic sources of variability **30**: 240
Abnormalities **33**: 36
    chromosomal **33**: 48
Abnormality of bivalve larvae at end of
    bioassay **37**: 55, 56, *57, 58*, 59-62
Absolute yield, sole **29**: 311-315
Absorption cross-section,
    photosynthesis **29**: 16
Absorption efficiency, *Donax* and *Bullia* **25**:
    226, 228, 229, 231, **231**, *233*, 238
Abundance
    and environmental gradients **26**: *189*
    and spatial distribution **26**: *193*
    in faunal data matrices **26**: 178, 180, 181
    of Atlantic halibut eggs **26**: *10*
Abundance (*Pandalus borealis*)
    38: 208-210
Abundance behaviour, dynamics of
    36: 123-137
Abundance distribution and biomass **26**: 210
Abundance of parasites **43**: *26*
    see also niches
Abundance of seaweeds
    dominance estimates **23**: 61
    regulatory factors **23**: 3, 4
    relationships at frontal systems **23**: 263,
    264
    relative **23**: 13
        in mixtures and monoculture
        23: 13-16
Abundance *see* Population biology
Abundance, predation **25**: 41, 42
Abundance, *see* Population
Abyssal and hadal zones **32**: 325-387
    Arctic Ocean **32**: 354-358
    Caribbean Sea **32**: 359, 360
    defined **32**: 323, 324

extent of faunal ranges **32**: 327-331
hadal (ultra-abyssal) fauna **32**: 72-74
    geographic distribution **32**: 367-373
macrobenthos, trophic structure
    32: 427-525
Mediterranean Sea **32**: 360, 361
near-bottom communities, macroplankton
    distribution **32**: 74-78, **77**
regions of World Ocean, faunal
    similarities **32**: 336-347
vertical zonation and geographic
    distribution **32**: 331-336
World Ocean **32**: *348*
zonation schemes **32**: 336-354
Abyssal environment **35**: 6, 7, 8-16,
    114, 115
    see also Deep-sea environment
Acantharia
    parasites of **25**: 125, 126, 127, 129
    symbionts **25**: 124
Acanthocephala, parasitic **25**: 151, 152
Acarids, hydrothermal vent dwelling **23**: 331
Acartia sp **36**: 36
Acartiidae **44**: 51-72
Accession Treaty 1972 **29**: 331
Accidental transfers **38**: 46
Acclimation, ecological principle of **36**: 234,
    235
Accumulation and production of substance
    and energy **36**: 142-162
Acetylation test **29**: **154**
Acetylcholine **21**: 177
Acidic mucopolysaccharide (AMPS),
    crustaceans **29**: 168, 199
Acoustic backscattering, fish eggs and
    larvae **28**: 15, 16
Acoustic sampling **43**: 189
Acoustic stimuli, larvae, predation **25**: 16,
    18, 19
Acoustic techniques **44**: 263

Acoustic telemetry data **44**: 261

Acoustico-lateralis system **20**: 137, 140, 145

Acquired characteristics of pollock **37**: 213, **214**, 215

Acrosome, crustaceans **29**: 180
  reaction **29**: **188**, 189, 190

Acrylic acid **43**: 206, 207

Actinomyxidea, parasitic **25**: 144, 146, 158

Activity
  budgets, *Donax* and *Bullia* **25**: 211–214, **213**, **217**, 218, 219
  larvae, predation **25**: 14, 23

Activity of clupeoid fish **20**: 91–93
  diurnal **20**: 93
  rhythmic **20**: *92*

Actograph **20**: 93

Acts of Parliament, fishing **29**: 323, 324, 331, 332

Actual fecundity **43**: 92

Acuity **20**: *135*, 138

ACW (Antarctic Circumpolar Wave) **43**: 174, 186, 240, 241

Adak Island, pollock near **37**: 217, *219*

Adaptation **36**: 1, 2, 7–58
  see also Genetics; Salinity; Temperature
  ecological principles **36**: 236, 240
  food and feeding **36**: 47–58
  oxygen level **36**: 28–43
  pressure and buoyancy **36**: 43–47
  strategies **36**: 59–87
    biological progress **36**: 59
    energy metabolism, divergences of **36**: 59–74
    plastic metabolism, divergences of **36**: 75–87

Adaptation and evolution, forces of **36**: 237–245
  energy optimum and structural functional homeostasis **36**: 243–245
  environmental factors **36**: 237–240
  motor activity **36**: 240–242

adaptation to temperature **36**: 7, *11*, 15

Adaptation, phytoplankton **29**: 16, 18

Adaptation, to background **27**: 188, 190, 192, 199–201

Additive variance **23**: 37

Adenosine triphosphate (ATP) **33**: 249, 250

Adenosine triphosphate, crustaceans **29**: 139

Adhesion, crustaceans **29**: 174–176, *177*, 178, 189, 199

ADONIS **43**: *189*, 190

Adriatic Sea **36**: 17, 30, 39, 48; **42**: 37

Adsorption systems, parasites **25**: 126, *128*, 131

Adult activity budget **20**: 238

Adult sea cucumber
  see also anatomy
  daily burrowing cycle **41**: 178–180, *179*
  feeding behaviour **41**: 157, 182–185
  habitat **41**: *150, 151*

Adults, as prey **27**: 359
  growth **27**: 240–250
  life-cycles **27**: 284–288
  migration **27**: 297, 312, 313
  morphology **27**: 7–45
  mortality rate **27**: **357**
  taxonomy **27**: 58–106
    key to families of Penaeoidea **27**: 58–66
    key to genera of Penaeidea **27**: 61–63
    keys to species of Penaeidea **27**: 66, 68, 72–78, 80–83, 85–87, 89, 90, 92, 93, 95–98, 103–105

Adults, sole
  see also Reproduction
  distribution **29**: 246, 247, *247*
  migration **29**: 248–251, 271, 272
  mortality **29**: 278, 295–298
  seasonal patterns **29**: 247, 248
  'stock' **29**: 249–251, *251*
  temperature **29**: 249

Advective and diffusive dispersal processes **28**: 41–45

Aerial photography **44**: 264

Aesthetascs **33**: 18, 23

Africa, coastal areas of: parasites **43**: 31, 53, 56
  diversity **43**: *10, 12*
  uses **43**: 64, 69, 70

Age
  see also body size; eggs; juveniles; larvae
  and maturity of Atlantic halibut **26**: 7, 8
  and ontogenetic variation in fish **26**: 150
  and size of Atlantic halibut **26**: 23, *24*, 25, 29–31
  of bivalve egg at beginning of test **37**: 51, 52
  of bivalve larva and toxicity reaction **37**: 118, 119

Age and fecundity of invertebrates **43**: 131–134, 152
  see also demography

at first maturity **43**: 134–136, 152, 153, 191
Age and growth in *Pandalus* **38**: 182–203
Age specificity **30**: 271–278
Age, predation **25**: 2
Age, seaweed
  class structure **23**: 39–44
    environmental variations **23**: 40, 41
    interpretational considerations **23**: 39, 40, 41
    irregular distributions **23**: 41
  growth ring method of estimation **23**: 39
Age, sole
  fecundity **29**: **284**, *285*
  harvesting **29**: 299–303, *300*, **301**, *302*, 309–311
  maturity **29**: 218, 275–277, *276*, 278, 279
Ageing and death **36**: 210, 211, 222, 249
  after spawning **36**: 102, 103, 125, 127, 237
  life cycles **36**: 90, 99, *100*, 120, *121*, 122–127, 130
    molecular and metabolic aspects **36**: 101–104
  substance and energy **36**: 159, 160
Aggregation, phytoplankton **29**: 3, 37–40, *39*, **40**, *41*
Aggregation, prey **25:** 53
Aggregations **33**: 424–426, 428
  and theoretical behavioural ecology **30**: 160–162
  as inappropriate strategy **30**: 197, 198
  benefits of **30**: 181–183, 191–199
  costs of **30**: 183, 184
  definitions **30**: 158
  density within **30**: 169–172
  exchange between **30**: 174, 175
  general features of **30**: 162, 163
  genetic relatedness in **30**: 186
  mathematical models of **30**: 175
  multispecies **30**: 174
  multispecies aggregations **33**: 429–432, **430, 431**
  protection from predators **30**: 191
  proximate **30**: 160
  sorting within **30**: 172–174
  ultimate **30**: 160
Aglantha sp **36**: 36
Agriculture, Fisheries and Food, Ministry of (MAFF), 350
Air, exposure to *see* osmotic shock
Airborne sensors *see* remote sensing
Akpatok Island **20**: 228

Alaska **42**: 272
Alaska pollock **36**: 15
Alaska *see* Gulf of Alaska
Albacore **20**: 62
Albacore and match/mismatch hypothesis **26**: 263
Albatross **20**: 295
  black-browed **20**: 295
  grey-headed **20**: 295
Alcian blue test **29**: 136, 154, **155, 156,157**
ALD (approximate lethal dose) **21**: 64
Aldehyde fuchsin test **29**: **155, 156**
Aldrin toxicity and bivalves **37**: 97, **103**
Aleutian Basin, pollock in **37**: 179, *180*, 184, 187
  population dynamics **37**: 194–199
  population structure
    genetic **37**: 216, 217, 219, *219*, 221, 223
    mechanisms **37**: 232–234
    metapopulation **37**: 225, 226, 228
    phenotypic **37**: 210, 211, 212, 213
  stocks **37**: **185**
Aleutian Islands, pollock near **37**: **185**
  population dynamics **37**: 194, 195, 196, 199
  population structure **37**: 216, 217, 219, 221, 223, 232
Aleutian Trench, hadal fauna **32**: 369, 370
Aleutians, biotic interactions **23**: 77–80
  kitchen midden animal analysis **23**: 80, 81
Alewife **20**: 34–41
  osmoregulation in **20**: 160
  passim **20**: 93, 124–126, 175
  spawning of **20**: 22, 24, 55
Algae **23**: 24, 54, 55; **36**: 37, 53
  see also Seaweeds
  barnacle growth and **22**: 219
  blue green **23**: 56, 104
  competition between **23**: 13–17
    encrusting and foliose forms **23**: 22
  coral infection **22**: 48
  depth distribution effects **23**: 142
    chromatic adaptation theory **23**: 142
  drought recovery **23**: 139
  elimination by predators **23**: 73
  encrusting forms
    brown **23**: 22, 116, 139
    competition between **23**: 116
    red **23**: 22, 54, 116

foliose forms **23**: 22, 104
    upright **23**: 87
grazer co-evolution **23**: 122
grazing pressure
    diversity relationship **23**: 77–87
    dominant communities established
        with **23**: 113
    midshore establishment **23**: 98, 99
    numerical classification **23**: 48, 49
        intertidal/subtidal **23**: 54–56
    percent cover-herbivore density
        relationship **23**: 97, 98
    physiognomic classification **23**: 62–65
    recruitment, interactions determining
        adult/juvenile **23**: 30
        effect of red algal turf **23**: 30, 31
    red turf **23**: 30, 31, 83
    reef **23**: 101–108
        distribution **23**: 101, 102
        exclusion **23**: 106
    sciaphilous, association of **23**: 68
    'sociations' **23**: 62
    species succession **23**: 114–116
        models **23**: 114
        role of physical disturbance **23**: 115,
            116
    spores, *see* Spores
    submergence/emergence **23**: 131, 132
    substrate colonization **22**: 19
    tidal and subtidal limits **23**: 127
    tidal pool **23**: 89
    zonation **23**: 111, 112, 116, 117
Algae, and plankton **26**: 251
Algae, Baltic Sea **29**: 87
    see also Eukaryotic algae
    distribution **29**: 87–89, *88*
    grazing **29**: 104, 105
    nutrients and temperature **29**: 101–104
    seasonal variation **29**: 90–92, *92*, 94, 98
Algae, *Bullia* **25**: 191, 237
Algae, toxins **21**: **71**, 86
    blue green **21**: 87
    green **21**: 92
    red **21**: 93
    yellow-green **21**: 90
Algae/dinoflagellates and bivalves **37**: 6
    bioassay procedures **37**: 64, 65, 66–68,
        69, 70, 72, 73
    larval rearing in laboratories **37**: 26, 38, 39
    toxicity **37**: 125, 126
Algal blooms and toxins, fish egg and larval
    mortality **28**: 79

Algelasine **21**: 100
Alimentary canal **33**: 39–41, *40*
Alimentary system, pycnogonid **24**: 17, 18
Alimentary tract *see* digestive system
Alimentary tract, sole **29**: 252
Alizarin red S stain
    coral growth **22**: 21, 23
Alkaline phosphatase **36**: *153*, 154
Alkaloid toxins **21**: 67
Allele frequency estimates **23**: 33
Allelic phylogenies method **35**: 57
Allochthonous organisms and sea ice **43**: 196
Allometric growth **33**: 338
Allometry hypothesis **43**: 120, 121
Allophycocyanin **43**: *283*
Allozyme electrophoresis **35**: 6, 28–30,
        31–42, 43, 112
    deep-sea fauna **35**: 30, **31–39**
    hydrothermal vent and seep fauna **35**: 30,
        40, 41, 43
    seamount fauna **35**: **42**, 43
Allozyme frequencies, sole **29**: 250
Allozyme studies of pollock **37**: 207, 215,
        216, **218**, 222, 223, 237
Alpha granules, crustaceans **29**: 175, 176,
        179
Alternative prey, predation **25**: 15, 27, 36,
        37, 58, 61
Aluminium **41**: *12, 122*
Aluminium toxicity and bivalves **37**: 87, 94
Alvinellidae
    development and dispersal **34**: 411
    distribution **34**: 410, 411
    gene flow **34**: 411, 412
    phylogenetic relationships **34**: 412, 413,
        413
Ambassiadae as predators **27**: **362**
Ambush predation **25: 5**, 12, 15, 23, *24*, 26
American fisheries
    Northeast coast **38**: 222
    Northwestern coast **38**: 224, 225
American shad,
    activity in **20**: *92*
    batch fecundity of **20**: 21
    decline in fecundity of **20**: 22
    demersal spawning in **20**: 24
    osmoregulation in **20**: 160
AMERIEZ (Antarctic Marine Ecosystem
        Research at Ice Edge Zone)
        43: 216, 217, 219–221
Amersham International **29**: 350
Amine sensitivity, *Donax* and *Bullia* **25**: 193

Amine toxins **21**: 67
Amines, posterior salivary glands,
  Hapalochaena **21**: **187**
Amino acids **33**: 249
  in sea anemones
    feeding activation **22**: 67–69
    metabolism **22**: 71, 72
    uptake **22**: 70–72, 87
Amino acids in fish eggs **26**: 83
Amino acids, crustaceans **29**: 139
Amino acids, dietary and ^{15}N **24**: 429,
  430
Amino acids, molluscs
  excretion **25**: 229
  sensitivity **25**: 192, 193
Amino acids, toxins, *Cerebratulus* **21**: **135**
Aminophenols, crustaceans **29**: 187
Ammonia excretion, molluscs **25**: 229,
  230, 231
Ammonia *see* Nitrogen/ammonia
Ammonia, excretion of **27**: 181
Amnesic shellfish poisoning (ASP) **31**: 310,
  311
Amoebophryidae, parasitic **25**: 126
Amoy (China) **43**: *12*
Ampharatid, hydrothermal vent
    dwelling **23**: 327, 328
Ampharetidae
  development and dispersal **34**: 414
  distribution **34**: 414
Amphidromes **23**: 168
Amphipoda
  eggs, prey **25**: 16, 29, 43, 47
  parasites of **25**: 135, 136, 137, 146
  parasitic **25**: 154, 156, *155*
  predators **25**: 8, 12, 16, 36, 38
Amphipoda, growth performance **35**: 173,
    174, **177**, 178
  see also Crustaceans
Amphipods **20**: 52, 53; **29**: 185, 253; **44**: 33
Amphipods, hydrothermal vent dwelling
    23: 329
Amphiporine **21**: 134
Amphizone **23**: 129, 130
Ampoule, crustaceans
  Anomura **29**: 146, *147*, 150, *151*, 152,
    159, 183, 197
  Macrura **29**: 164, 169, 171
Amundsen Sea (Antarctic) **43**: 185
Amylase
  in sea anemones **22**: 80, 83, 86, 87

α-Amylase activity, *Donax* and *Bullia*
    25: 192
Anadromous clupeoids **20**: 11
Anaerobic metabolism **35**: 27
Anaerobic metabolism, crustaceans **29**: 137,
    139
Anaerobic metabolism, *Donax* and
    Bullia **25**: 208, 210, 211, 213,
    223, 224, 229, 237
Anal fin, sole **29**: 221, **222**
Analytical methods in benthic
    studies **26**: 169–247
Anaphases, delayed in fish eggs **26**: 85
Anatomy **33**: 37–48
  see also Morphology; Reproductive
    system
  circulatory system **33**: 39
  digestive system **33**: 39–41, *40*
  endoskeleton **33**: 38
  excretory system **33**: 46
  muscular system **33**: 38
  nervous system **33**: 38, 39
  oil sac **33**: 47, 48
Anatomy and physiology
  Pleurotomarioidean, external *Plates 5 and
    6* **42**: 249–251, *249, 251*
    see also structure and morphology
      under Pleurotomarioidean
      gastropods
  Protobranch bivalves **42**: 4–17, 6, *10, 12*,
    30, 31
  shelled Opisthobranchs,
    external **42**: 91–94
anatomy of sea cucumber **41**: **140–148**
  see also morphology
  auto-evisceration and regeneration
    41: 142, 143
  internal **41**: 140–142
  ossicles **41**: 143–145, *144*
Anatomy, sole **29**: 217
  see also Morphology
Anatoxin-a **21**: 87, 88
anchoveta **38**: 3, 45, 47
Anchovy **36**: *108*; **20**: 17, *31*, 39, 40, 69,
    107, 133, 250–252, 263
  adaptation **36**: **60**, **77**, 78
  food and feeding **36**: 48, *49*, 53, 55,
    56–58
  temperature **36**: 17, 18
Argentine **20**: 38, 55
Azov **20**: **71**, 73
Bay **20**: 26–30

big-eye **20**: 28, **68, 70**
Black Sea **20**: **71**, 74
Californian **20**: 8
deep-bodied **20**: 131
differentiation and variability **36**: 224, *225*
ecological principles **36**: 230, 238, 242, 243, 247, 248
eggs and larvae, *see also* Fish eggs and larvae
field surveys, egg production **28**: 25–29
indicators of condition **36**: 217
lipids **36**: 206–211, **207**
Japanese **20**: 53, 55, **72**, 83, 113, 140
large-toothed **20**: 34
life cycles **36**: 101, 115
abundance, dynamics of **36**: 126, 127, 129, 130, *131*, 132
annual **36**: 105, *106*, 107, *108, 109, 111*, 112–114
interannual fluctuations **36**: 118, *119*
longnose **20**: 27
muscle fibres in **20**: 87
northern **20**: 8–78 *passim* **20**: 131, 133, 143, 246
passim **20**: **44**, **70**, **72**, 76
passim, **145**, *146*, *149*, **155**, **156**, *158*, **160**, **161**, *197, 198*
consumption, transformation and utilization **36**: **179**, 179–196
expenditure **36**: *164*, **164**, 164–174
passim, **166**, **168**, *169, 172, 173*, **174**
passim, *180, 181*, **185**, *187, 188*, **192**, **193**
trophic significance **36**: *199*, 200
Peruvian **20**: 3, *4*, 8, *9*, 32–60 *passim* **20**: 259, 264, **266**, 267
South African **20**: 35, 36, 240, 247
spawning peak in **20**: 10
substance and energy **36**: *144–145*, 144–162
Anchovy larva **20**: 32, 89
Anchovy, spawning **26**: 262, 263
Ancillary genital segment, crustaceans **29**: 175
and endogenous rhythms in fish **26**: *117*, 118
Andaman Islands, sea cucumber in **41**: 136, 153, 154, 155, 183, 191
Andaman Sea, fisheries **39**: 268–289, 276
androgenic gland **38**: 133
Anecdysis **27**: 213

Animal movement
data analysis **44**: 259–262
implications for conservation and resource management **44**: 268–271
patterns **44**: 256, 257
quantification **44**: 247–259
Animal Movement Analyst Extension (AMAE) **44**: 260
Animal–environment linkage, theoretical constructs of **44**: 236–245
Animal–environment relationships
application and tools **44**: 245–268
operational framework **44**: 245–247
Animalia, parasitic **25**: 146–157, *149, 155*
Annelids **21**: 137; **29**: 199, 253
Annelids, hydrothermal vent dwelling **23**: 326–328
Hirudinea **23**: 328
Polychaetes, *see* Polychaetes
Annelids, parasites of **25**: **152**
Annual cycles **36**: 2, 101–114, 127
see also Reproduction; Substance and energy
ecological principles **36**: 229, 232, 238, 239
indicators of fish condition **36**: *211*, 226
substance and energy **36**: 141–144, 147, 148, 156, 157
Annual fish **36**: 103
Annual variation, picoplankton **29**: 95–97, *96*
Annular gilthead/bream **36**: 25, 190
adaptation **36**: 64, 86
oxygen level **36**: *31, 33*, 34, *35*
pressure and buoyancy **36**: 44
Anomura, spermatophores **29**: **132**, 145, 146, 197
chemical composition **29**: 153–159, **154–158**
hardening **29**: 186
morphological diversity **29**: 146–148, *147*, 149–152, *151*
origin **29**: 148, 149, 152, 153, 178
transfer and dehiscence **29**: 159, 160
Anostraca **29**: 199
ANOVA approach **37**: 84, 221, 222
ANOVA, *see* Variance, Analysis of
Anoxia *see* Hypoxia and anoxia
Anoxia, Brachiopods **28**: 330
Anoxic water, Baltic Sea **29**: 97
Antarctic **36**: 107, 196; **43**: *50*
see also Southern Ocean sea ice

adaptation **36**: 8, 10, 13, 61, 73, 78
Bottom Water **43**: 176
Circumpolar Current, Southern Boundary
    of **43**: 230, 231
Circumpolar Wave **43**: 174, 186, 240, 241
ecological principles **36**: *233*, 236
invertebrates **43**: 116, 135, 141, 143,
    147–149 *passim*
Sea Ice Processes and Climate group
    43: 174
Antarctic Bottom Water (AABW) **30**: 20, 21
Antarctic fishes
    adaptive radiation of **24**: 333, 334
    buoyancy of **24**: 333, 334, 376
    continental drift and evolution of
        24: 323–325
    fishing techniques **24**: 328, 329, 332
    kinds of **24**: 328, 330, 331
    laboratory maintenance of **24**: 332, 333
    scientific importance of **24**: 325
Antarctic intermediate water (AIW)
        32: 154
Antarctic marine ecosystem, food web
        for **20**: 272
Antarctic Ocean
    bipolar distributions of abyssal
        fauna **32**: 365–367
    ice-water interface **33**: 493, 494
    polar and subpolar regions **32**: 24–29, *26,
        27*
Antarctic penguin, creching in **20**: 231
Antarctica
    Protobranch bivalves **42**: 34, 36, 37,
        40, 45
    Scaphopoda **42**: *170*, 272
Antarctica, fossils **27**: 148, 149
Antenna **33**: *18*, 19
    see also Swimming appendages
Antenna, morphology **27**: 11, 12, *11*, **47**, 48,
    49, 50, 49, 51, 52
antennae **38**: 62
Antennal glands, morphology **27**: 37
Antennata **29**: 197
Antennules **33**: 17–19, *18*; **38**: 62
    see also Swimming appendages
    copepodids **33**: 27
    feeding function **33**: 140
Antennules, morphology **27**: 10, 11, *10*, **47**,
    48, *49, 51*
Anterior mantle of Scaphopoda **42**: 191, 192
Anthomedusae, parasites of **25**: *128*, 131
Anthopleurin-A **21**: 125

Anthropocentric perspective of
    habitat **44**: 235, 236
Anthropogenic impact on community in
    benthic studies **26**: 190–192
Antibiotics and bivalves **37**: 42, 43, 54, 63
Antifreeze, biology of **24**: 376
    mode of action **24**: 337, 338
    molecular structure **24**: 335, 336
    renal retention **24**: 339
    synthesis **24**: 338
Antilles *see* Guadeloupe
Antimora sp **36**: 44
Antipredation and other activities **30**: 196,
    197
Aomori Aquaculture Center **20**: 310, 340,
    341, 359, 367, 378
Aplacophora **23**: 324
Aplysiatoxin **21**: 90, 181
Aplysin **21**: 180
Aplysinol **21**: 180
Apodemes **33**: 38
Apodinidae, parasitic **25**: **126**
Apolysis *see* Moulting
Apparent fecundity **43**: 92
Apparent Looming Threshold (ALT),
    predation **25**: 16
Appendages **33**: 15, *16*, 16, 18
    see also individual appendages
    feeding appendages **33**: 140–143
    nauplius **33**: 23, 24
Appendages, crustaceans **29**: 182
Appendages, morphology **27**: 7–18, **47**,
    48–52, *49, 51*
Appendicularia, parasites of **25**: 126, 127,
    128, 129, 133, 134, 143
Appendicularian tunicates **44**: 37
Appendicularians, feeding by **23**: 230
Appendicularians, role in plankton
    communities **32**: 64–68
Appendix musculina, crustaceans **29**: 140
Apterygotes **29**: 148, 199
Aquaculture **27**: 6
Aquaculture of Atlantic halibut **26**: 54–63
aquaculture of sea cucumber **41**: **196–201**
    see also fisheries
    brood stock collection and
        maintenance **41**: 197, 198
    polycultures **41**: 200, 201
    rearing **41**: 198, 199
    stock enhancement **41**: 201
Aquaculture *see* mariculture *under*
    exploitation

Aquaculture, crustaceans **29**: 189–191
aquaculture, *Pandalus* and **38**: 226, 227
Aquatic ecosystems, diversity in **26**: 199
Aquatic Environmental Monitoring Reports,
 MAFF Directorate of Fisheries
 Research, Lowestoft
 29: 350
Aquazone **23**: 129, 130
Aquiferous spaces, *Bullia* **25**: 200, 205, *205*,
 207
Arabian Sea **29**: *49*; **35**: 24, 25, 83; **36**: 30,
 39, 240; **43**: 56
Arachidonic acid (ARA) **44**: 303, 304, 309
Arachnids **29**: 198
Aral Sea **43**: 69
Archaean oceans **41**: *13*
Arctic
 parasites **43**: 51, 69
 sea ice **43**: 173, 179, 183, 236
  trophic processes **43**: 200, 214, 216, 218
Arctic and Arctic Ocean **36**: 8, 46, 79, 107,
 226
Arctic area **42**: 272
 Protobranch bivalves **42**: 29, 35, 37, 40
Arctic Ocean **32**: 19–22; **41**: *126*
 abyssal fauna distribution **32**: 354–358
 bipolar **32**: 365–367
 circular ranges **32**: 361–365
 ice-water interface **33**: 493, 495
 location of plankton stations **32**: *19*
 subarctic gyres **32**: 546
Arctic Oceans, *Pandalus*
 distribution **38**: 113, 114
Arctic skua **20**: 282, 285, 286, 293, 294
Ardtoe, plaice enrichment **38**: 15
Area temporalis **20**: 131, 133, 134, 135
Arendal Symposium on Sea Ranching
 (1993) **38**: 21
Argentina, clupeoids of **20**: **180**
Argentina, coastal areas of: parasites **43**: *11*,
 34, *50*
Argentine anchovy **20**: 38, 55
Argentine Basin **42**: 39
Argonautids, integument **25**: 104
Ariidae as predators **27**: **362**
Aristaeidae **27**: 3, 58, 115, 116, 118, 190
Arms, Cephalopoda **25**: 99, 100
Arrhenius activation energy (E_A) **24**: 358
arsenic **41**: *12, 120, 122*
Arsenic toxicity and bivalves **37**: 87, 88,
 90, 91
Artemia nauplii **26**: 53

Arthropods **29**: 130, 148, 178, 189, 197, 198;
 21: 139
Arthropods, hydrothermal vent dwelling
 23: 328–331
Artificial fertilization, Cephalopods **44**: 151
Artificial fertilization, of scallop
 20: 324–326
artificial incubations **44**: 41
Artificial insemination, crustaceans
 29: 189–193, **192**
artificial reefs **38**: 30–32
Artificial seawater quality and bivalves
 37: 43, **44**, 136
Ascomycetes, plankton **25**: 121
Ascorbic acid in fish eggs **26**: 81, 82
AseevSamyshev trawl, macroplankton,
 sampling **32**: 7, 8
Ash content **33**: 233–235, 235
Ash-free dry weight **33**: 233
Askö station, Baltic Sea **29**: **92**, 98, **99**
Aspartate, phytoplankton **29**: 34
ASPeCt (Antarctic Sea Ice Processes and
 Climate group) **43**: 174
Assemblage models **26**: 194
Assemblages *see* Spatial distribution
Assessment *see* Status, stock
Assimilation (see Digestion and assimilation)
Assimilation efficiency **33**: 205, 206, **206**
 net growth efficiency **33**: 205–207, **208**
Assimilation efficiency, *Donax* and
 Bullia **25**: 191, 192
Assimilation index **29**: *51*, 52
Associations **33**: 460–470
 diversity **33**: 461–464, **462, 463**
 dominance of species **33**: 464–466, *466,
 467*
 niches **33**: 466–470
Associations of Scaphopoda **42**: 210–216,
 211, 212
associations of sea cucumber with other
 species **41**: 149, 150, 151, 188,
 189
Assortative mating **31**: 2
Astacidea **29**: **132**, 136
Astaxanthin **33**: 251; **43**: *291*, 292
Asterids
 sea anemone predators **22**: 92
Asterosaponins **21**: **153, 165**
Asymmetry **33**: 36
Asymptotic length (L), 5, 20
Asynchronic spawning **20**: 11
Asynchronous oogenesis **43**: 89

Atlantic cod parasites **40**: 1–59
  and marine pollution **40**: 47–49
  as biological tags, population studies of
        cod **40**: 42–47, **43**
  Baltic Sea **40**: *46*
  checklists **40**: **3–9**
    see also Taxonomic index
    additions/deletions **40**: 10, 11
  host range **40**: 11–21
    cod and other gadoids **40**: 11–14
    cod and other groups **40**: **13**, 14–19
    cod only **40**: 11
    infestations considered accidental
        40: 19–21
  host specificity **40**: 21–24, **23**
  life cycles **40**: 24–29
    cod as definitive host **40**: **27, 28**
    cod as second intermediate host
        40: **25, 26**
  pathology **40**: 49–58
    farmed cod **40**: 54–57
    wild cod **40**: 49–54
    wild/farmed cod, interactions **40**: 57,
        58
  population dynamics **40**: 34–41
    age **40**: 34–38, **36**
    long-term changes **40**: 40, 41
    seasonal variations **40**: 38–9
  sexual differences in infections **40**: 41, 42
  zoogeography **40**: 29–34
    Arctic–Boreal species **40**: 30–32
    Boreal species **40**: 32, 33
    low Arctic species **40**: 30
    species of worldwide distribution
        40: 33, 34
Atlantic croaker, vertical migration
      of **26**: 121, 141
Atlantic fisheries **39**: 271, 272, 277, 286,
      266, 275
Atlantic halibut
  and pollution **26**: 42
  and salinity **26**: 53, 55
  aquaculture of **26**: 54–63
  biology of **26**: 1–70
  diseases of **26**: 41, 42
  distribution of **26**: 5, *6*, 7
  egg development of **26**: *12*
  exploitation of **26**: 42–53
  fisheries of **26**: *43*
  identity of **26**: 3–5
  immature phase of **26**: 22, *23*, 24–27
  larvae of **26**: 14, *15–18*

mature phase of **26**: 27–38
  parasites **26**: 38–41
  pelagic phase of **26**: 11–22, *20, 21*
  rearing of **26**: 53
  reproduction **26**: 7–11
  sexual maturity of **26**: *8*
  species status **26**: 2–4
Atlantic herring **20**: *4, 5,* 20–64
  daily ration of **20**: **70**
  demersal spawning in **20**: 23
  egg size in **20**: 15
  eye in **20**: 133
  fat content of **20**: 74
  fecundity in **20**: 21
  gut clearance rate in **20**: **68**
  oxygen uptake in **20**: 64
  passim **20**: 68, **72**, 74, 85, 133, 144
Atlantic herring larva
  feeding in **20**: 32
  mortality rate in **20**: **59**
  skin removal in **20**: **57**
Atlantic menhaden **20**: 5, 15, 21, 34, 36,
      66, 144
  egg size in **20**: 15
  fecundity of **20**: 21
  respiration rate of **20**: *66*
Atlantic menhaden larva **20**: 32
Atlantic Ocean **35**: 8, 22, 84; **36**: 217
  see also Eastern Atlantic; Southern
        Atlantic; Western Atlantic
  adaptation **36**: 3
    food and feeding **36**: *50, 51,* 52–55, 58
    oxygen level **36**: 36, 38
    pressure and buoyancy **36**: 45, 46
    salinity **36**: 21, 22, 26, 28
    strategies **36**: 61, 79, 80
    temperature **36**: **17**
  calcareous sediments **32**: 471, 481
  deep-water bottom fauna, taxonomic
        differentiation **32**: *337*
  differentiation and variability **36**: 221,
        224, 226
  ecological principles **36**: 231, 246, 249,
        251
  fisheries **38**: 221–224
  hydrothermal vents **35**: 17, 18, 20
  International Commission for Northwest
        Atlantic Fisheries (ICNAF) **28**: 30
  invertebrates **43**: 96, *97*, 143, 147
  life cycles **36**: 115
    abundance, dynamics of **36**: 126, 127,
        129, 130, 133, 134, *135, 136*, 137

annual **36**: 107, 114
  interannual fluctuations **36**: *121–123*
  ontogenesis **36**: *93*, 98, 102, 103
  macrobenthos, abyssal, trophic
      areas **32**: *478*, 479–484, **484**
  mid-Atlantic Ridge vents **43**: 96, 143
  Mid-Atlantic Ridge, hydrothermal vent
      sites **32**: 93–107
  North
    cross-section, bottom invertebrates
        distribution **32**: *338*
    plankton distribution **32**: 22–24
    Pandalus in **38**: 106–108, 113, 114
    parasites **43**: *50*, 51, *52*, 53, 59, *60, 61*
      diversity **43**: 4, 6, 9, *10–14, 18, 19*
      ecology **43**: 24, 34, 47
      uses **43**: 69, 70
    Protobranch bivalves **42**: 4, 32, 34, 39,
        40, 45
    salinity **36**: 26, 28
    Scaphopoda **42**: 147, 205, 207–210
    See also Mid-Atlantic Ridge; North
        Atlantic
    South **42**: 39, *273*
  substance and energy **36**: 149, 150, 152,
      153, 180, 181
  survey grids **28**: **31**
Atlantic Ocean, trace elements in
  and phosphates **41**: *126, 128*
  cadmium **41**: 72, *73, 75*, 76, *126*
  cobalt **41**: 77
  concentration **41**: *120–123*
  copper **41**: 63, *67, 124*
  indices of fractionation **41**: 80, *81, 82*
  iron **41**: 43, *44, 45*, 56
  nickel **41**: 78, *79, 128*
  silicon **41**: *124, 125, 128*
  vertical concentration profiles **41**: 9, 43,
      44
  zinc **41**: 67, *68, 125*
Atlantic salmon **38**: 5, 41
Atlanto–Scandian **20**: 121, 127, 174
Atlanto–Scandian herring **20**: 121, 127,
    174
Atlanto–Scandian spring spawners **20**: 127
Atmozone **23**: 129, 130
ATP (Adenosine triphosphate),
    spermatophores **29**: 139
ATP **34**: 10
ATP concentrations in seamounts **30**: 312
Atrazine-simazine toxicity and
    bivalves **37**: 66, 97, **104**

Atresia, sole **29**: 283, 286–288, 339, 341
Attachment disc, crustaceans **29**: 180,181
Attachment organs of parasites **43**: 41–44,
    43, 44
Attachment systems, parasites **25**: 126, *128*,
    134, 135, 156
Attack inhibition **30**: 192
"Attractin" (pheromone) **43**: 104
Auditory systems, larvae, predation **25**: 16,
    18, 19
Auk **20**: 231, 286
Auklet, Cassin's **20**: *244*, 246, 247
  rhinoceros **20**: 246
Australia **35**: 84–86, 109
  Pleurotomarioidean gastropods **42**: 272,
      273
  Protobranch bivalves **42**: 28, 35, 37
  Scaphopoda **42**: 146, 204, 205
Australia, coastal areas of **43**: 228
  see also Great Barrier Reef uses **43**: *63*,
      70
    western **43**: 4, *10*
    parasites **43**: 35
      distribution **43**: *50*, 53, 55, *56–58*,62
      diversity **43**: 4, 7, *10–12, 14, 15*, 16,17
      New South Wales and south-east **43**: 4,
          7, *10–12*, 14, *15, 50*, 53, *56–58*,
          62, *63*, 70
      north, north-east and north west **43**: 7,
          56–57
    pigmentation in coral reef organisms
        43: 285, 287, *288*, 290
Australia, sea cucumber in **41**: 134
  anatomy **41**: *138*, 145
  distribution and population
      structure **41**: 149, *150*, 154, 157
  fisheries and aquaculture **41**: 193, 195,
      198
  reproduction and spawning **41**: 159, *161*,
      163, *164*, 166, 176
Australian tuna **36**: 55
Autochthonous organisms and sea
    ice **43**: 196
auto-evisceration of sea cucumber **41**: 142,
    143
Automation, oyster culture **21**: 53
Autosynthetic vitellogenesis of invertebrates
    43: 89, 90, 141
Autotomy
  defensive **24**: 190, 191
    and dispersion of sexual products
        24: 191

as development or growth
    24: 191–193
of siphonophores **24**: 189–193
Autotrophic picoplankton *see*
    Cyanobacteria, Eukaryotic algae
AUVs (autonomous underwater vehicles)
    43: 180, 188, 189, 199, 225
availability *see* biological availability
Averaged General Quality Indicator
    (QIC) **26**: 79, 84
avoidance responses **38**: 34
Ayu
    egg quality **26**: 74, 98, 99
    egg viability **26**: 92
Azov anchovy,
    daily ration of **20**: **71**
    fat content of **20**: 73
Azov, Sea of **36**: 76, 159
    adaptation **36**: 44
        food and feeding **36**: *48, 49, 56,
            57*
        oxygen level **36**: 29, 30, **42**
        strategies **36**: **77**
    differentiation and variability **36**: 222,
        224, *225,* 227
    ecological principles **36**: 236, 237, 246,
        247, 248
    indicators of fish condition **36**: 213, 214,
        215, 217
    lipids **36**: 206, 208, 209, 210, 211
    life cycles **36**: 119
        abundance, dynamics of **36**: 127,
            129–136, *131*, 132, *135*
        annual **36**: 105, 107–109, *108*, 113,
            114
        ontogenesis **36**: *92*, 103

B spermatozoa, crustaceans **29**: 179
Bacillariophyceae **29**: *17*, 78, 79
Bacillariophyta *17*
Bacteria (see Parasites)
Bacteria **29**: 6, 21, 77
    see also Cyanobacteria
    carbon fixation **23**: 325, 326
    chemolithotrophs, *see* Chemolithotrophic
        bacteria
    coral infection **22**: 7, 48
    crustaceans **29**: 187, 189, 194, 195
    diet, *Donax* **25**: 185, 191, 227
    digestion by sea anemones **22**: 79, 80
    dissolved organic matter **29**: 22, 29,
        31–34, *33*, 41

ectosymbionts **33**: 524
frontal activity **23**: 229, 263, 264
gut, *Bullia* **25**: 192
hydrothermal vent dwelling **23**: 315–318,
    340
    anaerobic **23**: 318
    electron acceptors/donors **23**: 317
    isolation 23: 317
    metabolic rates **23**: 344
    productivity **23**: 318
    symbiotic **23**: 315, 321, 325, 327
in diet **33**: 154, 155
in faecal pellets **33**: 203
plankton **25**: 119, 120
sulphur oxidizing **23**: 316, 317, 325, 340
vertical mixing **29**: 43, 44, 55
Bacteria and bivalves **37**: 39, 40, 42, 43,
    125, 126
Bacterial growth in fish eggs **26**: 76
bacterial parasites of *Pandalus* **38**: 178,
    179
Bacteriophages, plankton **25**: 120
Bacterioplankton, Baltic Sea **29**: 115–119,
    118
    annual and seasonal variation **29**: 92–97,
        93, 95, 96
    distribution **29**: 97–101, **99–103**
    estimates **29**: 85–87
    nutrients **29**: 105–110, *107, 108, 110*,
        115–117
    predation **29**: 111–115, *112*, **113**, *115*
Bahamas **42**: *273*, 274, *275, 276*
Baikal, Lake **36**: 79, 252; **43**: 47
Bajkov's Equation, predation **25**: 39, 40,
    45
Ballycottin **29**: 223, *223*
Baltic **20**: 74, 127
    Banks **20**: 121, 122, 127
    binocularity in **20**: 131
    Buchan **20**: *16*, 17, *19*, 127
    Clyde **20**: 23
    Dogger **20**: *16*, 17, 121, 122
    Downs **20**: *16*, 17, *19*, 121, 122, 127
    East Anglian **20**: 50
    feeding success in **20**: 28
    flatiron **20**: 107, 128, 130
    gill area in **20**: 62, *63*
    Hokkaido–Sakhalin **20**: 5, 169
    Kamchatka **20**: 169
    Kiel **20**: *19*
    Maine **20**: 62
    Newfoundland **20**: 75

Norwegian **20**: 8, 15, *16, 19*
Onega **20**: 11
Pacific **20**: 20, 22, 25, 51, 61, 90, 143, 240, 241
round **20**: 247, 249, 252
Scottish **20**: 15
sinuous feeding posture in **20**: 27
spawning **20**: 10
summer spawning **20**: 17
tagging in **20**: 126–128
thread **20**: 34
winter spawning **20**: 10, 17
wolf **20**: 34
Baltic herring,
    fat content of **20**: 74
    summer spawners **20**: 127
Baltic Sea **29**: 52, 73–78, *74, 75*, 217, 275; **36**: 49
    see also Bacterioplankton, Cyanobacteria, Eukaryotic algae
    and vertical migration **26**: 138, 139
    differentiation and variability **36**: 221, 222, 226
    life cycles **36**: 101, 104, 107, 129
    recruitment of cod in **26**: 273, 274
    sole **29**: 217, 275
    substance and energy **36**: 141, 147
Baltic Sea: parasites **43**: 36, 55, 68
Baltic, plaice transplants (1893–1990) **38**: 11–13
Bandit cormorant **20**: 60
Banks herring **20**: 121, 122, 127
Barbs, *Physalia* **21**: *110*
Barents Sea **20**: 226; **36**: 50, 176, 226, 246; **43**: *11, 14*
    life cycles **36**: 101, 103, 117
Barents Sea ecosystem **34**: *267*
Barents Sea fisheries **38**: 208, 224
Barents Sea, halibut fisheries in **26**: 43, 45, 46
barium **41**: *11, 12, 120*
Barnacles **23**: 134, 135
    growth
        curve, equations **22**: 220, 221
        daily changes **22**: 213, 236, 237
        effects of
            breeding **22**: 238
            competing organisms **22**: 219
            food **22**: 216, 235
            light **22**: 216, 235
            lunar cycles **22**: 237
            moulting **22**: 235, 238

            orientation to current **22**: 217
            parasites **22**: 219
            population density **22**: 217–219
            reproduction **22**: 219, 220, 235
            surface contour **22**: 217
            temperature **22**: 212, 215, 235
            tidal levels **22**: 212, 213, 216, 234–236
        measurements, precision **22**: 238
        orientation during **22**: 207, 208
        primordial valves **22**: 204–207
        rates in various species **22**: 221, 222
        seasonal changes **22**: 211, 212, 237
        shell plates **22**: 203, 204
        tissue/shell weight ratio **22**: 215
    integument
        composition and origin **22**: 223–235
        cuticle formation **22**: 225–227
        shell deposition **22**: 225–228
    shape, effects of
        crowding **22**: 208, 209
        salinity **22**: 211
        substratum **22**: 209–211
    shell
        defence against predators **22**: 234
        deposition **22**: 225–228
        evolution **22**: 200–202
        structure **22**: 228–233
            microstructure in various species **22**: 231–233
        and limpets **23**: 117
        interactions with algae **23**: 99
Barnett Watson multiple corer **30**: 7
Barnstaple Bay **29**: 237
Baroclinic motion **23**: 204–207
    continental shelf spectrum of **23**: 206
Baroreceptor **20**: 115, 119
'Barrels', Crustacea **25**: 155, *155*, 156, 158
Barriers, oceanic, parasites as indicators of **43**: 55–59
Bartlett's test in benthic studies **26**: 230
Basidiomycetes, plankton **25**: **121**
Bass **36**: 26, *92*
Batch culture experiments **29**: 86, 87
Batch culture method **44**: 304
Batch size, sole **29**: 267
Batch spawner **20**: 127
Bathyal environment **35**: 6, 7, 8–16, 114, 115
    See also Deep-sea environment
Bathyal zone fauna **32**: 389–425
    classification of distribution **32**: 396

defined **32**: 389
latitudinal changes in relation to
    depth **32**: 394–400
management and conservation **32**: 413,
    414
methods **32**: 392–394
penetration to deep sea **32**: 404–407
relict and 'fossil' species **32**: 407–413
symmetry and asymmetry **32**: 400–404
Bathymetric distribution
    Pleurotomarioidean gastropods
        42: 274–277, *275, 276*
    Scaphopoda **42**: 204–207, *206*
Bathymetry **44**: 267
Bathymetry, Bristol Channel **29**: 223, 224,
    223
Bathypelagic zone **33**: 483
Batrachotoxin **21**: 102
Bay anchovy 26–30
    feeding success in **20**: 28
    passim **20**: 43, **44, 70, 72**, 76
    sinuous feeding posture in **20**: 27
Bay anchovy larva **20**: **72**
Bay of Bengal **35**: 24
Bay of Nigg, MA, hatchery **38**: 8
Beach environment, South Africa **25**:
    183–186
Beam-trawl fishing **29**: 216, 246, 247, 266
    regulation **29**: 332, 333
    surveys **29**: 237, 238, **240**
beche-de-mer, sea cucumber as *see*
    aquaculture; fisheries
"Becoculaire" structure **27**: 10
Behaviour
    abnormal, in rearing tanks **38**: 15
    escape response **24**: 369
    Pandalus
        individual **38**: 166–168
        larvae **38**: 156–159
        schooling **38**: 168, 169
    reared vs wild stocks **38**: 34, 35
    siphonophore **24**: 172–182
    protective **24**: 175–177
Behaviour *see* Learning; Feeding; Mating
    behaviour, Spatial distribution;
    Swimming activity; Vertical
    migration
Behaviour, copulation **27**: 260–262
    during moult cycle **27**: 221–223
    feeding **27**: 323–328
    jump response **27**: 376
    mating **27**: 257–262, *261*

migratory **27**: 297–314
parasite induced **27**: 379, 389
pressure changes **27**: 339
schooling **27**: 376, 377
Behaviour, sole **29**: 253, 267, 268, 351
Behaviourial
    factors, predation **25**: 11
    studies, Cephalopoda **25**: 107
Behavioural responses, bacterial
    infection **27**: 382
Birds (see Predators)
Blood composition and
    transport **27**: 173–177, **174**
    (see also Circulatory system)
    amino acids **27**: **174**, 175
    blood cells **27**: 31–33, 173
    carbohydrates **27**: **174**, 175, 176
    haemocyanin **27**: 33, 174, 184
    inorganic ions **27**: **174**
    ionic regulation **27**: 208, 209
    lipids **27**: **174**, 176, 177
    proteins **27**: **174**, 175
    volume **27**: 173
    endogenous rhythms **27**: 353–356
    food **27**: 355, 356
    light **27**: 192, 193, 353, *353*
    tides **27**: 355
    environment **27**: 333–356
    burrowing **27**: 333–354
    emergence and activity **27**: 334, 335,
        335
    substratum **27**: 341–349 (see also
        Substratum)
    light, diet cycle **27**: 336, 337
    moonlight **27**: 337
    turbidity **27**: 338
    oxygen concentrations **27**: 352
    salinity **27**: 351, 352
    temperature **27**: 349–351
    tides, currents and water depths
        27: 338–341
Bellingshausen Sea (Antarctic) **43**: 185,
    214, 226–228
Belt Sea, plaice transplants **38**: 12
Bengal, Bay of **43**: 84
Benguela Current **35**: 24
Benguela upwelling **32**: 36
Benthic animal mortality **21**: 75
Benthic assemblage, and biomass
    distribution **26**: 209
Benthic biology of seamounts **30**: 317–326
Benthic environments **39**: 39, 40

Benthic fauna and habitat **34**: 208-236
  degradation of reef habitats **34**: 232, 233
  direct effects of fishing gears
      34: 209-227
  drive netting **34**: 225-227
  effects of trawls and dredges on
      epifauna **34**: 219-222, **220**
  effects of trawls and dredges on
      infauna **34**: 213-219
  explosives **34**: 225-227
  indirect effects on habitat **34**: 227-233
  mobile epibenthic invertebrate fauna
      34: 288
  natural *vs.* fishing disturbance
      34: 233-235, *234*
  poisons **34**: 225-227
  potential effects of sediment
      resuspension **34**: 228, *228*
  static fishing gears **34**: 223-225
Benthic invertebrates, predators **25**: 47
Benthic mapping **44**: 263, 267
Benthic marine infaunal studies, analytical
      methods in **26**: 169-247
Benthic structure, collecting, classifying and
      mapping **44**: 265
Benthopelagic fauna
  hadal (ultra-abyssal) fauna **32**: 72-74
  hydrothermal ecosystems **32**: 78-82
  near-bottom communities **32**: 74-78, **77**
Benthopelagic habit **33**: 496, 497, **498-501,**
      501, 502
Benthos
  see also deep-sea
  coral reef *see* pigmentation in coral reef
Benthos, marine, structure, biotic
      interactions determining, *see*
      Biotic interactions
Benthos, sampling **30**: 317, 318
Bering Sea **29**: *56*, 57; **32**: 17, 18; **35**: 84;
      43: *11*
Bering Sea fisheries **38**: 225
Bering Sea, pollock in **37**: 178, 179, *180*,
      184
  see also Aleutian Basin
  fisheries and management **37**: 182, 183,
      237, 238, 239
  macroecology **37**: 187, 188, *189*, 190, 192
  population dynamics **37**: 193-199
  population structure
      genetic **37**: 215-218, *219*, 221-223
      mechanisms **37**: 228, 229, *230, 231,*
      232, 233, *234, 235,* 236

  metapopulation **37**: *226, 227*
  phenotypic **37**: 208-211, *212*, 213, *214*
  prey and predators **37**: 186, 187
  recruitment **37**: 199, *200*, 201, *202*, 203,
      204
  stocks **37**: **185**, *186*
Berkeley power station **29**: 237, 350
Bermuda **42**: 278
Berryteuthis anonychus **32**: 268-270
Berryteuthis magister **32**: 249-268
Bertalanffy, Von, growth parameters
      29: 258, 259, **259**, 301, 303,
      309, 347
Best's Carmine test **29**: **154**, **156**
Beta granules, crustaceans **29**: 176
Beverton-Holt Model, predation **25**: 59
Beyer's prey size selection model **20**: 28, 29
Big eye anchovy **20**: 28, **68**, **70**
Bigeye and match/mismatch
      hypothesis **26**: 263
Bigger is better hypothesis **31**: 171, 211
Bighead **36**: 227
Bile **36**: 214, 246
Biliverdin **43**: *291*
Bioassay of bivalves
  advantages and limitations **37**: 4, 5
  applications **37**: 74-82
      elutriate water **37**: 80-82
      pore water **37**: 80
      rationale for use of **37**: 76-82
      sediment quality **37**: 77-80
      toxicity tests with pure chemicals
          37: 74-76
      water quality **37**: 77
  broodstock availability and gamete
      production **37**: 135, 136
  methodology **37**: 40-50, 133-137
      broodstock **37**: 47, 48
      seawater quality **37**: 43, 44
      spawning **37**: 46-49
      species **37**: 135
Bioassay, toxinology **21**: 69
Biochemical indices **30**: 225-239, *258*
  digestive enzymes **30**: 237, 238
  lipids **30**: 230-232
  metabolic enzymes **30**: 239
  nucleic acids **30**: 232-237
  proteins **30**: 228-230
Biochemical measurements **30**: 259-265
Biochemistry *see* Chemical composition
Biochemistry *see* Elemental composition;
      Organic components

Biocide toxicity (herbicides, pesticides and
　　insecticides) and bivalves **37**: 63,
　　66, 95, **98–106**, 97–107, *133*
　see also in particular TBT
Biocoenoses **23**: 58
　in rocky shore communities **23**: 59, 61
　zonation **23**: 59
　　vertical **23**: 60
Bioeconomics, sole **29**: 319–321
Bioenergetic advantage, as
　　zeitgeber **26**: 136, 137
Bioenergetics equations **20**: 229–234, 236,
　　275, 276
Bioenergetics modelling **20**: 228, 233–250
　for Shetland seabird **20**: *235*
　passim **20**: 265, 275
BIOGAS programme **30**: 5, 21, 73
Biogeochemical cycling **44**: 3
Biogeographic barriers **35**: 6
Biogeography **34**: 355, 356, *356*
　see also ecology
　approaches **34**: 356, 357
　hydrothermal vent fauna **34**: 385–394
　Pleurotomarioidean
　　gastropods **42**: 277–279
　Scaphopoda **42**: 204–210, *206, 208, 209*
biological availability and uptake of trace
　　metals by phytoplankton **41**: 6, 7,
　　13–15, *13,* 23, 24–33
　see also primary production
　cadmium **41**: *26,* 27, 28, 29, *31*
　capture and sequestration **41**: 28–32, *29*
　cobalt **41**: *26,* 28, 29, 48
　copper **41**: 20, 27, 29, *31,* 66
　iron **41**: 25–32, 46–49, 47
　manganese **41**: *26,* 27, 29, 30, *31,* 48, 66
　nickel **41**: *26,* 27, 29
　theory **41**: 24–27, *25, 26*
　zinc **41**: *26,* 27, 28, 29, 30, *31,* 32, 48
'Biological clock' **36**: 114, 115, 127,
　　128, 239
Biological control of introduced pests **43**: 70
Biological factors affecting bivalves *see*
　　intrinsic
Biological markers *see* indicators
Biological Species Concept (BSC) **31**: 9
'Biological tagging' **24**: 300–302
Biology of Atlantic halibut **26**: 1–70
Bioluminescence **30**: 198, 199; **33**:
　　434–439, **435**; **20**: 93, 99, 100,
　　139, 185, 186
　luminescent behaviour **33**: 437–439

luminescent glands **33**: 434–436
luminous secretions **33**: 436, 437
Bioluminescence, pycnogonid **24**: 33
Biomass **23**: 3, 48; **33**: 393, 398, 399,
　　457–460
　see also Population biology
　and abundance **26**: 181
　coastal and shelf regions **33**: 457, 458
　distribution in benthic studies
　　26: 209–212
　importance value measurement using
　　23: 66
　inter-annual fluctuations **33**: 394
　oceanic water column **33**: 458, 459
　production/biomass (P/B) ratios **33**:
　　373–381, **375–379**
　reproductive **23**: 25–27
Biomass estimates
　see also Bloom conditions, Spawning
　　stock biomass, Stock-status
　bacterioplankton **29**: 86, 97, 101, **103**
　phytoplankton **29**: 2, 4, 5, 83–85
　size **29**: 19, 20
　sole **29**: 265, 266, 298
Biomass *see* Substance and energy
Biomass, *Donax* and *Bullia* **25**: 180,
　　182, 236
Biomass/production ratio **35**: 183–186, *187,*
　　187, 189, 190, **190**, *191*
　growth performance relationship **35**: 193,
　　194, *193*
　mortality relationship **35**: 192–194
Biomedical research, Cephalopoda **25**: 107
BIONESS **38**: 174
Bioremediation, oil spill sedimentary
　　environments **39**: 29, 30
BIOSIS bibliographic database **44**: 215
Biotic interactions determining marine
　　vegetation structure **23**: 72–126
　competition and/or herbivory in
　　isolation **23**: 108–117
　importance **23**: 72–77
　integrated studies of competition,
　　herbivory and carnivory **23**:
　　72–108
　　N.E. Pacific intertidal **23**: 72–77
　　N.E. Pacific subtidal **23**: 77–86
　　N.W. Atlantic intertidal **23**: 86–96
　　N.W. Atlantic subtidal **23**: 90–96
　　S.E. Australian intertidal **23**: 96–99
　synthetic ideas **23**: 117–126
　types **23**: 72

versus physiological competence **23**: 145
Biotic sources of variability **30**: 241
Biotoxicology **21**: 66
Biotoxins **21**: 66
BIOTRANS **30**: 5, 6, 19, 21, 35, 38, 52, 53,
    61, 64, 65, 73, 75
Bioturbation and Protobranch bivalves
    42: 32, 33
Biovolumes, plankton **29**: 83
Bird Rock platform **20**: 254, 255, 256
Birds **34**: 258
    changes in kittiwake and fulmar
        populations **34**: 285, *285*
    metazoan **25:** 147, 148, 149, 150, 152, 158
    oil effects
    parasites of **25:** 119, 122, 125, 129, 138,
        141, *142*
    predation **25:** 4, 32, 47, 64, 196
    prey removal **34**: 268–274, *269*
    seabirds, seaducks, shorebirds **39**: 64–70,
        65, *67*, **69**
    terrestrial birds **39**: 45–48, **46**
Birth rate **29**: 249
    see also Fecundity
Birth rates, seaweed **23**: 6
Biscay, Bay of **29**: 235, 236, 242, 298, 326,
    327, 351
    see also Vilaine, Bay of reproduction
        29: 271–273, 275, **289**
Bivalve embryos and larvae **37**: 1–175
    see also bioassay; laboratory rearing;
        pollutants; statistical methods;
        toxicity tests
    biology **37**: 10–25
        see also reproduction; species
    future research **37**: 137, 138
Bivalve molluscs, stress
    detection and responses to
        heavy and toxic metals **22**: 132, 133,
            143–146
        salinity **22**: 133, 134, 137–143, 145
        calcium ion dependence **22**: 138,
            139, 141
    heart rate **22**: 111–116
        copper effect **22**: 117–119
        temperature effect **22**: 130–132
        zinc effect **22**: 118–121
    heavy metal effects **22**: 104, 105, 129,
        132, 133, 143–146
    accumulation in tissues **22**: 162, 163
    behaviour and physiology
        changes **22**: 163–168

calcium role **22**: 176–179
    gametogenesis inhibition **22**: 166, 168,
        169, 173
    growth inhibition **22**: 167
    mitochondrial respiration and
        22: 176–181
    unfertilized eggs and **22**: 173–175
    methods for monitoring of
    "activity" **22**: 107, 108
    heart activity **22**: 105, 106
    shell growth **22**: 107
    valve movements **22**: 106, 107
    water pumping **22**: 106, 107
    mortality
        copper-induced **22**: 122–123T
        zinc-induced **22**: 122, 123T
    nature of stress **22**: 104, 105
    pollutants, threshold **22**: 104, 105
    pumping activity **22**: 112–116
        salinity effect **22**: 113, 116
    respiration
        anaerobic **22**: 146–150
        glycogen increase **22**: 149
        metabolism **22**: 148–150
        heart rate and **22**: 146–148, 155
        pH decrease **22**: 151–155
        valve movements and **22**: 146–148
    sensory receptors
        in inhalant siphon **22**: 134–136
        in mantle **22**: 134–136
    shell
        calcium reabsorption from **22**: 156–162
        physical protection **22**: 155, 156
        strength reduction **22**: 161, 162
    sublethal stress levels **22**: 128–130
        avoidance behaviour and **22**: 129, 130
    valve movements **22**: 108–112, 116
        effects of
            copper **22**: 117–119
            salinity **22**: 110, 113
            temperature **22**: 109
            zinc **22**: 118–121
        in epifaunal and infaunal species
            22: 121, 124–128
        oxygen consumption and
            22: 125–128
Bivalves **29**: 252; **34**: 418, 419
    burrowing rate, comparison with
        gastropods **28**: 421, 422
    clearance rates **28**: *311*
    comparisons with brachiopods
        28: 354–356

deep-sea species diversity **35**: 5
dimensions and mantle cavity
　　volumes **28**: *357*
genetic differentiation **35**: 81
genetic identity related to taxonomic
　　divergence **35**: 69
growth performance **35**: 167–173, **171**,
　　172, **173**, **174**, **175**
hydrothermal vents **35**: 18, 19
oxygen consumption **28**: *319*
oxygen-minimum zones **35**: 25
seeps **35**: 23
Bivalves, burrowing **25**: 198, 235
Bivalvia, hybridization bias in **31**: 39
Bjørnsen **29**: 86
Black Sea **35**: 24; **36**: 3
　　adaptation **36**: 44
　　　food and feeding **36**: 47–49, 56, 57
　　　oxygen level **36**: 29, 30, *31*, 32, *33*, 34,
　　　　35, 36, 41, 432
　　　temperature **36**: 7–9, *11*, 12–14, 18
　　adaptation strategies
　　　energy metabolism **36**: 60–74, **60**, **64**,
　　　　66, *67*, *73, 74*
　　　plastic metabolism **36**: 75–82, **75**, **77**,
　　　　81, 82, 85, 86
　　Calanus euxinus, vertical distribution of
　　　biomass **32**: *41*
　　ctenophores **32**: *42*
　　differentiation and variability **36**: 222,
　　　224, *225*, 226
　　ecological principles **36**: 232, 238, 239,
　　　241, 246, 248
　　indicators of fish condition **36**: 213, 218
　　　lipids **36**: **207**, 208–211, *212*
　　life cycles
　　　abundance, dynamics of **36**: 125, 129,
　　　　130, 132, 137
　　　annual **36**: 105, 107, *108*, 112–114
　　　daily rhythms **36**: 115–117
　　　interannual fluctuations **36**: 118, *119*
　　　ontogenesis **36**: 96, 101, 104
　　mesoplankton **32**: 40, 41
Black Sea anchovy **20**: **71**, 74
Black smokers **23**: 303, 308, 318; **35**: 17
　　faunal distribution and density **23**: 340
　　microbial productivity **23**: 318
Blackwater Estuary **29**: 293
Blastocones, Cephalopoda **25**: *92*, 93, 94
Blastodinida, parasitic **25**: **127**
Blastomeres, Cephalopoda **25**: *92*, 93, 94
Blastula phase, sole **29**: 229

Blastulation, Cephalopoda **25**: 86, 87, **108**
Bleached (and dead)　coral reef organisms
　　Plate 4, **43**: 280, 284, 292, 293,
　　306–308, 309
Bleak **36**: 183, 190, 227
Blenny **36**: 17, 66, 190
Blood **36**: 94, 213, 214, 221
　　see also circulation
　　adaptation strategies **36**: 62, *63*, **64**, 66,
　　　73, *84*
　　haemoglobin levels **24**: 350–356
　　oxygen affinity **24**: 353, 355, 356
　　Scaphopoda **42**: 185–188, *187*
Blood circulation, *Donax* and *Bullia* **25**:
　　200–203, 205, *206*
Bloom conditions **29**: 4, *59*, 60, 79
　　Baltic Sea **29**: 87, 92–97, *93*, *95, 96*,
　　　98–101, **99–101**
　　nutrients **29**: 103, 104, 106–110, *107*,
　　　108, 116, 117, *110*
　　dissolved organic matter **29**: 32, 35
　　fronts **29**: 50–58, *51*, *53, 55–58*
　　predators **29**: 20–22–24, *23, 26*
　　seasonal events **29**: 42–47, *43*, **43**, *46, 47*
　　sedimentation **29**: 36, 39, 40, *41*
　　wind events **29**: 47–50, *48, 49*
Bloom, planktonic, misuse of term **23**: 233
　　see also Plankton; Red tides
"Blooms" **37**: 125–127, **127**
Blooms **21**: 65, 70, 74; **43**: 197, 198,
　　231–233, 235, 308
Blooms, exceptional *see* Exceptional
　　phytoplankton blooms
Blue grenadier (hoki) fishery **35**: 107–109
Blue shark **36**: 78
Bluefish **36**: 77
Body axes, Cephalopoda **25**: **102**
Body density, sinking rate and **33**: 409, 414,
　　415–417
Body form **33**: *4*, 14, 15
　　see also Morphology
Body size of bivalve larva
　　see also age
　　and toxicity reaction **37**: 118, 119
　　shells **37**: *63*
Body size *see* Size
Body size, *Donax* and *Bullia* **25**: 216, 217,
　　217, **226**, 227, 228–230
Body size/length **43**: 213
　　and fecundity of invertebrates
　　　43: 131–134, 140, 148, 152
　　and parasites **43**: 31, *32*, 33, 36

and diversity **43**: 5, 8
and latitude **43**: 51
feeding organ sizes compared **43**: 29, 30
genital hook **43**: *12*
hamuli lengths compared **43**: *10, 13,15*
hosts of **43**: *15*, 33–35
Body weight **33**: 221–235, 333–337
ash weight **33**: 233–235, *235*
ash-free dry weight **33**: *227*, 233, **234**
dry weight **33**: 227–233, **228–222**, *334, 335*, **337**
wet weight **33**: 221–226, **223–225, 226**
Bogoslof Island, pollock near **37**: *180*, 183, 184
population dynamics **37**: 193, 194, 195, 196, *197*
population structure **37**: 217,*219*,221,233
Bohuslän herring period **20**: 127
Boothbay Harbor, ME, hatchery **38**: 8
Bootstrap methods in classification in benthic studies **26**: 219, 220
Bornholm Basin **29**: 76
Bothidae **29**: 219
Bothidae as predators **27**: **362**
Bothnian, Sea of **29**: *74*, 75, *75*, 76, 80, 86, 87, 116
autotrophic picoplankton **29**: *88*, 89
bacterioplankton **29**: 94, *96*, 98, **99, 100**, 111, **113**
Bottlenecks
heterozygosity and **35**: 64
hydrothermal vents **35**: 20, 21, 64
seeps **35**: 23
Bottom trawl surveys of pollock **37**: 193–196
"bottom-up" approach to resource evaluation **38**: 30, 39
Boudouresques's taxonomic synthesis **23**: 58–62
steps involved **23**: 58
Bounded hybrid superiority hypothesis **31**: 53
Bourrelets **23**: 190–195, 267, 268
and seawater chemistry **23**: 192–195
evidence for existence **23**: 189, 190
nitrogen reserves **23**: 193–195, 242
nutrient accumulation **23**: 194
nutrient mixing from **23**: 194–197, 241, 242, 267
tidal periodicity **23**: 194–197
oxygen minimums **23**: 192
seasonal effects **23**: 190

Boyle's Law **20**: 147
Bracco–Curti test **29**: **155, 156**
Brachiopods
articulate and inarticulate, World Ocean **32**: *403*
E and W Pacific distribution **32**: *398, 399*
latitudinal faunistic belts, World Ocean **32**: *395*
relicts **32**: **409**, *411*
vertical distribution **32**: *393*
Brachiopods, classification **28**: 179,359, 360
Brachiopods, comparisons with bivalves **28**: 354–356
dimensions and mantle cavity volumes **28**: *357*
Brachiopods, development **28**: 273–282
development type **28**: *265*
larva
early **28**: 278, 279
hatching **28**: 279–282
juvenile development after settlement **28**: 282
pigment spots and setae **28**: 248
mantle **28**: 185–187
oocyte cleavage **28**: 278
oocyte diameters **28**: *265, 282*
summary **28**: 277–282
times for embryological features to appear, comparisons **28**: *282*
Brachiopods, ecology **28**: 332–346
community ecology **28**: 345, 346
geographic distribution **28**: 346
life history strategies **28**: 333–337
growth rate **28**: 335–337
patterns **28**: 337
recruitment **28**: 333–335
survivorship **28**: 337–345
competition **28**: 339–341
disturbance **28**: 341
pathology **28**: 341, 342
patterns **28**: 342–345
predation **28**: 337–339
Brachiopods, genetics and biochemical systematics **28**: 347–352
classification **28**: 179, 359, 360
enzyme systems **28**: 347, 348
immunology **28**: 349–352
unweighted pair-group using averages (UPMGA) **28**: *351, 352*
morphometric analysis **28**: 349
palaeontological studies **28**: 177, 354
single-linkage cluster analysis **28**: *353*

terminology **28**: *178*
Brachiopods, physiology and
    metabolism **28**: 282–332
  digestion **28**: 312–316
    distribution of enzymes, selected
        species **28**: *315*
  energy turnover **28**: 355, 356, 358
  environmental tolerances
    anoxia **28**: 330
    salinity **28**: 329, 330
    temperature **28**: 329
    turbidity **28**: 331, 332
  excretion and ionic regulation
        28: 322–325
  feeding
    clearance rates **28**: 310–312
    feeding currents **28**: 292–295
    food sources **28**: 306–309
    nutritional sources **28**: 306–309
    particle capture **28**: 295–300
    particle rejection mechanisms **28**: 302,
        303
    particle retention efficiency **28**: 309,
        310
    particle transport **28**: 300–304
    quantitative physiology **28**: 309–312
    selectivity **28**: 304–306
  metabolic pathways **28**: 325–328
  respiratory physiology **28**: 316–322
    freeliving inarticulates **28**: 288–292
    oxygen exchange **28**: 320–322
    rates of oxygen consumption
        28: 317–320
    rates of water transport **28**: 316,
        317
    respiratory proteins **28**: 322
  sensory/neuromuscular physiology/
        behaviour **28**: 282–292
  sessile articulates **28**: 285–288
Brachiopods, reproduction **28**: 248–273
  adaptations **28**: 358, 359
  fertilization **28**: 277, 278
  gametogenesis **28**: 255–270
    oogenesis **28**: 258–270
    spermatogenesis **28**: 249, 255–258
  'genital pits' **28**: 192
  gonad shape, various brachiopods **28**: *253*
  hermaphroditism **28**: 265
  morphology **28**: 250–254
    genital lamella **28**: 251–253
    gonad development **28**: 253, 254
  oocyte diameters **28**: *282*

oocyte maturation **28**: 277, 278
  reproductive cycles **28**: 272, 273
    published data **28**: *274–276*
  sexual dimorphism **28**: 250
  spawning **28**: 270–272
    brooding species **28**: 271, 272
    free-spawning species **28**: 270, 271
    strategies **28**: *265*
  vitellogenesis **28**: 259–270
Brachiopods, structure and functional
    morphology **28**: 178–248
  coelomic and vascular system **28**:
        212–217
    amoeboid cells **28**: 215, 216
    blood cells **28**: 214
    coelomocyte types **28**: *213*, 214–217
    general structure **28**: 213, 214
    spindle bodies **28**: 216
  digestive system **28**: 226–236
    digestive cells **28**: 233, 234
    diverticula **28**: 231–234
    functional morphology **28**: 234–236
    intestine **28**: 229, 230
    type A secretory cells **28**: 232, 233
    type B secretory cells **28**: 234
  general characteristics **28**: 178, 179
  lophopore **28**: 217–226
    blood vessels **28**: 225
    coelomic epithelium **28**: 224, 225
    connective tissue **28**: 222–224
    epidermis **28**: 217–222
    myoepithelia **28**: 225
    nerves **28**: 224
  mantle **28**: 179–200
    caeca **28**: 194–200
    development **28**: 185–187
    gamete load **28**: 192
    general structure **28**: 182–185
    generative region **28**: 187–200
    inner mantle membrane **28**: 182,
        189–192
    mantle canal (sinus) **28**: 193
    outer mantle membrane **28**: 182, 183,
        193
    ultrastructure **28**: 187–189, 195, 196
  muscle arrangement **28**: 237–239
  muscular system **28**: 236–244
  myoepithelial cells **28**: 240, 241
  nerves and sensory structures **28**:
        244–248
    gap junctions **28**: 245
    setae (chaetae) **28**: 246, 247

statocysts **28**: 245, 246
paramyosin **28**: 243, 244
    smooth muscles **28**: 243
    striated adductor muscles **28**: 242
    ultrastructure **28**: 240-248
pedicle **28**: 200-212
    classification **28**: 210-212
    functional morphology **28**: 209, 210
    general structure **28**: 201-209
    muscular system **28**: 239
    regeneration **28**: 210
    rootlet and bulb structure **28**: 208, 209
    trunk **28**: 203-208
shell secretion **28**: 247
Brachyura, spermatophores **29**: 130, **133**,
        195
    chemical composition **29**: 136-139, **138**,
        139
    copulatory organs **29**: 139-141
    dehiscence **29**: 145
    morphology **29**: 133-135, 197
    origin **29**: 135, 136
    sperm receipt and storage **29**: 141-145,
        198, 200
    sperm transfer **29**: 139, 199
Brackish water **33**: 472-479
    estuaries **33**: *474*, 475-477
    fjords **33**: 349-351, **350**, 477-479, *478*
    river plumes **33**: 474, 475
Brackish water, parasites **25**: **146**
Brackish water, parasites in **43**: 55
Brain **33**: 38
Brandt's cormorant **20**: 241, 243, *244*
Bray-Curtis similarity coefficients in
        benthic studies **26**: 182, 215, 216, *218*
Brazil, coastal areas of **43**: *11, 50*
Breakage and repair of shells **42**: 152,
        281-286, *282-284*
Bream **36**: 25, 250
    annular *see* annular gilthead
    egg quality **26**: 74, 77, *84*, 98, 99, *101*
    vertical migration in **26**: 120, 121
Breeding biology *see* Reproduction
Breeding fish *see* Farming,
        fish; Reproduction
Breeding season, *Donax* and *Bullia* **25**: 194,
        236
Breeding seasonality, pycnogonid **24**: 43, 44
breeding sex ratio, *P. borealis* **38**: 211, 212
Brevetoxins **31**: 313
Bridgwater Bay **29**: 240, 243
Brightness, as *zeitgeber* **26**: 123, 124

Brightness, predation **25**: 15
Brine
    see also salinity
    channels **43**: 176, *177, 178*, 179, 181,182
        and trophic processes **43**: 198, 205,
            206, 208
    exclusions/pores **43**: 176
    frequency distribution **43**: 182
    gas concentrations **43**: 191, 200, 202
    release **43**: 177
Bristol Channel **29**: 218
    sole, adults **29**: 247, 266, 290
    sole, juveniles **29**: 237, 238, *239*, 240,
        240, 243
Bristol Channel, sole **29**: 216-219,
        348-352
    see Adults, Eggs, Exploitation,
        Harvesting, Juveniles, Larvae,
        Mortality rate, Reproduction,
        Stock
    classification and identification **29**: 219,
        220
    description and movements **29**: 221, 223
        currents **29**: 227-229, *228*
        depth and geomorphology **29**: 223,
            224, *223*
        salinity **29**: *226, 227*
        temperature **29**: *224, 225*
    description and related genera **29**: 220,
        221, **222**
    feeding **29**: 251-254
    future **29**: 348-350
    length at age **29**: 255-262, **257**, *258*,
        259
    size and growth **29**: 254, 255
    weight **29**: 262-266, *264*, **265**
Britain **42**: 26, 32, 37
British Columbia **20**: 240-243
British Columbian waters, pollock
        in **37**: 209, 210, 213, 214, *227*
Brixham **29**: 324-326, 330, 333
Broadcast fertilization, crustaceans **29**: 130,
        195
Broadcast spawning species of invertebrates,
        fecundity of **43**: 89, 92, 93, 118,
        129, 140
    quantification methodologies
        43: **100-109**
        chemically induced **43**: 101, 103, 104
        dissection of ovaries **43**: *95*, 105, 106,
            151

environmental shocks **43**: 101–103, *102*
histology **43**: 106–109, *108*
non-destructive weight **43**: *95*, 104, 105
Broadcaster copepods **44**: 10, 16, 41, 42, 51
Broken Spur site, MAR **32**: 79, 80, 100–102, **126–129**
Broken Spur vent fields **35**: 88, 89
Bromate toxicity and bivalves **37**: 96, **102**
bromine **41**: *12, 120*
Bromoaplysiatoxin **21**: 89
Bromophenol blue staining, crustaceans **29**: 136
Brood size **33**: 272–274, **275, 276**, 276–281, *281*
seasonal variation **33**: 281, 282, 371
Brood survival **20**: 42
Brooding species of invertebrates, fecundity of **43**: 89, 92, 114, 115, 130, 131
quantification methodologies **43**: **93–100**, *94*
Broodstock
management of fish **26**: 96–103
and hormonal treatment **26**: 102, 103
and induced spawning **26**: 102, 103
and nutrition **26**: 97–102, **98, 99**
of Atlantic halibut **26**: 54, 55
Broodstock/spawners, bivalve
and bioassay methodology **37**: 45, 46
availability and gamete production **37**: 134, 135
conditioning of **37**: 25–27
Brook trout **36**: 9, 136
Brotula **36**: 46
Brown bullhead **36**: 9
Brown rockfish **36**: 12
Brown tides **31**: 309, 361
Brown trout **36**: 9, 23
Brownian motion, phytoplankton **29**: 37
Brünnick's guillemot **20**: 228
Brunt-Väisälä oscillation **23**: 204–206
Bryozoans **21**: 141
B-sclerotization, crustaceans **29**: 187
Buccal cavity **33**: 39
Buccinacea **25**: 186
Buchan herring **20**: *16*, 17, *19*, 127
Buchan spawning area **26**: 252
Budget, substance and energy **36**: 196–199
Bulla **20**: 118, 119, *146*, 150–152
Bulla system **20**: 147–152, 155

Bulletin Statistique des Pêches Maritime **29**: 328
Bullhead **36**: 9, 19
Bullia digitalis
burrowing performance **28**: 413–418
adaptations **28**: 418–421
comparisons **28**: 421, 422, **425**
mechanism **28**: 404–407
burrowing, evolution **28**: 422–426
classification **28**: 392
comparisons with *Donax* **28**: 390, 421, 422
digging cycles **28**: 405
locomotion, comparisons **28**: *417*
structure **28**: 395–403
foot **28**: 394, 395–401
shell shape **28**: 401–403
Buoyancy **36**: 44–47
Actinomyxidea **25**: 146, 158
Cephalopoda **25**: 87
of Atlantic fishes **24**: 333, 334, 376
of siphonophores **24**: 185, 186, 193–199
and distribution **24**: 163, 164
Calycophora **24**: 199
Cystonectae **24**: 193, 194
Physonectae **24**: 194–199
Buoyancy control, phytoplankton **29**: 9
Buoyancy in fish eggs **26**: 76, *77*
and feed quality **26**: 98, 99
Burma **41**: *137*
Burned otolith technique **29**: 296
Burrowing **27**: 34, 35, 37, 187, 223, 333, 334
defence **27**: 375
temperature **27**: 350
Burrowing behavior of Scaphopoda **42**: 162, 163
burrowing by sea cucumber, daily cycle of **41**: **176–181**, *179*
Burrowing Rate Index (BRI) **30**: 103
Burrowing, *Donax* and *Bullia* **25**: 180, *181*, 182, 184, 187, *188*, 196, 197, 235, 236, 237
see also Locomotion
oxygen consumption **25**: 213, **217**, 218, 219, 224
'Burst' of activity **36**: 66, 68, 69, 70
Burst swimming **20**: 90, 187
burying behaviour in sole **38**: 34–35
Butterfish **36**: 190
Buzzards Bay (Massachusetts) **42**: 31, 32
^{13}C **24**: 430

Cadmium toxicity and bivalves **37**: 45, 66, 67
  tests **37**: 87, 88, **90–95**, 116
cadmium **41**: **72–77**, *120*
  biological availability and uptake **41**: *26*, 27, 28, 29, *31*
    impact on primary production **41**: *13*, 14, 23, 74–77, *75*
  cycling ratio **41**: *84, 85, 86*
  distribution in oceans **41**: 72–74, *73, 75, 76, 126, 127*
  functions and properties **41**: *16, 17*
  indices of fractionation **41**: *81, 82*
  inputs to system **41**: *123*
  interactions **41**: *34, 35*, 37, *126, 127*
Cadmium, *Donax* and *Bullia* **25**: 224
Cages, exclusion, design **23**: 112
Calanoid copepods
  parasites of **25**: 148
  predators **25**: 8
Calanoid copepods, life-history **44**: 297
Calanoidea **29**: **131**, 174, 175, 178–181
Calanus euxinus, vertical distribution of biomass **32**: *41*
Calanus hyperboreus
  Arctic basin **32**: 19–22
  North Atlantic **32**: 22, 23, *24, 25*
Calanus in match/mismatch hypothesis **26**: 272, 274, *275*, 276, *280*, 281
Calcarsaponin **21**: 154
calcium **41**: *11, 12, 84, 120, 186*
Calcium hypochlorite, *Donax* and *Bullia* **25**: 224
Calcium uptake of bivalve larvae **37**: 69, 70, 94
Calcium, bivalve molluscs
  concentration under stress **22**: 151
  in response to salinity **22**: 138, 139, 141
  mitochondrial respiration and **22**: 176–179
  reabsorption from shell **22**: 156–162
Calcium, crustaceans **29**: 137, **139**, 160, 168, 186
California anchovy **36**: 48
California Current fisheries **39**: 270, 271, 277, **266**, **275**
California sardine **36**: 55
California, Gulf of
  hydrothermal ecosystems **32**: 80
  oxygen minimum layer **32**: 45–48
Californian anchovy **20**: 8
Californian current **20**: 167, 243–247

Californian sardine **20**: *4*, 5, 8, *35*, 64, 131, 144
  feeding of **20**: *35*
  larva **20**: 32
  oxygen uptake in **20**: 64
  spawning in **20**: 8, 22
Caloric density **33**: 250
Calorie intake *see* Food
Cambrian
  Pleurotomarioidean gastropods **42**: 237, 238, 239, *242*
  Protobranch bivalves **42**: 3, 41, 42, 44
Camouflage, siphonophore **24**: 202
Canada **42**: 272
Canada, coastal areas of **43**: *11, 50*, 145
Canada, halibut fisheries in **26**: 43, 45, 46, 48, 49
Canada, *Pandalus borealis* stock limitation in **38**: 207
Canary Islands **42**: 39
Canberra–Metric measures in benthic studies **26**: 215, 216
Candaciidae **44**: 72
Cannibalism **20**: 26, 54–56, 170; **25**: 3, 48, 49, 50, 59, 64; **33**: 159; **38**: 7, 21, 37
  MacCall's model of **20**: 170
  northern anchovy **25**: 14, 15, 16, *18*, 25, 30, 40, 48, 61
Cannibalism of pollock **37**: 185, 202, 203
Canonical correlation in benthic studies **26**: 230, 231
Canonical discriminant analysis
  use in hybridization **31**: 7
Canonical variate analysis, *see* Multiple discriminant analysis
Capacitation, crustaceans **29**: 183
Cape cormorant **20**: 247, **248**, 252
Cape gannet **20**: 60, **248**, 252, 259
Cape of Good Hope, sole **29**: 217
Cape Thompson **20**: 226, 227
Capelin **36**: *17*, 152, 246
  and Atlantic halibut **26**: 33
  in match/mismatch hypothesis **26**: 257, 272
  vertical migration of **26**: 121, 130
Captacula of Scaphopoda **42**: 164–168, *166, 168*, 192
capture of trace metals by phytoplankton **41**: 28–32, *29*
Capture, predation **25**: 12, 16–22, *17, 18, 20*
Carangidae as predators **27**: **362**

carapace **38**: 60
Carassius sp **36**: 36
Carbohydrase, *Donax* and *Bullia* **25**: 192
Carbohydrate content **33**: 248
Carbohydrates and glycogen **36**: 5, 41, 56, 222, 230
　adaptation strategies **36**: 68, 72, **73**, *74*, 86
　indicators of fish condition **36**: 216, 217
　interannual fluctuations **36**: 119–123
　life cycles **36**: 90, *93*, 126, 136
　substance and energy **36**: 170, **179**, *180, 181*
Carbohydrates, crustaceans **29**: 137, 143, 144, 178, 189
Carbon **41**: *12*, 40, 42, 46, 120
　and community structure **41**: 90
　biochemical function **41**: *11, 12*
　cycling ratio **41**: *85, 86*
　dissolved organic-315
　fixing bacteria **23**: 325, 326
　interactions with trace metals **41**: *57*, 67, 71, 83
　internal economy and near-field chemistry **41**: *85, 86,* 87, 88
　particulate organic-315
Carbon content **33**: 236–240, **238, 239**; **44**: 33, 34
　carbon–nitrogen ratio **33**: 242
Carbon dioxide oversaturation **36**: 43
Carbon in Antarctic
　assimilation **43**: 199, 200
　consumption **43**: 205
　depletion **43**: 200
　isotopes **43**: 234
　sequestration **43**: 197
Carbon, plankton **29**: 31, 32, 35, 83, 101, **103**
　Baltic Sea **29**: 104, *107, 108*, 109, 110, *110*, **114**, 115–117, 119
Carboniferous **42**: 217
Cardiff **29**: 325–328, 330
Cardiotoxins **21**: 68
Caribbean area
　see also Gulf of Mexico
　Pleurotomarioidean gastropods **42**: 272, *273*, 274, 276, 277
　Protobranch bivalves **42**: 3, 35, 37
Caribbean corals, stress
　histopathology **22**: 48, 49
　loss of zooxanthellae **22**: 30, 31T
　metabolism **22**: 28
　mucus production **22**: 40T
　sediment shedding **22**: 44T

Caribbean Sea, abyssal fauna
　distribution **32**: 359, 360
Caribbean: pigmentation in coral reef organisms **43**: 282, 285, 289
Caridea, features **27**: **58**
Caridean sp. **26**: 131
Caridia **29**: **132**, 173, 174
Carlo Erba analyser **24**: 396, 397
Carmarthen Bay, sole **29**: *223*, 227, 349
　early fishing **29**: 323, 324, 326
　juveniles **29**: 237, 240, 245
Carnivores and carnivory, herbivory, competition and-, interactions **23**: 72–108
　identification **23**: 73–75, 91
　interspecific competition and **23**: 121
　'keystone' 73–75, 93
　-prey interactions, *see* Predator–prey interactions
Carnivorous gelatinous species **44**: 38
Carnivory *see* Predatory feeding
β-carotene **43**: *283*, 285
Carotenoids **33**: 251
　as dietary markers **33**: 151
Carotenoids and coral reef organisms **43**: *283*, 285, 289, 292, 293, 307
Carotenoids in fish eggs **26**: 80, *81*
Carp **36**: 227
　adaptation **36**: 12, 54, 80
　　food and feeding **36**: 58
　　oxygen level **36**: 41
　　temperature **36**: 8, 10, 12
　ecological principles **36**: 231, 250
　life cycles **36**: 117, 126, 129
　substance and energy **36**: 148, 190
Carp, egg quality **26**: 74
carrier complexes **41**: 28, 29
carrying capacity **38**: 7, 21, 30, 39, 40, 45
Carrying capacity *see* resources
Caspian Sea **43**: 69
　adaptation **36**: 37, 49
　ecological principles **36**: 246, 247, 248
　indicators of fish condition **36**: 210, 211, 214
　life cycles **36**: 107, 114, 129, 132
Castor oil fish **36**: 46
Castration, parasitic **25**: 134, 135, 148, 154
Catch of Atlantic halibut **26**: 42–52
　and over-exploitation **26**: 49, 50
　and recruitment and mortality **26**: 50, 51
　by area **26**: 45, *46*, 47

by nation **26**: 47, *48*, 49
  development of **26**: 42, 43, *44*, 45
  management of **26**: 51, 52
Catch per unit of effort (CPUE) **38**: 4, 208
Catch, sole **29**: 249, 312, **313–315**, 322,
    327–330, 333, 334
  see also Exploitation, Harvesting
  catch per unit effort (CPUE) **29**: 334, *341*,
    341–343, 345
  estimation of natural mortality **29**: 296,
    297
Catch, world total, of prawns **27**: 4
Catch-at-age data on pollock **37**: 194, 238
Catch-per-unit-effort **20**: 265, **266**, *267*, 290,
    291
Catfish **36**: 34, 77
Caudal
  fin, sole **29**: 219
  region, sole **29**: 266
Caulerpicin **21**: 92
Caulerpin **21**: 92
Cave habitat **33**: 488, **489**
CEFAS **44**: 253, 255
Cell, plankton
  see also Size
  membrane **29**: 30, 31
  size **29**: 101
  volume **29**: 101, **102**, 117–119, *118*
Cellulase activity, *Donax* and *Bullia*
    25: 192
Celtic sea **23**: 167; **29**: 34, *34*, 217, 218
  see also Bristol Channel
  cold bottom waters **23**: 190, 191
  fronts, horizontal exchanges **23**: 242,
    243
  plankton distributions **23**: 271, 272
  shelf break, *see* Celtic sea shelf break
Celtic sea shelf break tidal fronts **23**: 165,
    197–204, 205, 276
  biological characteristics **23**: 270–274
    physical causes **23**: 271–274
    plankton distributions **23**: 271, 272
  cold water band, models accounting
    for **23**: 197
  internal waves, *see* Internal waves
  upwelling interpretation **23**: 197, 198,
    202
  water column capsizing **23**: 198
  satellite scanning **23**: 270, 271
  secondary upwelling **23**: 271, 273
  tidal streaming **23**: 199, 200

Cenozoic and Pleurotomarioidean
    gastropods **42**: 237, 239, *242,
    268*, 281
  see also Recent
Cenozoic ridge **34**: 388, 389
Centrarchids **36**: 196
Centropagidae **44**: 72–79; **29**: 174, 175, 182
Centropomidae as predators **27**: **363**
Cephaloidophoridae, parasitic 25: 141
Cephalolobidae, parasitic **25:** 141
Cephalon, pycnogonid **24**: 4, 8
Cephalopod beaks **39**: 287
Cephalopod light-fishing **39**: 261–292
  catches **39**: **266, 275**
    and global biomass **39**: 286, 287
  DMSP-OLS images
    California Current **39**: 270, 271, 277
    China Sea Shelf **39**: 267, 268, 276
    Humboldt Current **39**: 271, 277
    Kuroshio Current **39**: 265–267, 275,
      276
    New Zealand **39**: 269, 270, 276, 277
    Southwest Atlantic **39**: 271–273, 277
    Sunda-Arafura Shelves **39**: 268–289,
      276
  other squid fisheries and unexploited
    stocks **39**: 273, 274
  predator consumption of cephalopods and
    fisheries yield, discrepancy
    between global and regional
    estimates **39**: 286–288
  predator diets **39**: 278
  predator interactions **39**: 278–288
    implications for precautionary
      measures **39**: 285, 286
  predators **39**: **279–281**
  predators of species exploited by light-
    fishing **39**: 278–285
  prey interactions **39**: 288, **289**
  use of predator data to identify new
    fisheries **39**: 285
Cephalopod non-light-fishing **39**: 277, 278
  catches **39**: **274**
Cephalopoda, parasites of **25**: 144
Cephalopods **29**: 195–197, 200; **36**: 36, 83;
    44: 143–203
  ambient temperature and rate of
    embryonic development **44**: 150
  arm buds **44**: 153–155
  artifical fertilization **44**: 151
  blastulation **44**: 156
  bodily outfit of hatchlings **44**: 161–164

cephalic organs **44**: 152, 153, 161–163
chorion micro-structure **44**: 148, 149
classification **44**: 146
defence **44**: 175, 176
digestive tract **44**: 163
dispersal and recruitment **44**: 179, 180
early embryonic stages **44**: 149–152
early juvenile life **44**: 182
egg size and formation of outer yolk
    sac **44**: 156, 157
embryogenesis **44**: 147, 181
embryology based on natural egg
    masses **44**: 150, 151
embryonic development **44**: 182
enzymatic activity of yolk
    syncytium **44**: 159
epibolic growth of gastrula **44**: 157
evolutionary developmental
    biology **44**: 182
feeding behaviour **44**: 173–175
funnel/mantle complex **44**: 153–155
gastrulation **44**: 156
hatching competence **44**: 159, 160
hatching gland **44**: 159
hatching stage **44**: 159
histological differentiations of body
    surface **44**: 152–156
inner yolk sac **44**: 157–159
integument development **44**: 155, 156
juvenile mortality **44**: 180, 183
larval features and paralarval
    adaptations **44**: 182
learning behaviour **44**: 172, 173
life style **44**: 176, 177
long-distance dispersal (LDD) **44**: 183
mantle and fins **44**: 164
microenvironment of embryo
    44: 147–159
morphological changes **44**: 164–168
morphological discontinuities
    44: 168–170
mortality rates **44**: 183
organogenesis **44**: 151–156
oviducal and nidamental jellies **44**: 147,
    148
paralarval specializations **44**: 168–170
phylotypic phase of development **44**: 151,
    152
post-hatching behaviour **44**: 171–179
post-hatching growth **44**: 164–168
prehensile organs: arms and
    tentacles **44**: 163, 164
protective envelopes or
    capsules **44**: 147–149
secondary head cover **44**: 153–155
shell field **44**: 155
social behaviour **44**: 177–179
staging systems **44**: 149, 150
suckers **44**: 164
swimming **44**: 176
symbiotic luminescent bacteria **44**: 170,
    171
systematics **44**: 145–147
taxonomy **44**: 145–147
terminology **44**: 168–170
tranquilliser effect of perivitelline
    fluid **44**: 160
transition from embryo to
    young **44**: 159–171
use of growth records in age
    determination **44**: 164–168
yolk absorption to food ingestion **44**: 160,
    161
yolk reserves **44**: 160
yolk storage, redistribution and
    absorption **44**: 156–159
Cephalopods as predators **27**: 359, 360
Cephalosome **33**: *15*, 16, 24, 25
Cephalotoxin **21**: 183
Cercariae, parasitic **25**: 146, 156
Cerebratulus A toxins **21**: **135**, 136
Cerebratulus B toxins **21**: **135**, 137
Cestoda, parasitic **25**: 150, 151, 158
CF *see* condition factor
Chaetognath **20**: 52, 159
Chaetognath prey **33**: 160
Chaetognatha **29**: 195, 196, 199; **44**: 3, 10,
    22, 24, 26–28, 30, 32–35, 39, 42,
    43, 119
    numbers of species, data points and
        ranges **44**: 11
    parasites **25**: 119, 122, 125, 129, 138,
        141, *142*
        metazoan **25**: 147, 148, 149, 150, 158
    predators **25**: 7, 11, 12, 14, 15, 23, 27, 29,
        36
Chaetognaths **36**: 36, 200
Challenger expedition **35**: 3
Channa sp **36**: 36
Chaos and metapopulations of parasites
    43: 36–41, *38–40*
Char, Arctic **36**: 107
Character species **23**: 58, 62
Charleston Lumps **42**: *276*

Chelicerata **29**: 197, 198
Chelifore, pycnogonid **24**: 4, 8, 9
Chemical composition, crustaceans **29**: 130,
    186, 187
  Anomura **29**: 153–159, **154–158**
  Brachyura **29**: 136–139, **138**, **139**
  Copepoda **29**: 178, 179
  Macrura **29**: 168
Chemical content and egg quality in
    fish **26**: 79–85
Chemical induction method of quantifying
    fecundity **43**: *94–95*, 101, 103,
    104
Chemical pollution, *Donax* **25**: 186, 197
Chemistry
  marine poisons **21**: 66
  paralytic shellfish poisoning **21**: 77
  sponge poisoning **21**: 97
  toxic echinoderms **21**: 151
  venomous cnidarians **21**: 112
  venomous molluscs **21**: 174
Chemocline **29**: 97
Chemolithotrophic bacteria **23**: 317
  oxidation reactions **23**: 316
    free energy **23**: 316
Chemoreception
  feeding responses **33**: 1*44*, 144–146,
    149
  reaction distances **33**: 432–434, **433**
  role in mate-seeking behaviour **33**: 257
Chemoreceptors
  prefeeding, sea anemones **22**: 69
Chemosensitivity **30**: 201, 202
Chemosensory detection
  Donax and *Bullia* **25**: 192, 193
  predation **25**: 7, 8, 10, 14, 15, **26**
Chemo-sensory perception, sole **29**: 252
Chemosynthetic nutrition in protobranchs
    42: 21–24
Chesapeake Bay **43**: 65, 70
Chesapeake Bay, striped bass in **38**: 28, 29
Chile, coastal areas of **43**: 149
China Sea Shelf fisheries **39**: 267, 268, 276,
    266, **275**
China, coastal areas of **43**: *12*
China, sea cucumber in **41**: 131, 134, *137,
    138, 150,* 190
Chironomids **36**: 236
Chi-squared goodness of fit models **26**: 194
Chitin **33**: 248
Chitin, crustaceans **29**: 187
  Copepoda **29**: 175, 178

Decapoda **29**: 136, 141, 142, 153, **155**,
    157, 178
  euphausiids **29**: 183
Chitosan test **29**: **155**, **157**
Chloride cells **20**: 164
chlorine **41**: *11, 12, 120*
Chlorine and chlorophenol toxicity and
    bivalves **37**: 3, 77, 96, 116, 123,
    132
  bioassay procedures **37**: 70–72
  in biocides **37**: **101**, **102**, **103**, 97
Chlorine pollution, *Donax* and *Bullia*
    25: 197
Chloriridovirus, plankton **25**: 119
Chloroform toxicity and bivalves
    37: 96–**105**
Chlorophyceae **29**: 78
Chlorophyll
  beyond Ushant front, temperature
    associations **23**: 249, 250
  internal wave effects **23**: 254–256
  Celtic sea shelf break **23**: 271, 272
  frontal content **23**: 236, 237, 274–276
  in eddies **23**: 244
  mathematical models simulating
    23: 239
  maximum subsurface **23**: 236, 239,
    250, 251, 259–261
  organism associations **23**: 264–266
  screening effect **23**: 250–252
  taxonomic composition **23**: 260–266
  thermoclinic association **23**: 236, 251,
    257
  vertical physical gradient
    associations **23**: 258
  oxygen correlations **23**: 259–261
Chlorophyll concentration, phytoplankton
    29: 4, 6, *6*, 16, *17*, 28, *28*, 34
  see also Biomass, Bloom
  fluorescence **29**: 82, *82*
Chlorophylls and coral reef organisms
  see also photosynthesis
  chl-*a* **43**: 282, *283*, 285, 293, 309
  chl-*b* **43**: *283*, 292, 293
  chl-*c* **43**: *283*, 285, 292, 293
  optical properties **43**: 282, *283*, 285, 286,
    289, 292, 293
  separability of spectra **43**: 306, 307,309,
    310
Chlorophyta **29**: *17*
Chloroplastic pigment equivalents
    (CPE) **30**: 19

CHN analyses **30**: 18
Cholesterol **36**: 18, 64, 90
Chondroitin sulphate, crustaceans **29**: 168
Chorioid gland **20**: 131
Chorion hardening in fish eggs **26**: 12, *13*, 76
Chorion, Cephalopoda **25**: 87, 88, *89*, 90, 105
Chorion, eggs, predation 25: 16, 38
Chromatography, high performance liquid **43**: 285, 309, 310
Chromatophores **27**: 199–201
Chromatophores, Cephalopoda **25**: 87, *96*, 98, 103
Chromatophores, sole **29**: 230
chromium **41**: *12, 13, 120, 122*
Chromium toxicity and bivalves **37**: 87, 88, **90–95**
Chromosomal aberrations and egg quality in fish **26**: *85, 86*
Chromosomes **33**: 48
  aberrations **33**: 48
  haploid number of **33**: **47**, 48
Chronobiological theory and vertical migration **26**: 148–150, 152
Chrysophyceae **29**: 78
Chukchi Sea **29**: *56*, 57
Chum salmon **36**: 171
Chytridiomycetes, parasitic **25**: 122
Cicatrix, Cephalopoda **25**: 108
Cichlids **36**: 133
Ciguareta poisoning **31**: 315
Ciguatoxin **21**: 127
Cilia
  cephalopod embryo **25**: 95, 97, 103, 105
  Donax **25**: *182*, 193
Cilia motion **44**: 34
Ciliary currents, predation **25**: 17
Ciliated epithelial field (CEF) **44**: 172
Ciliated epithelial receptors of Scaphopoda **42**: 191, 192–194
Ciliates **29**: 19, 20, 24, *25*, 42, 49, 50, 105, *106*
  ectosymbionts **33**: 525
  in diet **33**: 146, 156, 157
Ciliates, autotrophic, red tidal **23**: 221
Ciliates, bacteria **25**: 120
Ciliophora
  parasites and symbionts of **25**: 120, 124, 126, 129–131, *130*
  parasitic **25**: 137–139, 158
Circadian rhythms **26**: 116, 117, 119, 150
Circadian rhythms *see* Daily rhythms

Circatidal activity **26**: 137–139
  short persistence of **26**: 138
Circulation and circulatory system
  Pleurotomarioidean gastropods **42**: *251, 253*, 259, 260
  Scaphopoda **42**: 180–188, *183, 187*
Circulation pattern(s), tidal front **23**: 184, 185–189, 277, 278
  Langmuir- 221
Circulatory system **33**: 39
Circulatory system, blood cells **27**: 31–33
  morphology **27**: 29–33, *30*
  volume **27**: 173
Circulatory system, Cephalopoda **25**: 94, 95, *96*
Circulatory system, pycnogonid **24**: 18–20
  see also Blood
Cirrata, eggs **25**: 88
Cirripedians, hydrothermal vent dwelling **23**: 329, 337
Citation frequency **33**: 7
Citharidae **29**: 219
Citrate synthase (CS) **30**: 239
Citric acid cycle, crustaceans **29**: 143, 144
Cladocera **29**: 77, 199, 252
Cladocerans **23**: 225
  annual cycles and population dynamics **31**: 132–147
  biochemical and genetic analyses of populations **31**: 151, 152
  birth rates and life tables **31**: 141
  brood size **31**: 139–140
  development of gamogenic eggs before diapause **31**: 125–129
  developmental physiology **31**: 150, 151
  distribution and abundance **31**: 133–138, **134–5**
  grazing **23**: 230
Clam poison **21**: 75
Clambake I hydrothermal vent **23**: 333
  Mussel Bed **23**: 333, 334, 347
Clams *see* taxonomic index
Clams, hydrothermal vent dwelling **23**: 333, 334
Clams, toxicity **21**: **81**
Clarion–Clipperton Zone **35**: 93
Classical food chain **29**: 42, 49, 60
Classification **33**: 1, 2, **2**, 50, **51–53**
  floristic **23**: 57–71
  of pycnogonids **24**: 25–27, 76–78
  of siphonophores **24**: 110–121
  vegetation, *see* Vegetation classification

classification and feedbacks in
        system **41**: 86, 87
Classification in benthic marine infaunal
        studies **26**: 217–221
Classification, sole **29**: 219, 220
CLD (complete lethal dose) **21**: 64
Cleaners of hosts **43**: 35, 36
Clearance rates, *Donax* and *Bullia*
        25: 237, 238
Cleavage of bivalves **37**: 22–26, *36*
Cleavage, Cephalopoda **25**: 91–93, *92*
Climate
        see also temperature
        and pollock **37**: 203, 204
        and fish stock recruitment **26**: 256
        change and recruitment in fish
                populations **26**: 256
Climate *see under* seasonality
Climate, global, winter isotherms **27**: *128*
Climate, predation **25**: 50
Clinical problem
        cnidarians **21**: 128
        echinoderms **21**: 167
        molluscs **21**: 188
Closed areas, fishing **29**: 299
Close's Fishermen's Chart **29**: 246
Clupeidae, as predators **27**: **362**
Clupeids, predators **25**: 9
Clupeoid larva, and predation **20**: 56–58
        feeding in **20**: 26, 27
        feeding rate of **20**: 30–32
        growth rate of **20**: 77, 78, **79**
        mortality rate in **20**: 58
        prey of **20**: 27, 28
        swimming of **20**: 87, 88
Cluster analysis and classification in benthic
                studies **26**: 217, *219*, 220,229, 230
Clutch size *see* Brood size
Clyde herring **20**: 23
Clyde Sea **42**: 31
Clyde, Firth of **43**: 67
Cnidaria **44**: 3, 10, 21–24, 26, 27, 28, 30,
                32, 34, 40, 42, 121, 122
        numbers of species, data points and
                ranges **44**: 11
        parasites of **25**: 125, 126, 127, 131
                metazoan **25**: 148, 154, 156
        predators **25**: 5, 14, 27, 33, 36
        symbionts **25**: 124
Cnidarian stings, characteristics **21**: **132**
Cnidoblasts **21**: 108
Cnidocil **21**: 108

Coagulation theory, plankton **29**: 37–40,
        39, **40**, *41*
Coastal and Shelf Ecology of Antarctic
        Sea-Ice Zone Group **43**: 174
Coastal currents, dispersal processes
        28: 51–53
Coastal environments **33**: 470–480
        biomass **33**: 457, 458
Coastal jets and eddies, entrainment,
        dispersal processes **28**: 53, 54
Coastal lagoon fisheries **34**: 73–199
        anthropogenic constraints **34**: 134–151
        anthropogenic variables **34**: **137**
        biological characteristics **34**: **135**
        biological variables **34**: 115
        change in catch per unit effort accordingto
                number of fishermen per unit
                area **34**: *146*
        change in fishery yield
                according to mean depth **34**: *127*
                according to water area **34**: *126*
        collected lagoon data **34**: **184–199**
        correlation between area of immersed
                vegetation (*SG*) and *Res* **34**: *143*
        correlation between fishery yield and
                maximum depth **34**: *127*
        data description and collection
                34: 86–116, **87–109**, **111, 112**
        definitions of lagoon **34**: 77–79
        descriptive statistics **34**: 117–119, 134
        diversity of fishery data and fishery
                techniques **34**: 110
        effect of brushparks **34**: 823
        effect of devices and practices on
                catch **34**: 82
        effect of fish barrages **34**: 83
        effect of freshwater run-off on yield
                34: *139*
        effect of introduction of exotic fish
                species **34**: 83–85
        effect of tide height on yield **34**: *140*
        environmental constraints **34**: 134–151
        environmental data **34**: 114, 115
        fishing effort and catch per unit effort
                34: 144, 145
        fishing practices **34**: 116
        fishing pressure **34**: 150
        flushing index (*FI*) **34**: 115
        geographical constraints **34**: 117–134
        geographical data **34**: 110, 119–121, *121*
        geographical variables **34**: **124**

maximum sustainable yield (*MSY*) **34**: 130, 132, 148
mean and maximum depths **34**: 114
morphoedaphic index (*MEI*) **34**: 133, 148
morphometrical constraints **34**: 117–134
morphometrical data **34**: 113, 114, 121–128
morphometrical variables **34**: **124**
multivariate analysis **34**: 128, 129
overview **34**: 77–86
physical characteristics and variabilities **34**: 79
physico-chemical characteristics **34**: **135**
physico-chemical data **34**: 140–142
physico-chemical variables **34**: 115, **136**
productivity **34**: 149
significant relationships **34**: **124**, **136**
statistical analysis **34**: 110, 116, 117
statistical approaches **34**: 86
studentized residuals (*Res*) **34**: 117
tidal prism (*TP*) **34**: 115, 138
type I–V lagoons **34**: 78, 79, 118
under-estimation (*UE*) **34**: 110, **113**, 119
vs. wetland area **34**: *144*
water exchange
    characteristics **34**: **135**
    data **34**: 135–139
    variables **34**: **136**
yield **34**: 80–82, *80*, 110, **122**, 131–133, 138
    and anthropogenic characteristics **34**: **135**
    and latitude **34**: 119
    data **34**: **179–183**
    distribution **34**: *123*
    per unit area according to tidal prism (*TF*) **34**: *138*
    per unit area as function of fishing pressure **34**: *145*
    records **34**: 117, *118*
    vs. latitude **34**: *125*
    vs. mean nitrite–nitrogen concentration **34**: *142*
    vs. minimum salinity **34**: *141*
    vs. minimum temperature **34**: *141*
    vs. radiation balance **34**: *125*
    vs. under-estimation (*UE*), **34**: *123*
    vs. water volume **34**: *128*
    year-to-year variations **34**: *120*
Coastal water and plankton production **26**: 254
Coastal zone colour scanner (CZCS) **31**: 352

Coasts and estuary frontal systems **23**: 208, 209; *see also* River plumes
    h/u³ criterion **23**: 208
    heterotrophic accumulations **23**: 225, 226
    Loire, *see* Loire outflow
    zone colour scanning **23**: 271, 274
cobalt **41**: *12*, **77, 78**, *122*
    biological availability and uptake **41**: *26, 28, 29, 48*
    distribution in oceans **41**: 77
    functions and properties **41**: *15–17*, 19
    impact on primary production **41**: *13*, 14, 77, 78
    interactions **41**: 33, 34–37, *34–36*
Coccidia, parasitic **25**: 139, 144, 158
Coccoid cyanobacteria **29**: 89, 90
Coccolithophores, in diet **33**: 159
Cockburn, Bank of **29**: *223*, 224, 246
cod (Scandinavia) **38**: 18–23
Cod **36**: 3
    adaptation **36**: 28, 38
        food and feeding **36**: *50, 51*, 52, 53, 55
        pressure and buoyancy **36**: 45, 46
        temperature **36**: 11, 15, 17
    adaptation strategies
        energy metabolism **36**: 61, 68, 70
        plastic metabolism **36**: 78, 79
    and gadoid outburst **26**: 274–276
    and match/mismatch hypothesis **26**: 252, *270, 282, 283*
    and salinity anomaly **26**: *277*
    diel vertical migration of **26**: 120, 121, *126*, 141
    differentiation and variability **36**: 221, 222, 224–226, 227
    ecological principles **36**: 231, 246, 251
    egg quality **26**: 73, 74, 77
    egg viability **26**: 90, 92
    indicators of condition **36**: 214, 215, 217, *218, 219*
    lipids **36**: 206, 210, 211
    larvae of and plankton production **26**: 279–281
    life cycles **36**: 114, 115
        abundance, dynamics of **36**: 126, 127, 133, 134, *135, 136*, 137
        interannual fluctuations **36**: 118, 120, *121–123*
        ontogenesis **36**: 90, 98, 102, 103, 104
    prey composition of **26**: 130, *131*, 132
    recruitment of in Baltic Sea **26**: 272–274
    substance and energy **36**: 141, 170

accumulation and
    production **36**: 148–150, 152, 153
consumption, transformation and
    utilization **36**: 176, 180, 181, 190,
    196
Cod, Atlantic: comparison with
    pollock **37**: 182, 192, 237, 239
    population structure **37**: 207, 218, 220,
    221
Cod, Pacific **37**: 217, 218
Coelacanth **36**: 36, 46
Coelenterates, deep-sea, vertical
    distribution **32**: *333*
Coelenterates, hydrothermal vent
    dwelling **23**: 319, 336
Coelomocyte types, Brachiopods **28**: *213*,
    214–217
Coenobitidae **29**: **132**
Co-evolution **23**: 121, 127
Co-evolution of parasites and hosts **43**: 46
Cohesion Species Concept (CSC) **31**: 11
Coho salmon **36**: 23
Cohort approach **44**: 4, 9, 24
cohort competition **38**: 7
Cohort measurements **44**: 41
Cohort methods **44**: 39
Cold regions
    plankton distribution **32**: 11–29
    summary **32**: 29
Coleman analyser **24**: 396
Coleoids, arms **25:** 99, 100
Collagen and temperature **36**: 15
Collagen, crustaceans **29**: 166
Colonial species
    see also pigmentation in coral reef
        organisms
    of invertebrates, fecundity of **43**: 130
    life history **43**: 133, 134, 149, 150
    quantification methodologies
        43: 110, 111
Colonialism, siphonophore **24**: 109, 110
Colonization rate **23**: 28
Colonizations **33**: 505–508, **507**
Colour *see* pigmentation
Colour, skin **36**: 52, 225, 226
Colour, sole **29**: 220, 221, **222**
Colouration
    of pycnogonids **24**: 16
    of siphonophores **24**: 199–202
Comber **36**: 20, 21
Commensal species and vertical
    migration **26**: 145

Commensalism **33**: 503
Commensalism, pycnogonid **24**: 49–52
Commensals of Protobranch bivalves
    Plates 3 and 4 **42**: 36
Commensals, zooplankton **25**: 129, 138
Commercial exploitation of deep sea **35**: 5,
    6, 93–109, 116
    mineral extraction **35**: 96, 97
    oil **35**: 97–101
    radioactive waste **35**: 101, 102
Committee, Fisheries **29**: 324
Common Fisheries Policy (CFP) **29**: 331,
    333
common grey mullet **38**: 33
Communities
    see also ecology
    of parasites **43**: *30*, 31, 33–36
Communities *see* Associations
Community
    and biomass distribution in benthic
        studies **26**: 209–212
    anthropogenic impact on **26**: 190–192
    concept of benthic studies **26**: 171,
        186–192
    descriptive univariate indices
        26: 192–212
    structure in benthic marine infaunal
        studies **26**: 197–209
    vs continuum in benthic marine
        infaunal studies **26**: 186–190
Community ecology, seaweed **23**: 41–45
    'concept of the community', 42, 43
        holistic view **23**: 42
        individualistic/continuum view **23**: 42
    experimental design considerations
        23: 112, 113
    floristic classification **23**: 57–71
    structure
        keystone species **23**: 73
        models of mechanisms determining,
            see Population growth
        multivariate classification **23**: 133, 134
        physical disturbance roles in **23**: 115
        rock surfaces, *see* Rock surfaces
        variables influencing, *see* Biotic
            interactions; Physiochemical
            determinants
    vegetation; *see* Vegetation structure
community structure, influence of
    41: 89–92, *91*
    on iron **41**: *58–61, 59, 60*

Compact Airborne Spectrographic Imager
  Plate 3 **43**:
Comparative fecundity, sole **29**: 288–290, **290**
Competition **33**: 468, 469
  crustaceans **29**: 144, 145
  phytoplankton **29**: 18
  sole **29**: 252, 253
Competition and parasites **43**: 24–33
Competition for food **36**: 56, 57, 200, 202, 209, 217
Competition *see also* Resources
  coefficients **23**: 14
  effects on recruitment **23**: 30, 31
  hierarchies **23**: 110, 111
  interactions with carnivory **23**: 121, 122
  interactions with herbivory **23**: 113–117 and carivory **23**: 72–108
  interspecific **23**: 10, 12–23, 72–109, 110, 111, 121
    removal of keystone predator **23**: 73–75
    structuring dominance **23**: 82
  planktonic **23**: 223
    possible configurations **23**: 16
    ratio diagrams, input/output **23**: 16
  role in community organization **23**: 122–126
  structuring of zonation by **23**: 108, 109
  studies in isolation **23**: 108–111
  thinning by, *see* Thinning
  'triangular' examination technique **23**: 17
Competition, *Donax* and *Bullia* **25**: 189, 236, 237
Competition, in communities **26**: 198, 199
competitors of *Pandalus* **38**: 178, 179
Competitors, Japanese oyster culture **21**: 32
Complexation Field Diagram **41**: 8
Composition of sea ice *see* sampling
Compositional analysis **44**: 260
concentration of elements in seawater **41**: 5, 6, 120–123
  vertical profiles **41**: 7–9, *8*, 9, 43, *44*
Concentration of parasites in some clades **43**: 17–22
Condition factor **20**: 46, 47, 48, 49
'Condition factor', sole **29**: 263
Condition factors (CF) **33**: 336
Condition indices **30**: 219–241
  future research **30**: 286
Condition measures **30**: 217–303
  ideal **30**: 283

laboratory versus field
  estimates **30**: 265–271
processing time, costs, and requirements **30**: 281, 282
summary of key attributes **30**: 283, **284**
Condition trajectories **30**: *266*
Conditioning of spawners **37**: 25–27
Conducting system
  sea anemones **22**: 69
Conductivity temperature depth probes (CTDs) **31**: 174
Cone ellipsoid **20**: 135
Conger **36**: 77
Conidae, locomotion **25**: 198
Connell's models
  first model **23**: 118–120
    effect of physical conditions **23**: 119
    experimental testing **23**: 119
  second model **23**: 120
  third model **23**: 120, 121
    experimental testing **23**: 121
Consumption rates, *Donax* and *Bullia* **25**: 192, 225–229, **226**, 230, 231, **231**, 232–234, 233
Consumption, transformation and utilization of substance and energy **36**: 174–196
Contact predation **25**: *6*, 7
Contaminants, fisheries **29**: 350
Continental Shelf, European, tidal front, *see* European Shelf
Continental shelves **33**: 444, *478*, 479–482
  across-shelf gradients **33**: 480, 481
  biomass **33**: 457, 458
  upwelling **33**: 482, **484, 485**
Continuous Plankton Recorder (CPR) Surveys **28**: 129
Continuous Plankton Recorder Network **26**: 251, 252, 269, 270
Continuous tracking **44**: 252
Continuum, *vs* community in benthic marine infaunal studies **26**: 186–190
Contractile proteins and temperature **36**: 14, 15
Contrast, prey detection **25**: 15
Controlled approach **44**: 4
Convergences, *see* Fronts; Surface convergence
Convergent adaptation, *Donax* and *Bullia* **25**: 183, 198, 205, 235
Conway diffusion technique **24**: 397
Co-occurrence patterns **43**: 85

Cooperation in communities **26**: 198, 199
Co-ordinated group avoidance and evasion
    tactics **30**: 193
Coordination, siphonophore **24**: 182–189
Copepod nauplii **26**: 251, *283*
Copepoda **29**: 178, 179
  Macrura **29**: 187, 189
  parasites of **25**: 122, 144, 157
    bacteria **25**: 119
    ciliophoran **25**: 137, 138, 139
    dinoflagellate **25**: 122, 125, 127, *132*,
        133
    ellobiopsid **25**: 136, *136*
    fungi **25**: 121
    metazoan **25**: 147, 148, *149*, 150, 151
    sporozoan **25**: 141, 142, 143
  parasitic **25**: 152, 153
  predation **29**: 2, 19, 20, 77, 105
    bloom conditions **29**: *43*, 44–47, *48*,
        51, 52–54, *53*, *55*, 57
    cell size **29**: 22–24, *23*
    dissolved organic matter **29**: 33–35, *34*
    turbulence **29**: 24, 27–29, *27*, *28*
  predators **25**: 5, 6, 8, 12, 21, 22, 36, 38,
        48, 59
  prey **25**: 29, 36, 37, 42, 59; **29**: 252, 253
  spermatophores **29**: **131**, 174, 195
    changes during passage through
        male **29**: 179, 180
    chemical composition **29**: 178, 179
    morphological diversity **29**: 174, 175,
        177
    origin **29**: 175, 176, 178
    predation **29**: 33–35, *34*
    transfer and storage **29**: 180–183, 199,
        200
  Stomatopoda **29**: 184
Copepodid
  see also Development
  feeding **33**: 167
  morphology **33**: 24–28
  segmentation development **33**: 24–27,
        25, **27**
  swimming pattern **33**: 403–407
Copepodites **26**: 251
Copepods **20**: 25–30, *46*, 50; **30**: *29, 30, 38*,
        51, **57**, **58**, *62, 63*, 64, *68*, **71**, 170;
        34: 419, 420, *420*; **36**: 117, 200,
        217; **37**: 107, 185, 187; **44**: 32, 43
  adaptation **36**: 32, 36, 44, 45, 46, 55, 85
  as prey items for fish larvae **44**: 297
  broadcasting **44**: 10, 16, 41, 42, 51

broadcast-spawning **44**: 14
carbon content **44**: 33
culture, overview **44**: 296–298
data collection methods **44**: 4–8
egg production **44**: 35
food role in **44**: 36, 37
growth rates **44**: 4–8
  versus body weight **44**: 10, 12, 13
  versus temperature **44**: 18–21
in benthic studies **26**: 191
in match/mismatch hypothesis **26**: 251,
        262, 263, 269, 280
numbers of species, data points and
        ranges **44**: 11
  see also harpacticoids
results of investigation **44**: 10–12
sac-spawning **44**: 10, 15, 17, 41, 42, 51
temperature influence on growth rates
        in **44**: 7
Copepods, hydrothermal vent dwelling
        23: 328, 329, 336, 337
  Vestimentiferan associations **23**: 329
Copepods, parasitic **39**: 184
copper **41**: 22, **63–67**, *120*
  biochemical function **41**: *11, 12*
  biological availability and uptake **41**: 20,
        27, 29, *31*, 66
    impact on primary production **41**: *13*,
        14, 23, 64–67, *66, 67*
  cycling ratio **41**: *84–86*
  distribution in oceans **41**: 63, 64, 66,
        67, 124
  functions and properties **41**: *15–17*,
        20, 21
  indices of fractionation **41**: *81, 82*
  inputs to system **41**: *123*
  interactions **41**: 33, 34, *35*, 37–39, *124*
  internal economy and near-field
        chemistry **41**: 27, 29, *31, 67*, 88,
        89, 90, *91*
Copper toxicity and bivalves **37**: 3, 45, *54*,
        132
  tests **37**: 88, **90–95**, 116, 119, 122, 123
    larval physiology **37**: 66, 67, 68, 70, 72
Copper, bivalve molluscs
  accumulation in tissues **22**: 163
  behavioural responses to **22**: 117–119,
        164–168
  effects on
    gametogenesis **22**: 166, 168, 169, 172,
        173
    isolated siphons **22**: 163, 164, 167

mitochondrial respiration **22**: 176–180
  physiology **22**: 164–168
  unfertilized eggs **22**: 173–175
mortality **22**: 122, 123T
  sublethal levels **22**: 129, 130
Copper, crustaceans **29**: 137, **139**
Copulation (see Behaviour, reproduction)
Copulatory organs, crustaceans
  Brachyura **29**: 139–141
Coracidia **25:** 158
Coral **36**: 83
  see also Reef corals, adaptations of
  bleaching **31**: 254–257, 275, 276
  ecomorphs **31**: 225
  fish/coral-reef fish **36**: 66, 69
  hybridization in **31**: 53–4
Coral reef ecosystem **34**: 230, 231,
  231
Coral reefs **44**: 242
  changes monitoring by
    coral colony size measurements **22**: 4,
     5
    nonliving substrate assay **22**: 4
    quantitative technique
      plotless **22**: 13
      quadrant **22**: 13
      stereophotography **22**: 4
  future research **22**: 51–54
  natural disturbances **22**: 5–9
    in various geographical areas **22**: 6, 7T
  pollution effects **22**: 9–16
    juvenile coral mortality **22**: 19
    responses to various
      pollutants **22**: 10–12T
    tolerance in shallow water **22**: 14
  recovery rate **22**: 16–20
  temperature effects **22**: 11, 14–16, 26
Coral reefs *see* pigmentation
Coral reefs, deep sea **35**: 6
Coral reefs, *see* Reefs
Corals, stress
  behaviour **22**: 35–46
    feeding and mesenterial filament
      extrusion **22**: 35–39
    mucus production **22**: 39–42
    polyp retraction **22**: 38
    sediment shedding **22**: 43–46
  biochemical indexes
    lipid/protein ratio **22**: 50
    lysosomal glucosaminidase **22**: 50
    lysosomal hydrolases **22**: 50
  definition **22**: 9

growth rate **22**: 20–27
  alizarin red S stain **22**: 21, 23
  calcification **22**: 24, 27
  disturbance effects **22**: 22–24T
  fixed base line **22**: 21, 23
  X-radiography **22**: 21, 22, 25
histopathology **22**: 48–50
  algal infection **22**: 48
  bacterial infection **22**: 48
  shutdown **22**: 48
  tumours **22**: 49
  white band disease **22**: 48
loss of zooxanthellae **22**: 29–35, 39
metabolism **22**: 27–29
  photosynthesis **22**: 28, 29
  respiration **22**: 28, 29
reproduction **22**: 46–48
  laboratory studies **22**: 47, 48
  oil pollution and **22**: 46
  polyp bail-out **22**: 47, 48
Corals, toxins **21**: 104, **105**
Core sampling in Antarctic *Plate 2* **43**: 186
  damage caused by **43**: 189, 190
Core-satellite hypothesis **43**: 36
Cormorant **20**: 60, 231, 249
  bandit **20**: 60
  Brandt's **20**: 241, 243, *244*
  Cape **20**: 247, **248**, 252
  Cunay **20**: 166
  double crested **20**: 61, 245, 246
  pelagic **20**: 243
  Peruvian **20**: 260–265 *passim*
cormorants, predation of plaice **38**: 37
Cornish pilchard **20**: 17
Correspondence Analysis in benthic
    studies **26**: 222
Cortical reactions in fish eggs **26**: 73
Corycacidae **29**: 179; **44**: 79
Coryphaenoides sp **36**: 44
Cosmopolitanism of Protobranch
    bivalves **42**: 39
cost-benefit analysis of enhancement **38**: 43,
    44
Costs, Japanese oyster culture **21**: 29
Cotton tail symptom **27**: 379
Counting, picoplankton **29**: 81–83, *82*, 85
Coupling apparatus, crustaceans **29**: 175,
    176–178, *177*, 182
Cousin Reserve **34**: *250*
CPUE *see* catch-per-unit-effort
Crab **36**: 27
Crab(s)

algal grazing **23**: 114
hydrothermal vent dwelling, *see* Decapods
snail grazing by green- 88, 89
Crabs **29**: 130
see also Brachyura
Crabs, toxic **21**: 139
Crangonoidea **29**: **132**
Crash **20**: 262, 264, 265, **270**
Crepuscular spawning **20**: 22
Cretaceous
Pleurotomarioidean gastropods **42**: 237,
242, 266, 267, *268*, 281
Scaphopoda **42**: 215, 217
Critical period concept **31**: 170, 171
Crowding coefficient
individual **23**: 14
relative **23**: 14, 15
Cruise-searching larval predator **31**: 186
Cruising invertebrates, predators **25:** 23–26,
24, **26**
Crustacea **44**: 3, 10, 21–24, 26–28, 32, 34,
39, 40, 42, 122–128, 253
see also Amphipoda, Copepoda,
Decapoda, Euphausiids
as food **26**: 26, 27
as parasites **26**: 38
numbers of species, data points and
ranges **44**: 11
parasites of **25:** 135
bacteria **25:** 119
dinoflagellate **25:** 125, 127, 131, 133
metazoan **25:** 148, 150, 151, 158
sporozoan **25:** 141, 142
parasitic **25:** 152–156, *155*, 158
predators **25:** 5, 6, 8, 12, 21, 22, 36, 38,
48, 59
prey **25:** 29, 36, 37, 42, 59
Crustacea, parasitic **27**: 387–389
as predators **27**: 360
crustacean parasites of *Pandalus*
38: 181, 182
Crustaceans **29**: 253; **36**: 36, 46, 231; **37**: 97,
131, 132
see also Spermatophores
deep-sea species diversity **35**: 5
growth performance **35**: 173–175, *176*,
177, *178*, *179*
heterozygosity **35**: 59
hydrothermal vents **35**: 18, 19, 27
oxygen concentration influence on **35**: 27,
28
relict and 'fossil' species **32**: **409**

shrimps of hydrothermal vent
sites **32**: 107–135
small, in sea anemone diet **22**: 93
Crustaceans, hydrothermal vent
dwelling **23**: 328–331, 336, 337
Cryoinjury, crustaceans **29**: 189
Cryopreservation, crustaceans **29**: 187–189,
188
Cryoprotectants **29**: **188**, 189
Cryoprotective agents, *see* Antifreeze
Cryptophyceae **29**: 78, 82
Cryptotoxic animals **21**: 63
CS-EASIZ (Coastal and Shelf Ecology of
Antarctic Sea-Ice Zone Group)
43: 174
C-shaped contortion, larvae, predation
25: 19
CSIRO, Commonwealth Scientific Industrial
Research Organisation **27**: 5, 6
Cuticle **27**: 213–221, *215*, *216*, **217**
unpublished information on, fish as
predators **27**: 371
larvae in plankton **27**: 358
spawning migration **27**: 314
swimming speeds **27**: 313
C-start **20**: 39
"C-start" fright responses **38**: 34
Ctenophora **44**: 3, 10, 21–24, 26–28, 30,
32, 34, 35, 39, 40, 42, 128, 129
numbers of species, data points and
ranges **44**: 11
parasites of **25:** 147, 148, 149, 150, 154,
155
predators **25:** 5, 7, 8, 29, 30, 33, 36,
41, 47
symbionts **25:** 124
Ctenophore **36**: 56, 130, 200, 202, 209,
217, 246
Ctenophores **20**: 53
Black Sea **32**: *42*
role in plankton communities **32**: 64–68
Cubomedusae, predators **25:** 7
Culture of copepods, overview **44**: 296–298
Culture schemes, oysters, Hiroshima Bay
21: *24*
Cumaceans **29**: 185
Cunay cormorant **20**: 166
Current
barnacle growth and **22**: 216–218
Current edges, biological characteristics
23: 216

Current patterns, South African coast
  25: 185
Currents, Bristol Channel **29:** 218,
  227–229, *228,* 236, 340, 350
Currents, ocean, pollock and **37:** 190, *227,*
  229, *230, 231*
Cushing's match-mismatch hypothesis
  20: 7, 25
CUSTOMBLEN **39:** 29
Cuticles, eggs, predation **25:** 16
'Cyanellae' **25:** 120, 124
Cyanide toxicity and bivalves **37:** 96, 97,
  120, 132
α-Cyanoacrylate **29:** 189, 190
Cyanobacteria **21:** 87
Cyanobacteria, Baltic Sea **29:** 4, 13, 80,
  87, 98
  counting **29:** 81, 82
  distribution **29:** 89, 90, **89**
  grazing **29:** 104, 105
  nutrients and temperature **29:** 102, 104
  seasonal variation **29:** 87, **89,** 91, 92, 94,
    118
Cyanobacteria, N2-fixing reef **23:** 105
Cyanobacteria, plankton **25:** 120
Cyanobacterial toxins **31:** 315, 316
cycling ratio and fractionation of elements
  in oceans **41:** 83–86, *84–86*
Cyclopoid copepods, predators **25:** 5, 7, 8
Cyclopoida *29:* **131,** 182
Cydippid ctenophores, predators **25:** 5
Cynoglossidae **29:** 219, 220
Cytochalasin B, effect on Cephalopoda
  25: 93
Cytochrome C oxidase, crustaceans
  29: 138
Cytogenetics, siphonophore **24:** 100
Cytotoxins **21:** 68
CZCS (coastal zone colour scan) **43:** 196
Czekanowski coefficient in benthic
  studies **26:** 215
Czekanowski's coefficient **23:** 52

Daily rhythms **36:** 114–117, *115,* 230, 239
Damselfish
  and community structure control
    23: 104, 105
    territorial defence **23:** 104
  -sea urchin interactions **23:** 105
  -seaweed interactions **23:** 101
Daphnia **31:** 87, 97
  embryonic development **31:** 99, 102, 103

fertilization **31:** 125
genetic diversification **31:** 151
oogenesis in **31:** 98
sex ratio **31:** 109
temperature and **31:** 145
Daphnia magna
  cultivation **31:** 148
  developmental physiology **31:** 151
  reproductive behavior **31:** 122
  sex determination **31:** 111
Daphnia pulex
  developmental physiology **31:** 151
  embryonic development **31:** 99
  gamogenic egg development **31:** 126,
    127
  sex determination **31:** 115
Dassen Island **20:** 249, 257, 258
Dassen Island penguins **20:** 259
Data
  analysis of in benthic marine infaunal
    studies **26:** 178–232, **179**
  and statistical inference **26:** 212–214
  collection of in benthic marine infaunal
    studies **26:** 172–178
  multivariate analysis **26:** 214–231
  nature of in benthic studies **26:** 178–186
    reductions **26:** 181, 182
    transformations **26:** 182–186
  time series analysis **26:** 231, 232
Data analysis, animal movement
  44: 259–262
Data and images **39:** 293
Data storage (or archival) tags
  (DSTs) **44:** 253–255
Davis Station (Antarctic) **43:** 211
DCMU-treated incubation **29:** 85
DDT toxicity and bivalves **37:** 97, **101**
De Belgische Zeevisserij **29:** 328
Death
  rates, *see* Mortality rates
  risks **23:** 40
Death rate *see* Mortality
Death *see* Ageing and death
Deborah number **31:** 183
Debromoaplysiatoxin **21:** 88, 181
Debromoaplysin **21:** 180
Decabrachia **44:** 154
Decapod crustaceans, movements of
  44: 205–294
Decapoda **29:** 130, **131, 188,** 190, 194, 197,
  199; **29:** 160, 164, 168, 169, 171

see also Anomura, Brachyura, Lobsters, Macrura, penaeid shrimps
embryonic development **25:** 90, 91, 95, 99, 100, 101, 103
euphausiids **29:** 183, 184
parasites of **25:** 135, 136, 141, 142, 144, 151
Decapoda, classification **27:** 57, 58
fossil record **27:** 149–152
key to families,
adults **27:** 58–60
larvae and postlarvae **27:** 115–118
Decapods **44:** 33
Decapods, hydrothermal vent dwelling **23:** 330, 331, 336, 337
metabolic rates **23:** 330, 344
new superfamily **23:** 330
Decollation of Scaphopoda shells **42:** 149, *150, 151*
Decomposer pathway **23:** 263
Deep bodied anchovy **20:** 131
Deep sea hydrothermal vents, *see* Hydrothermal vents
Deep-scattering layers **24:** 164, 165
Deep-sea and benthic organisms
invertebrates, fecundity of **43:** 93, 106, 115, 116, 143, 147–150
parasites **43:** 6, 7, 54
Southern Ocean **43:** 180, 181, 185, 211, 220, 239, 241
Deep-sea environment **33:** 501, 502; **35:** 3–7, 8–16
adaptations to **35:** 26–28
fisheries **35:** 5, 6, 93, 102–109
life history strategies **33:** 390–392
See also Commercial exploitation of deep sea; Hydrothermal vent environment
species diversity **35:** 5
Defence Meteorological Satellite Program (DMSP) and Operational Linescan System (OLS) **39:** 263, 265
Defense, eggs, predation **25:** 16
Degeneration, crustaceans **29:** 193–195
Dehiscence, crustaceans **29:** 145, 159, 160
Dehydro-abyetilamines **37:** 97, *100*
Delayed invertebrate maturity **43:** 135, 136
Delipidation test **29: 154**
Delrad toxicity and bivalves **37:** 97, **100**
Demarsupiation, Amphipoda **25:** 154
Demersal fish, predators **25:** 9, 10, 48
eggs **25:** 9, 43, 47, 52

Demersal spawners **20:** 13, 22, 23, 24, 193
Demersal species **20:** 252, 284
Demographic analysis *see* Population biology
Demography theory and fecundity of invertebrates **43:** 112, *113*, 118–120, 135, 152
see also age; life history
Winemiller-Rose **43:** 120–123, *121,*152
Denaturing gradient gel electrophoresis (DGGE) **35:** 49
Dendograms
in benthic studies **26:** 217, 218, 228
of Atlantic halibut **26:** 4
Dendrobranchiata **27:** 2, 3; **29: 131**
distinguishing features **27: 58**
Dendrograms **23:** 53, 54
Denmark **42:** 26
Denmark, turbot studies in **38:** 23
Density and salinity **36:** 24
Density of population and resources *see* deterministic
Density *see* Population biology
Density, *Donax* and *Bullia* **25:** 180, 182, 189
Density, of prey
dependent mortality **25:** 49, 55, 55, 59–64, 60, 62, 63, 65
encounter rate **25:** 10, 11
feeding mode, effect on **25:** 9
predator response **25:** 31–36, *32,* 34, 35, 47
Density, plant **23:** 3, 10
-dependent effects, inter- and intra-specific, *see* Thinning
herbivore browsing effects on, *see* Grazing
-mortality relationship **23:** 10, 12, 13–23
-weight relationship **23:** 10, 11
Density, water **29:** 3
Baltic Sea **29:** 75, *75,* 76
Density-dependent growth **20:** 167
Density-dependent mortality **20:** 226
Density-dependent processes **28:** 126–128
Density-dependent regulatory mechanism **20:** 164
Density-dependent relocations **44:** 228
Density-dependent reproductive mechanism **20:** 8, 167
Dentate pedicellariae, urchins **21:** 148
Depletion curve **23:** 6
Deposit feeding by Protobranch bivalves **42:** 17–19, *18*
Depth **20:** 7
critical **23:** 241, 248, 274

theory of planktonic cycles **23**: 210–212
vegetation distribution **23**: 109, 142
   limits **23**: 142, 143
depth distribution (*Pandalus*) **38**: 114–118
Depth distribution **33**: 458, 459, *459, 460*, 486, 487, 504, 505
   see also Vertical migration
Depth gradients and parasites **43**: 4, 5, 14, 54, 55
   see also deep-sea
Depth, Bristol Channel **29**: 223, 224, *223*
   sole **29**: 237, 238, 240, **240**
Depth, predation **25:** 15, 53
Depth, speciation and **35**: 82
Dermatitis **21**: 88, 142, 168, 176
Descriptive univariate indices in benthic studies **26**: 192–212
Desiccation, intertidal species **25**: 184
Desiccation, sperm **29**: 130
Detection, predation **25:** 10, *10*, 13–15, 19, 37
Detergents and oil toxicity to bivalves **37**: 3, 90, 107, **109–115**, 115, 116, *133*
   and bioassay procedures **37**: 63, 64
   and sediments **37**: 77, 129
'Determinate' spawners **29**: 280–282, *281*
Deterministic (r-K selection) model of life history and fecundity of invertebrates **43**: 115–117, 152, 153
Detrended Correspondence Analysis in benthic studies **26**: 226
Detritivores (deposit-feeders), abyssal macrobenthos **32**: 456–466
Detritus feeding **33**: 153, 154
Detritus, in diet **27**: 316, 318, 319, **320**
Deuteromycetes, plankton **25**: 121
Devardo's alloy **24**: 397
Development **33**: 296
   see also Growth; Life history
   development time **33**: **300–309**, *318*
   eggs **33**: 298, 299, **310–313**, *314*, 314
   median development time (MDS) **33**: 322, 323
   food availability and **33**: 3*16*, 316
   models of **33**: 317–321, *319*
      equiproportional development **33**: 318, 319, *327*
      isochronal development **33**: 319
      non-conformist development **33**: 321

   sigmoidal development **33**: 319–321
   of cystonects **24**: 108
   of pycnogonids
      embryonic **24**: 39, 40
      post-embryonic **24**: 40–43
   of siphonophores **24**: 99–110
      of colonies of **24**: 109, 110
      post-planula **24**: 102–108
         of calycophores **24**: 106–108
         of physonects **24**: 102–105
         to planula **24**: 101, 102
   segmentation **33**: 24–27, **27**
   sexual differentiation **33**: 43, 44
   stage duration **33**: **317**, 317–323
   swimming legs **33**: **27**, 27, 55
Development *see* growth and development
Developmental stage *see* age
deviation method **38**: 192
Devonian **42**: 41, 217, *242*, 243
Diadinoxanthin and diatoxanthin and coral reef organisms **43**: *283*, 285, 286, 292
Diapause eggs **33**: 268–272, **269–271, 273,** 388, 389
   see also Life history
Diarrhoeic shellfish poisoning (DSP) **31**: 314, 315, *314*
Diastase test **29: 154**
Diatoms **29**: 4, 5, 22, 78, 79, 87
   abortive outbursts **23**: 215
   arctoboreal species **32**: *549*
   blooms **23**: 260
      spring **23**: 212
   Donax **25**: 185, 191
   ectosymbionts **33**: 525
   exudation **29**: 31, **31**
   in diet **33**: 145, 155, 290
   notalantarctic species **32**: *552*
   photosynthesis **29**: 16, *17*
   predation **29**: 19
   sedimentation **29**: 36, 37
   sinking **29**: 9, 12
   symbionts **25**: 120
   tropical species **32**: *550, 551*
Dichloronaphthoquinone *see* Phygon
Dichlorvos toxicity and bivalves **37**: 97, **102**
Diecdysis **27**: 213
Diel migrations, distribution of macroplankton **32**: 64
Diel movements **44**: 222, 223
Diel rhythms **20**: 32–34, 91
   see also Vertical migration

diet **33**: 169
feeding periodicity **33**: 170, 171, *442*,
    450, 452, 453
swimming activity **33**: 420, 421
Diel vertical migration **24**: 135; **20**: 110,
    111, 112, 115
and depth intervals **24**: 159–162
regional differences in **24**: 163, 164
sinusoidal patterns of, and sampling
    time **24**: 157–159
small-scale patterns **24**: 162
total depth distribution range **24**: 156, 157
Diel vertical migration in fish **26**: 115–168
and endogenous rhythms **26**: 122–150
and ontogenetic variation **26**: 148–150
as facultative process **26**: 154, 155
local environmental factors
    for **26**: 145–148
types of **26**: *119*, **120**, **121**
zeitgeber for **26**: 122–145
Diel vertical migrations **44**: 251
Dieldrin toxicity and bivalves **37**: 97, **102**
Diet **33**: 149–160
bacteria **33**: 154, 155
ciliates **33**: 156, 157
coccolithophores **33**: 149
detritus **33**: 153, 154
diatoms **33**: 155
diel changes **33**: 169
dietary requirements **33**: 168, 169
dinoflagellates **33**: 157, **158**
dissolved organic matter (DOM) **33**: 153
egg production and **33**: 290, 292
foraminiferans **33**: 159
metazoans **33**: 159, 160
^{15}N, organisms and **24**: 424–432
of siphonophores **24**: 227–229
Phaeocystis pouchetii **33**: 155, 156
protozoans **33**: 156–159
radiolarians **33**: 159
selectivity **33**: 164–166
Diet cycles **30**: 175–179
Diet *see* Feeding
Diet variations **30**: 240, 241
Diet, intraspecific variation **27**: 328, 329
(see also Food, nutrition)
species examined **27**: **317–318**
variation in **27**: 328–332
Diet/nutrition **43**: 286
of invertebrates **43**: 99, 128, 141–147
of parasites **43**: 33
and species richness **43**: 34, 35

feeding organ sizes compared
    43: 29, 30
Differentiation and variability **36**: 221–228
interpopulation **36**: 224–227
interspecific **36**: 227, 228
intrapopulation **36**: 222–224
Diffusion chamber experiments **29**: 111, **113**
Diffusion limitation, nutrient
    uptake **29**: 10–12, *11*
Digestion **33**: 41, 204, 205, 209
see also Assimilation efficiency
digestive enzymes **33**: 204
gut clearance rate constant (K) **33**: 188,
    192, *194*, 195, 196
gut evacuation rate **33**: 192, 195, 196
in pycnogonids **24**: 28
rates of, in siphonophores **24**: 234
Digestion and assimilation **27**: 159–163
Digestion rates, predators **25**: 30, **31**, 36, 37,
    39, 47
Digestive enzymes, biochemical
    indices **30**: 237, 238
Digestive gland **27**: 162–163, 218, 219
parasites **27**: 387
Digestive rates and vertical
    migration **26**: 136
digestive system (*Pandalus*) **38**: 173
Digestive system **27**: 159, 160; **33**: 39–41,
    40
assimilation efficiencies **27**: 163
digestive gland **27**: 162, 163
enzymes **27**: 160–163, **161**
foregut **27**: 22–24
gut, hormonal control **27**: 201
hindgut **27**: 29, *29*
midgut **27**: 25–29, *27, 28, 29*, 162, 163
cell types **27**: 25, 26, 28, 163
morphology **27**: 21–29, *23*
Digestive system and feeding
Pleurotomarioidean gastropods **42**: *251*,
    252–259, *253, 255–257*
diet **42**: 279, 280, *279*
Protobranch bivalves **42**: 14, 15,
    17–24, *18*
diet **42**: 19–21
Scaphopoda **42**: 163–179
alimentary tract **42**: 173–179, *174, 177*
behavior **42**: 168–173, *170, 171*
captacula **42**: 164–168, *166, 168*, 192
diet **42**: 163, 164
glands **42**: 178, 179
shelled Opisthobranchs **42**: 96–100, 106

Digestive tract, Cephalopoda **25:** 99
Digging cycle, *Donax* and *Bullia* **25:** 198,
    204, 205, 212, 213, *213*
Digitonin **21:** 157
Dihydromurexine **21:** 177
Dimethylsulphide (DMS), phytoplankton
    production of **31:** 344
Dimethyl-sulphoniopropionate **43:** 206,
    207, 238
Dimethylsulphonium propionate (DMSP),
    phytoplankton production
    of **31:** 344
Dimorphism
    sole **29:** 266, 267, 301
    spermatophores **29:** 148, 150, 179
Dingle Bay **29:** 326
Dinitrogen fixation **24:** 412, 418, 432, 433
Dinoflagellata
    parasitic **25:** 122, 125–135, *128*, *130*,
        132, 157
    symbiotic **25:** 125
    viruses and bacteria on **25:** 118, 119, 120
Dinoflagellate poison **21:** 75, 85
Dinoflagellates **29:** 22, 87
    dominance **23:** 215
    exudation **29:** 31, **31**
    in diet **33:** 157, **158, 159**
    luminescence influence on vertical
        migration **33:** 171, 449, 450
    motility **23:** 219; **29:** 13, 14
        vertical migration **23:** 219, 223
    parasitic dinoflagellates **33:** 373, 524
    predation **29:** 19, 20, 45
    prey **29:** 252
    red tide **23:** 217, 220
    Russell cycle associations **23:** 234, 235
    summer/autumn blooming **23:** 211, 215
    swimming speed and **33:** 419
    taxonomic difficulties **23:** 234
Dinophyceae **29:** *17*
Dinoterbe toxicity and bivalves **37:** 97, **102**
Dinoxanthin and coral reef organisms
    43: *283*, 285
Dioecious, Cephalopoda **25:** 87
Diogenidae **29:** **132**
Diptera, growth performance **35:** 176, *180*,
    181
Discards **34:** 283–290, *284*, 289, 313
Discount rate or factor, harvesting
    29: 320, 321
Disease
    crustaceans **29:** 193

fish **29:** 350
Disease and parasites **43:** 64, 68, 69
disease(s)
    of *Pandalus* **38:** 179–182
    transfer to wild fish stocks **38:** 41
Diseases in the wild **27:** 379
    (see also Parasites)
Diseases of Atlantic halibut **26:** 41, 42
diseases of sea cucumber **41:** 188
Disjunct species **43:** 6
Disko Bay, *Pandalus borealis* stocks
    in **38:** 207
Dispersal **33:** 505–508, **507**; **34:** 395, 396,
    401, *401*; **35:** 110
    barriers to **35:** 86, 87
    hydrothermal vent dwelling species
        23: 346
    hydrothermal vent fauna **35:** 19, 20, *19*,
        22, 90, 91
    seaweed spores and gametes **23:** 27, 28
    behavioural types **23:** 27
Dispersal processes
    advective and diffusive, separation
        28: 42–45
    Atlantic herring larvae **28:** 68–74
    coastal currents **28:** 51–53
    coastal jets and eddies, entrainment
        28: 53, 54
    diffusion across shelf **28:** 67
    dispersal in coastal and estuarine
        spawning areas **28:** 67, 68
    Eastern boundary current upwelling
        28: 57–61
    Eulerian and Lagrangian approaches
        28: 41
    model **28:** 42
    ocean boundary currents, eddies and
        meanders **28:** 54
    offshore Ekman transport **28:** 57–61
    onshore (cross-shelf) transport, coastal
        and estuarine waters **28:** 61–63
    onshore Ekman transport **28:** 63, 64
    selective tidal stream transport **28:** 61–67
    spawning, spatial distribution **28:** 43–45
    summary and conclusions **28:** 124
    Taylor column circulation **28:** 46
    terminology **28:** 41
    tidal circulation and fronts **28:** 46–51
    Western boundary currents, eddies,
        meanders and rings **28:** 54–57
Dispersal routes, ancient, parasites as
    indicators of **43:** 55–59

Dispersion measures and spatial
        distribution **26**: 195–197
Disseminules **23**: 25
  dispersal behaviour **23**: 27
Dissimilarity
  coefficient matrix in classification **26**: 217
  indices in benthic studies **26**: 216
Dissolution of Scaphopoda shells **42**: 149
Dissolved free amino acids (DFAA) **29**: 34,
        34
Dissolved organic material *see* DOM
Dissolved organic matter (DOM) **33**: 153
Dissolved organic matter (DOM),
        phytoplankton **29**: 3, 22, 29–35,
        31, *33*, *34*, 41
  Baltic Sea **29**: 76, 97, 110, 117
Distance, prey detection **25:** 14, 15, 19
Distribution
  see also in particular Atlantic Ocean;
        Indian Ocean; Pacific Ocean
  as biological markers **43**: 29, 59–62
    see also gradients *under* latitude;
        indicators
    depth gradients **43**: 4, 5, 14, 54, 55
    in brackish water **43**: 55
    longitudinal gradients and geographical
        ranges **43**: 14, 53, 54
  autotrophic picoplankton **29**: 87–90, *88*,
        89
  bacterioplankton **29**: 97–101, **99–103**
    major areas *see* Atlantic; Indian Ocean;
        Indo-Pacific; Pacific; Southern
        Ocean
    of parasites **43**: **45–62**
  Pleurotomarioidean gastropods
        42: 272–274, *273*, 274–277, *275*,
        276
  Protobranch bivalves **42**: 36–39, *37*, *38*
  Scaphopoda **42**: 204–207, *206*
  sole **29**: 217, 221, 351
    adults **29**: 246, 247, *247*
    eggs and larvae **29**: 231–236, *232–235*
    juveniles **29**: 237–242, *239*, **240**, *241*
Distribution of Atlantic halibut
  areas **26**: 5, *6*
  differentiation **26**: 6, 7
Distribution of prey **25**: 53
distribution of sea cucumber
  see also habitat
  and population structure and
        dynamics **41**: **149–158**
  major areas *see in particular* Australia;

        India; New Caledonia; Papua
        New Guinea; Solomon Islands
  map **41**: *135*
distribution of trace elements in oceans
  cadmium **41**: 72–74, *73*, *75*, 76, *126*, *127*
  cobalt **41**: 77
  copper **41**: 63–64, 66, *67*, *124*
  iron **41**: 42–6, *44–45*, 49–57, *52*, *54*, *60*
  manganese **41**: 61, 62
  nickel **41**: 78, *79*, *128*
  zinc **41**: 67–69, *68*, *125*
Distribution pattern, dynamics
        of **36**: 123–137
Distribution *see* Geographical distribution;
        Spatial distribution
Distribution, *Donax* and *Bullia* **25:** 180, 189,
        190
Distribution, sibling **27**: 147, 148
  southern continent **27**: **148**
  species by regions **27**: 140–148,
        142–147
  zoogeographical regions **27**: 127–140,
        130, **131**, *137*, 147, 148
Distribution, siphonophore
  see also Zoogeography
  and deep-scattering layers (DSL) **24**: 164,
        165
  benthic and pleustonic **24**: 154–156
  buoyancy and **24**: 163, 164
  diel vertical migration (DVM) patterns
        in **24**: 135, 156–164
  geographical **24**: 139–151
  light intensity and **24**: 163
  neritic **24**: 151–154
  salinity and **24**: 163
  vertical **24**: 133–139, 148–150
  water temperature and **24**: 163
distributions (*Pandalus*) **38**: 108–121
  Arctic Oceans **38**: 113, 114
  Atlantic Ocean **38**: 113, 114
  depth **38**: 114–118
  geographic **38**: 108–114
  Pacific Ocean **38**: 112, 113
Disturbance, space supply by **23**: 125–127
Diurnal variation, predation **25:** 5, 7, 8,
        9, 15
Diuron toxicity and bivalves **37**: 97, **104**
Diver, black-throated **20**: 241, 294
Divergence, *see* Upwelling
Divergences and upwellings, tropical
        oceans **32**: 34–37
Diversity **33**: 461–464, **462**, **463**

in communities **26**: 197–201
indices of **26**: 203–208
Diversity of parasites **43**: **3–22**
concentration in some clades **43**: 17–22
estimates of species richness **43**: 16, 17
problems in estimating species
richness **43**: 4–15
Diversity of Scaphopoda
gradients **42**: 207–210, *208, 209*
through geologic time **42**: 216–219, *218*
D-larvae, bivalve, growth tests and
37: 64–66
DMSP *see* dimethyl-sulphonio-propionate
DNA analysis of pollock **37**: 200–207, 216
see also mtDNA
DNA content **44**: 36; **33**: 249
RNA/DNA ratio **33**: 256
DNA damage in Antarctic **43**: 238, 239
DNA fingerprinting **35**: 44, 46–48
DNA sequencing **35**: 6, 49, 50
DNA–DNA hybridization **35**: 51
DOC release **44**: 35
Docosahexaenoic acid (DHA) **44**: 303, 304,
309, 311
Dogfish **36**: 46, 116
Dogger Bank, plaice enhancement **38**: 12, 13
Dogger herring **20**: *16*, 17, 121, 122
Dogger spawning area **26**: 251
Doliolid blooms **23**: 264
Doliolids **44**: 34, 35
Dolphin fish **36**: 79
DOM (dissolved organic material) in
Antarctic **43**: 191, 200, 201, 204,
205, 214, 234, 239
Dominance in communities **26**: 197, 201
Dominance of species **33**: 464–466, *466, 467*
Dominance types, zones of **23**: 44–46, 54
English school of definition **23**: 44
indicator species **23**: 45
Mediterranean school of definition
23: 44, 45
indicator species **23**: 45
redefinition **23**: 45, 46
Dominance variance **23**: 37
Dominants and biomass distribution **26**: 210
Donacidae **25**: 186
Donut Hole, pollock in **37**: *180*
fisheries **37**: 182, 183
population dynamics **37**: 194, 195
population structure **37**: 211, 217, *219*
DORA **30**: 5, *18*, 35, **64**
Dorsal fin, sole **29**: 221, **222**, 230

Dose-response curves and bivalves
37: 82–84
Douarnenez, Bay of **29**: 257, 261, 262, 269,
275
Double crested cormorant **20**: 61, 245, 246
Dove prion **20**: 295
Dover, Straits of **29**: 248
Dowcide toxicity and bivalves **37**: **103, 105**
Downs herring **20**: *16*, 17, *19*, 121, 122, 127
Downs spawning area **26**: 251, 271
Downwelling, and fish larvae
recruitment **26**: 260
Drag-nets **29**: 323
Draw-nets **29**: 323
Dredges **34**: 209–213
effects on epifauna **34**: 219–222, **220**
effects on infauna **34**: 213–219
Drift, sole **29**: 218
Drifter returns, Bristol Channel **29**: 229
Drifter tracking and pollock **37**: *234, 235*
Drifters **44**: 251
Drill cuttings **35**: 98, 99, 100
Drive netting **34**: 225–227
Drogues **44**: 251
Dromiacea **29**: 135, 141
Dromiidae **29**: **133**
Dromioidea **29**: 136
Drought, seaweed **23**: 137–141, 145
avoidance **23**: 138
differential species responses **23**: 138
neap tide damage **23**: 137, 138
recovery **23**: 139, 140
species distribution effects **23**: 137–140
Drum, red and black **37**: 220
Dublin Bay **42**: 26, 32
*Ductus ejaculatoriu*s **33**: 45, 254, 260
Dumas combustion **24**: 395–397, 399
Dutch age-length keys (ALK) **29**: 296–297
Dwarf salmon **36**: 103
Dynamic mosaics of patches **44**: 244, 245
Dynamics **30**: *252, 253*, 259, 285
Dynamics of abundance behaviour and
distribution pattern **36**: 123–137
Dynamics of population of pollock
37: 193–199

Eagle, sea **20**: 282
Ear **20**: 141–144
Ear hanging, of scallops **20**: 353–355
East Africa: parasites **43**: 31, 53, 56
East Anglian herring **20**: 50
East China Sea **42**: 37

East Pacific **35**: 24
East Pacific Ridge, spreading rate **32**: 148
East Pacific rise (21°N) hydrothermal
    vents **23**: 303, 310, 311, 332
  faunal distribution and density **23**: 336,
    337, 338
  geological features **23**: 308
East Pacific Rise (EPR) **34**: 354, *362*, 365,
    386, 387, 398, 399, 404, 405, 410,
    414–416, *420*, *422*
East Pacific rise (13°N) hydrothermal
    vents **23**: 303, 305–310, 311
  faunal distribution and density
    23: 334–337
    changes over two years **23**: 335, 336
  geological features **23**: 307
East Pacific Rise **35**: 16, 17, 78, 88, 90, 91
Eastern and North-Eastern Atlantic Ocean
  invertebrates **43**: *97*, 143, 147
  parasites **43**: 4, *18, 19, 50, 60, 61*
Eastern Pacific Ocean
  Barrier **43**: 59, *61*
  invertebrates **43**: 125, 150
  parasites **43**: 35, 69, 70
    distribution **43**: 53, *57–60*
    diversity **43**: *10, 11*, 14, *18, 19*
Ecdyses **43**: 85
Ecdysis, crustaceans **29**: 140, 141, 143, 144
Echinoderm heterozygosity at
    hydrothermal vents **35**: 63
Echinoderms **29**: 253; **36**: 36, 73
  deep-sea **32**: **350**
  Elasipoda (holothurians), distribution
    path **32**: *372*
  relicts **32**: **409**
Echinoderms, hydrothermal vent
    dwelling **23**: 335, 337
Echinoderms, toxic **21**: **144**
Echiuran worms, distribution **32**: 506
Ecological hypotheses **30**: 163
Ecological importance
  see also Synecology
  of siphonophores **24**: 241–244
Ecological neighbourhood **44**: 238–241
Ecological niche_see_niche
Ecological niches **33**: 466–470
Ecological principles **36**: 229–253, 238,
    239, 243
  see also Adaptation and evolution
  acclimation **36**: 234, 235
  death **36**: 90
  environmental specificity **36**: 231, 232

functional rhythmicity **36**: 232, 233
  functional state **36**: 233, 234
  integration **36**: 235, 237
  metabolism of substance and
    energy **36**: 201–203
  multidimension **36**: 229, 230
  niches and specialization **36**: 59
  protecting fish stocks **36**: 246, 247
  rationalization of fisheries **36**: 247, 248
  rearing marine fish **36**: 248–251
  species centricity **36**: 237
  taxonomic specificity **36**: 230, 231
  technology for using marine fish
    resources **36**: 251–253
Ecology
  see also biogeography *and under*
    Pleurotomarioidean gastropods
  see also environment
  and fecundity of invertebrates **43**: 90, 91
  frontal systems, *see* Fronts
  hydrothermal vents, *see* Hydrothermal of
    seaweeds, *see* Seaweeds vents
  of parasites **43**: **22–47**
    see also hosts
    and morphology **43**: 44, 45
    communities **43**: *30*, 31, 33–36
    ecosystem engineering **43**: 84
    evolution and speciation **43**: 45–47
    metapopulations and chaos **43**: 36–41,
      38–40
    niches **43**: 24–33
    site restriction **43**: 23, 24
  of sea ice *see* Southern Ocean sea ice
  Protobranch bivalves **42**: 31–36
  economics of enhancement **38**: 43, 44
Economics, sole **29**: 319–321
ECOPATH models **34**: 298
Ecosystem **20**: 227
Ecosystem processes **34**: 306–308
  modelling **34**: 296–299
Ecosystems **33**: 3–5, 519–525
  food webs **33**: 519–521
  human exploitation **33**: 522, 523
  in open oceans **26**: 259
  key species **33**: 528–530
  large marine ecosystems (LMEs) **33**: 509,
    510, 511, 510, 515, 516, 519
  parasites **33**: 523–525
    ectosymbionts **33**: 524, 525
    endosymbionts **33**: 523, 524
  perturbations within **33**: 526, 527
  predation **33**: 521, 522

water stratification as **26**: 254, 259
Ecosystems of pollock **37**: 185–187
Ecotone **26**: 189
Ecotones, hybridization at **31**: 45
Ecotoxicology vs field assessment **39**: 75–78
Ecotypes **23**: 38
  mineral nutrient concentration related
    23: 38
  other **23**: 38
Ectocochleate, shell **25**: 87
Ectoparasites **43**: 47, *48–50*, 52, 54, 85
  distribution **43**: *48–50*, 53, 55–59,
    56–58, 61
  ecology **43**: 25, *32*, 34, 36, 37, 47
Ectoparasites, *see* Parasites
Ectoparasitism, zooplankton **25**: 158
  Animalia **25**: 147, 152, 153, 157
  Protista **25**: 138, 139
    Dinoflagellata **25:** 125, 126, 127, 129,
      131, 133, 134, 135
Ectosymbionts **33**: 524, 525
Ectreposebastes sp **36**: 44
Ecuador, coastal areas of **43**: *10*
ED$_{50}$ (median effective dose) **21**: 64
Eddies **33**: 471, 472
Eddies, phytoplankton **29**: 14
Eddies, tidal front **23**: 177–179, 184
  development **23**: 177
  fertilizing potential of **23**: 243, 244
  real cases **23**: 178
  theoretical behaviour **23**: 178
Eel
  adaptation strategies **36**: 61, 66, 69, 77
  life cycles **36**: 96, 101, 126
  salinity **36**: 21, *22*, 27
Eelgrass communities **39**: 34–39
Eelpout **36**: 152
Eelpout, hydrothermal vent dwelling **23**: 332
Eels, vertical migration of **26**: 121, 130, 141,
  142, 149
EFA **44**: 310
"effective evolutionary time" **43**: 48
Efficiency, ecological **36**: 242
Effluents *see* urban and industrial
Effort, harvesting **29**: 299
Egestion, rate of **33**: 195, 196
  see also Faecal pellets
Egg **20**: 181–183
  predation on **20**: 51–54
Egg laying **33**: 266–268
  egg mortality (*Pandalus borealis*)
    38: 212–216

Egg number **33**: 256, 272–274, **275, 276**,
  276–281
  see also Fecundity
  egg production rates **33**: 273, 289, 290,
    291, 292
  food availability and **33**: 289–292
  seasonal variation **33**: 281, 282
  stored lipids and **33**: 292
Egg production **20**: 232
Egg production method (sampling) **28**: 27,
  28
Egg ratio method **44**: 6
Egg size **20**: *18*
Egger's foraging model **20**: 109
Egger's Model, predation **25**: 40
Eggs
  see also fertilization; spawning
  bivalve
    normal development **37**: *19*
    size **37**: *32*
  pollock **37**: *188*, 228
Eggs (see Embryology, Life histories)
  see also Fecundity, Spawning
  see also Predation
  Bullia **25**: 187, 194, 236
  Cephalopoda **25**: 87–90, 90, 108, 109
  copepods **29**: 22, 26, 23, 27–29, *28*
  Crustacea and fishes, parasites of **25:** 125,
    127, 133, 134, 135
  development time **33**: 298, 299,
    310–313, *314*, 314
  egg mass **33**: 267, 279, 280, 281, 283,
    299
  hatching **33**: 293, 294
  identification **33**: 23
  morphology **33**: 22, 22, 23
  mortality **33**: 292, 293, 366, 367
  movements of **44**: 213–215
  of Atlantic halibut **26**: 11, *12*
    in rearing experiments **26**: 53
    incubation and hatching **26**: 55–57
    quality in fish **26**: 71–113
    and fertilization **26**: 72, *73*, 74, 75
    and size **26**: 74, 75, 78, 79
    broodstock management **26**: 96–103,
      98, 99
    characteristics of **26**: 72–86
    chemical content **26**: 79–85
    chromosomal aberrations **26**: 85, 86
    criteria of **26**: **74, 75**
    factors of importance **26**: 91–103
    in wild fish **26**: 86–91, *88, 89*

morphology **26**: 74, 75, 76–88
overripening and storage **26**: 74, 75,
    91–96
physical and physiological
    properties **26**: 76
resting eggs in sediments **33**: 268–272,
    269–271, 388, 389
size **33**: 274–276, **275, 276**, *277, 278,
    279*, **280**, *280*
sole **29**: 217, 218
    description **29**: 229, 230
    distribution **29**: 231–236, *232–235*
    mortality **29**: 230, 291–293, *292*, 298,
        299
    stock status **29**: 334–342, *339, 347*,
        348, 352
    trajectory mapping **44**: 247–251
    viability of **26**: **92**
Eggs of invertebrates
see also gametogenesis
    counts **43**: *94*
    loss **43**: 98, 99
    number of *see* fecundity
    size and quality **43**: 91, 92, **123–131**
        see also quantification methods
        environmental factors **43**: 142, 143,
            145–149
            passim **43**: 153
        evolution, design barriers to
            43: 129–131
        optimal, models of **43**: 126–129
        quantification methods **43**: 96–98, *97*,
            105
        trade-off **43**: 124, 125
    yolk *see* lecithotrophic larvae;
        vitellogenesis
Eggs, bivalve molluscs
    heavy metal effects **22**: 173–175
Egypt, sea cucumber in **41**: 136, *164, 186*
Eicosapentaenoic acid (EPA) **44**: 303, 304,
    309
Ekman transport
    and mortality rates **28**: 130
    offshore **28**: 57–61, **58, 60**
    onshore **28**: 63, 64
    upwelling areas **28**: 130, 131, **132**
Ekman transport and fish larvae
    recruitment **26**: 260
Ekman, regional classification, tropical shelf
    fauna **27**: 127–131
El Niño **20**: 8, *9*, 11, 166, 169, 260, 264

El Niño southern oscillation
    (ENSO) **31**: 228, 231–233
El Niño Southern Oscillation **43**: 172, 174,
    185
El Niño, effect of **38**: 3, 45
Elbe estuary **43**: 68
Electivity index, predator–prey studies
    28: 22
Electromagnetic induction **43**: 179, 180
Electromagnetic telemetry **44**: 253
Electron microscopy **43**: 182
    of attachment organs of parasites
        43: 41–44, *43, 44*
Electron transport system (ETS) **33**: 210
Electrophoresis, *See* Allozyme
    electrophoresis; Denaturing
    gradient gel electrophoresis;
    Isoelectric focusing
Electroretinogram technique **20**: 116
Eledoisin **21**: 184
ELEFAN **38**: 190
ELEFAN, Electronic Length Frequency
    Analyses program **27**: 241, 245
Elemental composition **33**: 236–245
    atomic ratios **33**: 243
    carbon **33**: 236–240, **238, 239**
    carbon–nitrogen ratio **33**: 242
    hydrogen **33**: 242
    mineral composition **33**: 243–245, **244**
    nitrogen **33**: 240, **241**
    phosphorus **33**: **241**, 242, 243
Elemental composition and pollock **37**: 206,
    207
elements in oceans
    see also concentration; distribution;
        macronutrients
    inputs **41**: *123*
    list of **41**: *120–123*
        accumulated **41**: *120*
        recycled **41**: *120–2*
        scavenged **41**: *122, 123*
    periodic table **41**: *12*
    trace *see* micronutrients; phytoplankton
        and trace metals
Elephant Island **43**: 223
Elephant seal, northern **20**: 246
Elevated platforms **44**: 264
ELISPOT assay method **28**: 23
Elliot and Persson's Model, predation
    25: 39, 40
Ellobiopsidea, parasitic **25**: 135–137, *136*,
    158

Elopidae, as predators **27**: 363
Elops sp **36**: 233
Eltonian Pyramids **26**: 210
Elutriate water and bioassay of
    bivalves **37**: 80–82
Emaciated larvae **30**: 240
Embryo
    development of in fish eggs **26**: 76, 77
    of Atlantic halibut **26**: 11–13
Embryogenesis of bivalves **37**: 21–25
    and metals **37**: 87–89
    and toxicity **37**: **89–94**, 97, 99–106, 136
    success assessed **37**: 134, 135
Embryogenesis of invertebrates **43**: 90
Embryo-larval bioassay of
    bivalves **37**: 51–62
    age of fertilized egg at beginning of
        test **37**: 51, 52
    embryo stages **37**: *19*, 21, 22, **35**
    incubation temperature, duration and egg
        density **37**: 52, 53
    larva
        abnormality at end **37**: 55, 56, *57, 58*,
            59, 62
        mortality at end **37**: 54, 55
        stages **37**: 22, 23, 29, **35**
    volume of test vessels **37**: 53, 54
Embryology **27**: 267–281
    cleavage **27**: *273*
    development **27**: 268, 272, *274*, 275–277,
        278–280, 281
    eggs **27**: 267, 268, **269, 270**, 275–277,
        278–280, 281
    hatching **27**: 273–275
    pycnogonid **24**: 39, 40
    siphonophore **24**: 99–109
    embryonic development in
        Pandalus **38**: 155, 156
Embryonic development *see* Eggs
Embryonic development, Cephalopoda
    cleavage **25:** 91–93
    gastrulation **25:** 93
    organogenesis **25:** 94–105, *95, 96, 97,
        98, 100, 104*
Embryos, bivalve *see* bivalve; embryo-
    larval; embryogenesis
Emergence effects **23**: 131, 132, 143–145
Emigration, sole **29**: 217, 249, 250, 278
Empirical measurement of reefs *see in situ*
"Empty niche" and parasites **43**: 25–27, 31
Encapsulation, Cephalopoda **25:** 87–90

Encounter rates and feeding in larval
    fish **31**: 175–182
    discrepancies between laboratory and field
        results **31**: 181, 182
    encounter process **31**: 175, 176, *177*
    Gerritsen Strickler model **31**: 176–178
    Rothschild Osborn model **31**: 179, 1880
Encounter, predation **25**: 10, *10*, 11–13, *13*,
    23, *24*
Endeavour Ridge, hydrothermal
    ecosystems **32**: 80
Endemism **43**: 212
Endemism of Protobranch bivalves **42**: 39,
    40
Endocochlean, shell **25:** 87
Endocrine control, crustaceans **29**: 198
Endocrine system **27**: 195–202
    effectors **27**: 199–202
    gonadal maturation **27**: 197–199
    heart control **27**: 201
    metabolic regulation **27**: 202
    morphology **27**: 40, 41
    moulting **27**: 196, 197
    neurosecretory cells **27**: 195
    target gland control **27**: 195–199
Endocrine system *see* hormones
Endocytoplasmic bacteria, plankton **25:** 120
Endocytosis
    in sea anemones **22**: 78, 79
Endogenous rhythmicity **26**: 116, 117
    categories of in fish migration
        26: 119–122
    evidence for **26**: 122–150
    importance of **26**: 118
Endonuclear bacteria, plankton **25:** 120
Endoparasites **43**: 31, *32*, 34, 36, 47, *48–50*,
    52, 54
Endoparasitism, zooplankton **25:** 118, 158
    Animalia
        Crustacea **25:** 15, 2, 153
        Helmintha **25:** 147–152, *149*
    bacteria **25:** 120
    fungi **25:** 121
    Protista **25:** 122, 136–139, 144–146,
        145
        Dinoflagellata **25:** 125, 127, 129, *130*,
            131, *132*, 133, 134, 135
        Sporozoa **25:** 139–144, *140, 142*
Endoplasmic reticulum, crustaceans **29**: 135,
    166
Endopodite, crustaceans **29**: 139, 140
Endoskeleton **33**: 38

Endosymbionts **33**: 523, 524
Endrin toxicity and bivalves **37**: 97, **102**
Energetic cost of swimming **33**: 421–423
Energetics, *Donax* and *Bullia* **25**: 182, 183,
        221, 225–234, **226**, **231**, *232*, *233*,
        237, 238
    locomotion **25**: 208–211, **209**, **211**, 236
Energy and energy metabolism
    see also Substance and energy
    adaptation **36**: 22–24, 32, 33
        divergences **36**: 59–74
        oxygen level **36**: 59, **60**, 62
    high *see* 'Burst'
    life cycles **36**: 94–98, 116, 117, 127
    optimum and structural functional
            homeostasis **36**: 243–245
Energy anomaly, Bristol Channel **29**: 231,
        232, *234*
Energy consumption **20**: 227, 250
Energy content **33**: 250
Energy cost **20**: 250
Energy economics of Atlantic halibut **26**: 37,
        38
Energy requirement **20**: 238, **239**
Energy requirements, predators **25**: 40, 41
Energy reserves **30**: 225
Energy submodel **20**: 236
Energy turnover, Brachiopod **28**: 355, 356,
        358
Energy used in reproduction of invertebrates
        (reproductive effort) **43**: 91, 114,
        115, 136–140, *139*, 142
Enforcement, fishing **29**: 349
English Channel
    phytoplankton **29**: *8*
    sole **29**: 240, 246, 250, 349
        feeding **29**: 252, 253
        length **29**: 260
        natural mortality **29**: 293, 295, 298
        reproduction **29**: 269, 275, 286, **289**,
        290
English Channel frontal systems **23**: 170
    biological and climatic changes
            23: 234, 235
        red tides **23**: 234, 235
    low salinity influences **23**: 208
    non-classical plankton cycles
            23: 212–214
    thermal structures and budget **23**: 170,
        173
Engraulidae, as predators **27**: **363**
Engraulids, predation **25**: 9

Engraulis See anchovy
enhancement
    behavioural differences **38**: 34, 35
    carrying capacity **38**: 39, 40
    economics **38**: 43, 44
    genetic studies on released fish **38**: 41–43
    impact on wild stocks **38**: 40, 41
    rationale **38**: 5–7
    release strategy **38**: 37–39
    size and feeding **38**: 35, 36
    studies **38**: 7–33
    survival after release **38**: 36, 37
    viability of reared fish **38**: 33–36
Enoploteuthinae, eggs **25**: 88
ENSO events **39**: 271
ENSO *see* El Niño
Enteropneusts, hydrothermal vent
            dwelling **23**: 331, 337
Entrainment, dispersal processes **28**: 49,
        53–54
Environment
    gradients and abundance **26**: *189*
    habitat factors in benthic studies
            26: 226–228
    variations of in vertical
            migration **26**: 145–148
Environment and bivalve
        bioassays **37**: 125–130
    see also extrinsic
    algal and bacterial toxins **37**: 125, 126
    receiving waters **37**: 128, 129
    urban and industrial effluents **37**: 127, 128
Environment/environmental
    see also ecology
    factors and fecundity of invertebrates
            43: 113, **140–150**
        depth **43**: 96, 106, 115, 116,
            143,147–150
        ecology **43**: 90, 91
        food availability **43**: 99, 128, 141–147
        latitude **43**: 147–150
        temperature **43**: 99, 101, 102
    quality damaged *see* pollution
    shock methods of quantifying
            fecundity **43**: *94, 95*, 101–103,
            102, 151
Environmental conditions, sole **29**: 290,
        350, 351
    see also Temperature
Environmental data,
        quantifying **44**: 262–268
Environmental factors

see also Ecological principles; salinity;
Temperature
adaptation and evolution, forces
of **36**: 237–240
predation **25**: 11, 20, 27, 53, 59
recruitment variation **25**: 2
specificity **36**: 231, 232
Environmental heterogeneity, role in
promoting hybridization **31**: 13
environmental sex determination
(ESD) **38**: 123, 128
Environmental stressors **30**: 251
Environmental variability **30**: 241
Environmental, tolerance, *Donax* and
Bullia **25**: 196, 197
migration, effect on **25**: 187, 236
Environments *see* Pelagic environments;
Restricted environments
enzyme systems and trace metals **41**: *16*
Enzymes **27**: 160–163, **161**, 184; **33**: 252;
36: 94
adaptation
food and feeding **36**: 54, 55
salinity **36**: 22
strategies **36**: 61, 62, 68, **75**
temperature **36**: 7–14
adaptations to deep-sea environment
35: 26
adaptations to hydrothermal vents **35**: 28
differentiation and variability **36**: 229,
231–236
digestive **33**: 204
hydrostatic pressure effects on **35**: 11, 64
indicators of fish condition **36**: 215, *216*,
217
selection of alleles **35**: 63, 90
temperature effects on **35**: 11
Enzymes in Atlantic fishes **24**: 357–361
and protein variability **24**: 360, 361
denaturation of **24**: 360
functional properties of **24**: 357–360
Enzymes, *Donax* and *Bullia* **25**: 192
Eocene **42**: 217, *268*, 280
Eosinophils, crustaceans **29**: 137, 160, 161,
162, 165, 166
Epibacteria **25**: 119, 120
Epibionts, plankton **25**: 136, 137, 138, 152,
153, 158
Epibiosis on pycnogonid **24**: 53, 54
Epiboly **20**: 162, 163
Epibranchial organ **20**: 39
Epifaunal habit, prosobranchs **25**: 183

Epifluorescence microscopy **29**: 74, 81, 82,
82, 83, 85, 89, 90, 98, 101, **102**
Epifluorescence microscopy **43**: 182
Epipelagic invertebrates **44**: 40
Epipelagic metazoan zooplankton, weight-
specific growth rates of **44**: 8
Epipelagic zone **33**: 483
epipods **38**: 63
Epithelial receptors of Scaphopoda **42**: 191,
192–194
Epitokous spawning, Polychaeta **29**: 195
EPOS (European *Polarstern* Study) **43**: 216,
219
EPS (extracellular polymeric substances)
43: 205, 206, 208, 213
Equation,
bioenergetic **20**: 229–234
cost of moult **20**: 232
Kendeigh's **20**: 230, 231, 238, 286
logarithmic allometric **20**: 230
Von Bertalanffy's **20**: 20
equilibrium
of parasites, ecological **43**: 24–33
strategy and fecundity of invertebrates
43: 121, 122, *121*
Equinatoxin **21**: 121
Equiproportional development
33: 318–319, *327*
Escape reaction **33**: 406, 414, **418**, 419
response to predators **33**: 419, 420
Escape reactions **27**: 186
Escape response **24**: 369
Escape, predation **25**: 16–22, *17*, *18*, *20*, *22*
ESD (equivalent spherical diameter) **44**: 37
Esophagus of Scaphopoda **42**: 176, 177
Essential fatty acids (EFA) in fish eggs
26: 98–100
Estuaries **33**: 473, 474, *474*, 475–477
and tidal influence on vertical
migration **26**: 139–141
as ecotone **26**: 189, 190
Estuaries, hybridization in **31**: 4
Estuarine fronts, *see* Coasts and estuaries
Eucalanidae **44**: 79–81
Eucalanus inermis, oxygen minimum layer,
tropical oceans **32**: 41–48
Euchaetidae **44**: 81
Euclidian methods in benthic
studies **26**: 215, 217
Euglenians, symbiotic **25**: 124
eugorgiaenoic acid **43**: *291*

eukaryotes and evolution **41**: 3, 4, 5, 10, 14, 23, 94
  see also phytoplankton
Eukaryotic algae **29**: 77–80, 87
  analysis **29**: 81–85, *82*, **84**
  areal and vertical distribution **29**: 87–90, *88*
  grazing **29**: 104, 105, *106*
  nutrients and temperature **29**: 101–104
  seasonal variation **29**: 90–92, *92*
Euler-Lotka equation **43**: 118, 119
Eumalacostraca **29**: **131**
EUMELI **30**: 35
Euphausiacea **29**: 19, **131**, 183, 184
Euphausiids **44**: 33
  ellobiopsid **25**: 135, 136, 137
  metazoan **25**: 148, 150, 151, 152, 153, 154
  parasites of **25**: 133, 138, 139, 141, 142, 144
  predators **25**: 7, 27, 38, 39, 41, 42
Euphyllopoda **29**: 199
European Community, fishing **29**: 319, 331–333
European *Polarstern* Study **43**: 216, 219
European shelf tidal fronts **23**: 167–209
  biological characteristics **23**: 232–276
    high productivity hypothesis, *see* Phyto-plankton, high productivity hypothesis
    planktonic cycles **23**: 213–215
    red tides, *see* Red tides
  coasts and estuaries, *see* Coasts and estuaries
  physics and chemistry **23**: 167–209
  summer structures **23**: 166
    computed characteristics **23**: 181
    infrared recording **23**: 166
European Space Agency **43**: 185
Eustigmatophyceae **29**: 79
Eutrophic environment, plankton **29**: 4, 35
  Baltic Sea **29**: 92, 97, 104
Eutrophication **36**: 30, 236, 252; **43**: 68
  see also fitness
  see also pollution
  and fecundity of invertebrates **43**: 89, 90, 111–115, 118, 134, 151
    design barriers to egg size changes **43**: 129–131
  and parasites **43**: 44, 45
  "effective evolutionary time" **43**: 48
  phylogenetic trees **43**: *20–22*

speciation **43**: 45–47
evolution/natural selection
Evadne **31**: 86
  embryonic development **31**: 105
  female reproductive anatomy **31**: 93
  fertilization **31**: 124, 125
  measurements of body size in **31**: *84*
  oogenesis in **31**: 98
  reproductive anatomy **31**: 116
  resting embryos **31**: 86, 130
Evadne anonyx **31**: 90, 91
  embryonic development **31**: 102, 104, 106
  female reproductive anatomy **31**: 94
  fertilization **31**: 125
  gamogenic egg development **31**: 128
  reproductive anatomy **31**: 117, *118*, 119, 120
Evadne nordmanni **31**: 92, 152
  birth rates **31**: 141
  brood size **31**: 140
  cultivation **31**: 147, 148
  distribution and abundance **31**: 87, 136, 137
  embryonic development **31**: 101, *101*, 102, 103, 106
  female reproductive anatomy **31**: 93
  gametogenesis in **31**: 119, 120
  gametogenic egg development **31**: 128
  life cycle **31**: *88, 89*, 89, 90, 138
  male **31**: *109*
  male gametes **31**: 121
  nutrition **31**: 91
  optimum temperature and salinity **31**: *136*
  pædogenesis **31**: 138, 139
  population densities **31**: 133
  population dynamics **31**: *144*, 145, 146, 147
  population growth rates **31**: 145
  prenatal mortality **31**: 142
  reproductive anatomy **31**: 116, 119
  resting embryos **31**: 130
  sex determination **31**: 112, 115
  sex ratio **31**: 111
  sexual dimorphism **31**: 107
  temperature and **31**: 145
Evadne spinifera
  cultivation **31**: 148
  distribution **31**: 87, 137
  optimum temperature and salinity **31**: *136*
  temperature and salinity in **31**: 142
Evenness in communities **26**: 197, *204*

Evolution **27**: 127–157
  barnacles and their shells **22**: 200–202
  Cephalopoda **25:** 86, 87, 108
  of Antarctic fishes **24**: 323, 324
  of pycnogonids **24**: 76–78
  of siphonophores **24**: 121–133
  predation **25:** 51
Evolution and genetic diversity **27**: 152–157
Evolution of Protobranch bivalves **42**: 41–44
Evolution *see* Adaptation
Evolution, crustaceans **29**: 140, 198, 199
Evolutionary Species Concept
    (ESC) **31**: 9–11
Exceptional (noxious; nuisance) **31**:
  bloom dynamics **31**: 355–365
    bloom composition **31**: 358, 359
    bloom initiation **31**: 355–358
    bloom maintenance **31**: 363, 364
    bloom species succession **31**: 360–362
    bloom termination and fate of bloom
        products **31**: 364, 365
  causative organisms **31**: 307–310
  critical concentrations of
        phytoplankton **31**: 365, *366*
  definition **31**: 302
  effect of human activities **31**: 322–337
    eutrophication in selection for harmful
        species **31**: 330–335
    harmful phytoplankton
        species **31**: 326–330
    increase in harmful algal
        blooms **31**: 322, 323
    influence of cultural
        eutrophication **31**: 323, 324
    spreading of harmful
        species **31**: 335–337
  harmful effects **31**: 302–304
  history **31**: 305–307
  inhibitory effects on potential
        grazers **31**: 333
  phytoplankton blooms
  role of bacteria in toxin
        production **31**: 318, 319
  role of nutrient availability in toxin
        production **31**: 319–322
  seasonal bloom development
        31: 337–355, *338*
    arctic and antarctic
        regions **31**: 352–355
    spring bloom **31**: 338–342
    subsurface blooms in summer
        months **31**: 345–351

    summer blooms in offshore surface
        waters **31**: 342–345
    temperate regions **31**: 338–351
    tropical and subtropical
        regions **31**: 351, 352
  sudden collapse of **31**: 364, 365
  time series of data n **31**: *325*
  toxic algal blooms **31**: 310–318
Exclusion cages, design **23**: 112
Exclusive Economic Zones (EEZs), World
    Ocean **32**: 413
Excretion **27**: 180–182; **33**: 209–211, **213**,
    214
Excretion and excretory system
  Pleurotomarioidean gastropods **42**: *251,
        253*, 260
  Scaphopoda **42**: 188–190
Excretion rates, siphonophore **24**: 232–234
Excretion, *Donax* and *Bullia* **25**: 225, **226**,
    229, 230, 231, **231**, 232, **232**, **233**
"Excretory organs" **27**: 37
Excretory rates **27**: 181
Excretory system **33**: 46
Expenditure of substance and
        energy **36**: 162–174
Exploitation of Atlantic halibut **26**: 42–53
Exploitation of resources
  mariculture (fish farming)
        43: 62,64–66, 69
  Southern Ocean sea ice **43**: 229–231
    fishing **43**: 221, 229–231
    whaling **43**: 185, 229, 230, 243
Exploitation, Bristol Channel Sole **29**: 322
  catches from 1903 **29**: 327–330
  early fisheries **29**: 323, 324
  early quantitative information **29**: 326,
        327
  early trawl fisheries **29**: 324–326
  effort, harvesting **29**: 299
  evolution to modern fishery **29**: 330–334
Exploitation, prawn stock **27**: 289
Exploitation, *see* Commercial
        exploitation of deep sea;
        Fisheries
"Explosive cells" **27**: 32
Explosives **34**: 225–227
Export of biogenic material from Southern
        Ocean sea ice **43**: 231–235
Extent and thickness of sea ice **43**: 174–180,
    175
  sampling **43**: 184–186
External anatomy

Pleurotomarioidean gastropods
    Plates 5 and 6 **42**:, 249–251, *249, 251*
shelled Opisthobranchs **42**: 91–94
External fertilization of invertebrates
    see broadcast spawning
Extinction of parasites unlikely **43**: 37
Extinctions
    heterozygosity and **35**: 64
    hydrothermal vents **35**: 20, 21
    seeps **35**: 23
Extinctions *see* mortality
Extracellular polymeric substances *see* EPS
Extra-ovarian oogenesis **43**: 89
Extrinsic (environmental) factors affecting
        toxicity **37**: 119–123
    see also environment
    salinity **37**: 123
    temperature **37**: 122, 123
Exudation, dissolved organic matter,
        phytoplankton **29**: 3, 22, 29–35,
        31, *33, 34*, 41
    Baltic Sea **29**: 76, 97, 110, 117
Eye **33**: 3*1*, 31, **32**; **38**: 60–62
    Cephalopoda **25**: 87, *95, 98, 100*, 101
    Donax and *Bullia* **25**: 186, 193
Eye, of clupeoid **20**: 131–136
Eyes, biorhythms **27**: 192, 193
    morphology **27**: 9, 10, 9, *38*
    ommatidium **27**: *189*
    pigments **27**: 199–201
    structure and function **27**: 187–193, *189,
        191, 192*
Eyes, sole **29**: 219

F statistics **35**: 53, 55, 56, 71–73, *72*
    spatial patterns of genetic variation **35**:
        74, 78–91
Facilitation, interspecific **23**: 110
Factor Analysis in benthic studies **26**: 222,
        224
Factor analysis of correspondences **23**: 68
Facultative phytoplankton feeder **20**: 42
Facultative planktivores, predators **25**: 15
Faecal pellets **33**: 196–203
    fate of **33**: 525, 526
    production of **33**: 196–199, *197*, **198,
        200, 201**
    sinking rates **33**: 199–203, *202*
Faecal pellets and Southern Ocean sea ice
        organisms **43**: 214, 233
Faecal pellets, zooplankton **29**: 2, 32, 34, 36,
        46

Faecal production, *Donax* and *Bullia*
    25: 225, 229
Faeroe Islands, sole **29**: 217
Fager's index of Affinity in benthic
        studies **26**: 215
Falkland Islands fisheries **39**: 271, 272, 277
FAO (Food and Agriculture Organization of
        the United Nations) **38**: 4
FAO, Fisheries synopses **27**: 237
Faraday/Vernadsky Station **43**: 240
Farallon Islands **20**: 243, *244*, 245, 246
Farming, fish **36**: 3, 55, 177, 182, 247,
        248–251
Farne Islands **20**: 227, 287
Fast ice **43**: 183, 186
Fat content in fish eggs **26**: 83
Fatness *see* Lipids
Fatty acid content **30**: 281
Fatty acids **44**: 303, 304; **33**: 246, 247
    see also Lipids
    adaptation **36**: 16, **17**, 18, 26, 27, 53–55
    adaptation strategies
        energy metabolism **36**: 62, **64**, 66, 71
            passim, 80–82, 85, **85**
        plastic metabolism **36**: 75–85
    as dietary markers **33**: 151, 153
    life cycles **36**: 91, **92**, 124
    substance and energy **36**: 152, 153
Faunal data matrices **26**: 178–181
Faunal provinces **33**: 509, *510, 511*, 516, 517
FDF (fast-death factor) **21**: 87
Featherback **36**: 117
Fecundity **27**: 257; **33**: 272, 273, 282, 283,
        284–287, 288; **20**: 13–15
    see also Egg number
    and egg size **20**: 15–17
    batch **20**: 13, 15
    estimates of **27**: **258, 259**
    related to female body weight **20**: *14*
    related to length **20**: **169**
    relative **20**: 13
fecundity (*Pandalus borealis*)**38**: 212–216
Fecundity estimation, fish eggs and
        larvae **28**: 26, 27
Fecundity of bivalves **37**: 28
Fecundity of invertebrates **43**: **87–171**
    see also factors *under* environment;
            life history; quantification
            methodologies; reproduction
    definitions **43**: 91, 92, 150
    fecundity-time hypothesis **43**: 126–129

Fecundity of Protobranch bivalves **42**: 24, 25
Fecundity, *see* Reproduction
Fecundity, sole **29**: 217, 218, 279, 280
  atresia **29**: 286–288
  comparative **29**: 288–290, **289**
  determinacy **29**: 280–282, *281*
  estimates **29**: 282–286, **284**, *285*, 338, 339, 341
Feed/feeding
  level and egg quality **26**: 97, 100
  of Atlantic halibut **26**: 26, *27*, 33, *34*
  quality and buoyancy in fish eggs **26**: 98, 99
  windows and vertical migration **26**: *133*, 134
feedbacks in system and phytoplankton **41**: 4, **86–92**, 93, 95
  classification **41**: 86, 87
  community structure, influence of **41**: 89–92, *91*
  feeding behaviour of sea cucumber **41**: 156, 157, **181–185**
  internal economy and near-field chemistry **41**: 87–89
Feeding **30**: 181–186; **20**: 140
  see also Diet; Foraging; Omnivorous feeding; Particle feeding; Predatory feeding
  see also Nematocysts; Parasitism; Predation
  copepodids **33**: 167
  corals under stress **22**: 35–38
  detritus feeding **33**: 153, 154
  Donax and *Bullia* **25**: 180, 182, 189, 190–192, 215, 236, 237, 238
  energetics **25**: 212, 214, **217**, 218, 219, 225–229, **226**, 230, 231–234, *232*, *233*
  feeding history influence **33**: 172–174, *173*
  filter **20**: 27, 34, 35, 67, 86, 109, 129, 176
  herbivorous feeding **33**: 142, 143
  nauplii **33**: 166, 167
  of pycnogonids **24**: 44–47
  of siphonophores **24**: 203–239
    and diet **24**: 227–229
    associations **24**: 203–208
      parasitic **24**: 207, 208
      with fish **24**: 203–205
      with invertebrates **24**: 205–207
    chemosensitivity and **24**: 174

  day–night differences in **24**: 174, 227
  fishing behaviour **24**: 223–225
  ingestion, digestion and egestion **24**: 225–227, 234–239
  prey selectivity **24**: 229–231
  rates **24**: 244
  particulate **20**: 27, 34, *35*,41, 109
  phytoplankton **29**: 26, 27–30, *27*
  pollution, effect of **25**: 197
  predation **25**: 11, 62
  reproduction **25**: 194
  sea anemone behaviour **22**: 66–69, 81
  sole **29**: 242, 243, 251–254, 266
  tidal migrations **25**: 187
Feeding appendages **33**: 140–143
feeding behaviour
  in *Pandalus* **38**: 173–175
  reared vs wild stocks **38**: 35, 36
Feeding current **33**: 143, 144, 146–148, *147*
Feeding periodicity **33**: 167–172
  diel periodicities **33**: 170, 171, *442*, 450, 452, 453
  dietary requirements **33**: 168, 169
  gut fluorescence analysis **33**: 167, 168
  seasonal periodicities **33**: 171, 172
  short-term periodicities **33**: 170
  vertical migration and **33**: 170, 171
Feeding rate **33**: 176–193
  criticism of laboratory methods **33**: 176–178, *179*
  filtering/clearance rates **33**: 180–184, *181*, *182*
  ingestion rate **33**: 185–193, **186, 187, 188, 189–191, 192, 193**, 207
  predatory feeding **33**: **184**, 184
Feeding relationships, *see* Food webs
Feeding *see* diet
Feeding *see* digestive system and feeding
Feeding *see* Food
Feeding *see* nutrition
Feeding system **34**: 27–32
  mucus in **34**: 27–32, *29*, 46, 47
  scavengers **34**: 288–289
Feeding threshold **20**: 137
Feeding, behaviour **27**: 323–328
  diet **27**: 328–332
  foregut, clearance **27**: 325, 326
    volume **27**: 324, 325
  ingestion **27**: 324–326
  location and handling of food **27**: 323, 324
  periodicity **27**: 325–356

selectivity **27**: 326–328
starvation **27**: 328
female reproductive stages, classification, in
    Pandalus **38**: 135–139
Female spawning biomass **29**: 218
female specific protein (FSP) **38**: 147
Females, reproductive system **27**: 44, 45
    (see also Reproduction)
Femtoplanktonic algae **29**: 77
Fenitrothion toxicity and bivalves **37**: 97,
    105
    see also reproduction; spawning
fertilization
    and bioassay methodology **37**: 49
    bivalves **37**: 136
    laboratory rearing **37**: 30–33
    pollock **37**: 184
Fertilisation
    Cephalopoda **25**: 90, 91
    Donax and *Bullia* **25**: 193, 194
Fertility, age specific **23**: 23–25
Fertility, crustaceans **29**: 190, 191, 193
Fertilization **27**: 263–267; **29**: 195,
    198–200; **33**: 43
    and egg quality **26**: 72–75
    and overripening of fish eggs **26**: 93, 94
    Copepoda **29**: 180
    Decapoda **29**: 141, 145, 146, 148, 160,
        164, 165, 174
    sequence of events **27**: 264
    sperm-egg interaction **27**: *265*
    time course **27**: 267
fertilization by nutrients **38**: 17, 18
Fertilization of invertebrates
    see also reproduction
    external *see* broadcast spawning
    internal *see* brooding species
Fertilization tubes **33**: 43, 262, 263, 266
Fertilizers for bioremediation, oil spill
        sedimentary environments
    39: 29, 30
Fertilizers, agricultural **24**: 412, 418
Fick's first law **29**: 11
Field surveys, fish eggs and larvae
        28: 24–38
    see also Fish eggs and larvae, research
        methods
    daily egg production method **28**: 27, 28
    demersal egg surveys **28**: 28, 29
    egg production, estimation **28**: 25, 26
    fecundity estimation **28**: 26, 27
    larval abundance **28**: 32, 33

larval census surveys **28**: 29–33
larval production method **28**: 33–36
recruitment forecasts **28**: 36–38
Fiji, sea cucumber in **41**: 131, 136, *137*, 190,
    191, 194
Filter feeders **44**: 34
Filter feeding **20**: 27, 34, 35, 67, 86, 109,
    129, 176
    see also Particle feeding
    feeding appendages **33**: 141
    filtering/clearance rates **33**: 180–184,
        181, 182
    obligatory **20**: 34, 36, 40,
Filter-feeding
    Donax see Feeding
    high energy beaches **25**: 180
    predators **25**: 9, 15, 17, *24*, 25, **26**, 27, 29
Filtration of seawater **37**: 41, 42
Filtration rate, *Donax* **25**: 191
Filtration rate, hardening, Japanese oyster
        culture **21**: 21
Filtration, picoplankton **29**: 84, 85, **84**
finfish, worldwide catch **38**: 2–3, *3*
Finland, Gulf of **29**: *74*, 75, 76
    autotrophic picoplankton **29**: 80, *88*, 90
    bacterioplankton **29**: *93*, 94, 95, *95*, 98,
        99, 100, 110, 111, 116
Fins, sole **29**: 219–221, **222**, 230
Fish
    see also Freshwater fishes, Predation
    see also Vertebrates
    countershaded **20**: *129*
    diel vertical migrations of **26**: 115–168
    -egg frontal accumulation **23**: 185, 186
    egg quality in **26**: 71–113
    genetic identity related totaxonomic
        divergence **35**: 69, 70
    grazer interactions among **23**: 105
    heterozygosity **35**: 59, 61, 64
    hydrothermal vent dwelling **23**: 331, 332,
        337
    parasites of
        dinoflagellate **25**: 125, 126, 127, 134,
            135
        metazoan **25**: 147, 149, *149*, 150, 151,
            152, 158
    plankton production and year class
        strength in **26**: 249–293
    reef dwelling **23**: 105, 107
    sea anemone predator **22**: 92, 93
    silvery **20**: *129*

spatial patterns of genetic variation,74, 78,
    83–87
fish aggregating devices (FADs) **38**: 30
Fish community structure **34**: 236–257
    annual size spectra **34**: *247*
    biomass changes and biomass
        composition **34**: *243*
    changes in multispecies communities
        34: 248–250
    composition of non-cryptic diurnally
        active reef-associated fish
        community in Seychelles' marine
        reserves **34**: *250*
    diversity **34**: 237–246
    diversity loss and ecosystem stability
        34: 240–245
    indirect diversity losses **34**: 240
    intraspecific changes in life histories
        34: 251–254
    intraspecific diversity **34**: 245–246
    life history traits **34**: 248–256
    local extinctions and redundancy
        34: 237–240
    mean biomass of piscivorous length and
        mean species richness **34**: *242*
    relationships between structure of target
        fish communities in Fijian fishing
        grounds **34**: *251*
    reproduction **34**: 254–256
    sex ratios of fish populations **34**: 254–256
    size selective fishing **34**: 254–256
    size structure of **34**: 246, 247
    species extinction **34**: 240, 241
    temporal changes in estimated abundance
        of nine species **34**: *244*
Fish eggs and larvae **28**: 1–133
    definitions **28**: 3, 4
    density-dependent processes
        growth rates **28**: 111–114
        summary and conclusions **28**: 126–128
    dispersal processes *see* Dispersal
        processes
    growth and feeding **28**: 97–114
        density-dependence of growth rates
            28: 111–114
        diet **28**: 98–101
        feeding rates **28**: 101–104
        size frequency analysis **28**: 97
        summary and conclusions **28**: 125, 126
        temperature and food density effects on
            growth rates **28**: 106–111

temperature effects on growth rates
    28: 104, 105
match–mismatch, member–vagrant and
    population theories **28**: 128–131
natural mortality **28**: 74–97
    algal blooms and toxins **28**: 79
    disease and parasitism **28**: 80–82
    endogenous vs. exogenous
        processes **28**: 74–77
    physical damage **28**: 78, 79
    pollution **28**: 79, 80
    predation **28**: 85–97
    starvation **28**: 82–85
    stress, lethal **28**: 77–88
    summary and conclusions **28**: 125
otoliths
    chemical composition **28**: 22
    growth rings **28**: 20–22
recruitment process studies **28**: 38–41
research methods **28**: 5–23
    age determination and growth
        history **28**: 20–22
    alternatives to net sampling **28**: 15–17
    analysis techniques **28**: 17–22
    determination of nutritional
        condition **28**: 19, 20
    ELISPOT assay method **28**: 23
    future prospects **28**: 131, 132
    predator–prey studies **28**: 22, 23
    preservation and shrinkage **28**: 17, 18
    sampling systems **28**: 6–10
    size measurement **28**: 18–20
    statistics of sampling **28**: 14, 15
    types of samplers and deployment
        strategies **28**: 10–14
research objectives **28**: 4, 5
schooling and refuge behaviour **28**: 123,
    124
terminology **28**: 3, 4
vertical migration behaviour
    28: 114–123
    advantages **28**: 114, 115
    design of sampling programme
        28: 114–117
    hydrographic conditions **28**: 120
    light **28**: 122, 123
    physiological requirements **28**: 119,
        120
    prey capture and predator
        avoidance **28**: 117–119
    tidal characteristics **28**: 121, 122
    turbulence **28**: 120, 121

Fish epidermis **40**: 253–327
  (immuno)histochemistry
    epithelial cells **40**: 257–263
    goblet cells **40**: 264–268
  cell types and secretions **40**: 255–296
    chloride cells and recent models of ion
        transport **40**: 286–294, 323, 324
    club cells **40**: 279–282
    epithelial cells **40**: 257–263
    goblet cells **40**: 264–268
    ionocytes *see* chloride cells
    luminescent organs **40**: 294–296, 324
    Malpighian cells **40**: 255
    sacciform cells **40**: 283
    unicellular glands and bioactive
        substances **40**: 276–283, 326
  chemosensory organs **40**: 311–322
    basal cells **40**: 317–320
    'dark' sensory cells **40**: 315–317
    hair cells **40**: 300–303
    'light' sensory cells **40**: 315
    marginal cells **40**: 320
    neuromasts **40**: 301–303
    olfactory cells **40**: 311
    oligovillous cells **40**: 314
    photophores **40**: 294–296
    receptor area **40**: 320
    solitary chemosensory cells
        40: 311–314
    stereovilli **40**: 303
    synaptic regions **40**: 320, 321
    taste activity **40**: 321, 322
    taste buds **40**: 314, 315
  compounds
    adhesive structures **40**: 260
    adrenaline **40**: 295, 322
    agglutinins **40**: 269
    alarm pheromones **40**: 283
    alkaline phosphatase **40**: 288
    ATPases **40**: 288, 292, 308
    biogenic alkaloids, amines, peptides
        40: 276
    bombesin **40**: 279, 283, 302, 309, 319,
        322
    caerulein **40**: 279, 283
    calmodulin **40**: 260, 261, 268
    carbohydrates **40**: 269
    concanavalin A **40**: 259, 269
    cytokeratins **40**: 271–276, 280, 322
    endolectins **40**: 269
    endothelin **40**: 280
    enkephalins **40**: 279, 299, 300

    F-actin-based microfilaments **40**: 271
    FMRFamide **40**: 302
    glutamate **40**: 307, 308
    glycoconjugates **40**: 259
    glycoproteins **40**: 258, 264
    glycosaminoglycans **40**: 264
    G-proteins **40**: 321
    immunoglobulins **40**: 266
    intermediate filaments **40**: 271, 299
    keratinization **40**: 256
    lectins **40**: 259, 266, 269, 270
    mannose **40**: 271
    microtubules **40**: 271
    mucopolysaccharides **40**: 308
    mucus **40**: 266–271, **270**, 322
    N-acetylglucosamine **40**: 271
    neurofilament protein **40**: 311
    neuron-specific enolase **40**: 279, 298,
        299, 311, 319, 322
    neuropeptides **40**: 299, 300, 322
    nitric oxide **40**: 322, 325, 326
    noradrenaline **40**: 295
    NOS enzymes **40**: 267, 268, 322,325,
        326
    parvalbumin **40**: 261
    proteoglycans **40**: 259
    s-100 protein **40**: 261
    serotonin **40**: 279, 283, 295, 296, 319,
        322
    sialated glycoproteins **40**: 264
    sialic acids **40**: 259, 271
    somatostatin **40**: 279, 319, 322
    substance P **40**: 279, 300, 308
    tetrodotoxin (TTX) **40**: 283
  electroreceptors **40**: 303–309
    ampullary organs,
        teleosts **40**: 303–306, 309
    ampulli of Lorenzini, selachians
        40: 308, 309
    cytochemistry **40**: 308
    tuberous organs, teleosts **40**: 306, 307
  lateral line system **40**: 300–311, 325
    see also electroreceptors;
        mechanosensory lateral line
        receptors
  mechanosensory lateral line receptors
    40: 300–303
    spiracular organs, elasmobranchs
        40: 310, 311
    vesicles of Savi, elasmobranchs **40**: 309,
        310
  paraneurons **40**: 296–322

defined **40**: 296, 297
  Merkel cells **40**: 297–300, 319
  venom glands **40**: 283–285
Fish farms *see* Farming, fish
fish parasites *see* Atlantic cod parasites
Fish population biology **34**: 258, 259
Fish schools
  facultative **30**: 158
  obligate **30**: 158
Fish weight **20**: 13–15
Fish, as predators **27**: 361–371
        (see also Predation)
  Ambassiadae **27**: **362**
  Ariidae **27**: **362**
  Bothidae **27**: **362**
  Carangidae **27**: **362**
  Centropidae **27**: **363**
  Clupeidae **27**: **362**
  Elopidae **27**: **363**
  Engraulidae **27**: **363**
  Merlucciidae **27**: **363**
  Platycephalidae **27**: **363**
  Pomadasydae **27**: **363**
  Sciaenidae **27**: **363, 364**
  Serranidae **27**: **364**
  Sillaginidae **27**: **364**
  Sparidae **27**: **364**
  Synodontidae **27**: **365**
  Teraponidae **27**: **365**
  Trichiuridae **27**: **365**
Fisheries **27**: 6, 291, 370; **36**: *52*, 247, 248,
        251, 252; **38**: 218–226
  see also Bristol Channel
  see also Farming, fish
  blue grenadier (hoki) fishery **35**: 107–109
  deep sea **35**: 5, 6, 93, 102–109
  growth performance related to
        exploitation **35**: 178–180
  management **27**: 6
  overfishing **36**: 245, 246
  phytoplankton **29**: 2, 58, 60
  roundnose grenadier fishery **35**: 103–107
  sole **29**: 218
Fisheries of Atlantic halibut **26**: *43*
Fisheries *See* Coastal lagoon fisheries;
        Fish community structure
Fisheries, pollock **37**: 178, 182, 183, *191*,
        193, 196
  management **37**: 236–240
  time trends of catches **37**: *195*
fisheries, sea cucumber **41**: **188–196**
  see also aquaculture

  catches **41**: 193
  harvesting **41**: 191
  history and price fluctuations **41**: 190, 191
  management **41**: 194–196, *197*
  processing into beche-de-mer
        41: 191–193
Fishery biology
  Cephalopoda **25**: 106, 107
  recruitment variation **25**: 3
Fishery Convention Act 1843 **29**: 324
Fishes hybridization in **31**: 40, 41
Fishes, oil spill on shoreline habitat
        effects **39**: 48–60, **49**
Fishing and fish farming *see under*
        exploitation
Fishing effects on marine ecosystems
        34: 201–351
  see also Benthic fauna; Trophic
        interactions
  active fishing techniques **34**: 209–222
  degradation of reef habitats **34**: 232, 233
  drive netting **34**: 225–227
  ecosystem management **34**: 305, 306
  ecosystem processes **34**: 306–308
  effect of fishing effort **34**: 308
  explosives **34**: 225–227
  fish community structure **34**: 236–257
  global discards and landings **34**: *208*
  global estimates of primary
        production **34**: *204*
  indirect effects on habitat **34**: 227–233
  investigating marine food webs
        34: 295, 296
  management **34**: 303–311
  marine reserves **34**: 309, 310
  modelling ecosystem processes
        34: 296–299
  poisons **34**: 225–227
  population management **34**: 305, 306
  potential effects of sediment
        resuspension **34**: 228, *228*
  protected areas **34**: 308–310
  research **34**: 293
  research sites **34**: 299, 300
  reversibility **34**: 290, 291
  role of fisheries in marine ecosystems
        34: 303, 304
  small-scale fishing effects **34**: 301
  spatial and temporal scales of study
        34: 300–302
  species replacement **34**: 276–283, *277*,
        278

static fishing gears **34**: 223–225
statistical basis for correlative studies
    34: 284, 285
study methods **34**: 292–302
trophic interactions **34**: 306–308
vs. natural disturbance **34**: 233–235,*234*
Fishing effort **27**: 293
Fishing effort in shell-fisheries and mortality
    or emigration of shorebirds
    34: *274*
Fishing intensity
    and mean biomass of fishes recorded
    34: *241*
    vs. mean number of grouper species
    34: *238*
Fishing mortality rate **29**: 295
    see also Harvesting
Fishing, frontal relationships **23**: 216, 266
Fitness and fecundity of invertebrates
    43: 111–115, 118, 134
Fixation **33**: 10, 11
Fixation systems, parasites **25:** 126, *128*,
    134, 135, 156
Fixed base line
    coral growth **22**: 21, 23
Fixed-point samples **44**: 257
Fixed-point techniques **44**: 258
Fjordic environments **33**: 349–351, **350**,
    477–479, *478*
Flagellates **29**: 4, *5*, 13, 14, 19–21
    sinking **29**: 9
Flat-bed Passive Integrated Transponder
    (PIT) technology **44**: 253
Flatfishes
    hybridization bias in **31**: 41
    mating behavior in **31**: 33
Flatiron herring **20**: 107, 128, 130
Flatworms **21**: 133
Flødevigen, Norway, hatchery **38**: 9
Fleetwood **29**: 333
Flexibility and fecundity of invertebrates
    43: 140, 141
Flies, algal grazing **23**: 11
Flight, horizontal flapping **20**: 231
Flocculation, phytoplankton **29**: 37–40, *39*,
    40, *41*
Florida escarpment hydrothermal vents
    23: 304, 311
    faunal distribution and density
    23: 318, 319
Floristic classification **23**: 57–71
Flotation, *see* Buoyancy

flounder (Japan) **38**: 24, 25
Flounder
    adaptation **36**: 20, 21, 77
    differentiation and variability **36**: 223,
    224, 226
    egg quality **26**: 74, *87*, 88, 89
    egg viability **26**: 92, 93, *95*
    life cycles **36**: 90, 104
    substance and energy **36**: 141, 190, 196
    vertical migration of **26**: 116, 121, 139,
    140, 146
Flounder, arrowtooth **37**: 198
Flowmeters **28**: 9
    see also Sampling systems, fish eggs and
    larvae
Fluid-filled cavities, *Donax* and *Bullia* **25:**
    198, 235
Fluorescence of coral reef organisms *Plate* 4
    FPs (fluorescent proteins) **43**: 287–290,
    288
    reflectance **43**: 280, 297–302, *298*, *301*
    separability of spectra **43**: 308, 309
    tissue pigments **43**: 287–289
Fluorescence, plankton **29**: *51*, *55*, 82, *82*,
    111
fluorine **41**: *12*, *120*
Fluorochromes **29**: 82
Flying fish **36**: 77, 79
Foam, frontal convergences **23**: 227, 231
    aerial view **23**: 228
    metal enrichment **23**: 227
    organic enrichment **23**: 227
Focal level **44**: 243
Foliate pedicellariae, urchins **21**: 148
Folklore, marine toxins **21**: 61
Food **27**: 315–323
    see also Nutrition
    animal **27**: 321–323
    as *zeitgeber* **26**: 129–136
    availability of in vertical
    migration **26**: 147
    for Atlantic halibut **26**: 33
    microbial and detrital material
    27: 316–319
    percentage frequency occurrence of
    categories **27**: **320**
    plant **27**: 319–321
Food and feeding **27**: 315–332
    see also Carbohydrates; Consumption;
    Lipids; Proteins; Substance and
    energy
    adaptation **36**: 47–58

competition for **36**: 56, 57, 200, 202, 209, 217
differentiation and variability **36**: 224, *225*, 226
ecological principles **36**: 239, 240, 242, 250, 251
indicators of fish condition **36**: 207–220
life cycles
  abundance **36**: 126, 127–129
  annual **36**: 106–108
  daily rhythms **36**: 115, 116
  ontogenesis **36**: 94, 103, 104
temperature **36**: 107, 108, 126
Food availability **30**: 240
  as cue for diel vertical migration **33**: 450, 451
  development and **33**: 3*16*, 316
  egg production and **33**: 289–292
  population regulation **33**: 372
Food capture **33**: 143–149
  food detection **33**: 145, 146
  particle feeding **33**: 146–148
  predatory feeding **33**: 148, 149, 162–164, *163*
Food chain, classical **29**: 2
  see also Pelagic food webs
Food chain, frontal **23**: 263–266, 269
  temporal characteristics **23**: 278, 279
Food consumption **20**: 227, 229
Food density threshold **20**: 40, **44**
Food fish **20**: 289, 290
food preferences (*Pandalus*) **38**: 173–175
Food role in copepods **44**: 36, 37
Food *see* Diet
Food *see* nutrition
Food utilization efficiency **20**: 238
Food web
  Antarctic **20**: 272
  North Sea **20**: *283*, 284
Food web(s) **23**: 72
  Aleutian Island, prehistoric **23**: 78
  human disruption **23**: 78, 79, 84, 96
  hydrothermal vent **23**: 341, 342
  intertidal **23**: 72
  New England intertidal **23**: 96
Food webs **33**: 519–521
Food webs, oil spill effects **39**: 70–74
Food, barnacle growth and **22**: 216, 235
Food, *see* Feeding
Food, *see* Resources
Food-chain dynamics **24**: 423–441, 443
  estuarine-salt marsh **24**: 436

hydrothermal vent organisms **24**: 433, 434, 435
  intertidal organisms **24**: 434, 436
  marine food-web model of **24**: 439–441
  reef fish **24**: 439
  terrestrial/marine **24**: 437–439
Foot
  Protobranch bivalves **42**: 14
  Scaphopoda **42**: 161, 162
Foot, *Donax* and *Bullia* **25**: *181*, 186, 187, *188*, 198, *199*, 200, 201–203, 235
  see also Locomotion
  respiration **25**: 214, 215, *215*, 224, 237
Forage fishes **34**: 257
Foraging **30**: 181–186; **33**: 160–166, *161*
  group size models **30**: 184–186
  particulate feeding **33**: 161,162
  predatory feeding **33**: 162–164
  selectivity **33**: 164–166
Foraging species **39**: 57–60, *58*
Foraminifera **30**: *31, 32, 39*, 51, 52, **57**, **58**, *62, 63*, 65–71, *69*, **71**
Foraminiferans, in diet **33**: 159
Foraminiferans, parasites of **25**: 125, 129
Forecasting **36**: 247, 248
Foregut clearance **27**: 325, 326
  volume **27**: 324, 325
Formaldehyde preservation **29**: 81
Formalin **33**: 10, 11
Formations, physiognomic unit of **23**: 62–65
'Former', crustaceans **29**: 176, 178
Fossils **27**: 283
  Antarctic **27**: 148, 149
  divergence **27**: 150, 152
  records **27**: 149–152, *151*
Foula **20**: 227, 240, 269, *285*, 293, 294
Fovea **20**: 133
Fowlkes-Mallows test in benthic studies **26**: 228
FPs (fluorescent proteins) **43**: 287–290,*288*
Fractal analysis **44**: 260
fractionation of elements in oceans **41**: 80–86
  cycling ratio **41**: 83–86, *84–86*
  indices of **41**: 80–83, *81, 82*
Fractionation procedure, plankton **29**: 83–85, **84**, 111
FRAGSTATS ARC **44**: 266
Frazil ice **43**: *177*, 178, 179, 197
Free amino adds **33**: 249

Freeboard and infiltration in sea ice
    43: 183, 184, 198
Free-living parasites, problems in estimating
    species richness of **43**: 4–7
Free-living stages, parasites **25**: 146, 147,
    150, 153, 154, 158
Freezing caught fish **36**: 52, 251, 252
French Polynesia *Plate 3* **43**: 29, 37, 84
Frequency distribution models **26**: 193–195
Frequency of the dividing cells
    (FDC) **29**: 116
Fresh meat, monthly shipments, Hiroshima
    21: *48*
Freshwater fishes
    parasites **25**: 146
    predators **25**: 4, 9, 12, 14, 58, 59, 63, 64
Frogfish **36**: 196
Frontal organs **33**: *31*, 31, **32**
Fronts (general features) **23**: 163–281
    biological characteristics **23**: 164–167,
        210–276
        accumulation biotype, *see* Fronts,
            accumulation biotype
        ecosystem boundaries **23**: 215
        high productivity **23**: 164–165; *see
            also* Phytoplankton, high
            productivity hypothesis
        in the High Seas **23**: 216
        red tides, *see* Red tides
    definition **23**: 164
    -fishing relationship **23**: 216, 266
    N.W. European, *see* European shelf tidal
        fronts; Celtic sea
    types, *see* Tidal fronts; Celtic sea shelf
        break
Fronts **33**: 471, 472
Fronts, accumulation biotype of
    23: 225–235 *see also* Red tides
    bacterial activity **23**: 229, 263, 264
    chemical factors **23**: 227–229
    food chain structure at and around
        23: 263–266, 269
    microbubble generated particles **23**: 229
    outcomes, range of **23**: 232
    transport mediation **23**: 231, 232
Fronts, plankton **29**: 4, 50–58, *51, 53,
    55–58*, 60, 98
Fructose, spermatophores **29**: 139
Fucoids **39**: 17–29
Fucoxanthin **43**: *283*
Fulmar **20**: 226, 280–294 *passim*

Fulton-K condition factor **30**: 266, 267, 271,
    272
Fumarate reductase, spermatophores
    29: 137, 143
Functional ecology and morphology of
    parasites **43**: 41–44
Functional rhythmicity, ecological principle
    of **36**: 232, 233
Functional state, ecological principle
    of **36**: 233, 234
Functionally-meaningful pattern
    44: 264–266
Fungi (see Parasites)
Fungi, plankton **25**: 120, 121
Funka Bay **20**: 310–351 *passim*
Funka Bay, pollock in **37**: *180*, 182, 184
Funnel Complex, Cephalopoda **25**: 99, 100
Furospongin-1 **21**: 103
Furospongin-3 **21**: 103
Future bivalve research **37**: 137, 138
future research **31**: 147–152
    cultivation **31**: 147–150
    developmental physiology **31**: 150–152
    gamogenesis **31**: 107–124
        fertilization **31**: 124, 125
        in males and females **31**: 119–21
        male gametes **31**: 121
        mating **31**: 122–124
        reproductive anatomy **31**: 116–119
        reproductive behavior **31**: 121, 122
        sex ratios and sex
            determination **31**: 109–115
        sexual dimorphism **31**: 107–109
    generation times **31**: 102, *104, 105*
    horizontal and vertical
        distributions **31**: 87–89
    in neritic and oceanic systems **31**: **81**
    life cycles **31**: 89–90
    nutrition **31**: 91–93
    parthenogenesis in **31**: 93–106
        embryonic development of subitaneous
            eggs **31**: 99–103, *103*
        oogenesis in parthenogenetic
            females **31**: 98, 99
        origin of germ cells **31**: 98
        reproductive anatomy of
            females **31**: 93–98
    pedogenesis **31**: 138, 139
    physiology **31**: 90, 91
    population dynamics **31**: 138–147
    population growth rates **31**: 145, 146
    prenatal mortality **31**: 141, 142

release of neonates **31**: 103–106
release, diapause and hatching of
    gamogenic embryos **31**: 129–132
taxonomy and distribution **31**: 80–87
temperature and salinity **31**: 142–145
"Fuzzy chaos" *see* chaos

Gadoid fish, predators **25**: 9, 50
Gadoid outburst and
    recruitment **26**: 274–276
Galapagos **43**: 35
Galapagos Rift **34**: 354, **365**, 385–387,
    398, 399, 415; **35**: 16, 87, 88, 90,
    91
Galapagos rift hydrothermal vents **23**: 303,
    305, 311, 314
    biological features **23**: 306
        faunal distribution and density **23**: 333,
            334, 336, 340
        groups uncommon or absent **23**: 332
        pink fish **23**: 331
    depth gradient **23**: 306
    geological features **23**: 306
    hydrothermal mounds 20–30 km
        from **23**: 340
    soft sediments around **23**: 340
Galathea expedition **35**: 3
Galatheidae **29**: **133**, 149
Gamete release **44**: 8
Gametes **29**: 129, 130
        see also Spermatophores
    Protobranch bivalves **42**: 25
    Scaphopoda **42**: 196
Gametes, bivalve
    Ageing after spawning **37**: 32, 33
    gametogenesis **37**: 12, 13, 16–19
    production and bioassay
        methodology **37**: 135, 136
Gametes, *see* Spores and gametes
gametogenesis (sea cucumber)
    and environment **41**: 166, 167
    and spawning periodicity **41**: 160–166,
        162, 164
Gametogenesis **27**: 251–255
    oogenesis **27**: 255, 256
    ovarian maturation **27**: 256
    spermatogenesis **27**: 251–255, *252*
Gametogenesis of invertebrates (oogenesis
        and spermatogenesis) **43**: 89–91,
        101, 103, 132, 150
Gametogenesis, bivalve molluscs
    heavy metal effects **22**: 166, 168, 169, 173

Gamma distribution **26**: 208
Gammarids **29**: 253
Gamogony, Sporozoa **25**: 144
Gannet **20**: 249, 281–294 *passim*
    Cape **20**: 60, **248**, 252, 259
Ganymedidae, parasitic **25**: 141, 142
Gap layers and infiltration in sea ice
    43: 183, 184, 198
Garden of Eden hydrothermal vent **23**: 334
Gas exchange of Scaphopoda **42**: 156, 159,
    161
Gas in ice **43**: 191, 200, 202
Gas resources **35**: 6
Gastric evacuation times, predation
    assessment **25**: 38, 39, 40, 41
Gastropoda
    development and dispersal **34**: 418
    distribution **34**: 415, 416
    gene flow **34**: 418
    locomotion **25**: 198, 235
    parasitic **25**: 155
    patterns of endemism **34**: 416, 417
Gastropods
    burrowing habits
        adaptations **28**: 418–421
        burrowing performance **28**: 413–418
        comparisons with bivalves **28**: 421,
            422
        evolution **28**: 422–426
        mechanisms of burial **28**: 404–413
        principal events **28**: *405*
        recordings **28**: *408*
        research methodology **28**: 391–395
    burrowing rate index **28**: *415*
    burrowing rate, evolution **28**: 422–426
    classification **28**: 392, 393
    classification and BRI **28**: 392, 393
    functional anatomy **28**: 395–403
        foot **28**: 395–401
        shell shape **28**: 401–403
        ultrastructure **28**: 398
    genetic differentiation **35**: 81
    grazing by
        intertidal **23**: 87, 90
        resultant dominant algae **23**: 113
        tidal pool **23**: 88, 89
    hydrothermal vent adaptations **35**: 27
    hydrothermal vent dwelling **23**: 321–323,
        336
    locomotion **28**: 390, 391
        comparisons **28**: *417*
        pedal structure **28**: *399*

sea anemone predators **22**: 92
small, in sea anemone diet **22**: 93
turritelliform snails **28**: 402
Gastropods *see* Pleurotomarioidean gastropods;
Gastropods, relict and 'fossil' species **32**: 409, 410
Gastrula phase, sole **29**: 229, 230
Gastrulation, Cephalopoda **25**: 93
Gause's hypothesis **20**: 292
Gelatinous animals
counting from submersibles **32**: 9
macroplankton **32**: 64–68
Gelatinous layer, spermatophores **29**: 150, *151*, 152, **154–158**, 160, 161, **162**, **163**, 165, 169, *170*
Gelatinous organisms **44**: 36
Gelatinous species **44**: 35
Gelatinous zooplankton **44**: 33; **25**: 5, 8, 33, 38, 43–45
Gene flow **34**: 396, 398; **35**: 6, 81–91, 116
Alvinellidae **34**: 411, 412
barriers to **35**: 84, 86–90, 116
Gastropoda **34**: 418
levels of **35**: 71–73, 81, 116
measurement of **35**: 56–58
vestimentiferans **34**: 407
gene pool, wild, dilution of **38**: 41
"generalists" (parasites) **43**: 35
Generation time **23**: 25; **33**: 363, 374, *380*
Generation time, plankton **29**: 19–21, 104
Generations per year **33**: **355–362**
Generic key **33**: 57–60, *59*, **65**, 65–97
Genetic differentiation **23**: 35, 36, 37
in response to selection **23**: 38
Genetic distance (*D*) **35**: 53, 54
Genetic distance **23**: 35
Genetic diversity and evolution **27**: 152–157
Genetic drift **31**: 26; **35**: 63
Genetic factors **30**: 240
Genetic hypotheses **30**: 163
Genetic identity (*I*) **35**: 53–55, 116
taxonomic divergence and **35**: 67–71, *68*, **70**, 116
Genetic identity **23**: 35–37
matrix of estimates **23**: 36
Genetic identity of Atlantic halibut **26**: *4*
genetic markers, tracing released fish **38**: 42, 43
Genetic population structure, *See* Population genetics

Genetic structure of populations **23**: 34
genetic studies on released fish **38**: 41–43
Genetic variation **33**: 63, 64, 371
Genetic variation in seaweed populations **23**: 32–39
between populations **23**: 35–39
electrophoretic enzyme analysis **23**: 32–34, 36, 37
multiple loci **23**: 37, 38
single loci, *see* Polymorphism
within populations **23**: 32–35
Genetic variation, deep-sea fauna **35**: 58–93
future study prospects **35**: 109–114
gene flow levels **35**: 71–73, 116
genetic identity related to taxonomic divergence **35**: 67–71, *68*, **70**, 116
measurement of **35**: 53–58
mitochondrial phylogeography **35**: 91–93
pollutants and **35**: 100, 101, 102
See also Heterozygosity
spatial patterns of **35**: 74, 78–91
study methods **35**: 28–52, 109–112
Genetics **36**: 230, 231
see also Adaptation; differentiation
see also evolution
and coral reef organisms **43**: 286
and fecundity of invertebrates **43**: 111–123, 140
genotypic condition for natural selection **43**: 118
Genetics of pollock **37**: 181, 208, 215, **216**, 217–223, 237
gene flow **37**: 204, 205, 224, 239, 240
molecular *see* mtDNA
mutation **37**: 204, 205, 207
random drift **37**: 204, 205
genetics of sea cucumber **41**: 157, 158
Genetics, sole **29**: 250
Genets **23**: 8
Genital
flap, crustaceans **29**: 182
plate, crustaceans **29**: 182
segment, crustaceans **29**: 182
valve, crustaceans **29**: 182
Genital antrum **33**: 41, 43
Genital somite **33**: 16, 25–27, *42*, 43, 63, 267
spermatophore attachment **33**: *42*, 43, 262, *263*
Genitalia, external **27**: 41–43, *42*
Genotoxicity and bivalves **37**: 95, 96
Genotypic variance **23**: 37
Geographic basin model of pollock **37**: 22

Geographic distribution *see* distribution
Geographic position *see* Distribution
geographic range of sea
  cucumber **41**: 134–139
Geographical distribution
  bivalves **37**: 10–18
  pollock *see in particular* Aleutian
    Basin; Bering Sea; British
    Columbia; Gulf of Alaska;
    Japan; Okhotsk
Geographical distribution **33**: 504–517,
    512–515
  see also Spatial distribution
  faunal provinces **33**: 509, *510, 511*, 516,
    517
  introductions **33**: 505–508, **507**
  large marine ecosystems (LMEs) **33**: 509,
    510, 511, 510, 515, 516, 519
Geographical Information System
  (GIS) **44**: 266, 272
Geographical Information System (GIS) **39**:
    293, 294
Geographical range, *Donax* and *Bullia*
    25: 180, 236
Geographical ranges of parasites and
    longitudinal gradients **43**: 14,53,
    54
Geographical species *see* "Rassenkreise"
Geography *see* Antarctic; Arctic; Atlantic;
    Azov; Baltic; Barents; Black;
    Caspian; Mediterranean Sea;
    North; Pacific
Geologic time
  see also Cambrian; Mesozoic; Paleozoic;
    Recent
  diversity of Scaphopoda
    through **42**: 216–219, *218*
Geomorphology, Bristol Channel **29**: 223,
    224, 246
Georges Bank **34**: 219, *220*, 246, 279, 280,
    281, 306
Geostrophic balance **23**: 177, 178
German Bight **29**: 261, 275, **289**; **42**: 26;
    43: 68
germanium **41**: *81, 121*
Germinal site **33**: 42
Gerritsen Strickler model **31**: 176–178
GFP (green fluorescent protein)
    43: 287–289, *288*
GI *see* Gonad Index
Giant gourami **36**: 117
Giant squid **36**: 103

Gill rakers **20**: 36, **37**, 38, 41, 176
gills (*Pandalus*) **38**: 63
Gills and salinity **36**: 20, 21, 22
Gills of Protobranch bivalves **42**: 9–13,
    10, 12
Gills, blood circulation **27**: *35*
Gills, *Donax* and *Bullia* **25**: 214–216, *215*
Gilthead *see* Annular gilthead
Gizzard shad **20**: 24, 35, 140
Glacier Bay, Alaska **44**: 260
Glamorgan **29**: 326, 327
Glands
  Protobranch bivalves **42**: 16
  Scaphopoda **42**: 178, 179
Glandular activity, crustaceans **29**: 142, 143
Global patterns *see* distribution
Global spreading hypothesis **31**: 335
Global warming, fisheries **29**: 350
GLOBEC (Global Ocean Ecosystems
    Dynamics) **43**: 174
Globiferous pedicellariae, sea urchins
    21: 148
Gloucester, MA, hatchery **38**: 8
Glucosaminidases
  in corals under stress **22**: 50
Glucose, crustaceans **29**: 139
Glutamine, plankton **29**: 34
Glutaraldehyde preservation,
    plankton **29**: 81, 83
Glutinant nematocysts **21**: 107
Glycerol, cryopreservation **29**: **188**
Glycoconjugates **34**: 3
Glycogen
  in bivalve molluscs under
    stress **22**: 146–150
Glycogen reserves, *Donax* and *Bullia*
    25: 230
Glycogen *see* Carbohydrates and glycogen
Glycogen, spermatophores **29**: 137, **138**,
    139, 153, 154
Glycolysis, spermatophores **29**: 139
Glycoproteins **34**: 3, 5
Glycoproteins and temperature **36**: 18, 19
Glycosaminoglycans (GAGs) **34**: 3, 5, 7, 13,
    15
Gnathobase **33**: 19,20, 141
Goby
  adaptation **36**: 66, 68, 78, 86
  food and feeding **36**: 57
  oxygen level **36**: 42
  pressure and buoyancy **36**: 45
  temperature **36**: 7

ecological principles **36**: 236, 237, 241
indicators of condition **36**: 215, 216
life cycles **36**: *92*, 103, 127
substance and energy **36**: 148, 159, 190, 196
Goldfish **36**: 18, 34, 54, 227, 231
Golgi complex, vas deferens **29**: 135, 166, 175
gonad development, *Pandalus* **38**: 141–147
Gonad index of larval scallop **20**: *326–335 passim*
Gonad maturation **27**: 197, 198
gonad morphology of sea cucumber **41**: 159–161
Gonadosomatic index (GSI) **29**: 268–270, 272, *273*, 275
gonadosomatic index (GSI) **38**: 147
Gonads in Atlantic halibut development of **26**: 9
and energy economics **26**: 37, 38
Gonads, bivalve, stripping of **37**: 47, 135, 136
Gonads, Cephalopoda **25**: 98, 99
Gonads, hardening, Japanese oyster culture **21**: 21
Gonads, sole **29**: 218
Gonatid squids *see* Squids
Gonatopsis japonicus **32**: 275–278
Gonatopsis kamtschaticus **32**: 282, 283
Gonatopsis magister **32**: 270–275
Gonatopsis onyx **32**: 280–282
Gonatus spp. **32**: 284–289
Goneplacidae **29**: **133**
Gonochorists, Cephalopoda **25**: 87
Gonomere, Ellobiopsidea **25**: 135
Gonophores **24**: 186, 188
Gonopods, crustaceans **29**: 130, 139–141
Gonopores, crustaceans **29**: 144, *170*, 174, 182–184, *185*, 186
Gorda Ridge **35**: 89
Gotland Deep **29**: *75*
Gourami **36**: 117
Gower's method in benthic studies **26**: 222
Gradient
    analyses **23**: 65–70
        depth **23**: 66–69
        direct **23**: 65, 66
        importance values in **23**: 66
        indirect **23**: 65–70
        use **23**: 65
    shore, outcome of competition reversal **23**: 124, 125

Gradients and ranges of parasites
    see also under latitude
    depth **43**: 4, 5, 14, 54, 55
    longitude **43**: 14, 53, 54
Gradients, latitudinal diversity (Scaphopoda) **42**: 207–210, *208, 209*
Graham's law of diffusion **24**: 401
Granular endoplasmic reticulum **29**: 175
Grayanotoxin **21**: 102
Grayling **36**: 13, 227
Grazing **23**: 18–23 *see also* Herbivores
    and algal species diversity **23**: 77
    co-evolution with food source **23**: 122
    effect **23**: 113
        on seaweed recruitment **23**: 29
Grazing, plankton **29**: 3, 22–29, *23–28*, 36, 47, 60, 77, 104, 105, *106*
    bacterioplankton **29**: 111–115, *112*, **113**, *115*, 117–119, *118*
Grease ice **43**: *177*, 179, 215
Great Barrier Reef
    parasites **43**: *11*, *14*, 16, 35, 53
    pigmentation **43**: *288*, 290
Great Barrier Reef, Australia **44**: 216
Great Island **20**: 296
Great skua **20**: 236, *238*, 282, **285**, 286, 293, 294
Great Sole, Bank of **29**: 224, 246, 326
Grebe **20**: 241
Greenland waters
    fisheries **38**: 222, 223
    Pandalus borealis stock limitation in **38**: 207
Greenland, halibut fisheries in **26**: 43, 46, 47
Greenling **36**: 62, 133
'Greenness'
    and spawning distribution of species **26**: *270*
    as measure of plankton production **26**: 251, 265, *266–269*
Gregarinia, parasitic **25**: 139–143, *140, 142*, 158
    Polymastigotes **25**: 124, 125
Grenadier **36**: 46
grenadiers **38**: 3
Grey mullet **36**: 17, 116, 165, 166, 233
Grey shark **36**: 78
Grillotia and Atlantic halibut **26**: 40, 41
Gross growth efficiency (GGE) **44**: 33
Groundwater seeps **35**: 23

Growth **27**: 223–247; **33**: 296, 314–327,
      315, 320, *336*; **35**: 158, 159;
      36: 222, *223*, 224; **20**: 78
see also Development; Life history; Size
see also Substance and energy
adult **27**: 240–250
   seasonal variation **27**: 243
   tagging **27**: 240, 241, 242
allometric growth **33**: 338
compensation irradiences **23**: 143
data distribution **35**: 155, **156**
general concept of **33**: 326, 327
growth rate **33**: 316, 338, 339, *339, 340,*
      341–343, 344, *381, 382*
   seasonal variation **33**: 371
harvesting **29**: 301, 303, 309, 316
in clupeoid larva **20**: 77, *78,* **79**
intermoult period **27**: 225, 226
isocline, zero net, *see* Zero net growth
      isocline
juvenile **27**: 236–240, *240*
larval **27**: 235, 236, *236*
   development times **27**: **278–280**
length and weight parameters **27**: **228,
      229–232**
models **35**: 158, 159
models for **27**: 224, 225, 233, 234
morphometrics **27**: 227–233
moult increment **27**: 224, 225
moulting **33**: 315–317
   intermoult duration **33**: 317–323
   moult increment **33**: 323–326, *324*, **325**
net growth efficiency **33**: 205–207, **208**
of Atlantic fishes **24**: 348
of Atlantic halibut **26**: *25, 26, 31, 32,* 33
   experiments on **26**: 62, *63*
of larvae **26**: 281, 284
plankton **29**: *17,* 19–21, 26, *25,* 104, 119
rates of, in siphonophores **24**: 239–241
sole **29**: 218, 254–262, **257,** *258,* **259,** 347
von Bertalanffy's equation **27**: 234
weight **29**: 262–266, *264,* **265**
Growth and development
   Protobranch bivalves **42**: 26–28, *27*
   Scaphopoda **42**: 196–204, *198, 201–203*
growth in *Pandalus* **38**: 182–203
   determining factors **38**: 196, 197
   differences between sexes **38**: 199
   estimating size at age **38**: 190
   intraspecific variations **38**: 196, 197
   juvenile stages **38**: 197–199
   larvae **38**: 162–164

seasonal variations **38**: 199–201
species comparison **38**: 196
temperature and **38**: 201–203
time-tracing of cohorts **38**: 190–192
growth models in *Pandalus* **38**: 192–196
Growth of bivalve larvae **37**: 94
   and toxicity **37**: 97, 98–106
   tests **37**: 62–66
   inhibition with already
      unexposed **37**: 64–66
   inhibition with previously
      unexposed **37**: 63
growth of sea cucumber **41**: 176–178, *177*
Growth overfishing **20**: 5
Growth performance **35**: 159–180, *170,*
      184, 185
   Bivalvia **35**: 167–173, **171,** *172,* **173,
      174, 175**
   crustaceans **35**: 173–175, *176,* **177,** *178,*
      179
   exploitation and **35**: 178–180
   extended measures of **35**: 160, *161*
   insect larvae **35**: 175, 176, *179, 180,* **181**
   living mode and **35**: 177, 178, **182**
   measures of **35**: 159, 160
   productivity/biomass relationship
      35: 193, 194, *193*
   species specificity **35**: 160–164, **163,
      164, 165**
   temperature effect on **35**: 165, 166, **168,
      169**
growth performance index **38**: 197
Growth rate,
   of adult clupeoid **20**: 83–85
   of anchovy species **20**: *84*
Growth rates
   for copepods **44**: 4–8
      temperature influence on **44**: 7
   in marine epipelagic invertebrate
      zooplankton **44**: 1–142
Growth rates, predation **25**: 32, 33, 41, 54,
      55–57, **56,** 58, 59, **62,** 66
Growth, *Donax* and *Bullia* **25**: 194, 195,
      225, 227, 229, 230, 231–234, **231,**
      232, 233, 236, 238
Growth, hardening, Japanese oyster culture
      21: 21
Guadeloupe/Antilles **42**: 274, *275, 276,* 277
Guanay (Peruvian cormorant) **20**: 260–265
      passim
Guanine crystal **20**: 128
Guanine layer **20**: 136

Guano **20**: 247
Guano Administration **20**: 263
Guano birds **20**: 264, 268, **270**
    crash in **20**: **261**, 262
    fluctuations in **20**: *263*
Guano yield **20**: 253–258 *passim*
    at Lamberts Bay **20**: 257
Guaymas Basin (Gulf of California) **42**: 41
Gudgeon **36**: 43
Guillemot **20**: 282, *285*, 286, 293, 294
    black **20**: 282, **285**, 286, 293, 294
    Brünnich's **20**: 228
    common **20**: 228, 241, 243, *244*
    pigeon **20**: 243, *244*
Gulf menhaden **20**: 21, 34, 108
Gulf of Aden **35**: 17, 24
Gulf of Alaska **35**: 84
Gulf of Alaska, pollock in **37**: 178, 179, 184,
        192
    fisheries and management **37**: 182, 183,
        237, 238, 239
    population dynamics **37**: 193, 194, *195*,
        197, 198, *199*
    population structure
        genetic **37**: 215–218, 219, 221– 223
        mechanisms **37**: *227, 228, 229, 230,
            231*, 236
        phenotypic **37**: 208, 213
    prey and predators **37**: 186, 187
    recruitment **37**: 200–203, 204
    stocks **37**: **185**
Gulf of Maine **43**: 145
Gulf of Mexico **34**: 384; **35**: 23, 97, 100;
        42: 36, 37, *273*, 278; **43**: *11*, 65
Gulf of Mexico, red drum and spotted
        seatrout in **38**: 27, 28
Gulf of St Lawrence **35**: 83; **34**: 279
Gulf of Thailand fisheries **39**: 268, 269, 276
Gulf stream 'rings' **23**: 216
Gull **20**: 260, 282, 286, 293
    black-headed **20**: 281
    common **20**: 281, **285**, 294
    glaucous-winged **20**: 241
    great black-backed **20**: 281, **285**, 294
    herring **20**: 281, **285**, 294, 296
    lesser black-backed **20**: 281, **285**, 294
    western **20**: 243, *244*
Gulland's method **20**: 268
Gullmarsfjorden, *Pandalus borealis* stocks
        in **38**: 207
Guppy **36**: 54
Gustation of Scaphopoda **42**: 192

Gut capacity, predation **25**: 37
Gut clearance rate constant (K) **33**: 188, 192,
        194, 195, 196
Gut contents analysis **33**: 149–151,**151,152**
Gut evacuation rate **33**: 192, 195, 196
    see also Faecal pellets
Gut filling time **33**: 193
Gut fluorescence analysis **33**: 167, 168
Gut passage time **33**: 188, *193*, **194**
Gut *see* digestive system
Gut, hormonal control **27**: 201
Guthion toxicity and bivalves **37**: 97, **105**
Guyamas Basin hydrothermal vents **23**: 303,
        310, 311
    faunal distribution and density **23**: 338
        soft sediments adjacent **23**: 340
    geological features **23**: 338
Guyots **30**: 310
GXPOPS **38**: 203
Gymnodin **21**: 83
Gymnopleura **29**: 135, 136, 141

Habit, sole **29**: 216
Habitat
    anthropocentric perspective of **44**: 235,
        236
    concept of **44**: 237, 238
Habitat factors, comparisons of in benthic
        studies **26**: 226–231
habitat of sea cucumber **41**: 149, *150, 151*
    densities **41**: 152, 153, *152*
    distribution and size **41**: 153, 154
    substratum preferences **41**: 156, 157
habitat preference (*Pandalus*) **38**: 114–118
Habitat *see* ecology
Habitats **27**: 3
    see also distribution; ecology; niches
    Southern Ocean sea ice **43**: 180–184
Habitats *see* Pelagic environments;
        Restricted environments
Hadal zone *see* Abyssal and hadal zones
Haddock **36**: 141, 190, 224, 225, 231; **20**:
        291
    prey composition of **26**: 130–132
    vertical migration of **26**: 120, *126*, 147,
        148
Haddock, Atlantic **37**: 208, 220
Haematocrits **20**: 62
Haematology *see* Blood
Haemocoel, *Donax* **25**: *199*, 200, *200*, 201,
        202, 203, *206,*209, **209**, 237
Haemocyanin, *Bullia* **25:** 215, 216

Haemocytes **27**: *32*
Haemoglobin
  in Antarctic fishes
    and stress **24**: 357
    red-blooded **24**: 352–356
    white-blooded **24**: 350, 351
"Haemolymph" **27**: 29, 173
Haemolymph, *Bullia* **25**: 223
haemolymphic proteins **38**: 147, 148
Haemotoxins **21**: 68
Hagfish **36**: 19, 28
Hair cells
  kinocilium **20**: 142, 143
  stereocilium **20**: 142, 143
Halibut **36**: 101, 103; **37**: 198, 220
  see also Atlantic halibut
  eggs
    development of **26**: *78*
    viability **26**: 92
  stripping of eggs **26**: 91
Haline fronts **23**: 174
        see also Thermohaline fronts
  stratification **23**: 172, 209
Halocline **29**: 75, 76, 97, 225
Halocline responses **33**: 443
Halogen toxicity and bivalves **37**: *133*
  see also chlorine
Halogenated compound toxins **21**: 67
Halogenated metabolites, red algae **21**: 93
Halosauropsis sp **36**: 44
Hamburg Ship Model Basin ice tank **43**: *178*,
        193, *194, 195*
Hammer-head shark **36**: 78
Hanging culture, of scallops **20**: 310, 313,
        315, 316, 317, 321, 325, 344–355,
        373
  growth in **20**: 358–362
Hapalotoxin, properties **21**: **188**
Hardening, Japanese oyster culture **21**: 19
Hardening, spermatophore **29**: 130, 186,
        187, 198
Hardy-Weinberg equilibrium **23**: 34, 35
  deviation from, causes **23**: 35
Harmful phytoplankton blooms *see*
        Exceptional phytoplankton blooms
Harpacticoid **29**: **131**, 136, 178–181
Harpacticoids **44**: 298
  as prey for marine fish larvae **44**: 310, 311
  broodstock regulation of naupliar
        production **44**: 300, 301
  changes in behaviour **44**: 301
  culture, candidates for **44**: 299, 300

density effects **44**: 307, 308
  effects of diet on nutritional quality and
        population growth **44**: 301–304
  effects of diet on productivity **44**: 302
  effects of different substrata on
        populations **44**: 304, 305
  environmental effects on quality and
        growth **44**: 307–310
  essential food types for copepod
        culture **44**: 301, 302
  estuarine **44**: 308
  harvesting **44**: 307, 312
  life-history **44**: 297, 299–301, 311
  natural substrata **44**: 305
  potential as live feed for rearing marine
        fish **44**: 295–316
  preference for sand or mud strata **44**: 305
  rearing systems, food types and
        subsequent naupliar
        productivity **44**: 306
  salinity effects **44**: 308, 309
  sexual dimorphism **44**: 300, 301
  suitability of diet **44**: 304
  surface area to volutne ratio **44**: 304
  temperature effects **44**: 309
  turbulence effects **44**: 309, 310
Harvesting options, Bristol Channel
        Sole **29**: 299, 300
  absolute yields **29**: 311, 312, **312–315**
  appropriate fishery targets **29**: 321,322
  bioeconomics and dynamics **29**: 319–321
  spawning stock biomass per
        recruit **29**: 315, 316, *316*
  stock and recruitment
        relationship **29**: 316–318, *318*
  yield per recruit **29**: 300, 311, *300*, **301**,
        302, 306–309, 308
Harvesting, harpacticoids **44**: 307, 312
Harvesting, Japanese oyster culture **21**: 46
Hatching **33**: 293, 294
  of Atlantic halibut **26**: 55–57
  rate of and egg quality **26**: 98, 99
  hatching behaviour (*Pandalus*) **38**: 152
Hatching, Cephalopoda **25**: 105, *106*, 107
Hawaian corals, stress
  loss of zooxanthellae **22**: 30, 31T
Hawaii, threadfin studies in **38**: 29, 30
Hawaiian nehu **20**: 20
Hazmat study, fucoids **39**: 17–29
Heart **33**: 39
Heart and pericardium of Scaphopoda
        42: 180–184, *183*

Heart rate biotelemetry **20**: 229
Heart rate, bivalve molluscs
  effects of
    copper **22**: 117–119
    temperature **22**: 130–132
    valve closure **22**: 146–148, 155
    zinc **22**: 117–119
  methods **22**: 105, 106
  pumping activity and **22**: 111–116
  valve movements and **22**: 111–116
Heart, hormonal control **27**: 201
Heat stress, intertidal species **25**: **184**
  Donax and *Bullia* **25**: 184
Heavy metals
  deep-sea mining impact **35**: 96, 97
  hydrothermal vents **35**: 17
Heavy metals *see* metal toxicity
Heavy metals, bivalve molluscs
  accumulation in tissues **22**: 162, 163
  detection and response to **22**: 143–145
  effects on
    behaviour and physiology **22**: 163–168
    gametogenesis **22**: 166, 168, 169, 172,
      173
    growth **22**: 167
    life stages **22**: 168–175
    mitochondrial respiration **22**: 176–181
    unfertilized eggs **22**: 173–175
  toxicity **22**: 104, 105, 129, 132, 133
  zinc, *see* Zinc
Helgoland **43**: 23, 24
Heliozoa, parasites of **25**: 126
Helminth parasites **29**: 143, 144
Helmintha, parasitic **25**: 147–152, *149*, 157
Helminths **36**: 57
Hemocytes and hemolymph of
      Scaphopoda **42**: 185, 186, 188
"Hepatopancreas" **27**: 22
  red disease **27**: 380
hepatosomatic index (HIS) **38**: 147, 148
Herbicides *see* biocide toxicity
Herbivores of seaweeds **23**: 77–108 *see also*
      Grazing
  behavioural responses **23**: 75, 76, 94, 95
  -competition interactions **23**: 113–117
    with carnivory **23**: 72–108
  coral reef, *see* Reefs
  studies in isolation **23**: 111–113
Herbivorous feeding **33**: 142, 143
  see also Diet; Phytoplankton feeding
Heritability **23**: 37
Hermaphrodites **44**: 35

Hermaphroditism **37**: 17, **18**
  protandric **38**: 121–123, 128, 143
  sequential **38**: 121, 128
Hermit crabs **29**: 146–149, *147*
Heron Island (Great Barrier Reef) **43**: 14, 16
Herring **20**: 11, *12*, 13, *31*, 38–55
  adaptation **36**: 61, 62, 69, 70, 77
  food and feeding **36**: 52, 55
  temperature **36**: 9, 10, 17
  and match/mismatch hypothesis
      26: 265–269, 271, 272, 276, *277*
  Atlantic **20**: *4*, 5, 20–64 *passim* **20**: 68,
      72, 74, 85, 133, 144
  differentiation and variability **36**: 224,
      225, 226
  egg quality **26**: 74
  egg viability **26**: 92
  in retention areas **26**: 257, 258
  life cycles **36**: 116, 118, 129
    annual **36**: 104, 107
    ontogenesis **36**: 101, 104
  light as *zeitgeber* for **26**: 122, 123
  passim **20**: 117, 247, 279, 291
  spawning of **26**: 251, 252, *253*, 256, 271
  substance and energy **36**: 141, 190
  vertical migration patterns **26**: 120–122,
      151, 152
Herring egg **20**: 15
Herring larva **20**: 26, *33*, 48
Herring, eggs and larvae dispersal
      processes **28**: 41–74
  see also Field surveys; Fish eggs and
      larvae; Sampling
    Atlantic herring larvae **28**: 68–74
Herring, Pacific: comparison with pollock
      37: 190, 217, 218, 223, 239
Heteroduplex analysis **35**: 49
Heterosis, crustaceans **29**: 193
heterosynthetic vitellogenesis of
      invertebrates **43**: 89, 90, 141
Heterotrophic picoplankton *see*
      Bacterioplankton
Heterotrophs, micro-, *see* Zooplankton
Heterozygosity **35**: 30, 58–67
  faunal category relationships **35**: 61, 62,
      62, 115
  life history and **35**: 61–67
  polymorphic loci and **35**: 59–61, *60*, 115
  radioactive pollutants and **35**: 102
  reproductive strategy and **35**: 65, 67, **66**,
      115
  See also Genetic variation

Heterozygote estimates **23**: 33, 34
Hierarchical model **44**: 244
Hierarchical structure of the
    environment **44**: 244
Hierarchy theory **44**: 241, 242
High latitude life history strategies
    33: 384–390
    overwintering copepodids **33**: 385–388,
    396
    resting eggs in sediments **33**: 388, 389
High performance liquid chromatography
    43: 285, 309, 310
High-performance liquidchromatography
    (HPLC) **35**: 52
Hills **35**: 25
Himerometra sp **36**: 36
Himmerfjärden, Baltic Sea **29**: 92
Hindgut **33**: 39, *40*
Hinkley Point power station **29**: 237, 238,
    243, 244, 350
Hiogi scallop **20**: 325
Hippidae **29**: **133**, 197
Hippoglossus, biology of **26**: 1–70
Hiroshima Bay **21**: 13
Hiroshima rack, oyster culture **21**: *14*
Hiroshima rafts, oyster culture **21**: *14*, *25*, *26*
Hiroshima, Japanese oyster culture **21**: 6, 41
Hirudinea, hydrothermal vent dwelling
    23: 328
Hirudinea, parasitic **25**: **147**
Histological condition **30**: 270
Histological criteria **30**: 240
Histological indices **30**: 223–225, 279
Histological measurements **30**: 256–259,
    269
Histological method of quantifying
    fecundity **43**: 106–109, *108*
Histological scoring **30**: 281
Histopathology
    corals under stress **22**: 48–50
History, Japanese oyster culture **21**: 2
History, marine toxins **21**: 61
"Hitchhiker" parasites **43**: 85
Hjort's critical period concept **31**: 170
Hoki fishery **35**: 107–109
Hokkaido–Sakhalin herring **20**: 5, 169
Holocene **42**: 215
Holostanol **21**: **161**
Holothurians **29**: 253
    distribution path **32**: *372*
    Pacific Ocean **32**: 497–499
Holothurigenin **21**: 158

Holothurin A **21**: *157*, **166**
Holothurins **21**: 152–167, **159**, **164**, **165**
Holotoxin **21**: 159
Homarid lobsters **29**: 160–165, **162**, **163**
Homarine **21**: 151
Home range **44**: 238–241, 262
    estimation **44**: 259
    movements **44**: 215–223
    relocations of **44**: 226–228
Homeostasis, structural functional, energy
    optimum and **36**: 243–245
Homeoviscous cold adaptation **24**: 368
Homolidae **29**: **133**
Hong Kong, sea cucumber in **41**: 126, *137*,
    190
Hoplocarida **29**: **131**, 184
Hormonal control, egg production **29**: 279
Hormonal treatment and broodstock
    management **26**: 102, 103
Hormones **36**: 24–26, 86, 96, 128
Hormones and gametogenesis **43**: 103, 104,
    129
Hormones, chromatophorotropins **27**: 200,
    201
    gonad maturation **27**: 197–199
    metabolic **27**: 202
    moulting **27**: 195–197
    neurohormones **27**: 201
Horn's Information Theory in benthic
    studies **26**: 215
Horse mackerel, vertical migration
    of **26**: 120, 127, 145
Horse-mackerel **36**: 222
    adaptation
        food and feeding **36**: 47, 49
        oxygen level **36**: *31*, *33*, 34, *35*
        temperature **36**: 7, 9, *11*, 12, 14, 17
    adaptation strategies
        energy metabolism **36**: **60**, 62, **64**, 65,
        66, *67*, 68–70, 72, *73*, *74*
        plastic metabolism **36**: **75**, 76, **77**, 78,
        79, *80*, *81*, *82*, 86
    ecological principles **36**: 238, 243
    indicators of condition **36**: *212*
    life cycles **36**: 132
        annual **36**: 105, 107, 112
        daily rhythms **36**: *115*, 117
        ontogenesis **36**: *99*, *100*, *102*, 104
Horseshoe effect in ordination in benthic
    studies **26**: 225
Host range **24**: 265, 270
Host specificity **24**: 265, 271–279, 299, 306

indices **24**: 271–273
Hosts of parasites **43**: 14, *15*
  body size/length **43**: *15*, 33–35
  cleaners of **43**: 35, 36
  co-evolution with parasites **43**: 46
  effects of **43**: 44, 45
  latitudinal gradients in **43**: 52, 53
  migration monitored **43**: 55
  phenotype changed **43**: 34, 45
  phylogeny **43**: 33, 34
  populations and migrations **43**: 59–62
  ranges **43**: 22, 23, 52, 53
  specificity **43**: 5, 22, 23, 52, 53
Hotate gai scallop **20**: 310, 322, 323, 366
Hotate goten **20**: 362
HPLC (high performance liquid
        chromatography)
        43: 285,309–310
HSVA *see* Hamburg Ship Model
Human exploitation **33**: 522, 523
Human use of resources *see* exploitation
Humboldt current **20**: 259
Humboldt Current fisheries **39**: 271, 277,
        266, 275
Humpback salmon **36**: 213, 219, 222
Hurricane
  coral reefs and **22**: 6
Hybrid dysgenesis **31**: 28
Hybrid indices, use in hybridization **31**: 7
Hybrid swarm **31**: 53
  definition **31**: 2
Hybrid unfitness **31**: 2, 29, 30
Hybrid vigour (heterosis) **31**: 30
Hybrid zone **31**: 55–6
  definition **31**: 2
  shape **31**: 35–37
  theory **31**: 56
  types **31**: 2, 3
  width **31**: 33–35
Hybridization
  crustaceans **29**: 168, 190, 193
  sole **29**: 221
Hybridization, marine **31**: **70–8**
  as a fisheries and conservation
        problem **31**: 49–52
  bias **31**: 39–42
  definitions **31**: 2–4
  ecological differentiation **31**: 45–6
  experimental **31**: 17, 18
  factors preventing **31**: 11, 12
  factors promoting **31**: 12, 13
  fitness of individuals **31**: 28–31

fossil record **31**: 46–49
frequency of **31**: 13–15
geographical distribution **31**: 43–45, *44*
human disturbance and **31**: 22–3
intergenetic versus interspecific **31**: 42, 43
levels of **31**: 5, 6
natural disturbance **31**: 23–27
prospects for the future **31**: 55, 56
recognition of **31**: 5–7
relative parental abundance and **31**: 27, 28
reproductive biology and frequency
        of **31**: 31–33
social structure and **31**: 28
speciation and **31**: 52–55
species concepts and **31**: 9–11
systematics, taxonomy and **31**: 7–9
techniques for identifying **31**: 6, 7
temporal variability and stability in rates
        of **31**: 19–22
timing of reproduction and **31**: 15–17
Hybrizymes **31**: 52
Hydroacoustic surveys of pollock **37**: 193,
        197
Hydrocarbon seeps **35**: 23
Hydrodynamic characteristics, Bristol
        Channel **29**: 218
Hydrodynamic features **33**: *422*, 422, 423
hydrogen **41**: *11, 12, 15*
Hydrogen content **33**: 242
Hydrogen sulfide
  hydrothermal vents **35**: 17, 18
  oxygen-minimum zones (OMZs) **35**: 25
  seeps **35**: 23
  transport of **35**: 27, 28
Hydrogen sulphide **36**: 30
Hydrographic conditions and spawning of
        Atlantic halibut **26**: 10, 11
Hydrographic conditions, Bristol
        Channel **29**: 351
Hydroids, toxins **21**: 104, **105**
Hydrolases
  corals under stress **22**: 50
Hydrological connectivity **44**: 215
Hydromedusae **44**: 29
Hydromedusae, predator **25:** 7, 29
Hydrostatic pressure, deep-sea
  adaptations to **35**: 26
        environments **35**: 11–16, 64
Hydrostatic skeleton, *Donax* and *Bullia*
        25: 200, 202
Hydrothermal ecosystems
  Gulf of California **32**: 80

local influence **32**: 78–82
plumes **32**: 79
Hydrothermal vent communities
    30: 323–326
Hydrothermal vent environment **35**: 16–22,
    115
adaptations to **35**: 27, 28
allozyme electrophoresis studies **35**: 30,
    40, 41, 43
dispersal strategies **35**: 19, 20, *19*, 22, 90,
    91
heterozygosity **35**: 61–64, 115
plumes **35**: 18
spatial patterns of genetic variation
    35: 74, 78, 79, 87–91, 116
temperature **35**: 8, 10, 17, 18, 28
Hydrothermal vent fauna
Atlantic **34**: 391, 392
biogeography **34**: 385–394
causes of diversity **34**: 423–425, *425*
composition **34**: 363–379
dispersal **34**: 395, 396, 401, *401*
distribution **34**: 385–386
distribution patterns of taxa **34**: 402–420
ecological controls **34**: 424, 425
extinction **34**: 423
gene flow **34**: 396–398
generic/specific distributions
    34: 402, 403, *403, 404, 405*
global species differences **34**: 422, 423
gradients **34**: 421, 422
high sulphide environments **34**: 379–381
Indian Ocean **34**: 392
local to regional-scale processes
    34: 394–398
Mid-Atlantic Ridge **32**: 93–144
bivalves **32**: 121, 122
geographic patterns, vent and seep
    fauna **32**: 125–135
named sites **32**: **126–129**
shrimps **32**: 109–121
special features **32**: 119–135
trophic structure **32**: 122–125
Northeast Pacific **34**: 386, 387, *388, 389*
organic remains **34**: 384, 385
overall composition **34**: *381*
patterns in diversity **34**: 420–425, *422*
Phanerozoic **32**: 121
phyletic composition **34**: 364–379
pogonophorans **34**: 404–410
problems **34**: 357
regional studies **34**: 386–392

regional to global-scale processes
    34: 398–401
relict and 'fossil' species **32**: 408

representation of major invertebrate
    phyla **34**: *382*
reproductive strategies **34**: 395
ridge characteristics **34**: 399
seep/whale/vent relations **34**: 392–394
seeps **34**: 382–384, *383*
Southern Pacific **34**: 390, 391
speciation **34**: 423, *424*
species currently known **34**: 364,
    365–381
sulphide-rich habitats **34**: 393, 394
taxon listings **34**: 363, 364
vagrant taxa **34**: **380, 381**
vestimentiferans distribution **34**: *406*
vicariance **34**: 399–401
Western Pacific **34**: 387–390, *389*
Hydrothermal vent organisms **24**: 433–435
Hydrothermal vents **23**: 302–347; **33**: 502,
    503; **34**: 353–442
age **23**: 310–314
chemical composition **23**: 312, 313,
    314–315
of ambient waters **23**: 312, 313
of fluids **23**: 312, 313
of gases **23**: 314, 315
definition **34**: 358, 359
discovery **23**: 302, 303, 304
distribution **23**: 301–310
distribution of major sites **34**: *360*
distribution on a ridge **34**: *362*
faunal distribution and density
    23: 333–340
common organisms **23**: 315–332
microbes **23**: 315–318
relative consistencies between
    vents **23**: 332, 333
faunal temperature conditions **23**: 302,
    303, 318
formation **23**: 304
geographical sites
cold seeps **23**: 303–314
E. Pacific Rise (21°N), *see* E. Pacific
    Rise (21°N)
E. Pacific Rise (l3°N), *see* E. Pacific
    Rise (13°N)
Florida Escarpment **23**: 304, 311,
    338–339
Guyamas Basin, *see* Guyamas Basin

Juan de Fuca ridge, *see* Juan de Fuca
    ridge
Louisiana Continental Slope **23**: 339
Oregon subduction zone **23**: 339
Red sea, barrenness of **23**: 310
20°S vent, *see* 20°S vent (under Twenty)
habitat character **34**: 361–363
habitat distribution **34**: 359–361
locations **23**: 311
primary sites of biological collections
    34: **358**
ridge-crest venting **34**: 359
survey systems **23**: 304, 305
zoogeography and ecology **23**: 341–347
    difficulties in comparing different
        vents **23**: 341
    dispersal **23**: 346, 347
    food webs **23**: 341, 342
    fossil history **23**: 343
    instability relative to reefs **23**: 332, 333
    metabolic rates **23**: 343–345
    metal tolerance **23**: 342, 343
Hydroxysaxitonin **21**: 79
Hyperbenthic habit **33**: 496–501, **498–501**
Hyperiidae **26**: 131
    parasites of **25**: 137, 141, 142
    parasitic **25**: 154–156, *155*, 158
Hyperparasitic, peridinians **25**: 129
Hypnocysts **21**: 74
Hyponome, Cephalopoda **25**: 99, 100
Hypothesis testing in benthic marine infaunal
    studies **26**: 212
Hypoxia and anoxia **36**: 232, 236
    see also Oxygen level
    adaptation **36**: 30–41
    passim, 31, 38, 74

IASs (ice-active substances) **43**: 206
ice
    see also Southern Ocean sea ice
    forming and melting **43**: 175–180, *177*,
        183, 184, 198
    stations **43**: 187, 188, 197, 198, 211, 218,
        219, 240
Ice Age, Bristol Channel **29**: 223
Ice conditions, Baltic Sea **29**: 76, 94
Ice scour **23**: 135
Icefish **36**: 13, 236
Iceland **29**: 331
Iceland, halibut fisheries in **26**: 43, 45, 46, 48
Icelandic waters
    fisheries **38**: 223

Pandalus borealis stock limitation
    in **38**: 207, 208
ICES assessments **29**: 335, 336
    mark–recapture estimates **29**: 342, 343
    stimulation of population
        trajectories **29**: 343–347,
        344–347
ICES *see* International Council for
    Exploration of the Sea
Ice-water interface **33**: 490–496, **493**
    Antarctic Ocean **33**: **493**, 494
    Arctic Ocean **33**: **493**, 495
    ice cover at lower latitudes **33**: 496
    ice-edge zones **33**: 495, 496
Ichaboe Island **20**: *256*
Ichthyocrinotoxic fish **21**: 65; **21**: 65
Ichthyootoxic fish **21**: 65
Ichthyoplankton *see* Fish eggs and larvae
Ichthyoplankton, prey **25**: 7
Ichthyotosarcotoxic fish **21**: 65
Identification **33**: 57–64, 516
    eggs **33**: 23
    genera **33**: 65–97
    molecular genetics **33**: 63, 64
    pore signatures **33**: 61–63, *62*
    species **33**: 60, 61, 516
Identification, prey **25**: 38, 39
Identification, sole **29**: 219, 220
Image analysis **43**: 107–109, *108*, 182
Image-searching behaviour, predator **25**: 15
Immature phase of Atlantic halibut
    26: 22–27
    age/size composition **26**: 23–25
    distribution **26**: 6, 7
    feeding **26**: 26, 27
    growth **26**: 25, 26
    occurrence and migration **26**: 22, 23
    sex ratios **26**: 23–25
Immigration, sole **29**: 217, 249, 250, 261,
    294
Immunoassay, prey identification **25**: 38, 39
Immunology, Brachiopods **28**: 349–352
in estuaries, mesohaline or hypersaline
    systems **31**: **82–83**
In situ and *in vivo* spectral measurements of
    coral reef organisms **43**: 280, *283*,
    284, 285, 289, **293–303**, *298, 301*,
    305, 310
Incirrata, embryonic development **25**: 88,
    89, *89*, 105
Incubation approaches **44**: 39

Incubation of Atlantic halibut **26**: *55*,56, 57, *59*
Incubation of bivalve eggs **37**: 52, 53, 136
'Indeterminate' spawners, sole **29**: 280
India, sea cucumber in **41**: 132, 136, 185
  anatomy **41**: *138, 139*, 143
  associations **41**: *189*
  biochemistry **41**: *186*
  burrowing cycle **41**: 178
  common names **41**: *137*
  development **41**: 171, 173, 174
  distribution and population structure **41**: 149, *150, 152*, 153, 154
  fisheries and aquaculture **41**: 190, 191, 193, 194, 195, 198, 199
  reproduction and spawning **41**: *161*, 163, *164*, 165, 167, 168
Indian Ocean **34**: 392; **35**: 5, 8, 11; **36**: 15, 36, 170; **42**: 37, 147, 205
  see also Indo-Pacific Ocean
  deep-water bottom fauna, taxonomic differentiation **32**: *337*
  equatorial, meso- and macroplankton biomass distribution **32**: **59**
  hydrothermal vents **35**: 17
  invertebrates **43**: 143
  parasites **43**: 53, *57, 58, 60, 61*
    diversity **43**: 6, 11, *18, 19*
  Pleurotomarioidean gastropods **42**: 272–275, *273, 275*, 278, 281
Indian Ocean, sea cucumber in *see* distribution of sea cucumber
Indian Ocean, trace elements in
  and phosphates **41**: *127, 128*
  cadmium **41**: *75, 127*
  copper **41**: 63, *124*
  nickel **41**: 78, *79, 128*
  silicon **41**: *124, 125, 128*
  vertical concentration profiles **41**: 9
  zinc **41**: 67, *68, 125*
Indian oil sardine **20**: 35, 36
Indicator species analysis **23**: 49–51
Indicator threshold **23**: 50, 51
Indicators and biological markers, parasites as **43**: 29, 55–62 *passim*
indicators of fish condition **36**: 205–220
  lipids **36**: 206–212
  other substances **36**: 213–220
Indonesia **42**: *273*
Indonesia, sea cucumber in **41**: 132, 136
  anatomy **41**: *139*

associations **41**: *189*
distribution and population structure **41**: 149, *150, 152*
fisheries and aquaculture **41**: 190, 194, 198
reproduction and spawning **41**: 163, *164*, 165, 168
Indo-Pacific Ocean
  invertebrates **43**: 149, 150
  parasites
    distribution **43**: 47, *50*, 53, 55, *57–58*
    diversity **43**: 6, 9, *18*
IndoWest Pacific Region, Nazca and Sala y Gómez submarine ridges **32**: 145–241
Industrial waste, deep-sea dumping of **35**: 6, 93
Industry *see* urban and industrial
Infaunal habit
  Donax and *Bullia* **25**: 183
  gastropods **25**: 198
Infections, sole **29**: 298
Infiltration in sea ice **43**: 183, 184, 198
Infrared radiometry **23**: 166, 174
  sea surface **23**: 174, 176
    airborne measurement **23**: 176
    satellite measurement, *see* Satellites
Ingestion rate *see* Feeding rate
Ingestion rates, *Donax* and *Bullia* compared **25**: 192, 225–229, **226**, 230, 231, **231**, 232, 234, *233*, 237, 238
Ingestion rates, predation **25**: 42, 43
INIPOL **39**: 29
Injuries, predation **25**: 51
Innervation of Scaphopoda **42**: 194, *195*
Inorganic components of fish eggs **26**: 82, 83
Inorganic nutrients **43**: 199, 200
Inorganic nutrients, exudation **29**: 30
Inquilinism, Amphipoda **25**: 156
Insect algal grazing **23**: 112
Insect larvae, growth performance **35**: 175, 176, *179, 180*, **181**
Insecta, spermatophores **29**: **188**, 198, 199
Insecticides *see* biocide toxicity
Integration, ecological principle of **36**: 235–237
Integument **33**: 28–35
  see also Moulting
  eye **33**: *31*, 31, **32**
  frontal organs **33**: *31*, 31, **32**
  moulting **33**: 32–35, *33*, **34**
  pore pattern signatures **33**: 30

setae **33**: 28–30
subintegumental glands **33**: 29, 30
Integument, Cephalopoda **25:** 95–98, *96,* *97,* 103, 104, *104*
Integument, pycnogonid **24**: 23–25
Interaction between toxicants **37**: 123–125
interactions of trace metals **41**: 33–39, *34–35, 36,* 61, 64, 74–78, *125–128*
  with other elements **41**: 56–58, 62–64, *67, 68,* 71–76, *73, 75,* 78, 79, 83, *124–128*
"Interactive" communities **43**: 33
Interannual fluctuations **36**: 118–123
  carbohydrates **36**: 119–123
  lipids **36**: 118, 119
Intercanicular striae, sole **29**: 220, 221
Interfish communication **20**: 100
Interfish distance **20**: 95, 96, 101
Intermediate culture, of scallop **20**: 310, 341–344
  standard **20**: **343**
Intermediate forms of parasites **43**: 9–11
internal economy
  see also biological availability;seawater recipe
  and near-field chemistry **41**: **23–39**
    feedbacks **41**: 87–89
    interactions **41**: 33–39, *34–36*
  for essential trace metals **41**: **9–23**
    establishment of **41**: 9–15, *10–13*
    functions **41**: *11,* 15–23, *15, 16–17*
Internal fertilization of invertebrates *see* brooding species
Internal waves (internal tides) **23**: 198–204, 276, 277
  and phytoplankton production, *see* Phytoplankton, high productivity hypothesis range **23**: 199, 202
  between shelf break and tidal fronts **23**: 204–207
  cooling linked to **23**: 201
  energy, world-wide distribution **23**: 273
  high frequency **23**: 204–207, 280
    data detecting **23**: 207
  methodological problems **23**: 254–256
    spurious chlorophyll peaks **23**: 254, 255
  nutrient pulse generated **23**: 269, 270
  on-shelf propagation **23**: 200, 202, 205, 269, 270
  relevant model **23**: 201

standing wave mechanisms **23**: 198, 199
International Commission for Northwest Atlantic Fisheries (ICNAF) **28**: 30
International Council for Exploration of the Sea (ICES) **28**: 30, 33; **29**: 217, 218, 223
  see also Virtual population analysis
  sole stock **29**: 334, 344, 345, 348, 351
International Council for the Exploration of the Sea (ICES) **38**: 4, 57
International Whaling Commission **43**: 229, 230
Intersexes **33**: 37
Intertidal species **25**: 182
Intertidal zonation **44**: 266
Intertidal zone, hybridization in **31**: 12
Intertidal zone, oil spill effects **39**: 13–34
Intra-ovarian oogenesis **43**: 89
Intra-specific scaling **44**: 29
Intra-taxa scaling and comparisons **44**: 30–36
Intrinsic (biological) factors affecting toxicity **37**: 118, 119
  age and developmental stage **37**: 118, 119
  body size **37**: 119
  species **37**: 118
Introductions **33**: 505–508, **507**; **43**: 69, 70
introductions, definition **38**: 5
Introgression **31**: 37, 38
Invertebrate aggregations **30**: 156
  internal structure **30**: 169
  shape and density of **30**: 163–169
Invertebrates **36**: 9, 14, 27, 28
  alternative prey **25**: 36, 37
    population vulnerability **25**: 27, 30, **31**, **45,46**
  dynamics; Productivity
  genetic identity related to taxonomic divergence **35**: 69
  heterozygosity **35**: 61–66, 115
  Molluscs; Mortality;
  performance; Insect larvae;
  Polychaetes; Population
  predators **25**: 4, 5–8, *6,* 12, 13, 14, 16, 22, 27
    prey identification **25**: 38
  See also Crustaceans; Growth
  spatial patterns of genetic variation **35**: 74, 78–83
Invertebrates *see* fecundity of invertebrates
iodine **41**: *12, 121, 122*

Iodine values **36**: 15, **42**, 76, **77**, 78, 219, 220
Ireland, coastal areas of **43**: 68
Irish Sea **29**: 34
  adults **29**: 248, 250
  growth **29**: 260–262
  juveniles **29**: 237, 238, 240, 246
  natural mortality rates **29**: 295, 298
  reproduction **29**: 272, 275, 278
  sole **29**: 235, 349
  weight **29**: 263
Irish sea fronts, horizontal exchanges **23**: 243
Iroise front **23**: 225
  red tides **23**: 232, 233
iron **41**: 22, **42–61**, 88, 89, *120, 121*
  biochemical function **41**: *11, 12*
  biological availability and uptake **41**: 25, *26, 27,* 28, 29, 30, *31,* 32, 46–49, *47*
  community effects **41**: 58–61, *59, 60*
  cycling ratio **41**: *84, 85, 86*
  distribution in oceans **41**: 42–46, *44, 45,* 49–57, *52, 54, 60*
  functions and properties **41**: *15–17,* 18–19
  impact on primary production **41**: *13,* 14, 23, 49–55, *54*
  in sea cucumber **41**: *186*
  indices of fractionation **41**: *81, 82*
  inputs to system **41**: *123*
  interactions **41**: 33–*35*
    with macronutrient cycles **41**: 56–58
Irradiance **43**: 286, 296, 303, 309
  see also reflectance
  and radiance, differences between **43**: 294, 295
Irradiance, attenuation of and plankton production **26**: 254
IsaacsKidd trawl, macroplankton, sampling **32**: 7, 8
Island model **35**: 56
Islay front **23**: 171, 172, 187
  nutrient-phytoplankton relationship **23**: 242
Isle of May **20**: 296
Isochronal development **33**: 319
Isoelectric focusing (IEF) **35**: 43
Isoenzymes **29**: 137
Isolumes **26**: 125–127
Isopoda, parasitic **25**: 153, 154
Isopods **29**: 185, 186, 199
  deep-sea, vertical distribution **32**: *334*

relict and 'fossil' species **32**: **409,** 410
Isopods, hydrothermal vents **23**: 330
Isotopic fractionation **24**: 394, 395, 396, 398, 401–408, 417–419, 442
  and denitrification **24**: 405, 406
  and diet **24**: 430, 431
  and phytoplankton **24**: 406, 407, 414–416
  during amino acid transamination **24**: 443
  magnitude **24**: 404–408
  theory **24**: 401–403
'Isotopic memory' **24**: 430, 431
Isovalerylcholine **21**: 178
Israel: mariculture **43**: 64, 69
Itaya gai scallop **20**: 325
Iteroparity **43**: 118, 129; **20**: 22
Ivasi **36**: 129, 132, 248
Iwate and Miyagi, scallop production in **20**: 321
IWC (International Whaling Commission) **43**: 229, 230

Jaccard coefficient **23**: 52, 70
Jaccard's index in benthic studies **26**: 215, 216
Jackass penguin **20**: 60, 247, **248**, 249, 252, 259
Japan **21**: 1–57
  Pleurotomarioidean gastropods **42**: 239, 267, 272, *273*
  Protobranch bivalves **42**: 35, 36, 37, 39, 40
  Scaphopoda **42**: 215, 216
Japan Sea **36**: 114
Japan, coastal areas of
  parasites **43**: *10,* 65, 69
  whaling **43**: 229
Japan, sea cucumber in **41**: 136, *137, 150,* 183
Japan, Sea of/Japanese waters, pollock in **37**: *180,* 182, 184
  fisheries and management **37**: 183, *191,* 237
  population structure **37**: *227,* 236
    genetic **37**: 216, 218, 221
    phenotypic **37**: 209, 212, 213, *215*
  stocks **37**: **185**
Japanese anchovy **20**: 53, 55, **72**, 83, 140
Japanese currents,
  Kuroshio **20**: *314,* 321
  Oyashio **20**: *314,* 316, 320, 321
  Riman **20**: *314,* 316
  Tsushima **20**: *314,* 316, 317, 320, 321
Japanese eel **36**: 21, 22

Japanese gizzard shad **20**: 140
Japanese oysters **21**: 1–57
 characteristics **21**: 4
 culture areas **21**: 6
 culture problems **21**: 31
 harvesting **21**: 46
 larval development **21**: 5
 marketing **21**: 46
 production **21**: 38
Japanese round herring **20**: 140
Japanese sardine **20**: *4*, 5, 8, 36, 140, 258
Japanese scaled sardine **20**: 140
Japanese spotline sardine **20**: 140
Japanese waters fisheries **38**: 226
Jelly cat **36**: 46
Jellyfish
 sting pattern **21**: *129, 131*
 toxins **21**: 104, **105**
Jellyfish *see* Medusa
Jones, Bank of **29**: *223*, 224
Juan de Fuca Ridge **34**: 386, 387; **35**: 17, 78,
   89
Juan de Fuca ridge hydrothermal vents
   23: 310, 311, 346
 vent separation **23**: 346
Juan de Fuca Ridge, hydrothermal
   ecosystems **32**: 79, 80
Jump response **27**: 376
Jurassic **42**: 217
 Pleurotomarioidean gastropods **42**: 237,
   239, *242*, 267, 278, 280
Juvenile (plant) survivorship **23**: 12
Juvenile fish *see* Reproduction
Juvenile pollock **37**: 188, *189*, 192, 193,
   197, 199, 200–203, 226
juvenile sea cucumber **41**: **154, 155**
 daily burrowing cycle **41**: 180–181
 feeding behaviour **41**: 156, 157, 182–185
 habitat **41**: *150, 151*
 rearing in aquaculture **41**: 198, 199
juvenile stages, growth in, in
   Pandalus **38**: 197–199
Juveniles, as prey **27**: 357, **357**, 358
 diet **27**: 320, 321
 growth **27**: 236–240, *240*
  rate estimates **27**: **238**
 life cycles and histories **27**: 284–288,
   296, 297
 migration **27**: 297, 303–311
 response to tides **27**: 338
 salinity **27**: 351, 352
 species identification **27**: 126

 taxonomy **27**: 107–126, **108–112**
 temperature and habitat selection **27**: 350
Juveniles, *Donax* and *Bullia* **25**: 187, 189,
   190
Juveniles, sole **29**: 217, 218, 237
 distribution **29**: 237–242, *239*, **240**, *241*
 mortality **29**: 291, 294, 295, 299
 movement **29**: 242–246, *245*, 261, 294

Kagoshima Bay, Japan, red sea bream
   enhancement **38**: 25, 26
Kamchatka herring **20**: 169
Kamchatka, pollock near
 population structure **37**: 208, 209, 213,
   215, 221, *227*
 stocks **37**: **185**, *186*
Karyotyping **35**: 51, 52
Kattegat **29**: 22, *23*, *24*, 42, 44, *46*, *48*, 49,
   50, *51*, 52, 76
Kelp
 densities and biomass **23**: 20, 21, 80–82,
   91–96
 destruction **23**: 91–96
  feedback relationships **23**: 95
 giant **23**: 20, 21, 24
 patches, maintenance **23**: 18
 recolonization **23**: 91
 recruitment **23**: 20–30
 sea urchin grazing **23**: 91–96
  effect of removal **23**: 80–82
 spore production **23**: 21
 transplant experiments **23**: 137
Kelp communities **39**: 40–43
Kelp, South African coast **25**: 180, 185
Kelvin wave propagation **23**: 197, 204
Kendeigh's equation **20**: 230, 231, 238, 286
Kenya **41**: 136
Keratinization, crustaceans **29**: 187
Kermadec Trench **32**: *50*, 51
 hadal fauna **32**: 369
Keta-ami scallop dredge **20**: *357*
'Key and lock' relationship, crustaceans
   29: 182
Key species **33**: 528–530
Key to genera **33**: 57–60, 59, **65**, 65–97
Keys **27**: 4, 55
 Atypopenaeus spp. **27**: 66
 families, adults **27**: 58–66
 families, larvae and
   postlarvae **27**: 115–118
 Funchalia spp. **27**: 68
 genera, adults **27**: 61–63

genera, larvae and postlarvae
  27: 119-123
Metapenaeopsis spp. 27: 72-78
Metapenaeus spp. 27: 80-83
Parapenaeopsis spp. 27: 85-87
Parapenaeus spp. 27: 89, 90
Penaeoidea 27: 58-60
Penaeopsis spp. 27: 92, 93
Penaeus spp. 27: 95-98
to adult *Pandalus* shrimps 38: 64-66
to larvae of *Pandalus* shrimps
  38: 105-108
Trachypenaeus spp. 27: 103-105
Kiel Bight 29: 96, 98, 99, 101, 104, 113
  station 29: 89, 92, 92
Kiel herring 20: 19
Kilka 36: 159, 248
  adaptation 36: 49, 53, 76, 77
  indicators of condition 36: 208, 210, 211
  life cycles 36: 129, 132
Killifish 36: 19, 20, 21, 25, 28
Kinsale Head 29: 223, 223
Kittiwake 20: 281, 284, 285, 286, 293, 294
Kjeldahl-Rittenberg procedure 24: 394, 395,
  397, 399
Knolls 35: 25
Kokanee salmon 36: 117
Kölliker organs, Cephalopoda 25: 103, 104,
  105
Kolmogorov length 29: 14
Kolmogorov scale 31: 174, 175
Kolmogorov-Smirnov test 26: 194
Kombu 20: 320
Kosrae 41: 136, 150
Krebs cycle 29: 143, 144
Krill 20: 271, 273, 274, 275, 279, 295
  consumption of 20: 276, 277, 278
Kruskal-Wallis test in benthic
  studies 26: 228
K-selection 20: 260
K-strategist species 38: 6
K-strategist, *Bullia* 25: 194, 236
Kurile-Kamchatka region of NW
  Pacific 32: 11-19, 48-55, 51
  hadal fauna 32: 369-373
  meso- and macroplankton biomass
    distribution 32: 59
  vertical distribution, biomass of
    taxa 32: 56
Kuroshio Current fisheries 39: 265-267,
  275, 276, 266, 275

L_∞ 20: 5, 20, 21, 84, 169
L{S} − 1(s) components 44: 243, 244
L{S} + 1{s} level 44: 243, 244
Labadie, Bank of 29: 223, 224, 247
Labelling, bacteria 29: 111
Labels 33: 11, 12
Labium 33: 19, 39
Laboratory culture 33: 218, 297, 298
Laboratory rearing of bivalve
  larvae 37: 25-40
  see also physical requirements
  conditioning of spawners 37: 25-27
  fertilization 37: 29-33
  nutritional requirements 37: 37-40
  spawning 37: 27-30
Laboratory sampling of Southern Ocean
  sea ice 43: 192-196
Laboratory studies
  Donax and *Bullia* 25: 182
  mesocosm experiments 25: 43-45, 44,
    50, 51, 57, 58, 67
  predation 25: 33-37, 34, 35
Labrador Current
  Calanus hyperboreus 32: 23, 48, 49
  frontal with Gulf Stream
    macroplankton 32: 60
  meso- and macroplankton biomass
    distribution 32: 59
  zooplankton biomass distribution 32: 49
Labrador Sea 36: 50
Labral glands 33: 19, 39
Labrum 33: 19, 39
Lactate dehydrogenase (LDH) 30: 239
Lactate dehydrogenase (LDH),
  spermatophores 29: 137, 143, 189
Lactic acid, spermatophores 29: 139
Lady-fish 20: 34
lagoon fisheries 38: 32-33
Lagoons, see Coastal lagoon fisheries
Lake Erie 20: 297
Lake Quarun 29: 269, 272
Laker's theory 20: 9
Lallemand Fjord 42: 170
Lambert's Bay 20: 257
Lamellibranchs 29: 252
Laminarinase
  in sea anemones 22: 80
Laminarinase, *Donax* and *Bullia* 25: 192
Lampfish 36: 79
Lamprey 36: 19
Landing size, sole 29: 260, 299
Landsat Thematic Mapper data 44: 264

Landscape ecology **44**: 244, 245
Langranian measurements **23**: 187–189
Lanosterol **21**: 160
Lantern fish **36**: 44; **20**: 249
Lantern net **20**: 310, 351, *352*, 354, 360,
          362, 368
Large marine ecosystems (LMEs) 509,
          510, 511, 510, 515, 516, 519
Large-toothed anchovy **20**: 34
Larvacea **44**: 3, 10, 22–24, 26, 27, 28, 30,
          32, 35–37, 42, 129–132
    numbers of species, data points and
          ranges **44**: 11
Larvacean prey **33**: 160
Larvae
    see also Predation
    abundance and recruitment **26**: 271–273
    and members/vagrant
          hypothesis **26**: 256–258
    and upwelling areas **26**: 259–261
    bivalve *see* bivalve; embryo-larval;
          growth
    Donax and *Bullia* **25**: 193, 194, 236
    growth of and food supply **26**: 258
    movements of **44**: 213–215
    of Atlantic halibut
        and post-larval stage **26**: 18–20, *20,
          21*
        growth of **26**: *15, 16*
        in rearing experiments **26**: 53
        occurrence **26**: 11, 12, *13*
        rearing **26**: 57–60
        start-feeding **26**: 60–62
        yolk-sac **26**: 14–18, 57–60
    of cod and plankton production
          26: 279–281, *282, 283*
    pollock **37**: 184, 185, *188, 189*, 201–203
        migration **37**: 228, 229, 231, 232
    size of and egg size **26**: 78, 79
    survival of **26**: 264
        and egg quality **26**: 98, 99, 101
    time of release of **26**: 251, 252, 255
    trajectory mapping **44**: 247–251
Larvae of invertebrates **43**: 91, 93, 128
    nutrition (lecithotrophic or planktotrophic)
          43: 90, 92, 100, 124–126, 144,
          146, 147, 246, 247
larvae of sea cucumber **41**: 171, *172,* 173,
          174
    feeding behaviour **41**: 181, 182
    rearing in aquaculture **41**: 198, 199
Larvae, as prey **27**: 357–358, **357**

duration of development **27**: **278–281**
    growth **27**: 235, 236, *236*, **278–280**
    lens transition in **27**: 190
    life cycles **27**: 283–297
    life histories **27**: 283–314
    migration and dispersal **27**: 297–302
    morphology **27**: 48–53, **47**, *49, 53*
    spatial variation in distribution **27**: *301*
    stages, internal anatomy **27**: 53, 54
        morphology **27**: 45–54, **47**
        mysis **27**: 52
        nauplius **27**: 48–50
        nomenclature **27**: 45–48, **48**
        protozoea **27**: 50, 51, *51*
    taxonomy **27**: 107–26, **108–112,** *116,
          120, 122*
        key, to families **27**: 115–118
        key, to genera **27**: 119–123
    vertical distribution **27**: *300*
Larvae, condition measures, *see* Condition
          measures
larvae, *Pandalus*
    behaviour **38**: 156–159
    biology **38**: 156–166
    classification and identification
          38: 101–108
    growth **38**: 162–164
    nutrition **38**: 159–162
    survival **38**: 164–166
Larvae, sole **29**: 217, 218, 230
    distribution **29**: 231, *233*, 235, 236
    feeding **29**: 251, 252
    growth **29**: 255, 256
    mortality **29**: 293, 294, 298, 299
Larval abundance indices **28**: 33, **34**
Larval development
    Protobranch bivalves **42**: 26–28, *27*
    Scaphopoda **42**: 197–204, *198*
        shells **42**: 200–204, *201, 202, 203*
Larval development, Japanese oysters **21**: 5
Larval drift **20**: 50, 51
Larval predation **20**: 51–58
Larval production method (sampling)
          28: 33–36
larval stages, *Pandalus* **38**: 101–105
Larval starvation **20**: 45–50
Larval survival **20**: 17–20
LAS (linear alkylate sulphonates) toxicity
          and bivalves **37**: 107, **109, 110,**
          115
Lasker events and plankton
          production **26**: 261

Latency **30**: 252, *253, 255,* 256, 257, 285
Lateral line **20**: 146, 154, 155
Lateral line neuromast **20**: 151
Lateral line stimulus **20**: 98, 186
Lateral line, sole **29**: 219–221
Latitude
  and fecundity of invertebrates
    43: 147–150
  changes in sea-ice edges *Plate 1*
  gradients and parasites
    host ranges and specificity **43**: 52, 53
    reproductive strategies **43**: 51, 52
    species richness **43**: 11–14, 17, *19,* 34,
      47–51
Latitude and circadian rhythms in
    fish **26**: 151
Latitudinal diversity gradients of
    Scaphopoda **42**: 207–210, *208,
    209*
Lau Basin vent system **35**: 90
Law's method **20**: 278
LD (lethal dose) **21**: 64
LD$_1$ **21**: 64
LD$_{100 \times 2}$ **21**: 64
LD$_{50}$ (lethal median dose) **21**: 64, 69
LD$_{99}$ **21**: 64
Leach's storm petrel **20**: 60, *61,* 243, 245,
    282, **285**, 294
lead **41**: 7, 14, 43, *123*
Lead toxicity and bivalves **37**: 77, 88, 89,
    90–95, 116
Learning **33**: 174
Learning, predation escape **25:** 16, 17
Lecithotrophic larvae of
    invertebrates **43**: 90, 92, 100,
    124–126, 144,146, 147, 246,
    247
Lecithotrophic reproductive strategy **35**: 65,
    66, 79, 91
Legal requirements, fishing **29**: 260
Leitz Laborlux D epifluorescence
    microscope **29**: 85
Lemon sole **36**: 28
length *see* body size
Length, *see* Size
Length, sole **29**: 218, 255–262, **257**, *258,
    259,* 301
  fecundity **29**: 283–286, **284**, *285,* 288
  maturity **29**: 275–278
  relationship to weight **29**: 262–266, *264,*
    265

length-frequency distributions
    (LFD) **38**: 190–192
Leptostracans, hydrothermal vent
    dwelling **23**: 329
"Lessepsian migration" **27**: 139
Lethality **21**: 63
Liassic **42**: 217
Life cycle
  of herring parasites **24**: 279–285
  of pycnogonids **24**: 33–44
  of siphonophores **24**: 170–172
life cycle and development of sea
    cucumber **41**: **171–174**, *173*
  see also adult; juvenile; larvae
  embryonic **41**: *171*
  growth **41**: 176–178, *177*
  migration **41**: 175, 176
  settlement **41**: 174, 175
Life cycles (see Life histories)
Life cycles **44**: 209–213, 268, 269
  movement patterns in time and
    space **44**: 208–232
  movements **44**: 270
  trajectories **44**: 245, 272
Life cycles, molecular and metabolic aspects
    of **36**: 89–137
  see also Interannual fluctuations;
    Ontogenesis
  annual **36**: 101–114
  daily rhythms **36**: 114–117
  dynamics of abundance behaviour and
    distribution pattern **36**: 123–137
life cycles, schematic (*Pandalus* spp.)
    38: 124–127
life histories (*Pandalus*) **38**: 121–166
Life histories **27**: 3, 4, 283–314
  dynamics of **27**: 288–297
    equatorial pattern **27**: 289–291, *290*
    temperate pattern **27**: 294–297, *295*
    tropical/subtropical pattern
      27: 292–294, *295*
  life cycles **27**: 283–288, *285, 286,
    287,288*
Life history **33**: 347, 348, 352, 353
  deep-sea patterns **33**: 390–392
  high latitude patterns **33**: 384–390,
    391, 392
    overwintering copepodids **33**: 385–388
    resting eggs **33**: 388, 389
    resultant life histories **33**: 389, 390
  strategies **33**: 381–392
  temperate patterns **33**: 383, 384

tropical and subtropical patterns
  33: 382, 383
life history and fecundity of invertebrates
  43: 91, 92, **111–140**, 151
  see also demography; eggs of
    invertebrates; models, life history;
    reproduction
  age at first maturity **43**: 91, 134–136
  female size and age **43**: 131–134
  reproductive effort **43**: 91, 114, 115,
    136–140, *139*, 142
  theory **43**: 111–115
Life history strategies, deep-sea fauna **35**: 6
  genetic variation and **35**: 79
  heterozygosity and **35**: 61–67
Life history traits **34**: 248–256
  intraspecific changes **34**: 251–254
Life span, *Donax* and *Bullia* **25**: 236
Life tables, seaweed **23**: 6–8
  distribution curves produced **23**: 7
Light (and ultraviolet radiation)
  coral pigments as protection from **43**: *288*,
    289, 290
  fecundity **43**: 102
  sea ice organisms **43**: 199, 200, 235–240
Light **36**: 115, 128, 129
  as cue for diel vertical migration
    33: 446–449, *448*
  barnacle growth and **22**: 216, 235
  phytoplankton **29**: 3, 4
    cell size **29**: 15–18, *17*, *18*, 21
    diatom bloom **29**: *23*, 42
    predation **25:** 27
    response to **33**: 217, 218
    sensitivity, *Donax* and *Bullia***25**: 193
    sole **29**: 243, 252
Light as *zeitgeber* **26**: 122–129
Light attenuation coefficient **23**: 250–252
Light effects
  gametophyte death **23**: 143
  on plant competition **23**: 15, 17, 142
  subtidal **23**: 142
    colour modification **23**: 142
    decreases **23**: 142
Light intensity threshold **20**: **138**
Light, and predation **27**: 376
  responses to **27**: 336–338
Light-fishing, cephalopods, remote
    sensing **39**: 261–292
  California Current **39**: 270, 271, 277
  catches, light-fishing vs no light-
    fishing **39**: **266**, **274**, *275*

China Sea Shelf **39**: 267, 268, 276
DMSP–OLS **39**: 263, 265–272
Humboldt Current **39**: 271, 277
interactions with oceanography and other
    fisheries **39**: 265–275
Kuroshio Current **39**: 265–267, 275, 276
New Zealand **39**: 269, 270, 276, 277
other squid fisheries and unexploited
    stocks **39**: 273–275
predator interactions **39**: 278–288
prey interactions **39**: 288, **289**
Southwest Atlantic **39**: 271, 272, 277
Sunda-Arafura Shelves **39**: 268, 269, 276
Limpets
  displacement by seaweeds **23**: 98
  grazing by **23**: 97, 98, 113
    exclusion experiments **23**: 117
    on crustose algae **23**: 22, 113
    on fucoids **23**: 19, 20, 21
    shore zone variation in **23**: 97, 98
  hydrothermal vent dwelling **23**: 321, 322
    dense packing **23**: 322
Lindane toxicity and bivalves **37**: 97, **102,
    103**
Linear alkylate sulphonates *see* LAS
Line-intercept transect sampling **23**: 48
Linfjord, Denmark, plaice transplants **38**: 11,
    43
Lingula anatina
  development **28**: 274, 276–279, **280**
  genetics **28**: 347
  physiology and metabolism **28**: 289–292,
    308, 312, 314, 322, 323, 329, 330
  structure
    coelom and vascular system **28**: 212,
      213, 215
    digestive system **28**: 227, **228**, 229,
      231, 232, 234–236
    lophophore **28**: 217, 221, 222, 226
    mantle **28**: **180**, 189, 200
    muscular system **28**: **238**, 240, 241, 243
    nerves and sensory structures **28**:
      244–246
    pedicle **28**: **202**, 203, 209, 210
    reproductive system **28**: 248–250, 255,
      258, 271
  synonyms **28**: 178
Linkage rules and classification in benthic
    studies **26**: 217
Lipid content **33**: 245–248
  stored lipid influence on egg
    production **33**: 292

Lipid formation excess in Antarctic
  43: 201, 202
Lipid toxins **21**: 67
Lipid/protein ratio
  corals under stress **22**: 50
Lipid-iodine value, in Newfoundland
  herring **20**: 75
Lipids **36**: 4, 5
  see also fatty acids
  adaptation
    buoyancy **36**: 44–47
    food and feeding **36**: 48, *49*, 50–55, *56, 57*
    oxygen level **36**: 32, **42**
    salinity **36**: 26, 27
    temperature **36**: 15–18
  adaptation strategies
    energy metabolism **36**: 62, **64**, 66, 68–73
    plastic metabolism **36**: 75–85
  biochemical indices **30**: 230–232
  composition **30**: 279, 280
  composition of in Atlantic halibut
    26: *35*, 36, 37
  content in fish eggs **26**: 83, *84*, 85
    and overripening **26**: 96
  differentiation and variability **36**: 222, 224, *225, 226*–228
  ecological principles **36**: 230, 237, 239, 244, 245, 248, 251, 253
  indicators of fish condition **36**: 206–212, 214, 215, 217, 218
  interannual fluctuations **36**: 118, 119
  life cycles **36**: *97, 119*
    abundance, dynamics of **36**: 124–136
    ageing **36**: 101, *102*
    annual **36**: 106–112, *106, 108, 109*
    ontogenesis
    passim, 125, 131
    reproduction **36**: 90–93, *91, 92*, **92**
    sex differences **36**: *97*, 98, *99*
    passim, 80–82, 85, **85**
  substance and energy **36**: 143–201 *passim*
    accumulation and production **36**: *149, 150*, **155**
    consumption, transformation and utilization **36**: **179**, *180, 181*, **192**
    expenditure **36**: *173*, **174**
Lipids, crustaceans **29**: 137, 138, **138**, 143, 178, 189
Lipoproteins, crustaceans **29**: 187
Little Sole Bank **29**: 326

Littoral zone, predation **25**: 59
Liver **36**: 214, 215, 222
  adaptation
    buoyancy **36**: 46
    food and feeding **36**: 50, 54, 55
    strategies **36**: 68, 70, 76–78
    temperature **36**: 9, *11*
  glycogen *see* Carbohydrates and glycogen
  life cycles **36**: 105, 112, 120, 121, *122*, 126
Liver, in Atlantic halibut **26**: *39*
Lizard Island (Great Barrier Reef) **43**: 16
LKB–Wallac 1209 RackBeta liquid
  scintillation counter **29**: 86
L_m **20**: 20
Lobster **36**: 27, 39
Lobsters
  deficient areas **23**: 96
  effects of overfishing **23**: 91–96
    controversies **23**: 96
    density decline **23**: 92, 93
    kelp destruction **23**: 93
  feeding rates **23**: 93
  preferred food **23**: 92
Lobsters, spermatophore **29**: 130, 187, 199
  see also Spiny lobsters
  chemical composition **29**: 168
  morphology **29**: 160–165, **162, 163**
  origin **29**: 165–168, *167*
Local action systems **24**: 185–189
Loch Etive, Scotland **33**: 349–351, **350**
Loch Ewe, plaice enrichment **38**: 15, 37
Loch Fyne, MA, hatchery **38**: 8
  juvenile catches in **38**: 9, 9
Locomotion
  see also Swimming
  of pycnogonids **24**: 30–32
  of siphonophores **24**: 177–182
  Protobranch bivalves **42**: 14
  Scaphopoda **42**: 161–163
  shelled Opisthobranchs **42**: 105, 106
Locomotion *see* Motor activity
Locomotion, *Donax* and *Bullia* **25**: 183, 191, 192, 197–214, *200, 202, 204, 205, 206*, **209**, **211**, **213**, 235, 236, 237, 238
Locomotor activity
  and circatidal activity **26**: 137, 139
  and vertical migration **26**: 135
Log normal distribution **26**: 205, 206, *207, 208*
Log normal frequency, and biomass

distribution **26**: 209
Logatchev, 14–45 site MAR **32**: 103, 104, **126–129**
Loire Estuary **29**: 242, 252
Loire outflow **23**: 274–276
  salinity front **23**: 208, 209
  surface chlorophyll map **23**: 274–276
Lokta-Volterra equations **23**: 18, 19, 22, 23
Loliginid squids **25: 101**
Loligo sp **36**: 36
London Fisheries Convention 1964 **29**: 331
Long Island Sound, USA **29**: *5*
Long line **20**: 345–350
Longevity **33**: 339–344, **345**
Longevity of Protobranch bivalves **42**: 28–30
Longevity, of Atlantic halibut **26**: 31–33
Longfin cod **36**: 79
Longhurst/Hardy Plankton Recorder (LHPR) **33**: 9
Longhurst–Hardy plankton recorder **28**: **12**, 13, 14
Longitudinal gradients and geographical ranges of parasites **43**: 14, 53, 54
long-jawed mudsucker **36**: 21
Longnose anchovy **20**: 27
Looming threshold, predation **25**: 19
Lophopore, Brachiopods **28**: 217–226
Lophotoxin **21**: 113
Louisiana continental slope, hydrothermal vents **23**: 339
Loxophyceae **29**: 79
Luminescence *see* Bioluminescence
Luminescence, siphonophore **24**: 202, 203
Luminescent glands **33**: 434–436
Lunar cycles
  barnacle growth and **22**: 237
Lundy Island **29**: *223*, 227, 231, 236, 324, 325
  Marine Nature Reserve **29**: 333
Lungers, predation **25: 9**
lungfish **36**: 202
Lures, predation **25: 5**
Lutein **43**: *283*
Lyngbyatoxin A **21**: 89
Lyon, Gulf of **29**: 273

MacCall's model of cannibalism **20**: 170
Mackerel **36**: 101; **20**: 249, 252, 279, 297
  see also Horse-mackerel
  adaptation **36**: 17, 45, 49, 61, 70, 77
  and match/mismatch hypothesis **26**: 260
  egg quality **26**: 74

horse **20**: 247, 249, 258
  survival rates **26**: 264, 265
  vertical migration of **26**: 149
Mackerel tuna **36**: 17
Macquarie Trench, hadal fauna **32**: 369, 371
Macrobenthos, abyssal
  definitions **32**: 429, 430
  deposit-feeders, influence on benthic sediments **32**: 497–500
  feeding groups, concept and methods **32**: 443–470
    carnivores **32**: 445–456
    detritivores (deposit-feeders) **32**: 456–466
    sestonophages (suspension-feeders) **32**: 466–470
  food sources **32**: 438–443
  methods **32**: 431–433, **434–437**
  trophic areas **32**: 471–506
    Atlantic Ocean **32**: 479–484, *478*, **484**
    general description **32**: 472–479
    Indian Ocean **32**: *477*, 485–497
    Pacific Ocean **32**: 485–497, **486, 487**
    species composition **32**: 500–506
  volume of material **32**: 433–438
Macroecology of pollock **37**: 187–193
Macrofauna and biomass distribution in benthic studies **26**: 209, *211*
macronutrients **41**: *12*, 39–41
  cycling
    ratio **41**: *84–86*
    reaction of iron with **41**: 56–58
Macroplankton **32**: 1–91
  defined **32**: 2
  gelatinous animals **32**: 64–68
  near-bottom communities **32**: 74–78
    influence of hydrothermal ecosystems **32**: 78–82
  patterns of distribution, methods of evaluation **32**: 7–9
  vertical distribution **32**: 58–68
    composition and dependence on depth **32**: 62–64
    diel migrations **32**: 64
    ontogenetic migrations **32**: 63, 64
    pelagial zone **32**: 68–74
    quantitative **32**: 58–62
'Macroscale time' 232, 233
Macrosetellidae **44**: 81, 82
Macrura, spermatophores **29**: 136, 146, 178, 197
  see also Lobsters, penaeid shrimps

caridean shrimps **29**: **132**, 173, 174
Maculae **20**: 141, *142*, 143, 152
  anterior **20**: 143
  posterior **20**: 143
Maculotoxin **21**: 185
  isolation **21**: *186*
  properties **21**: **188**
Madagascar, sea cucumber in **41**: 132, 136,
  137, 147, 149, 176, *189*, 196, 201
magnesium **41**: *11, 12, 120*
Magnesium, spermatophores **29**: **139**
Magnetic resonance imaging **43**: 181
Magnetic stimuli **43**: 41
Maine herring **20**: 62
Majidae **29**: 133, 144
major *see* carbon; nitrogen; phosphorus;
  silicon
Malacostraca **29**: **131**, 184
Malacostracans, hydrothermal vent
  dwelling **23**: 329–331
Malathion toxicity and bivalves **37**: 97, **106**
Malaysia, sea cucumber in **41**: 136, *137,
  150,* 193, 195, 196
Maldives, sea cucumber in **41**: 136, 198
Males, reproductive system **27**: 43–45
Mallory's triple stain **29**: 136, 137
Mammals (see Predators)
Mammals **34**: 258
  prey removal **34**: 274–276
Mammals, parasites of **25**: 152
Mammals, spermatozoa **29**: 137
Man, effect on mollusc distribution **25**: 180,
  186, 197, 215, 216, 255
Management of Atlantic halibut **26**: 51, 52
Mandible **33**: *18*, 19, 20, 141
  see also Swimming appendages
  edge index, **33**: *141*, 141, 142
  moult cycle stages **33**: 35
manganese **41**: 22, **61–63**, 88, *123*
  biochemical function **41**: *11, 12*
  biological availability and uptake
    41: *26*, 27, 29, 30, *31*, 48, 66
  impact on primary production **41**: 13,
    14, 23, 62, 63
  cycling ratio **41**: *84–86*
  distribution in oceans **41**: 61, 62
  functions and properties **41**: *15–17,
    16–18*
  indices of fractionation **41**: *81*
  inputs to system **41**: *123*
  interactions **41**: 33, *34, 35*, 37–39
Manganese nodules **35**: *93–95*

Manganese toxicity and bivalves **37**: 88, **90,
  91,** 123
Manganese, spermatophores **29**: **139**
mangrove ecosystems **40**: 82, 83, 113–192,
  192
  see also mangroves
  associated fauna
    amphibians **40**: 154, 155
    ascidians **40**: 131–134
    barnacles **40**: 138, 139
    birds **40**: 107, 155–157
    copepods, planktonic **40**: 129
    crabs **40**: 139–144
    epibenthos, infauna and meiofauna
      40: 134–136
    epifauna **40**: 131–134
    fish **40**: 149–154
    hydrozoans **40**: 134, 135
    insects **40**: 107, 108, 144–147
    mammals **40**: 107, 157, 158
    molluscs **40**: 147–149
    nematodes **40**: 135, 136
    other crustacea **40**: 136–144
    reptiles **40**: 154
    shrimps and prawns **40**: 136–138
    sponges **40**: 131–134
    zooplankton **40**: 129–131
  associated flora
    bacteria **40**: 113–118
    fungi and fungus-like protists
      40: 118–122, **119**
    macroalgae **40**: 124–126
    microalgae **40**: 122–124
    phytoplankton **40**: 122–124
    saltmarsh plants **40**: 128, 129
    seagrasses **40**: 127, 128
  coastal and seaward views **40**: *85, 86*
  defined, and terminology **40**: 87, 88
  ecological roles
    food webs and energy
      fluxes **40**: 186–189
    litter decomposition **40**: 183–185
    litter production **40**: 111–113, **114**
  global distribution **40**: 88, **89**
  responses to environmental stress
    anthropogenic stress **40**: 83, 84, 163,
      179–181
    coastal changes **40**: 163–165
    gases **40**: 160–162
    global changes **40**: 179–183
    light changes **40**: 105, 106, 158, 159
    metal pollution **40**: 170–173

oil pollution **40**: 174–176
organic pollution **40**: 173, 174
pests **40**: 176–179
salinity **40**: 103, 104, 169, 170
soil conditions **40**: 167–169
tidal gradients and zonation **40**: 166, 167
wind **40**: 162, 163
mangroves **40**: 81–113
see also mangrove ecosystems
biochemistry **40**: 105–107
biomass **40**: 111, **112**
defined, and terminology **40**: 87, 88
evolution **40**: 90–93
food webs and energy fluxes **40**: 186–189
genetic variation **40**: 97, 98
global distribution **40**: 88, **89**, **94–96**
historical background **40**: 89, 90
litter decomposition **40**: 183–185
litter production **40**: 111–113, **114**
morphology
leaf anatomy **40**: 101, 102
root anatomy **40**: *86*, 99, 100
seed/seedling anatomy **40**: 102, 103
wood anatomy **40**: 100, 101
physiology
photosynthesis **40**: 104, 105, 160–162
salt regulation **40**: 103, 104, 169, 170
reproduction **40**: 107–111
dispersal and establishment
40: 108–111
pollination biology **40**: 107, 108
propagules **40**: *86*, 108–110
species, and taxonomic authorities
40: **94–96**
taxonomy **40**: 93–97, **94–96**
tissue culture **40**: 98, 99
Manhattan metric in benthic studies **26**: 215
Mann-Whitney U-test **30**: 58
Mantle and mantle cavity
Pleurotomarioidean gastropods **42**: *251*,
252
Protobranch bivalves **42**: 13, 14
Scaphopoda **42**: 153, 155–158, 157, 158,
160, 191, 192
shelled Opisthobranchs **42**: 94–96
Mantle cavity, *Donax* and *Bullia* **25**: 201,
205, *205*, *206*, 235
Mantle, Brachiopods, structure and
functional morphology
28: 179–200
Mantle, Cephalopoda **25**: *96*, *98*, 100, 101
Manus Basin **34**: 387; **35**: 96

Manx shearwater **20**: 282, **285**, 294
Mapping techniques **44**: 263, 264
Marginal ice zone **43**: 176, *177*
Mariana Trough **34**: 387
Mariculture production **20**: 312
Mariculture *see under* exploitation
Marine animal ecology, critique of
traditional
approaches **44**: 232–236
Marine epipelagic copepods **44**: 3
Marine epipelagic invertebrate zooplankton,
growth rates in **44**: 1–142
Marine fish, movements of **44**: 205–294
Marine food webs **34**: 295, 296
Marine mammals, shoreline habitats, oil
effects **39**: 60–64
marine pollution
and Atlantic cod parasites **40**: 47–49
mangrove ecosystems **40**: 170–176
Marine reserves **34**: 309, 310
Mark recapture methods **31**: 34
Marketing data **27**: 6
Marketing, Japanese oyster culture
21: 46, **49**
Markets, sole **29**: 333
Mark-recapture assessment **29**: 218, 335,
342, 343
Marquesan sardine **20**: 143
Marsupium **29**: 185
Marteiliidea, parasitic **25**: **146**
Masked greenling **36**: 62
Mass mortalities **33**: 369, 370
Mass mortality, Japanese oyster culture
21: 34
Mass mortality, of scallops **20**: 310, 314, 367
MAST I project **30**: 2
MAST II project **30**: 2
Mastigomycetes, parasitic **25**: 122
Match mismatch hypothesis **31**: 170, 171
Match/mismatch hypothesis and fish
populations **26**: 250–256, *255*,
284
in low latitude waters **26**: 259–263
testing of **26**: 264, 265
Match–mismatch hypothesis **28**: 128, 130
Maternal care, *Bullia* **25**: 236
Mathematical modelling **44**: 272
mating
frequency in *Pandalus*
borealis **38**: 211–212
in *Pandalus* **38**: 148–151
Mating behaviour **27**: 257–262

Mating behaviour **33**: 256–266
  see also Reproduction
  attraction of the sexes **33**: 256–258
  interspecific mating **33**: 258
  mating position **33**: 258–261
  spermatophore attachment **33**: 261–263
    number attached **33**: 261, 262
    placement position **33**: 262, 263, *263*
  spermatophore production **33**: 264–266
  spermatophore transfer **33**: *258*,
    258–260
Mats, microbial **23**: 316, 317
Mature phase of Atlantic halibut **26**: 27–38
  age/size composition **26**: 29–31
  body composition **26**: 35–37
  distribution **26**: 6, 7
  energy economics **26**: 37, 38
  feeding **26**: 33–44
  growth **26**: 31–33
  longevity **26**: 31–33
  occurrence and migration **26**: 27–29
  sex ratios **26**: 29–31
Maturity, sexual, in Atlantic halibut **26**: 7, 8,
    51
Maturity, sole **29**: 218, 275–277, *276*
  age **29**: 278, 279
  length **29**: 277, 278
  spawning **29**: 272, *273*
Maud Rise sea mount (Antarctic) **43**: 180
Mauthner cell **20**: 56, 57
Mauthner cells, larvae **25**: 19
Mauthner neuron **20**: 94
Mauthner-initiated startle **20**: 108
Maxilla **33**: 18, 20
  filtering setae **33**: 142, 143, 148
Maxillary glands **33**: 46
Maxilliopoda **29**: **131**
Maxilliped **33**: *18*, 20
Maxillule **33**: *18*, 20
Maximum economic yield (MEY) **29**: 319,
    320
maximum sustainable yield (MSY) **38**: 4
Maximum-likelihood method **35**: 56, 57
McLachlan Scale **25**: 234
McMurdo Sound and ice shelf **43**: 211, 236
Meadows **44**: 242
Mean beam trawling intensity **34**: 212
Mean cellular haemoglobin concentration
    (MCHC) **24**: 352, 353
Mechanical locomotory cost, burrowing
    Donax and *Bullia* **25**: 208,
    209–211, **209**, **211**, 213, 214

Mechanisms of population structure of
    pollock **37**: 228–236
Mechanoreception
  feeding responses **33**: *144*, 144–146, 149
  reaction distances **33**: 432–434, **433**
  role in mate-seeking behaviour **33**: 257
Mechano-reception, predation **25**: 6
  invertebrates **25**: 5, 7, 8, 12, 13, 14, 15,
    23, **26**
  larvae **25**: 16
Medaka **36**: 196
Median development time (MDS) **33**: 322,
    323
Mediterranean Sea **35**: 8; **42**: 37
  adaptation **36**: 15, 17, 30, 44, 48, 49, 69
  differentiation and variability **36**: 224,
    225, 226
  indicators of fish condition **36**: 207, 208,
    213, 214
  invertebrates **43**: 148
  life cycles **36**: 107, *108, 109*
  parasites **43**: *58*
    diversity **43**: *10, 12,* 17
    uses **43**: 64, 66, 67, 84
Mediterranean Sea, abyssal fauna
    distribution **32**: 360, 361
Mediterranean, sole **29**: 217, 220, 236, 271,
    273
Medulla oblongata **20**: 149
Medusa **36**: 56, 200, 202, 209, 217, 246,248
Medusae **20**: 53
  parasites of **25**: 149, 150, 152, 153, 154,
    157
  predators **25**: 23, 25, 29, 30, 42
Medusae, role in plankton
    communities **32**: 64–68
Meiobenthos of the deep Northeast
    Atlantic **30**: 1–88
  bathymetric trends **30**: 22–35
  collection and processing **30**: 6–15
  comparison with other regions **30**: 44–46
  deep bottom water masses **30**: 20
  environmental variables **30**: 16–21
  faunal composition **30**: 29–35, 38–41
  faunal densities **30**: 22–29, 35–38
  food supply **30**: 17–20
  historical background **30**: 5
  horizontal spatial patterns **30**: 21–53
  latitudinal trends **30**: 35–41
  multivariate analysis **30**: 41–43
  near-bottom currents **30**: 21
  oxygen profiles **30**: 17

patterns at different taxonomic
    levels **30**: 46–52
physiographic setting **30**: 4, 5
sampling areas **30**: 6
sediment type **30**: 16, 17
small-scale spatial patterns **30**: 52
suprabenthic microhabitats **30**: 53
temporal patterns **30**: 74, 75
vertical distribution within the
    sediment **30**: 53–74
vertical spatial patterns **30**: 53–74
Meiofauna
  and biomass distribution **26**: 209
  in polluted sites **26**: 191, 192
Meiofauna of the northeast Atlantic **30**: **8, 9**
  densities for samples collected **30**: **10**–**15**
Meiosis, invertebrate **43**: 89, 90, 103, 104
Meiosis, sole **29**: 279
Melanin, predation studies **25**: 43
Melanization, spermatophores **29**: 140, 168,
    194, 195, 200
Melanophore **20**: 128
Melting *see under* ice
Member/vagrant hypothesis on fish
    spawning **26**: 256–259
Member–vagrant hypothesis **28**: 44, 124,
    129
Menhaden **20**: 13, 30, 39, 40, 41
  Atlantic **20**: 5, 15, 21, 34, 36, *66*, 144
  burst swimming in **20**: 90
  daily ration for **20**: **71**
  evacuation rate of **20**: 67, **68**
  gill area in **20**: *63*
  Gulf **20**: 21, 34, 108
  larva **20**: 32
Mercury toxicity and bivalves **37**: 3
  sub-lethal and lethal effects **37**: *132*
  tests **37**: 88, **90–95**, 95, 116, *120*, 123, 125
    larval physiology **37**: 66, 68, *69*, 70, 73,
    74
Meristics and pollock **37**: 206, 207, 208–211
Merluciidae, as predators **27**: **363**
Meromictic lakes **29**: 90
Mesenterial filaments, corals
  extrusion, feeding and **22**: 35–39
Mesh size **29**: 218, 260
  harvesting options **29**: 299, 300, 303,
    305–309, *306–309*, **308**, 321, 322
Mesocosm, predation **25**: 43–45, *44*, 50, 51,
    57, 58, 67
Mesocosms **44**: 39
Mesopelagic zone **33**: 483

Mesoplankton **32**: 1–91
  defined **32**: 2
  patterns of distribution **32**: 9–58
    cold vs tropical regions **32**: 37, 38
    methods of evaluation **32**: 4–7
    open ocean water column,
      quantitative **32**: 48–53
    small-scale vertical spatial
      variability **32**: 38–48
    stratification types **32**: 39, 40
Mesozoic
  Pleurotomarioidean gastropods **42**: 237,
    242
    classification **42**: 262, 266, 267, *268*
    ecology **42**: 272, 278, 280, 281
    structure and morphology **42**: 243, 244,
      248
  Protobranch bivalves **42**: 41
  Scaphopoda **42**: 215, 217
Mesozooplankton **29**: 2, 22
  see also Copepoda
  blooms **29**: 20, 21, 25, 42, 44–46
  dissolved organic matter **29**: 32, 33–35,
    34
Metabolic cold adaptation
    (MCA) **24**: 340–345
  and Ca^{2+}-ATPase activity **24**: 359
  and enzymes **24**: 357, 358
  and proteins **24**: 359, 360
Metabolic enzymes, biochemical
    indices **30**: 239
Metabolic faecal loss (MFL) **34**: 43
Metabolic rate, deep-sea fauna **35**: 11
  adaptations to deep-sea environment **35**:
    26, 27
Metabolic rates, and vertical
    migration **26**: 137
Metabolism **27**: 174–184; **33**: 211–215;
    36: 1, 2
  see also Substance and energy
  acid-base system **27**: 184
  adaptation
    food and feeding **36**: 47, 53
    oxygen level **36**: 29, 30, 32, 33, 41
    pressure **36**: 43, 44
    salinity **36**: *23*, 24, 25
    temperature **36**: 8–14
  amino acids **27**: 179–182
  and enzymes and temperature **36**: 8–14
  antennal glands **27**: 180
  basal **20**: 230
  carbohydrates **27**: 182–184

corals, stress effects **22**: 27–29
digestive gland **27**: 183
excretion **27**: 180–182
existence **20**: 230
hormonal control **27**: *202*
life cycles
　　see also Reproduction
　　annual **36**: 109, *110*, 113, 114
　　daily rhythms **36**: 115, 116
　　ontogenesis **36**: 94, 95, 104–106
lipid **27**: 182–184, *183*
metabolic rates **27**: 177–179
nitrogen **27**: 179–182
　　catabolism **27**: 180–182
　　compounds **27**: 180
of Antarctic fishes **24**: 339–347
　　and cold adaptation
　　　(MCA) **24**: 340–345
　　and *in vivo* protein synthesis **24**: 345,
　　　346
　　in vitro measurements of **24**: 347
　　in vivo **24**: 339, 340
　　routine rates of **24**: 340
　　stress **24**: 357
rates of, in siphonophores **24**: 232–241
standard **20**: 230
Metabolism, pelagic food webs **29**: 2
Metacercariae, parasitic **25:** 146, 147–150,
　　149, 158
Metal toxicity and bivalves **37**: 3, 88–96,
　　90–95, 132, *133*
　　see also in particular cadmium; copper;
　　　lead; mercury; silver; zinc
　　embryogenesis **37**: 87, 88
　　genotoxicity **37**: 95, 96
　　growth, larval **37**: 94
　　mortality, larval **37**: 94
　　respiration **37**: 95
　　settlement **37**: 94, 95
　　tests **37**: 88, 89, **90–95**, 95, 116, *120*, 123,
　　　124
Metal(s)
　　foam enrichment **23**: 227
　　hydrothermal vent **23**: 312, 313
　　　biological tolerance **23**: 342, 343
Metals, *See* Heavy metals; Mineral resources
metals, trace *see* phytoplankton andtrace
　　　metals
Metamorphosis of bivalve embryos and
　　　larvae **37**: 23, *24*
　　and toxicity **37**: 70, 71, 95
　　physiology tests for **37**: 70–74

Metamorphosis, sole **29**: 230, 236, 255, 256
Metapodium, *Donax* and *Bullia* **25:** *200*,
　　204, 207, 210
Metapopulation of pollock **37**: 224–228
Metapopulations of parasites, chaos and
　　　43: 36–41, *38–40*
Metasome **33**: *15*, 16, 24, 25
Metazoan density **30**: *24, 26, 27, 36*
Metazoan meiofauna **30**: *47, 55, 56*, **57**, 58,
　　58, *59, 60, 66*, **71**
　　percentage total abundance of **30**: 57
Metazoans **44**: 3
Metazooplankton **29**: 77, 105, 199, 252
　　see also Copepoda
Methoxychlor toxicity and bivalves **37**: **103**,
　　107
Metridiidae **44**: 81
Metridiolysin **21**: 123
Microbes, hydrothermal vent dwelling
　　　23: 315–318 *see also* Bacteria
　　mat forming **23**: 316, 317
　　Microbio loop **23**: 263, 264
　　productivity **23**: 318
Microbes, in diet **27**: 316, 318, 319
Microbial loop **29**: 2, 77, 104, 111
　　see also Pelagic food webs
Microchemical analyses of body
　　　parts **44**: 258
Micro-electrode technology **43**: 191
Microencapsulated diets for bivalve
　　　larvae **37**: 40, 41
Micronekton, defined **32**: 4
micronutrients (trace metals) **41**: *12*
　　functions of **41**: *11*, 15–23, *15, 16, 17*
　　major *see* cadmium; cobalt; copper;iron;
　　　manganese; nickel; zinc
Microsatellite analysis of pollock **37**: 207,
　　216, 222, 223
'Microscale time' 233, 234
Microscopy *see* electron microscopy
Microsite analysis **35**: 44, 46–48
　　RAPD analysis **35**: 46–48
Microsporidia, parasitic **25**: 144
Microvilli
　　long **20**: 140
　　short **20**: 140
Microwave scanning **43**: 185
Mid-Atlantic Ridge **34**: 399, 405, 416, 421;
　　　35: 17, 20, 21, 105, 106
　　hydrothermal vent sites **32**: 93–107
　　　general structure **32**: 107–109
Middle Atlantic Bight **34**: 228, *228*, 255

Midgut **33**: 39–41, *40*
Migration **44**: 228–232; **27**: 297–314 ;
   36: 171
  see also Motor activity
  adaptation strategies
    energy and energy metabolism **36**: 61,
     71, 72
    plastic metabolism **36**: 79–81
  adults **27**: 297, 312, 313
  'biological tagging' and **24**: 300–302
  diel vertical in fish **26**: 115–168
    environmental factors **26**: 145–148
    model of **26**: 123, 124
  distances **27**: 313
  ecological principles **36**: 232, 237, 240,
    241, 247, 248
  indicators of fish condition **36**: 211, 214
  juveniles **27**: 297, 303–312
    cues for **27**: 303, 306–309
    timing of **27**: 309–311
    transport mechanism **27**: 311
  larval and postlarval **27**: 297–302
    vertical **27**: 298–289, *300*
  life cycles **36**: 112–115, 117
    abundance **36**: 127–137
    passim, 131
  non-spawning **44**: 229, 230
  of Atlantic halibut
    immature phase **26**: 22, *23*
    mature phase **26**: 27–29, *28*
  of hosts, parasites as biological markers
    of **43**: 59–62
  population dynamics **25:** 2
  predators **25:** 27
  siphonophore
    deep-scattering layers **24**: 164, 165
    diel vertical (DVM) **24**: 135, 156–164,
     227
    light sensitivity and **24**: 173, 174,227
  Southern Ocean sea ice organisms
    43: 213, 214, 217, 222
  spawning **44**: 230–232
  spawning **27**: 397, 314
  tidal, *Donax* and *Bullia* **25**: 187–189,
    188, 196, 207, 208, 212, 213, **213**,
    235, 236
  travel distances **27**: **313**
Migration of clupeoids
  anticlockwise trend in **20**: 121
  horizontal **20**: 119–128
  speed of **20**: 123, 124
  vertical **20**: 110–118, 138

Migration of pollock **37**: 184
  And population structure **37**: 206, *214,*
    215, 224, *234, 235*
  larvae **37**: 228, 229, 231, 232
  management implications **37**: 236,
    237–240
  vertical **37**: 190, 191
migration of sea cucumber **41**: 175, 176
Migration, ontogenetic **33**: 391, 391, 392,
    392, 445, **446**
  see also Vertical migration
Migration, sole **29**: 236, 261, 351
  adults **29**: 248–251, 271, 272
  juveniles **29**: 242–244, *245,* 261, 294
migrations (*Pandalus*) **38**: 132, 169–173
  nocturnal vertical **38**: 171–173
Milankovitch effect **31**: 231
Milford Haven **29**: 328–330, 333
  early fisheries **29**: 323, 325–327
Milkfish **36**: 26, 27
Mimicry of siphonophores **24**: 203
Mine Head **29**: 223, *223*
Mineral composition **33**: 243–245, **244**
Mineral resources, deep-sea **35**: 5, 6, 93
  extraction of **35**: 96, 97
Minerals
  see also Salinity
  balance **36**: 19–22
  in diet **36**: **179**, *180, 181,* 186
Minerals, and feed quality in fish
    eggs **26**: 98, 99, 101
Minimum convex polygon (MPC) **44**: 261
Minimum mapping unit (MMU) **44**: 265
Minnow **36**: 23
minor **41**: *11–13, 15, 84, 120, 186*
Miocene **42**: 217, *268*
Mismatched scale in sampling
    strategies **44**: 234–236
Mitochondria crustaceans **29**: 135, 137, 138,
    166
Mitochondria, bivalve molluscs
  respiration, heavy metal
    effects **22**: 176–181
    cation transport and **22**: 176–179
    in eggs and sperm **22**: 180
Mitochondria, *Bullia* **25**: 223, 237
Mitochondrial DNA analysis *see* mtDNA
Mitochondrial phylogeography **35**: 91–93
Mitosis
  Cephalopoda **25**: 91–93, *92*
  peridinians **25:** 126
MIX **38**: 190, 191

Mixed water **29**: 2
see also Pelagic food webs
Miyagi
Japanese oyster culture **21**: 10, 45
rafts **21**: *29*
MIZ (marginal ice zone) **43**: 176, *177*
MLD (minimum lethal dose) **21**: 64
Mobility *see* locomotion
Mobility *see* Motor activity
Modelling techniques **44**: 250, 251, 269
Models
frequency distribution **26**: 193–195
of assemblage **26**: 194
Models of life history and fecundity of
invertebrates **43**: **115–122**
see also demography theory
deterministic (r-K selection)
43: 115–117, 152, 153
egg size **43**: 126–129
stochastic **43**: 117, 118, 140, 152
Models, growth **27**: 224, 225, 233, 234
migration **27**: 299, 301
morphometric **27**: 227
Modular components
mortality rates **23**: 8, 9
recruitment **23**: 31, 32
Modular organisms *see* colonial species
Mohn's Ridge, MAR **32**: 119
Mole crabs **29**: 146, 149–153, *151*, **154**,
155, **156**, **157**, **158**, 197
Molecular clock **35**: 54
Molecular genetics **33**: 63, 64
Molecular genetics *see* mtDNA
Molecular studies of parasites **43**: 5, 6
Molecular-based classification of
Pleurotomarioidean
gastropods **42**: 269–272, *270*
Molluscs **29**: 196, 197, 199, 252, 253;
21: 170, **175**; **36**: 39, 40, 73, 207
see also Squids
aplacophera **23**: 324
bivalve **23**: 320, 321, 323, 335, 336
mussels, see Mussels
gastropods, *see* Gastropods
hydrothermal vent fauna
heterozygosity **35**: 62, 63
locomotion **25**: 198
metabolic rates **23**: 330, 344, 345
radiometric measurement **23**: 344
shell dissolution measurement
23: 344, 345
mortality data **35**: 155, **156**

parasites of **25**: 139, 142, 143, 148, 150,
152, 153, 154
parasitic **25**: 157
pollution, effect of **25**: 180, 186, 197, 215,
216, 255
relicts **32**: **410**
See also Bivalves; Gastropods
molybdenum **41**: *11–13, 16, 17, 120*
Molybdenum toxicity and bivalves **37**: 89,
123
Mongolian grayling **36**: 227
Monod equation **23**: 38
Mont St-Michel, du, Bay of **29**: 261
Monte Carlo technique **20**: 236, **237**
Monterey Bay **35**: 23
Moray eel **36**: 66
Morisita's Index in benthic studies **26**: 215
Morphological indices **30**: 281
Morphological measurements **30**: 240
Morphological variation **33**: 36, 37
morphology (*Pandalus*) **38**: 60–64, 61
Morphology **27**: 7–54 ; **33**: 14–28, *15, 16*
see also Anatomy; Integument; Size
adults **33**: 16–21
and egg quality in fish **26**: 74–78
and reflectance of coral reef organisms
43: 303
antennae **27**: 11, 12, *11*, **47**, 48–53, *49,
51*
antennal glands **27**: 37
antennules **27**: 10, 11, *10*, **47**, 48–53, *49,
51*
appendages **27**: 7–18, **47**, 48–53,*49, 51*
asymmetry **33**: 36
body form **33**: *4, 14, 15*
circulatory system **27**: 29–33
copepodid **33**: 24–28
crustaceans **29**: 183, 185, 186
Anomura **29**: 145–148, *147*, 149–152,
151
Brachyura **29**: 133–135, 197
Copepoda **29**: 174, 175, *177*
Macrura **29**: 160–165, **162**, **163**,
168–171, *170*, **172**, **173**
digestive system **27**: 21–29
egg **33**: 22, 23
endocrine system **27**: 40, 41
external **27**: 7–18, *8*
larval and postlarval **27**: **47**, 48–53, 49,
51, 53
eyes **27**: 9, 10, *9*, **47**, 48–53
integument **27**: 18–21, *19*

larval stages **27**: 45–54
malformations **26**: *88, 89*
moulting **27**: 213–217
mouthparts **27**: 12–14, *12, 13*, 47, 48–53, *49, 51*
muscular system **27**: 21
nauplius **33**: 23, 24, *24*
nervous system **27**: 37–40
of invertebrates **43**: 113, 129, 130
of parasites **43**: 6, 41–44
pereopods **27**: *14, 15, 15, 17*, **47**, 48–53, *49*
pleopods **27**: 17, **47**, 48–53
reproductive system **27**: 41–45
respiratory system **27**: 33–37
setae **27**: *16*
sole **29**: 217
uropods **27**: 18, *18*, **47**, 48–53, *53*
Morphology and structure
    see also under Pleurotomarioidean
        gastropods
    Protobranch bivalves **42**: 4–17, *6, 10,12*
    Scaphopoda **42**: 139–141, *140, 141*
    shelled Opisthobranchs **42**: 73–81, *74, 76–80*
morphology of sea cucumber **41**: *138, 139*, 140
    gonads **41**: 159–161
    morphotypes **41**: 145, 146, *145*
    sexual dimorphism **41**: 158, 160
Morphology, pycnogonid **24**: 3–17
    abnormalities **24**: 14, 15
    adaptations **24**: 16, 17
    colour **24**: 16
    sexual **24**: 13, 14
    size **24**: 14–16
Morphometric indices **30**: 219–223, 268
Morphometric measurements **30**: 252–256, 266, 281
Morphometrics of pollock **37**: 206, 208, 209–211, *212*
Mortality **27**: 357–359 ; **33**: 366–370; **35**: 180–194
    bivalve larvae **37**: 94
        and toxicity **37**: 98–106
        at end of bioassay **37**: 54, 55
    pollock **37**: 198, 200, 231, 237
        extinctions **37**: 224
    see also Population biology
    corals *see* bleached
    data distribution **35**: 155, **156**
    eggs **33**: 292, 293, 366, 367

invertebrates **43**: 135, 136
    mass mortalities **33**: 369, 370
    models **35**: 183
    natural mortality rate **35**: 186–190, **188**, *189*
    of Atlantic halibut **26**: 50, 51
    of larvae **26**: 281, 284
    parasites as cause **33**: 523, 524
    productivity relationship **35**: 183–186, *187*, **187**, 190, 192, 193
    rates **27**: 357, **357**
    taxonomic position and **35**: 190–192, *192*
mortality (*P. borealis*) **38**: 216–218
Mortality rate, sole **29**: 217, 218, 250, 290, 291, 298, 299
    see also Harvesting
    adults **29**: 278, 295–298
    eggs **29**: 230, 291–293, *292*, 298, 299
    juveniles **29**: 291, 294, 295, 299
    larvae **29**: 293, 294, 298, 299
Mortality rates, seaweed; *see also* Death; Survivorship
    density relationships **23**: 12, 13–23
        effects of reduced density **23**: 12, 14
        neighbour distance **23**: 12, 13
    determinants **23**: 9–23
        plant density **23**: 10–12
    long life span problems in studying **23**: 6
    macroscopic **23**: 3, 4–8
        annual seaweeds **23**: 5, 6
        perennial seaweeds **23**: 5
    microscopic **23**: 29, 30
        versus macroscopic **23**: 28, 29
    of modular components **23**: 8, 9
    size dependent **23**: 6
Mortality *see* Ageing and death
Mortality, fish eggs and larvae **28**: 74–97
Mortality, predation, assessing **25:** 37–45, *44*
Mosaic patterns at spatial extents **44**: 268
Motility, phytoplankton **29**: 12–14, **13**, **14**, 29
Motion, prey **25**: **16**
Motor activity **36**: 214
    see also Migration; Plastic metabolism
    adaptation **36**: 8, 30
        and evolution, forces of **36**: 240–242
        passim, 81, 82
        strategies **36**: 59–61, 64, 76–87
    ecological principles **36**: 233, 234, 243–245

life cycles **36**: 12, *95, 110*, 115, 116, 134, 135
  oxygen level **36**: 30
  substance and energy **36**: 165–167, 183, 184, 194, 195
  temperature **36**: 8
Moult cycle, anecdysis **27**: 213
  behavioural changes **27**: 221–223
  diecdysis **27**: 213
  mating **27**: 257
Moulting **44**: 35; **27**: 179, 181, 182, 183, *183*, 193, 213–223 ; **33**: 32–35, *33*, **34**, 296, 315–317; **43**: 85
  calcium metabolism **27**: 219, 220
  diet **27**: 330
  endocrine control **27**: 195–197, 226,
  growth **27**: 224–226
  intermoult duration **33**: 317–323
  moult increment **33**: 323–326, *324*, **325**
  physiology of stages **27**: 218–221
    ecydysis **27**: 220, 221
    postecdysis **27**: 221
    proecdysis **27**: 218–220
  stages **27**: 213–221, *215, 216*, **217**
  Y-organ **27**: 196
Moulting, crustaceans **29**: 140, 141, 143, 144, 159, 183
  artificial insemination **29**: 190, 191
Moulting, pyconogonid **24**: 24, 25, 40–43
Mouth **33**: 19, 39
Mouthparts **33**: 140–143; **38**: 62
Mouthparts, morphology **27**: 12–14, *12, 13*
Movement, sole *see* Migration
Movements of marine fish and decapod crustaceans **44**: 205–294
Movements over time **44**: 240
Mozambique, sea cucumber in **41**: 136, *139, 150, 176, 189*, 193
MRI (magnetic resonance imaging) **43**: 181
MtDNA (mitochondrial DNA analysis) of pollock **37**: 206, 207, 208, 216, 217, 219, **220**, 221–223, 237
Mucopolysaccharide toxins **21**: 67
Mucopolysaccharides **34**: *5*
Mucopolysaccharides, spermatophores **29**: 200
  Copepoda **29**: 179
  Decapoda **29**: 145, 153–159, **154–158**, 168, 186
  Stomatopoda **29**: 184
Mucoprotein, spermatophore **29**: 136

Mucous production, molluscs **25**: 210, 229, 230, 231
Mucus
  adhesive nature **34**: *30, 47*
  biochemical content **34**: **6**
  composition **34**: 4–8
  ecology **34**: 43–50
  energy-rich **34**: 26
  fate of **34**: 43–45
  from marine molluscs **34**: 1–71
  functions **34**: 2, 3, 9, 10, 23–35
  high viscosity **34**: 10
  in biological interactions **34**: 45–50
  in distribution of species **34**: 48
  in energy budgets **34**: 35–43, *38*, **41**, 46, 47
  in feeding **34**: 27–32, *29*, 46, 47
  in gardening behaviour **34**: 47, 48
  in locomotion **34**: 23–25, 40
  in non-molluscan groups **34**: 3
  in protection **34**: 32–34
  lubricatory **34**: 18
  nomenclature **34**: 3
  overview **34**: 50, 51
  production **34**: 9–23, *20*, 37, 41, 50
  production rates **34**: 19
  properties **34**: 8, 9
  structure **34**: 7
  trails **34**: 22, 23, 25–27, **25**, 45–47
  viscosity **34**: 7
Mucus, corals
  production, stress effects **22**: 39–42
Mucus-secreting cell **34**: 13
Mud skipper **36**: 15, 23, 26
Mudsucker **36**: 21
mullet (Hawaii) **38**: 29, 30
Mullet **36**: 116
  see also Red mullet
  adaptation **36**: 13, 17, *23*
  grey **36**: 17, 116, 165, 166, 233
multi species virtual population analysis (MSVPA) **38**: 203
Multidimension, ecological principle of **36**: 229, 230
Multidimensional scaling methods in benthic studies **26**: 221–226
  and non-metric scaling **26**: 221, 225
  and ordinal scaling **26**: 221
  and Principal Component Analysis **26**: 222–224
  and Principal Co-ordinates Analysis **26**: 222, 224, 225

Multidirectional hydrophone arrays **44**: 252
MULTIFAN **38**: 190
Multiline system, of scallop culture
    20: *346*
Multilocus DNA fingerprinting **35**: 44, 46
multiple **26**: 152, 153, *154*
Multiple discriminant analysis in benthic
    studies **26**: 228, *229*, 230
Multiple isotope analysis **24**: 423
Multiple Opening/Closing Net and
    Environmental Sensing System
    (MOCNESS) **33**: 9
Multiple spawning **20**: 8, 11, 13, 17
Multispecies assemblages **33**: 429–432,
    430, 431
Multispecies virtual population analysis
    (MSVPA) **34**: 305
Multivariate community analysis **26**: 172
Multivariate data analysis in benthic marine
    infaunal studies **26**: 214–231
  classification **26**: 217–221
  habitat factors, comparisons **26**: 226–231
  ordination **26**: 221–226
  similarity indices **26**: 215, 216
multivariate structure **23**: 57
Multivariate structure of vegetation
    23: 54–56, 57, 133, 134
Murexine **21**: 176
Murve, common **20**: 60, *61*
Muscle (red and white)
  adaptation
    food and feeding **36**: *51, 52, 58*
    oxygen level **36**: 28, 38, 39
    salinity **36**: 20
    temperature **36**: 8–10, *11*, 14, 15
  adaptation strategies
    energy and energy
      metabolism **36**: 60–74
    oxygen level **36**: 66, 69
    passim, **64**, *67, 73, 74*
    plastic metabolism **36**: 76, 84, 86, 87
  indicators of fish condition **36**: 213, *218,*
    219
  life cycles
    abundance, dynamics of **36**: 133, 134,
      135, 136
    interannual fluctuations **36**: 119, 120,
      121, 122
    ontogenesis **36**: 95, 112, 114
  substance and energy **36**: *149, 150*
Muscle function
  in saccadic eye movements **24**: 371–373

  in swimming **24**: 368–71
Muscles **27**: 186
  innervation **27**: 186
  ionic concentration **27**: 208, 209
  morphology **27**: 21, *22*
Muscular system **33**: 38
Musculature, Cephalopoda **25**: 104, 105
  Donax and *Bullia* **25**: 198, *199*, 200, *200*,
    201, 202, *204*, 205–207, *206*, 224,
    235, 236
Mussel **36**: 39, 83, 114, 117
Mussel beds **44**: 242
Mussel poisons **21**: 75
Mussel(s); *see also* Molluscs
  community interactions **23**: 87
  dwelling annelids **23**: 327
  hydrothermal vent dwelling **23**: 320–324,
    333, 335, 336, 344, 345
    age estimation **23**: 345
    dispersal **23**: 346, 347
    geographical distribution **23**: 323
    maturation **23**: 320
    reproduction **23**: 320, 321
    symbionts of **23**: 321
    starfish grazing on **23**: 73, 74
    zonation **23**: 17–18
Mussels *see under* taxonomic index
Mutation of pollock **37**: 204, 205, 207
Mutations **43**: 46
  see also evolution
Mutsu Bay **20**: 310–373 *passim*
  potential standard growth curve **20**: *361*
Mutsu Bay cooperatives **20**: *316*
Mxysids **36**: 27
'Mycelians', Polymastigotes **25**: 124
Mycosporine amino acids (MAAs) **31**: 268
"Myopic" selection **43**: 45
Myotomes, sole **29**: 230
Myotoxins **21**: 68
Mysid aggregations **30**: 158
Mysidacea, parasites of **25**: 135, 136, 137,
    142, 147, 151
  Crustacea **25**: 153, 154
Mysideae **29**: **131**, 184, 185
Mysids **44**: 33
Mysis, morphology **27**: **47**, *52, 53*
Mystacocarida **29**: **131**, 186
Mytilotoxin **21**: 75, 77
Mytilus edulis **31** : 7, 10, 17, 21, 35
  allozyme variation in **31**: 19
  domoic acid in **31**: 311
  effect of phytoplankton bloom on **31**: 317

hybrid zone shape **31**: 36, 37
hybrid zone width **31**: 35
rate of hybridization **31**: 21
size and potential for hybridization **31**: 17
stability in hybridization **31**: 17
Mytilus galloprovincialis **31**: 7, 10
  allozyme variation in **31**: 19
  hybrid zone shape **31**: 36
  hybrid zone width **31**: 35
  rate of hybridization **31**: 21
  size and potential for hybridization **31**: 17
Myxosporidia, parasitic **25**: 144–146, *145*

^{15}N
  and food-chain dynamics **24**: 423–441
  and sample preparation **24**: 394–398
  and terminology **24**: 398, 399
  calculations and precision **24**: 399–401
  historical aspects of **24**: 391, 392
  isotopic fractionation **24**: 394–396, 398,
      401–408
    magnitude of **24**: 404–408
    theory **24**: 401–404
  measurement of **24**: 392–394
  replicate analyses of **24**: 400, 401
  source studies of **24**: 408, 416–423
  variations of, in nature **24**: 408–416,
      431–432
    atmospheric **24**: 408–411
    freshwater and estuarine
        24: 408,412–413
    marine **24**: 408, 414–416
    terrestrial **24**: 408, 411, 412
Nanoflagellates **29**: **43**, 44, 105, 111–115,
    112, **113**, 117
Nanoplankton
  see also Pelagic food webs
Nassariidae **25**: 186, 198
Natal homing of pollock **37**: 234, 236, 237
Natality **23**: 6
Naticidae, locomotion **25**: 198
National production, Japanese oyster culture
    21: **4**, 38, **40**, **41**, **44**, **47**
Natural feeding condition **44**: 38
Natural gregarious behaviour **30**: 159, 160
Natural increase, intrinsic rate of
    see density of population
Natural seawater quality and
    bivalves **37**: 41–43
Natural selection **31**: 26; **37**: 204, 205
Natural selection *see* evolution
Natural water approach **44**: 4, 24

Nauplius
  see also Development
  feeding **33**: 166, 167
  morphology **33**: 23, 24, *24*, *403*
  sources of descriptions **33**: 26
  swimming pattern **33**: 402, 403
Nauplius, morphology **27**: **47**, 45–50, *49*
Navaga **36**: 11
Navassa **42**: *276*, 277
Nazca and Sala y Gómez submarine
    ridges **32**: 145–241
  benthic communities **32**: 167–185
    population stability **32**: 178–180
    species list **32**: **168–174**
    trophic structure **32**: 180–185
  benthopelagic communities **32**: 185–199
    fish distribution **32**: 185–199
    size and age structure **32**: 198, 199
    species lists and predominating
        foods **32**: **191, 194–197**
    stability and trophic
        structure **32**: 190–198
  biogeographic boundaries **32**: 201–209
  biogeographic classification **32**: 221–224
  biogeographic history **32**: 224–227
  bottom and near-bottom fish and
      invertebrates **32**: 156–167, **159**,
      160–166
  endemism **32**: 209–213
  faunistic relationships **32**: 213–221
    dendrograms of faunal
        similarities **32**: *205, 207*
  geographic position and
      geology **32**: 150–153
  location maps **32**: *151, 222*
  overlying water hydrology and
      plankton **32**: 153–156
  pelagic communities **32**: 200, 201
Near-bottom communities, macroplankton
    distribution **32**: 74–78
Nearest neighbour distances (NND) **33**: 425,
    428, 434
Nearest neighbour distances (NND) **30**: 162,
    170
near-field chemistry *see under* internal
    economy
'Nebenkern', Rhizopoda **25**: 122
Neburon toxicity and bivalves **37**: 97, **104**
Necrosis, spermatophores **29**: 194
Nectophores **24**: 181, 182, 188
Nectophores, siphonophores **25**: 158

Negative binomial distribution
    model **26**: 193, 194
Nehu, Hawaiian **20**: 20
    larva of **20**: 32
    tropical **20**: 13
Nekton
    effect of spatial pattern on **44**: 267
    spatial patterns **44**: 225
N_em statistic **35**: 56–58, 71–73, *72*
Nemathelminthes, parasitic **25**: 151
Nematocysts **21**: 107; **24**: 209–223
    Chironex **21**: **118**
    development of **24**: 209, 212, 213
    discharge of **24**: 218–221
    Physalia **21**: *112*
    toxinology of **24**: 221–223
    types of **24**: 213–217
    use of, in feeding **24**: 217, 218, 230
nematode parasites of *Pandalus* **38**: 181
Nematodes **30**: *28, 37*, 48–51, **50**, *55*, **57**,
        58, *58, 59, 60*, 61–64, **71**; **36**: 57
Nematodes (see Parasites)
Nematodes in benthic studies **26**: 191
Nematodes, hydrothermal vent dwelling
        23: 319
Nemertine **21**: 134
Neomysis sp **36**: 36
Neosaxitoxin **21**: 79
Neoxanthin **43**: *283*
Nepheloid layer, benthic **32**: 74, 75
Nephropidae **29**: **132**, 160–165, **162**, **163**
Nereid worms **29**: 253
Nereistoxin **21**: 138
Nerve function
    in Antarctic fishes **24**: 361–368
        and quantal release of neurotransmitter
            24: 363
        membrane fluidity and synaptic
            events **24**: 363–368, 376
        physiology of peripheral nerve
            24: 361–363
    in pycnogonids **24**: 32, 33
Nerve performance, pressure and **35**: 16
Nervous system **33**: 38; **24**: 182–185
    local action systems **24**: 185–188
        integration of **24**: 188, 189
    of pycnogonids **24**: 20, 21
    of siphonophores **24**: 182–189
    Pleurotomarioidean gastropods **42**: *253*,
        261, 262
    Scaphopoda **42**: 194, *195*
    shelled Opisthobranchs **42**: 103–105

Nervous system, central **27**: 185–187
    chemosensory perception **27**: *17*,194, 195
    eyes **27**: 187–193
    ionic regulation **27**: 208
    light perception **27**: 187–193
        (see also Eyes)
        apposition images **27**: 188, 191
        superposition images **27**: 188–190
    light-dark adaptation **27**: 188, 190, 191,
        191, 192
    mechanoreception **27**: *17*, 193, 194
    morphology **27**: 37–40, *39*
        motor control **27**: 185–187
        neuropiles **27**: 185
    neuromuscular organisation **27**: 186, 187
    neurotransmitters **27**: 186, 187
    sensory perception **27**: 187–195
Nervous system, Cephalopoda **25**: 98, *98*,
        108
Nested spatial hierarchy **44**: 242
Nestedness of parasite communities **43**: 31
Net
    drift **20**: 185
    gill **20**: 185
    monofilament nylon **20**: 186
    standard trap **20**: 112
Net avoidance **20**: 140
Net growth efficiency **33**: 205–207, **208**
Net to gross displacement ratio (NGDR)
        33: 407
Nets *see* Sampling systems, fish eggs and
        larvae
Nets, predation study **25**: 41, 42
Neural intergration **27**: 184–195
        (see also Nervous system)
        motor control **27**: 185–187
Neuromast organs, prey **25**: 17, 18, 19
Neuromuscular system **27**: 186, 187
Neurosecretory cells **27**: 195
Neurotoxic shellfish poisoning
        (NSP) **31**: 313, 314
Neurotoxins **21**: 68
Neuse River estuary, North Carolina,
        USA **44**: 228
Neustonic environment **33**: 488–490, **491**
Neutral theory **35**: 112–114, 115
New Caledonia **42**: *284*, 285; **43**: 33
New Caledonia, sea cucumber in **41**: 136
    see also Holothuria scabra versicolor in
        taxonomic index
    anatomy **41**: *139*, 146
    common names **41**: *137*

distribution and population structure
    41: 149, *151, 152,* 154, 155
fisheries and aquaculture **41**: 193–195,
    201
reproduction and spawning **41**: 158, *161,*
    163, *164,* 165, 168
new habitats, introduction of parasites
    into **43**: 69, 70
new species, introduction of **38**: 17
New Zealand **35**: 84–86, 109; **42**: 37
New Zealand fisheries **39**: 269, 270, 276, 277
New Zealand, coastal areas of: parasites
    43: *10, 11,* 34, 35, 62, *63, 64*
Newfoundland cod fishery, failure of **38**: 23
Newfoundland herring **20**: 75
Newlyn **29**: 224, 270, *271,* 333
Newman–Keuls test in benthic
    studies **26**: 228
News Bay, Japan, red sea bream
    enhancement **38**: 25
Niche, concept of **44**: 237, 238
Niches **33**: 466–470
Niches of parasites
    "empty" **43**: 25–27, 31, 48, 49
    saturation **43**: 24–33, 47, 48
nickel **41**: **78–9**, *121*
    biochemical function **41**: *11, 12*
    biological availability and uptake **41**: *26,*
        27, 29
        impact on primary production **41**: *13,*
        79
    cycling ratio **41**: *84–86*
    distribution in oceans **41**: 78, *79, 128*
    functions and properties *16, 17, 19*
    indices of fractionation **41**: *81, 82*
    inputs to system **41**: *123*
    interactions **41**: 33, *128*
Nickel toxicity and bivalves **37**: 87, 88,
    90–95, 116
Nicobar Islands, sea cucumber in **41**: 136,
    191
Nidamental glands and envelope,
    Cephalopoda **25**: 88, 90
NIPR-1 sampler **43**: 190
Nitrilotriacetate *see* NTA
nitrogen **41**: *12,* 39–41, 42, *121*
    and community structure **41**: 89, 90
    biochemical function **41**: *11, 12*
    biological availability and impact on
        primary production **41**: *13,* 14
    cycling ratio **41**: *85, 86*

interactions with trace metals **41**: 56–58,
    62, 63, 79
Nitrogen and nitrates
    particulate organic- **23**: 315
    supply **23**: 193–195, 242
        features determining **23**: 262
        regenerated **23**: 268
Nitrogen content **33**: 240, **241**
    carbon–nitrogen ratio **33**: 242
Nitrogen excretion **33**: 210, **213**
Nitrogen, Baltic Sea **29**: *107, 108,* 109, 110,
    110, 119
Nitrogen/ammonia (and excretion of)
    36: 215
    adaptation **36**: 32, *33,* 34, 39, 41, *51, 69*
        nitrogenous compounds **36**: 27, 28
    substance and energy **36**: 162, 170, 171,
        176
Nocturnal spawning **20**: 22
NOEC (no observed effect concentration)/
    ANOVA approach **37**: 85
Nomenclature **27**: 1–3
Non-conformist development **33**: 321
Non-destructive weight method of
    quantifying fecundity **43**: *95,* 104,
    105
Non-linearity in ordination in benthic
    studies **26**: 225
Non-passerine **20**: 230, 231
Non-random feeding **20**: 55
non-scleractinian corals **43**: 290–292, *291*
Non-upwelling regions **44**: 40
Nord Trøndelag, cod release **38**: 21
Normalized Expected Species Shared
    (NESS) in benthic studies **26**: 215,
    216
NORMSEP **38**: 190
North America **43**: 60, *121*
    see also Canada; United States
North Atlantic **29**: 46, 217, 219, 220; **35**: 5,
    83, 84
    roundnose grenadier fishery **35**: 104–107
North Atlantic Deep Water (NADW) **30**: 20
North Atlantic Fisheries Organization
    (NAFO) **38**: 57
North East Atlantic Deep Water
    (NEADW) **30**: 20
North Fiji vent system **35**: 90
North Pacific **35**: 5, 84
North Pacific Marine Sciences Organization
    (PICES) **38**: 57
North Pacific *see* pollock

North Sea **35**: 97, 99, 100; **36**: 15, 61, 147, 231; **42**: 32; **20**: 279–292
  differentiation and
      variability **36**: 224–226
  fish catches **38**: 4, *5*
  fisheries **38**: 221, 212
  halibut fisheries in **26**: 43, 45, 46
  larval abundance indices **28**: 34
  life cycles **36**: 103, 104, 107, 123
  managed stocks **38**: 4, 5
  Pandalus borealis stock limitation in **38**: 207
  plaice transplants (1893–1990) **38**: 11–13
  plankton **29**: 6, 27, 28, *28*, *41*, 42–44, *43*, **43**, 46, 52, *58*, 76
  sole **29**: 321, 334, 349
    adults **29**: 248, 249
    eggs and larvae **29**: 235, 236
    feeding **29**: 251, 252
    growth **29**: 255, 261
    juveniles **29**: 240, 246
    mortality rate **29**: 293, 295, 296–298
    reproduction **29**: 270–272, 277, 279, 286, **289**, 290
    Southern Bight **29**: 251, 272, 275
    weight **29**: 263, 265
  survey grids **28**: **30**
  time of spawning in **26**: 251, 252, *253*, 254
North Sea: parasites **43**: *10*, *12*, 17, *58*, 64, 66, 67, 84
North West Bank **29**: 247
North-eastern Pacific Ridge **35**: 16, 89
Northern anchovy **20**: 7–80 *passim* **20**: 131, 133, 143, 246
  and startle response **20**: 56
  daily ration **20**: **70**
  eggs in **20**: 15, 17
  excretion products of **20**: 176
  feeding success in **20**: 28, *35*, 43
  gut clearance rate in **20**: **68**
  histological characteristics of **20**: 49
  mortality rate in **20**: **59**
  muscle fibres in **20**: 87
  pelagic spawning of **20**: 24
  reproduction cost in **20**: *16*
  school **20**: *175*
  school profile **20**: *104*, 106
  sinuous feeding posture in **20**: 27
  spawning in **20**: 13, 23, 109, 143, 144, 172
Northern anchovy larva **20**: 61

  burst speed in **20**: 90
  feeding behaviour in **20**: 26
  food of **20**: 27
  growth rate of **20**: *81*, *82*
  heart length in **20**: 84
  minimum prey size **20**: 30
  patchiness of **20**: *101*
Northern elephant seal **20**: 246
northern pike **36**: 196
Norway (cod and plaice, 1882–1969) **38**: 9–11
Norway, halibut fisheries in **26**: 43, 45, 46, 47, 48
Norwegian Deep, *Pandalus borealis* stock limitation in **38**: 207
Norwegian herring **20**: 8, 15, *16*, *19*
Norwegian Sea **32**: 22, 23; **35**: 24; **36**: 107, 226; **42**: 40
Norwegian Sea–Barents Sea ecosystem **34**: 266
Nostril, sole **29**: 221, **222**
Notodelphid copepods **29**: 179
Notostaca **29**: 199
notothenids **36**: 196
  adaptation **36**: 19, 45, 61, 77, 78
  life cycles **36**: 101, 107
Novaya Zemlya **20**: 228
NTA (nitrilotriacetate) toxicity **37**: 107, **112**
Nucleic acid sequencing **35**: 6, 49, 50
Nucleic acids **33**: 249, 250
  adenosine triphosphate (ATP) **33**: 249
Nucleic acids, biochemical indices **30**: 232–237
Nuclepore filters **29**: 81, 84–86
Nucleus, gregarines **25**: 143
Number of generations **33**: **355–362**
Numerical classification of seaweeds **23**: 47–57
  agglomerative polythetic **23**: 48, 49, 50–57
    assessing sample resemblance **23**: 50–52
  association analysis **23**: 49
  divisive
    monothetic **23**: 49
    polythetic **23**: 49, 50
  sampling, *see* Sampling
Nursery grounds, sole **29**: 217, 236, 237–246, 349–351
  Bristol Channel stock **29**: 249, 250
  feeding **29**: 252, 253
Nutricline **29**: 14

Nutrients
   as a cause of red tides, *see* Red tides
   pulses **23**: 269, 270
   surface depletion **23**: 192–194
Nutrients, plankton **29**: 3, 6, 21, 60, 77,
      101–104
   bacterioplankton **29**: 105–110, *107, 108,
      110*, 115–117
   motility **29**: 12–14, **13**
   spring bloom **29**: 21, 45–47, *46*, 54
   turbulence **29**: 14, 15, **15**
   uptake kinetics **29**: 10–12, *11*
Nutrition **27**: 163–172
   bivalve larvae **37**: 20, 37–40, 66, 67
   pollock **37**: 185, 187, 202, 231
   amino acids **27**: 164, **165**, 169, 170
   artificial diets **27**: 167–169, 171, 172
   carbohydrates **27**: 166, 167, 168, 171, 172
   energy requirements **27**: 167–172
   lipids **27**: 164, 165, **165**, 168, 170–172
   minerals and trace elements **27**: 167
   natural food **27**: 168
   protein requirement **27**: 168, **168**,169,
      170, 172
   vitamins **27**: 165, 166, **165**
nutrition of *Pandalus* larvae **38**: 159–162
Nutrition *see* diet
Nutrition *see* digestive system and feeding
Nutrition, and broodstock management of
      fish **26**: 97–102
Nutrition, *see* Feeding
Nutritional requirements **44**: 29; **33**: 168, 169
Nymphe Bank **29**: 223, *223*, 235, 242, 324
Nyquist criterion for time-series
      analysis **26**: 177

Obligate planktivores, predators **25**: 15
Obligatory phytoplankton feeder **20**: 42
Ocean boundary currents **28**: 54–61
   Eastern boundary current upwelling
      28: 57–61
   Western **28**: 54–57
Ocean sunfish **36**: 55
Oceanic water column **33**: 483–487
   vertical distribution **33**: 486, 487
Ocelli **33**: *31*, 31
Octobrachia **44**: 154
Octopoda, chorion and its relation to
      brooding **44**: 149
Octopoda, embryonic development **25**: 88,
      97, 100, 101, 103, 104
Octopodidae, eggs **25**: 89

Octopodinae, eggs **25**: 89
Octopus sp **36**: 36
Ocular tubercle, pycnogonid **24**: 4, 8
Ocypodidae **29**: 133
Odhner, Nils **42**: 111
Oegopsid squids, embryonic
      development **25**: 101
Oesophagus **33**: 39, *40*
Oesophagus of Scaphopoda **42**: 176, 177
Oil globules, sole **29**: 229, 230
Oil pollution
   bivalve molluscs and **22**: 129
   coral reefs and **22**: 11, 16
   coral stress
      mesenterial filament extrusion **22**: 36,
         37
      reproduction and **22**: 46
Oil pollution, South African Coast **25**: 186
Oil resources **35**: 6, 97–101
   consequences of pollution to
      deep-sea fauna **35**: 99–101
   sources of pollution from oil
      exploration **35**: 97–99
Oil sac **33**: 47, 48
Oil spill **20**: 191
Oil spill effects **39**: 1–83
   ecotoxicology vs field
      assessment **39**: 75–78
   food webs **39**: 70–74
   history and fate of *Exxon Valdez*
      oil **39**: 6–12, **9**, *10*
   intertidal zone **39**: 13–34
      dominant taxa, changes **39**: **18**
      fucoids, Hazmat study **39**: 17–29, *19*,
         27
      oligochaetes **39**: *30*
      rocky intertidal zone
         communities **39**: 16, 17, *22, 23*, **24**
      sedimentary environments **39**: 29–34,
         32, *33*
   location map **39**: *8*
   recovery delay **39**: 79–81
   shoreline habitats **39**: 43–70
      fishes **39**: 48–60, **49**, *51, 53, 56, 58*
      marine mammals **39**: 60–64, **61**
      shorebirds, seaducks,
         seabirds **39**: 64–70, **65**, *67*, **69**
      terrestrial birds **39**: 45–48, **46**
      terrestrial mammals **39**: 43, 44
   subtidal zone **39**: 34–43
      deeper benthic environments **39**: 39, 40

eelgrass zone communities **39**: 34–39, *36, 38*
  kelp communities **39**: 40–43, **42**
  summary **39**: 81–83
Oil toxicity and bivalves **37**: **113–115**
  see also detergents and oil
Oithonidae **44**: 82–86
Okhotsk Sea **36**: 114; **32**: 18
Okhotsk, Sea of, pollock in **37**: *180*, 188, 199, 237
  see also Kamchatka
  population structure **37**: 208, 209, 212–214, 216, 217, 226
  stocks **37**: **185**, *186*
Oldbury power station **29**: 237, 239, 350
Olfactory lobes, sole **29**: 252
Olfactory organs **27**: 194
Olfactory rosette **20**: 140
Oligocene **42**: 44, *268*
Oligochaetes, parasites of **25**: 146
Oligotrophic environment **29**: 2, 35, 42, 92
  see also Pelagic food webs
Olson and Boggs Model, predation **25**: 40
Ombre **36**: 116
Ommastrephid squids **25**: 88, 95, 100, 102
Omnivorous feeding **33**: 142, 143, 150
  see also Diet
  feeding appendages **33**: 141
One-finned greenling **36**: 133
Onega herring **20**: 11
Ongrowing, Japanese oyster culture **21**: 23, 27
Oniscid isopoda **30**: 89–153
  burrowing **30**: 100–107
  crawling **30**: 99, 100
  dispersal **30**: 134, 135
  distribution on the beach **30**: 130–134
  energetics **30**: 116, 117
  growth and survivorship **30**: 127, 128
  impact of man **30**: 136–139
  locomotion **30**: 98–107
  nutrition **30**: 112–114
  population densities **30**: 129, 130
  reproduction **30**: 122–127
  respiration **30**: 114–116
  rhythmic activity and behaviour **30**: 93–98
  sensory physiology and orientation **30**: 107–112
  water and heat relationships **30**: 117–122
Oniscidea, classification **30**: *91*

Ontogenesis **36**: 2, 89–104, 243
  see also Reproduction
  ageing and death **36**: 101–104
  sex differences **36**: 96–104
Ontogenetic changes **44**: 223–226
Ontogenetic migrations **33**: *391*, 391, 392, *392*, 445, **446**
Ontogenetic variation and vertical migration **26**: 148–150
Oocytes **29**: 279; **33**: 42, 43
  see also Fecundity
Oocytes, *Bullia* **25**: 194
Oogenesis **29**: 269, 279–282, *281*; **33**: 42, 43, 273
  oogenesis in *Pandalus* **38**: 153, 154
Oogenesis *see* gametogenesis
Oogenesis, nature of **20**: 14
Oomycetes, parasitic **25**: 122
Oozoids, tunicates **25**: *155*, 156, 158
Open-access fishing **29**: 319
Operculum, *Bullia* **25**: 200, 201
Operculum, crustaceans **29**: 182
Ophiocephalous pedicellariae, urchins **21**: 148
Ophiuroids **29**: 253
Opisthobranchs *see* shelled Opisthobranchs
Opportunist feeding, *Donax* and *Bullia* **25**: 191
Opportunistic strategy
  and fecundity of invertebrates **43**: 121, 122, *121*, 131, 132, 142, 144
  and Southern Ocean sea ice organisms **43**: 222, 223
Optic nerve **20**: 131
Optic tecta **20**: 131
Optical imaging, fish eggs and larvae **28**: 16, 17
Optical lobes, sole **29**: 252
Optical properties of coral reef benthos **43**: **281–293**
  see also bleaching; zooxanthellar pigments
  coral **43**: 284–293
  macroalgae **43**: 281–284, *283*
  skeletal pigments **43**: 290–293, *291*
  tissue pigments **43**: 287–290
Optimal digestion theory **29**: 33
Optimal fishing **29**: 320, 321
Optimal foraging strategy **20**: 29
Ordinal scaling in benthic studies **26**: 221
Ordination (MDS) **30**: *42, 43, 44, 45*, 65, *67, 68, 69*

Ordination
   in benthic studies **26**: 221–226, *223*
   non-linearity problems in **26**: 225
Ordination analysis of seaweed samples
      23: 67–70
   factor- **23**: 67
   multidimensional scaling **23**: 71
   polar- **23**: 68
   principal components- **23**: 70
   principal coordinate- **23**: 69, 70
      undimensional **23**: 70
   reciprocal averaging- **23**: 68, 69, 70
      species division and scoring **23**: 68, 69
Ordovician **42**: 217
   Pleurotomarioidean gastropods **42**: *242*,
      243
   Protobranch bivalves **42**: 3, 9, 42, 217
Oregon **20**: 269
Oregon subduction zone, hydrothermal
      vents **23**: 339
Oreochromis sp **36**: 10, 36
Organ systems, reproductive
      (Scaphopoda) **42**: 195
Organic components **33**: 245–250
   adenosine triphosphate (ATP) **33**: 249, 250
   carbohydrates **33**: 248
   chitin **33**: 248
   enzymes **33**: 252
   free amino acids **33**: 249
   lipids **33**: 245–248
   nucleic acids **33**: 249
   protein **33**: 245
Organic components of fish eggs **26**: 83–85
Organism-based tracking devices **44**: 251
Organism-environment linkage **44**: 237, 238
Organogenesis, Cephalopoda **25**: 94–102,
      95, 96, 97, 98, 100
Organogenesis, sole **29**: 230
Organo-phosphate toxicity and
      bivalves **37**: 97, 116
Organs of Bellonci **33**: *31*, 32
Organs of Gicklhorn **33**: *31*, 31, 32
Orientation of Scaphopoda **42**: 192
Origin, spermatophores
   Anomura **29**: 148, 149, 152, 153, 178
   Brachyura **29**: 135, 136
   Copepoda **29**: 175, 176, 178
   Macrura **29**: 165–168, *167*, 171–174
Orkney **20**: 281
Orkney Trench, hadal fauna **32**: 367, 371
Orloci's index in benthic studies **26**: 215
Ornithine, plankton **29**: 34

Orthoptera **29**: 184, 199
Oscillatoxin **21**: 89, 90
Oslo Commission, pollution **29**: 350
Oslofjord, Norway, enhancement **38**: 10, 12
Osmo-phototrophic physiology,
      peridinians **25**: 126
Osmoregulation **33**: 215, 216
Osmosis in fish eggs **26**: 76
Osmosis, spermatophores **29**: 130, 145
Osmotic regulation **27**: 37, 203–207, 208,
      204, *205, 206,* 209–211
Osmotic shock **43**: *94*, 103, 190
Osmotrophic physiology, peridinians **25**:
      126, 127, 135
Osphradium, *Bullia* **25**: 192
ossicles of sea cucumber **41**: 143–145, *144*
Ostracoda **29**: 199
Otic bulla, larvae **25**: 19
Otolith **20**: 78, 80
Otolith measurements **29**: 255
Otoliths **28**: 20–22
Otters, *see* Sea otters
Otter-trawl fishing **29**: 216, 330, 331,
      342–345
Ovarian cycle
   crustaceans **29**: 142, 143, 145, 146, 159,
      182
   sole **29**: 279
Ovary **33**: 41–43, *42*
   crustaceans **29**: 184
   ovarian development **33**: 254–256, **255**
   sole **29**: 263, 266, 267, 269, 272, 282, 283
Ovary dissection method of quantifying
      fecundity **43**: *95*, 105, 106, 151
Overall growth performance (OGP) **35**:
      159*See also* Growth performance
Over-culture, oyster production **21**: 52
Over-exploitation of Atlantic halibut **26**: 49,
      50
Over-exploitation of marine animals **44**: 272
Over-fishing **29**: 316, 319; **36**: 245, 246
Overripening of eggs
   and chemical changes **26**: 94, 96
   and lipid content in fish eggs **26**: 96
   and storage **26**: 74, 75, 91–96, *93*
   and stripping **26**: 91–93
Overwintering stages **33**: **386, 387**
   copepodids **33**: 385–388
   diapause eggs **33**: **267–271**, 268–272,
      273, 388, 389
Oviduct, crustaceans **29**: 141, 143, 164, 174,
      182, 200

Oviducts **33**: 41, *42*, 43
ovigerous period in *Pandalus* **38**: 154, 155
Ovigers, pycnogonid **24**: 4, 9
oviposition, *Pandalus* **38**: 151, 152
Ovoviviparous reproductive strategy **35**: 65, 67, 79, 115
Ovulated eggs, viability of **26**: 92
ownership, enhancement and **38**: 43, 44
Oxidative phosphorylation, crustaceans **29**: 138
Oxycline responses **33**: 443, 444
Oxyconformers, *Donax* **25**: 222, 223, 237
oxygen **41**: *12*, 40, 63
  see also Respiration
  beach environment **25**: 183
    Donax and *Bullia* **25**: 213
  biochemical function **41**: *11*, *12*
  -chlorophyll correlations **23**: 259–261
  content of bottom waters **23**: 192
  measurement and phytoplankton production **23**: 259
    pelagic **23**: 259
  saturation
    correlation with plankton biomass **23**: 259, 260
    super-, of surface waters **23**: 195, 259
  vertical distribution **23**: 259
Oxygen affinity, blood **24**: 353,355, 356
  and stress **24**: 357
Oxygen availability and respiration of bivalves **37**: 81
  and laboratory rearing **37**: 34, 37
  and metals **37**: 95
  physiology tests **37**: 67, 68
Oxygen concentration, deep-sea environments **35**: 11, 27, 28
  hydrothermal vents **35**: 18
  seeps **35**: 23
Oxygen consumption **33**: 210, 211
  see also Respiration
  pycnogonid **24**: 29, 30
    during swimming **24**: 31, 32
  rates of **24**: 252
  resting rates of **24**: 352
Oxygen consumption, deep-sea fauna **35**: 11
Oxygen level **36**: 28–43, 240
  see also Hypoxia and anoxia
  and energy metabolism **36**: 32, 33
  and salinity **36**: 23
  and temperature **36**: 8, 10–12
  differentiation and variability **36**: 222, 227, 228

energy and energy metabolism **36**: 59, **60**, 62
energy metabolism **36**: 59, **60**, 62
life cycles **36**: 94, *95*
substance and energy **36**: *164*, 165, 167, 168, 171, 172, 176, 194, 195
Oxygen minimum layer, tropical oceans **32**: 41–48
Oxygen transport
  and responses to stress **24**: 357
  gas exchange surfaces **24**: 348, 349
  in Antarctic fishes
    red-blooded **24**: 352–356
    'white-blooded' **24**: 349–352
Oxygen, atmospheric
  consumption in bivalve molluscs **22**: 125–128
Oxygen, concentration of in vertical migration **26**: 147
Oxygen, concentration, and behaviour **27**: 352
  consumption **27**: 177–179, *178*
Oxygen, response to **33**: 217
Oxygen-minimum zones (OMZs) **35**: 11, 24, 25, 115
  adaptations to **35**: 27
Oxyregulators, *Bullia* **25**: 222, 223, *222*, 237
Oxyrhynchan **29**: 134
oyster **36**: 83
'Oyster basket' **24**: 18, 25, 44
Oyster culture areas **21**: 6
  Hiroshima **21**: 6
  Japan **21**: *3*
  Miyagi **21**: 10
Oyster culture grounds
  Hiroshima Bay **21**: *8*
  Kobuchi Bay **21**: *10*
  Mangoku-Ura **21**: *10*
  Matsushima Bay **21**: *10*
Oyster culture problems **21**: 31
  competitors **21**: 32
  mass mortality **21**: 34
  parasites **21**: 33
  predators **21**: 31
  water quality **21**: 36
Oyster culture techniques **21**: 13
  culture costs **21**: 29
  hardening **21**: 19
  ongrowing, Hiroshima **21**: 23
  ongrowing, Miyagi **21**: 27
  spat collection **21**: 13

Oyster culture, Japan **21**: 47
Oyster meat production
  Hiroshima **21**: *42*, **43**
  Japan **21**: **4**, **41**
  value, Japan **21**: *39*
  world **21**: **39**
Oyster meat values, Japan **21**: **49**
Oyster meat, Hiroshima **21**: *47*
Oyster production, Japan **21**: **4**
Oyster seed, production and
    distribution **21**: *7*
Oyster shell utilization **21**: *51*
Oyster stock, Matsushima Bay **21**: *17*
Oyster yield, Japan **21**: *45*
Oysters **21**: 1–57
Oysters *see under* taxonomic index
Ozonation and bivalves **37**: 70, 71, 72
Ozone depletion and sea ice **43**: 235–240

Pacific herring **20**: 20, 22, 25, 51, 61, 90,
    143, 240, 241
  demersal spawning in **20**: 23, 24
  fecundity of **20**: 21
  sperm density of **20**: 23
Pacific herring larva **20**: **72**
Pacific Ocean **35**: 8, 22, 84; **36**: 39, 211
  see also Eastern Pacific; Indo-Pacific
      Ocean; Southern Pacific;
      Western Pacific
  see also western Pacific
  see *also* World Ocean
  abyssal ocean floor faunistic
      regions **32**: *364*
  adaptation **36**: 15, 30, 39, 55
  adaptation strategies **36**: 48, 55
    energy metabolism **36**: 62, 71
    plastic metabolism **36**: 79–81, *80*, 84
  benthic invertebrates, vertical
      zones **32**: **339**
  Central **42**: 39
  deep-water bottom fauna
    coefficients of similarity **32**: **341**, **343**
    coordination of geographical
        affinities **32**: *342*
    endemics **32**: *336*
    invertebrates **32**: *340*
    taxonomic differentiation **32**: *337*
  E and W, brachiopod distribution **32**: *398,*
      399
  East **42**: 37, 40, 41
  Eastern Barrier **43**: 59, *61*

  ecological principles **36**: 237, 247, 248,
      249, 251
  equatorial, meso- and macroplankton
      biomass distribution **32**: **59**
  eutrophic/oligotrophic areas **32**: 428
  fisheries **38**: 224–226
  gelatinous animals, distribution **32**: 66
  hydrothermal vents **35**: 16, 17, 18, 20
  Indo-West Pacific Region **32**: 221–227
  invertebrates **43**: 102, 125, 148–150
  life cycles **36**: 96, 102, 103, 107, 118
    interannual fluctuations **36**: 125, 129,
        132, 137
  mineral extraction **35**: 96
  Nazca and Sala y Gómez submarine
      ridges **32**: 145–241
    fish distribution occurring in other
        areas **32**: **218**, *219*
  North **42**: 26, 40, 210, 216
  northwest *see* Kurile–Kamchatka region
  oligotrophic areas **32**: 491–497
  Pacific Insular Province **32**: 223
  Panama-Pacific **42**: 37, 205, 210
  Pandalus **38**: 106, 112, 113
  panthalassic E and W faunal
      distributions **32**: *214*
  parasites
    distribution **43**: 53, *57, 58,* 59, *60, 61*
    diversity **43**: *10–14 18, 19*
    ecology **43**: 29, 35, *37, 38*
    uses **43**: 69, 70
  Protobranch bivalves **42**: 26, 29, 37, 39,
      40, 41, 45
  Scaphopoda **42**: 147, 164, 205, 210, 216
  seamounts **35**: 25
  See also East Pacific; East Pacific
      North-eastern Pacific Ridge;
      Rise; North Pacific;
      South Pacific
  South **42**: 210
  species shared with Atlantic and Indian
      Oceans **32**: **344**
  squids, gonatid, northern North
      Pacific **32**: 243–324
  substance and energy **36**: 141, 152, 196,
      202
  trophic areas, macrobenthos,
      abyssal **32**: 485–497, **486, 487**
Pacific Ocean, sea cucumber in *see*
      distribution of sea cucumber
Pacific Ocean, trace elements in
      and phosphates **41**: *127, 128*

cadmium **41**: 72, *73, 75*, 76, *127*
cobalt **41**: 77
concentration **41**: *120–3*
copper **41**: 63, 66, *67, 124*
indices of fractionation **41**: 80, *81, 82*
internal economy and near-field
    chemistry **41**: 37–8
iron **41**: 43–46, *44, 45*, 49–55, *52, 54*, 57,
    60
nickel **41**: 78, *79, 128*
silicon **41**: *124, 125, 128*
vertical concentration profiles **41**: 9, 43, *44*
zinc **41**: 67, *68*, 69, *125*
Pacific salmon **38**: 5
Pacific sardine **20**: 10, 20, 61
    daily ration **20**: **70**
    first maturity in **20**: 21
    gut clearance rate in **20**: **68**
    pelagic spawning in **20**: 24
    spawning in **20**: 8
Pacific sardine eggs **20**: 25
Pacific *see* pollock
'Paddle cilia', Cephalopoda **25**: 103
Padstow **29**: 327, 333
Pagothenia sp **36**: 44
Paguridae **29**: **132**, 146–149, *147*, 152
Painted comber **36**: 20, 21
Palaemonetes sp **36**: 36
Palaemonidae **29**: **132**
Palau, Micronesia **44**: 231
Palau, sea cucumber in **41**: *137, 139, 151*
Paleogene **42**: 44, 215
Paleogeography **27**: 147–157
Paleontology, *see* Evolution
Paleozoic and Pleurotomarioidean
    gastropods **42**: 237, 239, *242*
    ecology **42**: 272, 274, 280, 286
    structure and morphology **42**: 243, 248,
        262
Palinura **29**: **132**, 160–165, **162, 163**
Palpons **24**: 186
Palps, pycnogonid **24**: 4, 9–11
Palytoxin **21**: 127
Panacene **21**: 182
Panama-Pacific **42**: 37, 205, 210
Pancake ice **43**: *177*, 179
Papua New Guinea, sea cucumber
    in **41**: 131, 132
    anatomy **41**: *139*
    associations **41**: *189*
    distribution and population structure
        41: 149, *151, 152*, 153, 154

fisheries and aquaculture **41**: 190,
    193–195
reproduction and spawning **41**: 158, 163,
    164, 170
PAR (photosynthetically active radiation)
    43: 199
Paracalanidae **44**: 86–93
Paralichthyidae **29**: 219
Parallel communities hypothesis **26**: 186,
    187
Paralytic shellfish poisoning (PSP) **31**: 311
    13, 312
Paralytic shellfish poisoning **21**: **71**, 75, 170,
    178
Parameter sensitivity value **20**: 236
Paramyxidae **25**: *145*, 146
Parapeneini **27**: 150
Parasites **27**: 379–390; **33**: 373, 523–525;
    43: **1–86**
    see also diversity; hosts; uses *and under*
        distribution *and* ecology
    as biological markers **27**: 389, 390
    bacteria **27**: 382
    behavioural changes to **27**: 389
    Crustacea **27**: 387–390
    ectosymbionts **33**: 524, 525
    endosymbionts **33**: 523, 524
    fungi **27**: 383
    herring **24**: 263–319
        'accidental' **24**: 273–275
        and herring population differences
            24: 299–303
        and host sexual differences **24**: 298,
            299
        as pathogens **24**: 303–307
        'essential' **24**: 275–278
        host range of **24**: 265, 270
        host specificity and **24**: 265, 271–279,
            299, 306
            definitive host **24**: 284, 285
            paratenic or reservoir host **24**: 284
            second intermediate host **24**: 284
    in aquaculture **26**: 54, 55
    larval **24**: 42, 43
    nematodes **27**: 387
    of Atlantic halibut **26**: 38–41
    of digestive tract *see* endoparasites
    of head and gills *see* ectoparasites
    of pycnogonids **24**: 49, 52, 53
    on siphonophores **24**: 207, 208
    platyhelminthes **27**: 385–387
    protozoans **27**: 383–385

symptoms of **27**: 379, 380
  viruses **27**: 380, 381
parasites of *Pandalus* **38**: 179-182
Parasites of pollock **37**: 206, 207
Parasites of Protobranch bivalves **42**: 35, 36
Parasites, Japanese oyster culture **21**: 33
Parasitism **36**: 57, *58*, 206, **207**
'Parasitoids', hyperiids **25**: 154, 158
Parasitology, zooplankton **25**: 118-159
  Animalia **25**: 146-157, *149*, *155*
  fungi **25**: 120, 121
Parathion toxicity and bivalves **37**: 97, **105**
parent stock (*Pandalus borealis*)
        38: 210-216
Paris Commission, pollution **29**: 350
Particle capture and transport,
        Brachiopods **28**: 304-310
Particle feeding **33**: 146-148
  see also Filter feeding
  feeding appendages **33**: 141
  foraging tactics **33**: 161, 162
  selectivity **33**: 165, 166
Particulate feeding **20**: 34, 35, 41, 93, 109
Particulate organic nitrogen (PON) **24**: 443
Passerine **20**: 230, 231
Patagonia toothfish **36**: 78
Patch Analyst **44**: 266
Patchiness of food **20**: 43-45, *46*
Patchiness of larva **20**: *101*
Patchy distribution **33**: 8
Pathogenic parasites, plankton **25**: 118, 119,
        120, 121, 144, 158
Pathogens of herring **24**: 303-307
  on adult host **24**: 305, 306
  on larval and juvenile host **24**: 303-305
Pathology, crustaceans **29**: 193-195
Pause travel predators **31**: 178, 187
PCP (pentachlorophenol) toxicity and
        bivalves **37**: 97, **101**
PCR (polymerase chain reaction) **37**: 206,
        222
PCR amplification **35**: 49
PCS(Amersham) scintillator **29**: 86
Pearl net **20**: 310, 321
Peclet number **29**: 14, **15**
Pectinacea, growth performance **35**: 167,
        171-173, *175*
Pectinidae, stress on **25**: 235
Pectoral fin, sole **29**: 220, 221, **222**
Pedal *see* Foot
Pedestal, crustaceans **29**: 146, *147*, 148, 197
Pedicellariae

echinoderms **21**: 148
  sea urchins **21**: *149*
Pedoncule, peridinians **25**: 126, *128*, 131
Peduncle
  Anomura **29**: 146, *147*, 148, 150, *151*,
        153, 159, 197
  euphausiids **29**: 183
  Macrura **29**: 164, 197
Pelagic
  Cephalopoda **25**: 88, 89
  fish
    eggs **25**: 9, 11, *25*,43, **52**, 53
    predators **25**: 9, 39, 41, 47, 48, 50
    prey **25**: 54
Pelagic armourhead **30**: 335, 336
Pelagic communities over
        seamounts **30**: 315-317
Pelagic ecosystems **30**: 311-317
Pelagic environments **33**: 470-487
  brackish water **33**: 472-479
  estuaries **33**: *474*, 475-477
  fjords **33**: 349-51, **350**, 477-479, *478*
  river plumes **33**: 474, 475
  coastal and shelf habitats **33**: *478*,
        479-482
    across-shelf gradients **33**: 480, 481
    biomass **33**: 457, 458
    upwelling **33**: 482, **484, 485**
  eddies **33**: 471, 472
  fronts **33**: 471, 472
  oceanic environment **33**: 483-487
  vertical distribution **33**: 486, 487
Pelagic food webs **29**: 2, 3, 60, 61
  fate of primary production **29**: 22-41,
        23-28, **31**, *33*, *34*, *39*, **40**, *41*
  turbulence, water column structure and
        phytoplankton cell size **29**: 4-21,
        5-8, **9**, *11*, **13-15**, **17**, **18**
  vertical mixing **29**: 41-59, *43*, **43**,
        46-49, *51*, *53*, *55-59*
Pelagic invertebrates, social aggregation
        in **30**: 155-216
Pelagic marine spawners **20**: 11, 13, 22, 181
Pelagic metazoan zooplankton **44**: 3
Pelagic phase of Atlantic halibut **26**: 11-22
Pelagic seabird numbers **20**: 226
Pelagic *see* deep-sea
Pelagic species **20**: 252, 284
Pelican **20**: 262, 263, 264
  brown **20**: 60, 260, **265**
Penaeid shrimps, spermatophore **29**: **131**,
        174, 187, 189

morphology **29**: 168–171, *170*, **172**, **173**
origin **29**: 171, 172
Penaeidae **27**: 59, 115, 117, 118, 119
 characteristics of **27**: 60, 61
 key to genera, adults **27**: 61–63
  larvae and postlarvae **27**: 119–123
 keys to species (see Keys, see Taxonomy)
Penaeoid prawns **29**: 189, 191, 193, 194
Penaeoidea **27**: 3
 morphology **27**: *59, 60*
Peneini **27**: 150
Penes, crustaceans **29**: 139, 140, 185, 199
Penetrability, sand, burrowing macrofauna
  25: 184, 190, 235, 236
Penetrant nematocysts **21**: 107
Penguin **20**: 249, 260, 271
 Adelie **20**: 278
 Antarctic **20**: 231
 chinstrap **20**: 278
 Dassen Island **20**: 259
 jackass **20**: 60, 247, **248**, 249, 252, 259
 King **20**: 273, 295
 macaroni **20**: 275
 royal **20**: 296
Penguin egg harvest **20**: 257, 258
Penilia **31**: 86, 90, 92, 151, 152
 cultivation **31**: 150
 development and hatching of
   embryos **31**: 132
 embryonic development **31**: 102
 female reproductive anatomy **31**: 93, 98
 gamogenic egg development **31**: 129
 gamogenic embryo release **31**: 129
 measurements of body size in **31**: 84
 oogenesis in **31**: 98, 99
 reproductive anatomy **31**: 116
 resting embryos **31**: 130, 132
 sexual dimorphism in **31**: 107–109
 spermatozoa in **31**: 121
Penilia avirostris *31* : 85, 90, *110*
 brood size **31**: 140
 cultivation **31**: 147, 149
 developmental stages **31**: *100*, 101
 distribution and abundance **31**: 133, 136,
   137
 embryonic development **31**: 106
 female reproductive anatomy **31**: 93
 gamogenic egg development **31**: 126, *127*
 life cycle **31**: 138
 optimum temperature and salinity **31**: *136*
 pedogenesis **31**: 139
 population densities **31**: 133

population growth rates **31**: 146
prenatal mortality **31**: 141
reproductive anatomy **31**: *117*
reproductive behavior **31**: 122
resting embryos **31**: 130, *131*
temperature and **31**: 145
Pennsylvanian **42**: *242*, 248, 286
Pentachlorophenol *see* PCP
Peptide mapping **35**: 52
Peracarida
 anatomy **39**: 115–127
 copulation and fertilization **39**: 153–156
  ovulation and fertilization **39**: 156
  sperm storage **39**: 154–156
  sperm transfer into marsupium **39**: 153,
   154
 development **39**: 156, *157–161*, **195–219**
  cleavage **39**: 162
  differentiation of mesoderm **39**: 164,
   165
  egg size, brood size and
   duration **39**: *186, 187, 189, 190*,
   195–219
  eggs produced/day of
   incubation **39**: *193*
  external differentiation **39**: 168–170,
   169
  fate mapping, gastrulation and germ
   layer formation **39**: 163, 164
  larval stages and moulting **39**: 170, 171
  organogenesis **39**: 165–168, *167*
  salinity and temperature effects **39**: *191,
   192*
 gametogenesis **39**: 128–142
  hormonal control **39**: 141, 142, 145
  oogenesis **39**: 137–141
  origin of spermatogonia and
   meiosis **39**: 132–136
  ovarian cycle **39**: 138, 139
  pre-vitellogenic development **39**: 137,
   138
  sperm structure **39**: 128–132, *129, 130*
  spermatogenesis **39**: 128–136
  spermatophore formation **39**: 136, 137
  vitellogenesis **39**: 139–141
 life cycles **39**: 143, 144
  larval stages and moulting **39**: 70, 71
  schemas **39**: *144*
 marsupial brooding pattern **39**: 172–194
  brood chamber structures **39**: 157–161
  brood mortality and brood
   parasitism **39**: 183, 184

brood size **39**: 187, 188
brooding behaviour and parental
    care **39**: 179–181
brooding in gelatinous zooplankton
    (hyperiid amphipods) **39**: 182,
    183
brooding in tubes **39**: 181, 182
egg size **39**: 185–187
evolutionary trends **39**: 190–194
extramarsupial brooding **39**: 181–183
incubation time **39**: 188
internal brooding **39**: 176–179, *179*
maternal care **39**: 181
oöstegite and marsupial
    structure **39**: 173–176, *174–177*
temperature and salinity **39**: 188–190
pairing and precopulatory
    behaviour **39**: 146–153, *148*
chemical recognition **39**: 146–149
complex mating systems **39**: 152, 153
cruising and non-cruising
    males **39**: 150, 151
planktonic pairing **39**: 151, 152
precopulatory pairing **39**: 149–153
protracted pairing and mate
    guarding **39**: 149, 150
reproductive behaviour **39**: 145–156
complex mating, spongocoels **39**: 153
reproductive cycles **39**: 143–145
reproductive system **39**: 116–127
female anatomy **39**: 116–121, *117, 119*
female histology **39**: 120, 121
male anatomy **39**: 121–125, *122, 123*
male histology **39**: 125–127
sex determination **39**: 108–114
environmental factors **39**: 111–114
genetic aspects **39**: 110, 111
hermaphroditisn **39**: 114
sex determining
    hormones **39**: 108–110
sexual dimorphism and secondary sexual
    characteristics **39**: 115, 116
Peracarids **29**: 186
"percent net risk" (PNR) **37**: 137
Perch **36**: 77, 116, 148, 190, 196
Perch, Pacific **37**: 190, 218
Pereiopod, crustaceans **29**: 182
pereiopods **38**: 62, 63
Pereopods, morphology **27**: 14, 15, *15, 17*
Performance **35**: 160
    See also Growth performance
Pericardium *see* heart

Peridinians *see* Dinoflagellata
Peridinin and coral reef organisms **43**: *283*,
    285, 292, 293, 307
Periodic acid Schiff test **29**: 143, 153, 154,
    156, 162, 163, 164, 166
Periodic strategy and fecundity of
    invertebrates **43**: 120, 121, *121*
Periodicity **27**: 353–356
Periodicity, reproductive (Protobranch
    bivalves) **42**: 25
Perivitelline fluid **44**: 160
Perivitelline fluid, Cephalopoda **25**: 105
Perivitelline space **33**: 22, 23
Perivitelline space, sole **29**: 229
Periwinkle, *see* Gastropods
Permeability, sand, burrowing
    macrofauna **25**: 184, 190
Permian **42**: 44, 217, *242*, 243
Perturbation effect on community
    diversity **26**: 199
Peru Countercurrent **32**: 155
Peru current **20**: 259–269
Peruvian anchovy **20**: 3, *4*, 8, *9*, 32–60
    passim **20**: 259, 264, **266**, 267
    growth of **20**: 83
    pelagic spawning of **20**: 25
Peruvian booby **20**: 166
Peruvian brown pelican **20**: 166
Peruvian cormorant **20**: 260–265 *passim*
Peruvian sardine **20**: 10, 65, 90, 91
Peruvian Shelf, fisheries **39**: 271, 277
Peruvian upwelling ecosystem **34**: 278, *278*
Pesticides *see* biocide toxicity
Pests, introduced, biological control
    of **43**: 70
Petasma, crustaceans **29**: 183
Petrel **20**: 231, 282, **285**, 293
    ashy storm **20**: 243, 245
    blue **20**: 295
    burrowing **20**: 260
    common **20**: 295
    Leach's storm **20**: 60, *61*, 243, 245, 282,
        285, 294
    South Georgia diving **20**: 295
Petroleum resources **35**: 93
    See also Oil resources
Petroleum *see* detergents and oil
pH **36**: *51*, 52, 219
    in bivalve molluscs under
        stress **22**: 151–155
    life cycles **36**: 120, *121*, 122, 123
PH and bivalves **37**: 37, 63, 81, 120

Phagocytosis
    in sea anemones **22**: 74, 76–79, 88
Phagotrophy, peridinians **25**: 126, 131, 134, 135
Phanerotoxic animals **21**: 63
Phanerozoic **42**: *242*, 281
Pharmacology
    marine poisons **21**: 66
    molluscs **21**: 174
    paralytic shellfish poisoning **21**: 77
    Porifera **21**: 95
    Protista **21**: 77
    sponge poisoning **21**: 97
    tentacles and nematocysts, *Chironex* **21**: **118**
Pharyngeal pockets **20**: 38
Phases of migration in fish **26**: 150, 151
Phenol oxidase, crustaceans **29**: 140, 194, 195
Phenol, respiratory response, *Donax* and *Bullia* **25**: 224
Phenolase, crustaceans **29**: 153, 187
Phenolic tanning, crustaceans **29**: 186, 187, 200
Phenols, spermatophores **29**: 136, 194
Phenotype
    condition for natural selection **43**: 118
    shift of hosts by parasites **43**: 34, 45
Phenotype variance
    genotypic components of **23**: 37, 38
    Photon flux density **23**: 142, 143
    Photosynthesis, *see also* Light
        compensation irradiences **23**: 143
        depth limitations 23: 141–143
        measurement at various tissue water contents **23**: 141
        recovery after drought **23**: 139–141, 144
    single loci, *see* Polymorphism
Phenotypic population structure of pollock **37**: 208, **209–211**, 212–215, 237
Pheromones **33**: 256, 257
Pheromones and spawning **43**: 104
Pheromones, crustaceans **29**: 181, 182, 185
Philippine Sea Plate **34**: 389
Philippine Trench, hadal fauna **32**: 369, 370
Philippines, coastal areas of **43**: *10, 12*, 65
Philippines, sea cucumber in **41**: 132, 136, *151*
    anatomy **41**: *139*
    common names **41**: *137*

reproduction and spawning **41**: *161, 164, 165*, 168
Phillipines **42**: 29
    Pleurotomarioidean gastropods **42**: 272, 273, *273*
    Trench **42**: 37
Philopathy of pollock **37**: 234, 236
Phoresy, plankton **25**: 118, 133, 134, 135, 138
Phosphagens, echinoderms **21**: 151
Phosphate
    coral reef response to **22**: 16
phosphorus (and phosphates) **41**: 22, 88, *121*
    biochemical function **41**: *11, 12*
    cycling ratio **41**: *84, 85, 86*
    in sea cucumber **41**: *186*
    indices of fractionation **41**: *81, 82*
    interactions with trace metals **41**: 56–58, 67, 72–76, *73, 75*, 78, *126–128*
Phosphorus content **33**: *241*, 242, 243
Phosphorus, Baltic Sea **29**: *107, 108*, 109, 110, *110*
Phosphorylation, *Donax* and *Bullia* **25**: 224
Photoinhibition, plankton **29**: 16
Photon flux density, plankton **29**: 16, *17*
Photoperiod, sole **29**: 268, 272
Photosynthesis **43**: 286
    see also chlorophylls
    in corals, stress effects **22**: 28, 29
    in sea anemones **22**: 91, 92
    in sea ice **43**: 197, 199, 200, 218
    photosynthetically active radiation (PAR) **43**: 199
Photosynthesis, phytoplankton **29**: 15, 16, *17, 18*
Phototaxis **21**: 74; **20**: 138, 140
Phototropism, pycnogonid **24**: 32
Phronima sp **36**: 36
Phycobilins and coral reef organisms **43**: *283*, 307
Phycobilins, picoplankton **29**: 82
Phycocyanin, picoplankton **29**: 82, *82*, 90
Phycoerythrin **43**: *283*, 307
Phycoerythrin, picoplankton **29**: 81, 82, *82*
Phycomycetes, parasitic **25**: 122
Phycotoxins **31**: 302
Phygon (dichloronaphthoquinone) toxicity and bivalves **37**: 97, **100**
    oxygen **37**: 34, 37
    pH **37**: 37
    physical requirements of bivalve larvae reared in laboratory **37**: 33–37

salinity **37**: 34, 36
temperature **37**: 36–6, 36
turbidity **37**: 37
Phylogenetic diversity **34**: 261
Phylogenetic hypotheses **30**: 163
Phylogenetic Species Concept (PSC) **31**: 10
Phylogenetics, crustaceans **29**: 168, 197, 200
Phylogeny **33**: 2, *3*, 49–56, *54*, 55, 56
   and taxonomic history of shelled
        Opisthobranchs **42**: *89,* 107–117,
        109, 114, 116
   Scaphopoda **42**: 141–145, *142, 143, 144*
Phylogeny *see* evolution
Phylogeny, siphonophore **24**: 121–133
   Cystonectae **24**: 124, 125
   of Brachystele families **24**: 125–128
   of Calycophorae **24**: 129, 130
   of diphyomorph families **24**: 131–133
   of Macrostele families **24**: 128, 129
   of prayomorph calycophores **24**: 129–131
   Physonectae **24**: 125, 127
Phylogeography **35**: 92
   mitochondrial phylogeography **35**: 91–93
Physalitoxin **21**: 115
Physical and physiological properties and
        egg quality in fish **26**: 76
Physical characteristics, Bristol Channel
        29: 218
Physical environment of Southern Ocean sea
        ice **43**: 175–184
   see also seasonality
   habitats **43**: 180–184
physical structure **23**: 99, 100
   sand plains adjacent **23**: 100
   seaweed depth distribution **23**: 100
   stability relative to hydrothermal
        vents **23**: 332, 333
   subtidal experiments **23**: 106
Physiochemical determinants of marine
        community structure **23**: 115, 116,
        126–145
   correlation studies **23**: 127–135
   early studies **23**: 126, 127
   experimental studies **23**: 135–145
      transplants **23**: 135–137
   tide levels, *see* Tide levels
   versus biotic interactions **23**: 145
Physiology **27**: 159–211
   (see also Metabolism)
   blood composition and
        transport **27**: 173–177
   digestion and assimilation **27**: 159–163

endocrine system **27**: 195–202
   metabolism **27**: 177–184
   moulting **27**: 217–221
   neural integration **27**: 184–195
   nutrition **27**: 163–172
   osmotic and ionic regulation **27**: 203–211
Physiology *see* anatomy
Physiology tests for bivalve larvae **37**: 66,
        67
   calcium uptake **37**: 69, 70
   feeding activity **37**: 66, 67
   metamorphosis **37**: 70–74
   respiration **37**: 69
   swimming activity **37**: 68, 69
Phytoplankton (general features)
        23: 210–215; *see also* Diatoms;
        Dinoflagellates
   Celtic Sea Shelf Break
        distribution **23**: 271, 272
   classical annual cycle **23**: 210, 211, 277
      arctic **23**: 211
      critical depth theory **23**: 210–212
      internal wave related **23**: 253, 254
      species succession **23**: 212
      temperate **23**: 211
      tropical **23**: 211
      variation and departures from **23**:
      212–214
   European tidal front accumulation **23**:
        232–236; *see also* Red tides
   high productivity hypothesis, *see* Phyto-
        plankton, high productivity
        hypothesis
   fortnightly cycle **23**: 183, 241, 268, 278,
        280, 281
   growth **23**: 245–266
      limiting factors **23**: 214, 215
      permanent vertical mixing **23**: 234,
      235, 246–248
      'replicative' **23**: 245
   in red tides, *see* Red tides
   light variations **23**: 249, 250–253
      attenuation model **23**: 250–252
   nutrient preferences and requirements
        23: 223
   photoinhibition **23**: 247, 248
   recycling **23**: 278
   regenerated production **23**: 261, 262, 268
   Russell cycle associations **23**: 234, 235
   sinking **23**: 257, 258, 268
   strategic ordering **23**: 224

temporal factors in the ecology of **23**:
    278–280
Phytoplankton **37**: 38, 39, 70; **20**: 34, 36, *37*,
    40
and dinitrogen fixation **24**: 432, 433
and isotope fractionation **24**: 406, 407,
    414–416
as food for larva **20**: 27
biotopes and ocean gyres **32**: 545–556
endemism **32**: 558–560
expatriation of species **32**: *557*
geographical distribution **32**: 527–563
    arctoboreal type **32**: 536, 537
    bipolar type **32**: 535
    boreal types **32**: 538–540
    borealnotal type **32**: 541–543
    cosmopolitan type **32**: 535
    methods **32**: 529
    ranges and environmental
        factors **32**: 556–558
    tropical type **32**: 535
    tropicnotal Antarctic type **32**: 541–543
growth season of **20**: 7
World Ocean, distribution
    maps **32**: *530–543*
phytoplankton and trace metals **41**: **2–128**
see also elements in oceans;
    feedbacks; fractionation; internal
        economy; macronutrients;
        seawater
    global imprint *see* Redfield Ratios
    major metals *see* micronutrients
    resilience of ocean system **41**: 92–95, *94*
Phytoplankton 'blooms' **36**: 30
Phytoplankton feeding **33**: 6–8
    feeding rate **33**: **187**
    feeding responses **33**: *144*, 144
Phytoplankton *see* Cyanobacteria,
        Eukaryotic algae, Pelagic food
        webs
Phytoplankton, high productivity
        hypothesis **23**: 236–266
    accumulation view **23**: 243, 256, 257
    active (convergent), 256, 257
    passive **23**: 257, 267
    [14]C uptake measurement **23**: 213, 214,
        245–248
    methodological problems **23**: 246–280
    detection of high plankton
        concentrations **23**: 236–238
    in situ growth view **23**: 221, 222
        239–243, 256, 261, 267

Holligan's mechanism **23**: 239
one-dimensional approach **23**:
    240–242
internal wave effects **23**: 249–256, 276,
    277
    light variations **23**: 250–253
    periodicity related **23**: 253, 254
mixing effects **23**: 246–249
nutrient availability **23**: 241–243;
        see also Red tides
questioning **23**: 244–266
water mixing factors
    horizontal **23**: 241, 242, 243
    vertical **23**: 240
Phytoplankton, *see* Plankton
Phytoplankton, South African coast **25**: 185
Donax **25**: 191
Phytosociology **23**: 57–65
    hierarchies **23**: 58–259, 65
    Zurich–Montpelier classification, *see*
        Zurich–Montpelier classification
Pickerel
    adaptation **36**: 7, *11*, 47, **60**, 64, **77**, 86
    ecological principles **36**: 239, 240
    life cycles **36**: 105, *115*
    substance and energy **36**: 200, 201
        accumulation and production **36**: 143,
            147, 148, 150, 151, **155, 156, 160,
            161**
        budget **36**: *197, 198*
        consumption, transformation and
            utilization **36**: **179**, 179–196
        expenditure **36**: *164*, **164**, 164–174
        passim **36**: *166, 169, 173*, **174**
        passim, 181, **185**, *188*, **192, 193**
Pickfordiateuthid squids, embryonic
        development **25**: 101
Picoplankton **29**: 4, 6, 77, 78
    see also Bacterioplankton, Eukaryotic
        algae, Cyanobacteria, Pelagic
        food webs
Pigmentation in coral reef organisms,
        biological and remote sensing
        perspectives **43**: **277–317**
    see also in situ; optical properties;
        reflectance; separability of spectra
    benthic categories **43**: 281
Pigmentation, prey **25**: 15
Pigments and quality in eggs **26**: 80, 81, 98,
    99, 102
Pike **36**: 231
    adaptation **36**: 66, **85**

life cycles **36**: *95*, 101, 103
substance and energy **36**: 148, 190, 196
Pike-perch **36**: 116
Pilchard **36**: 49; **20**: *12, 31*, 252, 258
　biomass of **20**: 258
　Cornish **20**: 17
　European **20**: 133
　South African **20**: 5, 15, 24
Pilsbry, Henry **42**: 68, 107–110, 129
Pineal organ **20**: 139
Pink salmon
　adaptation strategies **36**: 78, 79, *80, 85*
　life cycles **36**: 124, *125*, 133
Piquero (Peruvian booby) **20**: 260, 262, 263,
　　264, **265**
Plaice **36**: 62, 78, 141, 176
　and match/mismatch hypothesis **26**: 251,
　　252, 257, 270
　North Sea and Baltic
　　(1893–1990) **38**: 11–13
　rearing in UK (1957–1967) **38**: 14–16
　Scotland (1894–1920) **38**: 8, 9
　vertical migration of **26**: 121, 133, 138,
　　141, 152
Plains killifish **36**: 20
Planktivores **34**: 265
Planktivorous fish, predators **25**: 9, 11
Planktobacteria **25**: 119
Plankton
　classical annual cycle **23**: 211–215
　frontal accumulations **23**: 209–232
　　evidence of surface convergence
　　　23: 186
　phyto, *see* Phytoplankton
　zoo, *see* Zooplankton
Plankton and pycnogonids **24**: 75, 76
Plankton production and year class strength
　　in fish populations **26**: 249–293
Plankton sampling *see* Sampling systems,
　　fish eggs and larvae
Plankton, vertical distribution **27**: *300*
Planktonic food webs, predation **25**: 5
Planktotrophic larvae of invertebrates
　　43: 90, 92, 100, 124–126, 144,
　　145, 147, 246, 247
Plastic metabolism, divergences
　　of **36**: 75–87
　see also Motor activity
Plasticity and invertebrates **43**: 140, 141,
　　144, 146
Platycephalidae, as predators **27**: **363**
Platyhelminthes (see Parasites)

Platyhelminths, parasitic **25**: 147–152, *149*,
　158
pleonites **38**: 63
Pleopis polyphemoides **31** : 84, 86, 90, 92,
　93
　birth rates **31**: 141
　brood size **31**: 140
　cultivation **31**: 147, 148
　distribution and abundance **31**: 136, 138
　embryonic development **31**: 101, 102,
　　103
　gamogenic egg development **31**: 128
　nutrition **31**: 91, 92
　optimum temperature and salinity **31**: *136*
　pedogenesis in **31**: 139
　population densities **31**: 133
　population dynamics **31**: *143*, 145, 146,
　　147
　population growth rates **31**: 145
　prenatal mortality **31**: 142
　reproductive anatomy **31**: 119
　reproductive behavior **31**: 122
　resting embryos **31**: 130
　sex determination **31**: 112, *113*, 115
　sex ratio **31**: 111
　sexual dimorphism **31**: 107
　temperature and **31**: 145
Pleopods, crustaceans **29**: 139, 140, 159,
　　183, 185
Pleopods, morphology **27**: 17, 52, 53
Plerocercoid larva, Cestoda **25**: 150, 151, 158
Pleuronectidae **29**: 219
Pleuronectiformes **29**: 219
Pleuronectoidei **29**: 219
Pleurotomarioidean gastropods **42**: 237–294
　see also (in taxonomic index)
　　　Pleurotomarioidea
　ecology **42**: 272–286
　　bathymetric distribution and
　　　zonation **42**: 274–277, *275, 276*
　　biogeography **42**: 277–279
　　diet **42**: 279, 280, *279*
　　geographic distribution **42**: 272–274,
　　　273
　　predators and chemical
　　　defenses **42**: 281–286, *282–284*
　structure and morphology **42**: 243–262
　　circulatory system **42**: *251, 253*, 259,
　　　260
　　classification based on **42**: 262–268,
　　　263, 268

digestive system **42**: *251*, 252–259, *253, 255–257*
excretory system **42**: *251, 253*, 260
external anatomy *Plates 5 and 6* **42**: 249–251, *249, 251*
mantle cavity organs **42**: *251*, 252
nervous system **42**: *253*, 261, 262
reproductive system **42**: *251, 253*, 260, 261
shells **42**: 239–241, *240*, 243–249, *244, 247*
systematic relationships and classifications **42**: 262–272
molecular-based **42**: 269–272, *270*
morphology-based **42**: 262–268, *263, 268*
Plocamodiene A **21**: 93
Ploidy states, seaweed **23**: 38, 39
relative frequencies **23**: 39
Plymouth **29**: 324–326
Pneumatic duct **20**: 118
PNR ("percent net risk") **37**: 137
PNR *see* point-of-no-return
Pocilloporin **43**: 288, 290
Pocket net, for scallop culture **20**: 352, 362
Podon 31 : 86, 92
embryonic development **31**: 102
fertilization **31**: 125
measurements of body size in **31**: *84*
reproductive anatomy **31**: 116
resting embryos **31**: 130
Podon intermedius 31 : 85, 86, 92
brood size **31**: 140
cultivation **31**: 148
distribution and abundance **31**: 133, 137
optimum temperature and salinity **31**: *136*
pedogenesis **31**: 139
Podon leuckarti 31 : 85, 86, 152
cultivation **31**: 148
distribution and abundance **31**: 137
embryonic development **31**: 101
gametogenesis in **31**: 119, 120
male gametes **31**: 121
optimum temperature and salinity **31**: *136*
pedogenesis **31**: 139
population growth rates **31**: 145
reproductive anatomy **31**: 117, 119
resting embryos **31**: 130
sex ratio **31**: 111
temperature and **31**: 136, 145
Poecilogony **43**: 125, 146
see also Polychaeta *in taxonomic index*

Pogonophora **29**: 196
Pogonophorans **34**: 404–410
Point-of-no-return **20**: 43, 48, *49*
Poison apparatus, echinoderms **21**: 148
Poisoning
annelids **21**: 137
arthropods **21**: 139
bryozoans **21**: 142
ribbon worms **21**: 134
turbellarians **21**: 133
Poisonous marine animals **21**: 59–217
Poisonous marine plants **21**: 59–217
Poisons **34**: 225–227
Poisson distribution model **26**: 193, 194
Poisson–Inverse Gaussian distribution **26**: 208
Pojo-bay, Baltic Sea **29**: 90
Polar cod **36**: 11
Polar regions **43**: 96
see also Antarctic; Arctic
Polaria sp **36**: 36
Polinices josephinus
burial mechanism **28**: 407–409
burrowing performance, comparisons **28**: *417*
digging cycles **28**: 405
functional anatomy **28**: 397
Polinices strangei, see Uber strangei
Pollock **36**: 15, 132
Pollock, *see* walleye pollock
pollutants **41**: 38, 39; **20**: **188, 189, 190**
Pollutants and bivalves **37**: 3, 4, 6–10, 87–118
see also bioassay; biocide; detergents and oil; metal; toxicity
biocides **37**: 96–107, **98–106**
detergents and oil **37**: 107–116, **109–115**
Pollution **24**: 66, 67
and Atlantic halibut **26**: 42
and community stability **26**: 200,206, 207
and diversity **26**: 206, 207
and egg quality **26**: 87–89
as habitat factor in benthic studies **26**: 227, 228
indicators of **26**: 190, 191
Pollution and parasites
in mariculture **43**: 62, 65, 66
indicators **43**: 29
monitoring **43**: 66–68
Pollution and toxicity **36**: 3, 37, 245–247, 252, 253

Pollution effects
  on bivalve molluscs, *see* Bivalve molluscs,
    stress
  on corals, *see* Corals, stress
Pollution, deep sea
  heavy metals **35**: 96, 97
  oil pollution **35**: 97–101
  radioactive waste **35**: 101, 102
Pollution, fish egg and larval mortality **28**:
    79, 80
Pollution, fisheries **29**: 350
Pollution, Japanese oyster culture **21**: 2,
    9, 36
Pollution, South African coast **25**: 186
  Donax and *Bullia* **25**: 197, 215, 216, 255
Polychaeta **44**: 10, 21–24, 26–28, 42, 132;
    29: 195, 252, 253
  numbers of species, data points and
    ranges **44**: 11
Polychaeta, parasites of **25**: 135, 146, 150,
    158
  Dinoflagellata **25**: 125, 127, 131
  Sporozoa **25**: 139, 140, *140*
Polychaetans, hydrothermal vent
    dwelling **23**: 326–328, 336
  bacterial symbionts **23**: 327
  mussel dwelling **23**: 327
  newly identified species **23**: 328
Polychaetes **30**: 320, 321; **35**: 27
  deep-sea species diversity **35**: 5
  heterozygosity at hydrothermal vents **35**:
    62
  in sea anemone diet **22**: 93
Polychelidae **29**: **132**
Polymastigotes, parasitic **25**: **124, 125**
Polymerase chain reaction (PCR) **37**: 206,
    222
Polymetallic nodules **35**: 6, 93
Polymorphic loci **35**: 30
  heterozygosity and **35**: 59–61, *60*, 115
  See also Genetic variation
Polymorphism and pollock **37**: 206, 207,
    216, 222
Polymorphism, genetic, seaweed **23**: 32–34
  interspecies comparisons **23**: 33
Polymorphism, spermatophore **29**: 134, 135
Polynemidae, as predators **27**: **363**
Polynesia *Plate 3* **43**: 29, 37, 84
Polynyas **43**: 180
Polyphemus
  brood size **31**: 140
  female reproductive anatomy **31**: 95, 97

fertilization **31**: 124, 125
reproductive behavior **31**: 122
Polyphemus pediculus **31**: 80, 86
  embryonic development **31**: 99, 100, 106
  female reproductive anatomy **31**: 94, 95,
    96, 97
  gamogenic embryo release **31**: 130
  germ cells in **31**: 98
  mating **31**: 123
  nutrition **31**: 92
  oogenesis in **31**: 98
  reproductive anatomy **31**: 116, 117, 118
  reproductive behavior **31**: 121, 122, *123*
Polyploid forms **29**: 279
Polyps *see* colonial species
Polysaccharide
  acid, secretion by sea anemones **22**: 86
Polysaccharides **29**: 179
  see also Chitin
polysaturated fatty acids (PUFAs) in larval
    diets **38**: 36
Polyspermy and bivalves **37**: 30, 32
Polytypy and invertebrates **43**: 140, 141,
    144, 146
  see also flexibility; plasticity
Polyvinyl lactophenol **33**: 12
Pomadasydae, as predators **27**: **363**
Pompano, as predator **20**: 129
Pontellidae **44**: 93; **29**: 174, 175, 182
Population
  dynamics of herring
    parasites **24**: 292–298
    age **24**: 292–294, 299, 302, 307
    long-term changes **24**: 296–298
    seasonal variations **24**: 294–296, 303,
      307
  of siphonophores **24**: 165–172
    ecological importance of **24**: 241–244
    hydrography and **24**: 169
    impact on prey populations **24**: 244
    life cycles and **24**: 170–172
  pycnogonid **24**: 55
Population biology **33**: 348, 352–381
  see also Biomass; Life history
  breeding seasons **33**: 354, 363
  density **33**: 457–459
    coastal and shelf regions **33**: 457, 458
    oceanic water column **33**: 458, 459, *459*
  generation time **33**: 363, 374, *380*
  mortality **33**: 366–370
  population maintenance **33**: 392–398
    annual fluctuations **33**: 393

inter-annual fluctuations **33**: 394–398, **395–397**
population regulation **33**: 370–373
  brood size seasonal variation **33**: 371
  food availability **33**: 372
  genetic aspects **33**: 371
  growth rate variation **33**: 371
  parasites **33**: 373
  predation **33**: 372, 373
  production **33**: 373–381
  seasonal changes in stage structure **33**: 353, 354
  sex ratio **33**: 363–366, *365*
  size frequency **33**: 354
Population biology, sole **29**: 218, 251, 343–347, *344–347*
population discrimination in *Pandalus* **38**: 203–208
Population dynamics **25:** 2*see also* Predation
Population dynamics **27**: 6; **35**: 154
  data distribution **35**: 155, 156, **156**, *157*
  data evaluation **35**: 155
  data sources **35**: 154, 155
  See also Growth; Growth performance; Mortality; Productivity
  statistical analysis **35**: 156
population dynamics in *Pandalus* **38**: 203–218
Population ecology of pollock **37**: 187–204
  dynamics **37**: 193–199
  macroecology **37**: 187–193
  recruitment **37**: 199, *200*, 201, *202*, 203, 204, 238, 239
Population ecology, seaweed **23**: 2–41
  equation **23**: 2, 3
  growth and resource depletion *see* Population growth and resource depletion
  minimum components of study **23**: 4
  primary parameters **23**: 4–32
  secondary parameters **23**: 32–41
Population energy demand **20**: 236
population genetics of sea cucumber **41**: 157, 158
Population genetics, deep-sea fauna **35**: 6, 116
  future study prospects **35**: 109–112
  See also Genetic variation
  study methods **35**: 28–53, 109–112

Population growth and resource depletion **23**: 117–126
  at equilibrium **23**: 117, 118, 121–126
    competitor co-existence **23**: 124, 125
  at non-equilibrium **23**: 118
  models **23**: 118–121
Population size **20**: 238
Population structure of pollock **37**: 204–236
  see also genetics
  estimation methods **37**: 206–208
  mechanisms **37**: 228–236
  metapopulation **37**: 224–228
  phenotypic **37**: 208, **209–211**, 212–215, 237
Populations of hosts, parasites as biological markers of **43**: 59–62
Porcellanids **29**: 150
Porcupine Seabight **30**: 6
Pore pattern signatures **33**: 30, 50, 55, 56, 56
  species identification **33**: 61–63, 62
Pores in ice **43**: 176, 181, 187, 198
Porgy **36**: 69
Portunidae **29**: **133**, 187
Positive binomial distribution model **26**: 193
Posterior mantle of Scaphopoda **42**: 192
Postlarvae, abundance patterns **27**: **304, 305**
  diet **27**: 321
  life cycles and histories **27**: 284–289, 294, 296
  migration **27**: 297, 302, 303
  morphology **27**: **47**, 52, 53, *53*
  response to pressure changes **27**: 339
  salinity **27**: 351, 352
  species identification **27**: 125, 126
  taxonomy **27**: 107–126, **108–112**, 125, 126, *117, 122, 123*
    Key, to families **27**: 115–118
    to genera **27**: 119–123
  vertical distribution **27**: *300*
Post-larval stage of Atlantic halibut **26**: 18–21
post-larval stages, *Pandalus* **38**: 166–203
Post-planktonic juveniles and adults **44**: 251–259
potassium **41**: *11, 12, 120, 186*
Potassium toxicity and bivalves **37**: 72, **106**
Potassium, crustaceans **29**: 137, **139**
Potential fecundity **43**: 92
Powell Basin (Antarctic) **43**: 225
Prasinophyceae **29**: 79
Prawns **29**: 130, 189–191, 193, 194
Precious corals **30**: 338

Predation **27**: 357–377; **34**: 258, 280;
     43: 286; **20**: 51–62
see also Copepoda
and die vertical migration **26**: 127
and match/mismatch hypothesis **26**: 284
avoidance as *zeitgeber* **26**: 143, 144
bacterioplankton **29**: 111–115, *112*, **113**,
     115
birds **27**: 371, 372
by humans *see* exploitation
cephalopods **27**: 359–360
chemical detection **27**: 377
coelenterates **27**: 359
Crustacea **27**: 360
defence mechanisms **27**: 373–377
eggs and larvae **25:** 1–67
     assessment of **25:** 37–45, *44*
     population studies **25:** 45–51, *46*, *49*
     predator diversity **25:** 4–10, *6*
     predator–prey interactions **25:** 10–22,
          10, *13*, *17*, *18*, *20*, *22*
     reproductive strategies **25:** 51–66, **52**,
          56, *60*, **62**, *63*, *65*
     vulnerability **25:** 23–37, *24*, **26**, *28*, **31**,
          32, **34**, **35**
fish **27**: 361–371, **363–365**
mammals **27**: 372, 373
on *Donax* and *Bullia* **25:** 182, 187, 195,
     196, 236
on parasites **43**: 35, 36
on pycnogonids **24**: 48, 49
on siphonophores **24**: 208–211
     invertebrates **24**: 208, 209
     vertebrates **24**: 208
penaeids as prey **27**: 357–359
phytoplankton **29**: 3, 6, 45, 47, 60
     cell size **29**: 18–23, 41
     turbulence **29**: 23–29, *25–28*
reptiles **27**: 371
sole **29**: 238, 239, 251–254, 295
Southern Ocean sea ice **43**: 225–229,
     241, 274
Predation risk **20**: 108
Predator–prey interactions of pollock
     37: 185–187, 190, 202, 203
Predator removal **34**: 259–265, *260*
Predator–prey interaction **20**: 227; **23**:
     18–23 *see also* Carnivores
density regulation equation
     of predator **23**: 18
     of prey **23**: 18
     oscillation cycles **23**: 18, 19

phase diagrams **23**: 19–21
     partial **23**: 21
strength of **23**: 72, 73
Predator–prey relationships **34**: 260, 262,
     295
Predator–prey studies **28**: 22, 23
electivity index **28**: 22
fish egg and larval mortality **28**: 85–97
prey capture and predator avoidance **28**:
     117–119
Predators
as cue for diel vertical migration **33**: 449,
     450
Pleurotomarioidean gastropods
     42: 281–286, *282–284*
population regulation **33**: 372, 373, 470
predation effects **33**: 521, 522
Protobranch bivalves **42**: 35
responses to **33**: 419, 420
Predators of cephalopods *see* cephalopod
     light-fishing
predators of *Pandalus* **38**: 175–178
predators on sea cucumber **41**: 187
Predators, Japanese oyster culture **21**: 31
Predatory feeding **33**: 142, 143
see also Diet
ecological effects of **33**: 521
feeding appendages **33**: 140, 141, 142
feeding rate **33**: **184**, 184, **188**
food capture **33**: 148, 149, 162–164, *163*
foraging tactics **33**: 162–164
selectivity **33**: 165
Preservation of samples **33**: 10, 11
Preservation, and shrinkage, in sampling
     28: 17, 18
Preservation, picoplankton **29**: 81, 83
Presettlement oyster larval groups
     Matsushima Bay **21**: *17*
     Sendai Bay **21**: *18*
Pressure and buoyancy **36**: 43, 47
Pressure, response of postlarvae **27**: 339
Pressure, response to **33**: 218
Pressure, *See* Hydrostatic pressure
Prey
     availability of **26**: 127, 129, 130
     movements of **26**: 133, 134
Prey populations **34**: 264
Prey release **34**: 262
Prey removal **34**: 265–276
     birds **34**: 268–274, *269*
     fishes **34**: 265–268
     mammals **34**: 274–276

Prey *see* Diet; Predatory feeding
Prey species **34**: 262
Prey velocity **29**: 28
Primary lid, Cephalopoda **25**: 101
Primary oocytes, sole **29**: 279
primary production and biological
  availability **41**: 94
  cadmium **41**: *13*, 14, 23, 74–77, *75*
  cobalt **41**: *13*, 14, 77, 78
  copper **41**: *13*, 14, 23, 64–67, *66, 67*
  iron **41**: *13,* 14, 23, 49–55, *54*
  manganese **41**: *13*, 14, 23, 62, 63
  nickel **41**: *13*, 79
  zinc **41**: *13,* 23, 69–71, *70*
Primary production on Southern Ocean sea
  ice **43**: 196–202
Primary production, phytoplankton **29**: 2
Primary sensory modality **20**: 98
Primitive streak, sole **29**: 229, 230
Prince William Sound **34**: 270
Prince William Sound, pollock in **37**: 179,
  180, 196, 222, 224
Principal Component Analysis
  and plankton production **26**: 256
  in benthic studies **26**: 222–224
principal components analysis (PCA)
  38: 206
Principal components analysis
Principal Co-ordinate Analysis in benthic
  studies **26**: 222, 224, 225
Private alleles method **35**: 56, 71–73, *72*
Proboscis, pycnogonid **24**: 4, 5, 7, 8
  anatomy of **24**: 18
Procedures **37**: 50–74
  spermiotoxicity test **37**: 50, 51
  see also embryo-larval bioassay;
    physiology tests; tests *under*
    growth
  statistical evaluation **37**: 136
Proceedings Under Acts Relating to Sea
  Fisheries, Annual Reports **29**: 328
Procercoid larva, Cestoda **25**: 150, 151, 158
Processed meat exports, Japan **21**: *40*
Pro-coelomic gas duct **20**: 160
Production **33**: 373–381
  see also Egg number
  estimates, picoplankton **29**: 83–85, *84*,
    86, 91, 92, **92**
  production/biomass (P/B) ratios **33**:
    373–381, **375–379**
  rate, sole **29**: 230
Production data **27**: 6

Production rates, *Donax* and *Bullia*
  25: 194, 195, 236
Production to biomass ratio (P:B) **44**: 9
Productivity
  data distribution **35**: 155, **156**
  mortality relationship **35**: 183–186, *187*,
    187, 190, 192, 193
  production/biomass ratio **35**: 183–186,
    187, **187**, 189, 190, **190**, *191*,
    192–194, *193*
Productivity and biomass **26**: 210
Productivity, oyster culture **21**: 53
Productivity/production
  defined **36**: 139, 142
  somatic and generative *see* Substance and
    energy
Profit, sole **29**: 348
prokaryotes **41**: 4, *10,* 30, 94
  see *also* algae *and* bacteria in taxonomic
    index
Pro-otic bulla **20**: 112, 115, 141, 142, 145,
    147, 153
Pro-otic fenestra **20**: 144, 148, 151
Propagule
  digestion survival **23**: 20–22
  emersion/submersion studies **23**: 143, 144
Propodium, *Bullia* **25**: 193, 210
Prosobranchs
  epifaunal **25**: 183
  respiration **25**: 214
Prosome **33**: *15*, 16
  musculature **33**: 38
Prostaglandins **27**: 198
Protavine, picoplankton **29**: 82
Protease
  in sea anemones **22**: 82–85
Protecting fish stocks **36**: 246, 247
Protective responses, siphonophore
    24: 175–177
Protein **24**: 359, 360
  and feed quality in fish eggs **26**: 98–101
  in Atlantic halibut **26**: 35, 36
  variability **24**: 360, 361
Protein content **33**: 245
Protein electrophoresis, use in
    hybridization **31**: 6
Protein metabolism **20**: 75
Protein store **20**: 74
Protein venoms **21**: 67
Protein, spermatophores
  Anomura **29**: 153
  Brachyura **29**: 136, 137, **138**, 139, 143

Protein–polysaccharide complexes **34**: 5
Proteins
　adaptation
　　food and feeding **36**: 50, 53
　　oxygen level **36**: 32–34, *35*, 39–41
　　strategies **36**: 68, 69, 73, 74, 86
　　temperature **36**: 14, 15, 18, 19
　differentiation and variability **36**: 221,
　　223, 224, *225*, 226
　ecological principles **36**: 230, 239, 244,
　　245, 251
　indicators of fish condition **36**: 214, 215,
　　218
　life cycles **36**: 126
　　annual **36**: 105, 106, *111*, 114
　　ontogenesis **36**: 98, 99, 101, *102*
　　reproduction **36**: 90, *91*, 93–95, 98
　　substance and energy **36**: 148–201
　　　passim, *173*
　　accumulation and production **36**: **155,
　　　156, 160, 161**
　　consumption, transformation and
　　　utilization **36**: **174, 179,** *180, 181,*
　　　192, 193
Proteins, biochemical indices **30**: 228–230
Proteoglycans **34**: 5
Proteolytic enzymes **30**: 281, 282, 285
Protista **25**: 121–146, *123, 128, 130, 132,
　　136, 140, 142, 145*
　viruses and bacteria **25**: 118–120
Protista, definition of **25**: 121
Protistan toxins **21**: 70, **71**
Protobranch bivalves **42**: 1–65
　see also (in taxonomic index) Nuculoida;
　　Solemyoida
　anatomy and morphology **42**: 4–17, *6,
　　10, 12*
　bioturbation **42**: 32, 33
　characteristics and classification **42**: 2–4
　cosmopolitanism **42**: 39
　development **42**: 26–28, *27*
　digestive system and feeding **42**: 14, 15,
　　17–24, *18*
　ecology **42**: 31–36, 41
　endemism **42**: 39, 40
　evolution **42**: 41–44
　feeding **42**: 17–24, *18*
　foot **42**: 14
　gills **42**: 9–13, *10, 12*
　glands **42**: 16
　global patterns **42**: 36–39, *37, 38*
　habitat **42**: 31, 32, 41

　labial palps **42**: 13
　mantle **42**: 13, 14
　physiology **42**: 30, 31
　predators, parasites and commensals
　　42: 35, 36
　recruitment **42**: 34, 35
　reproduction **42**: 24–26
　resource partitioning **42**: 33, 34
　respiration **42**: 30
　salinity tolerance **42**: 30, 31
　sense organs **42**: 16, 17
　shells **42**: 4–9, 6
　soft tissues **42**: 9–17, *10, 12*
　zoogeography **42**: 36–40, *37, 38*
protoctist parasites of *Pandalus* **38**: 180
Protopodite, crustaceans **29**: 139
Protozoa, grazing **29**: 22, 25, 105
　see also Ciliates, Nanoflagellates
Protozoa, hydrothermal vent dwelling
　　23: 319
Protozoa, in diet **33**: 156–159
Protozoans (see Parasites)
Protozoans, toxic **21**: 70, 77
Protozoea, morphology **27**: 47, 50, 51, *51*
Protozooplankton **29**: 77
"Proventriculus" **27**: 23, 24, *24, 25, 26*
Prydz Bay (Antarctic) **43**: 228
Prymnesin **21**: 83, 90
Prymnesiophyceae **29**: 79
Psammophilies, tidal migrations **25**: 187
Psettodoidei **29**: 219
Psettoidae **29**: 219
Pseudevadne tergestina **31**: 86, *108*
　development and hatching of
　　embryos **31**: 132
　distribution and abundance **31**: 87,
　　133, 137
　embryonic development **31**: 102, 106
　optimum temperature and salinity **31**: *136,*
　　142
　p*æ*dogenesis **31**: 139
　reproductive anatomy **31**: 116
　reproductive behavior **31**: 122
　resting embryos **31**: 130
　sex ratio **31**: 111
　temperature and **31**: 145
Pseudocalanidae **44**: 93–100
Pseudodiaptomidae **44**: 100–109
Pseudo-oil based muds (POBMs) **35**: 98
Pseudosexual eggs **31**: 125
Pteropoda **44**: 10, 21–24, 28, 30, 32, 34, 42,
　　132

numbers of species, data points and
    ranges **44**: 11
Pteropods, parasites of **25**: 138, 139, 150,
    151, 152, 153
Pterotic bulla **20**: 144
Pterotracheidae, parasites of **25**: 142, 143
Puffin **20**: 282, **285**, 286, 294, 296
    tufted **20**: 61, 245, 246
Puget Sound, pollock in **37**: *180*, 183, 184,
    187, 188, 224, 225
Pull-seining **34**: 225–227
Purification, toxin, *Goniopora* **21**: *128*
Purse seine fishing **20**: 116, 120, 185, 186,
    247, 249, 254
Pursuers, predation **25**: 9
Push net **29**: 237, 239
Pusules, peridinians **25**: 134
Pycnocline **29**: 54, 90, 97; **20**: 320
Pycnogonida, parasitic **25**: 152
Pycnogonids
    sea anemone predators **22**: 92
Pyrosomidea, parasites of **25**: 138, 143

Q-mode cluster analysis in benthic
    studies **26**: 220
Q-spermatozoa **29**: 179
Quadrat sampling **23**: 47, 48, 74
    analytical method **23**: 48
    minimal size **23**: 47, 48
Qualitative assessment of environmental
    samples **37**: 84, 85
Qualitative genetics, determination of
    phenotype composition by **23**:
    37–38
Quality
    of seawater and bivalves **37**: 41–43, **44**,
    77, 137
    of sediment and bioassay of
    bivalves **37**: 77–80
Quality of invertebrate eggs *see* size *under*
    eggs
Quanidine base toxins **21**: 67
Quantification methodologies for fecundity
    of invertebrates **43**: **92–111**, 105,
    112, *113*
    see also under broadcast spawning
    brooding species **43**: 93–100
    colonial species **43**: 110, 111
    shocks **43**: *94, 95*, 101–103, *102*, 151
Quantum efficiency of photosynthesis
    29: 16, *17*, *18*
Quarun, lake **29**: 349

Quaternary ammonium compound toxins
    21: 67
Quinone toxins **21**: 67
Quinones, crustaceans **29**: 187

Radiance *see under* irradiance
Radiation within subclass on Protobranch
    bivalves **42**: 43, 44
Radioactive
    contaminants, fisheries **29**: 350
    tracers **29**: 74
Radioactive waste, deep-sea dumping of
    35: 6, 93, 101, 102
Radioisotope injection **20**: 229
Radiolarians
    bacteria **25**: 119
    parasites of
        Crustacea **25**: 153, 154, 155, 156
        Dinoflagellates **25**: 126, 127, 128, *128*,
        135
    symbionts **25**: 123, 124, 125
Radiolarians, in diet **33**: 159
Radiotracers **43**: 192
Radula of Scaphopoda **42**: 173–176, *174*
Raft culture (Ikada) of scallops **20**: *315*, 321,
    346–347
    Miyagi **20**: 347
Rafting **43**: 179
Rainbow site, MAR **32**: 119
Rainbow trout **36**: 126, 170, 249, 250
    adaptation **36**: 22–25 *passim*, 27, 54
"Rami" **27**: 17
Rance Estuary **29**: 252
Random encounter, predation **25**: 26
Random sampling, *see* Sampling, random
Ranges of hosts of parasites **43**: 22, 23, 52, 53
Rangiroa Atoll *Plate 3*
Raninidae **29**: 133, 141
RAPD analysis **35**: 46–48
Raptorial predation **25**: 6, *24*, **26**
    fish **25**: 9, 17, 25, 27
    invertebrates **25**: 5–7, 8
Rarefaction and dominance **26**: 201–203,
    206
Rasbora **36**: 34
"Rassenkreise" (geographical species) **43**: 4,
    5
Rationalization of fisheries **36**: 247, 248
Rat-tails **36**: 44
rattails **38**: 3
Ray **36**: 116
Razorbill **20**: 282, **285**, 293, 294

RCF *see* relative condition factor
Reabsorption and Scaphopoda **42**: 189, 190
Reaction distance, predation **25**: 10; **29**: 28
Realized fecundity **43**: 92
Reared fish **20**: 84–86
Rearing fish *see* Farming, fish
Rearing of Atlantic halibut **26**: 53, 57–60
Receiving waters and bivalve
          bioassays **37**: 128, 129
Recent Pleurotomarioidean
          gastropods **42**: 237, 239, 241
     classifications **42**: 267, *268, 270*
     ecology **42**: 273, *275*, 280
     structure and morphology **42**: *240*,
          244, *251*
Recognition Species Concept (RSC) **31**: 10,
          11
recruitment (*Pandulus borealis*)
          38: 210–216
Recruitment **30**: 217
     and climatic change **26**: 256
     and gadoid outburst **26**: 274–276
     and larval abundance **26**: 271–273
     and salinity anomaly of
          1970s **26**: 276–278
     and tests of match/mismatch
          hypothesis **26**: 264
     and upwellings **26**: 250, 259–261, 284
     in Baltic Sea **26**: 273, 274
     in fish populations **26**: 250, 284
     in low latitude waters **26**: 259–263, 284
     of Atlantic halibut **26**: 50, *51*
Recruitment and Protobranch
          bivalves **42**: 34, 35
Recruitment of pollock **37**: 199, *200*, 201,
          202, 203, 204, 238, 239
Recruitment overfishing **20**: 5
Recruitment, processes affecting **28**: 5,
          38–41
Recruitment, seaweed **23**: 28–32
     cohort **23**: 5, 7
     intra- and inter-specific interactions
          23: 30, 31
     rate **23**: 7
          frond **23**: 32
          lateral branches **23**: 31
     seasonal cycle **23**: 32
     using artificial and natural surfaces
          23: 29, 30
Recruitment, *see* Predation
Recruitment, sole **29**: 278, 279, 287, 351
     see also Harvesting

future **29**: 348, 349
     natural mortality rate **29**: 297
Recruits per stock biomass **29**: 298, 299
"Red disease" **27**: 279, 280
red drum enhancement (Gulf of Mexico)
          38: 27, 28
Red mullet
     adaptation **36**: 48, 49
          strategies **36**: *60*, 64, 70, **77**, 82
     ecological principles **36**: 234, 238, 243
     indicators of condition **36**: 215, *216*
     life cycles **36**: 105, 132
     substance and energy **36**: 200, 201
          accumulation and
               production **36**: 147–162 *passim*,
               153, **155, 156**, *158*, **160, 161**
          budget **36**: *197, 198*
          consumption, transformation and
               utilization **36**: **179**, 179–196
               passim, *181*, **185**, *188*, **192, 193**
          expenditure **36**: *164*, **164**, 164–174
               passim, **166, 168**, *169, 173*,
               174
Red muscle **20**: **86**, 87, 89
Red Sea **35**: 8; **42**: 37
Red sea bream **36**: 250
red sea bream enhancement (Japan) **38**: 24,
          25
Red Sea, sea cucumber in **41**: *164*
Red Sea: parasites **43**: 6, *57, 58*, 66, 67
Red tide **20**: 314, 315, 321, 369, 370, 371
Red tides **31**: 303, 323
     definition **23**: 217
     ecological components **23**: 217–225
     European tidal front **23**: 232–226
          locations **23**: 232, 233
          mortalities caused **23**: 233
     hydrographic views of development
          23: 219–221, 225
          circulation mechanism **23**: 220, 221
     in situ growth view, *see* Phytoplankton,
          high productivity hypothesis
     mortalities **23**: 217, 219
     nutrient theory of development **23**:
          217–219, 223
          other views **23**: 218, 219
          palaeontological view **23**: 217, 218
     seeding **23**: 222, 223
     senescent **23**: 224, 229, 268
     taxonomic difficulties **23**: 234
"Red tides" **37**: 125, 126, **127**
Redfish **26**: 33, 34

vertical migration of **26**: 120, 123, 132
Redfield Ratios **41**: **39–42**
    correlations for trace metals **41**: *124–128*
    essential trace metals **41**: 41, 42
    macronutrient elements **41**: 39–41
Reef corals, adaptations of **31**: 221–277, 223
    environmental fluctuations in geological time **31**: 225–230
    environmental fluctuations on modern reef **31**: 230–238
        daily fluctuations **31**: 236–8, *240*
        effect of latitude **31**: 234, *235*
        interannual/annual fluctuations **31**: 231–234
        ENSO 231–233
    enzymatic adaptations **31**: 247–250
        enzyme concentrations **31**: 248
        enzyme isoforms and alloforms **31**: 248, 249
        enzyme kinetics **31**: 249, 250
    extinction profile **31**: 227
    fluctuating abiotic influences **31**: 238–245
        irradiance **31**: 242–244
        temperature **31**: 244, 245
        water flow **31**: 241, 242
    heat-stress proteins **31**: 250–252
    homeoviscous adaptations **31**: 252, 253
    hurricanes **31**: 233, 234
        monsoon variability **31**: 233
        variations in solar radiation **31**: 231
        seasonal fluctuations **31**: 234–236, *238–239*
    inter-regional comparisons of thermal tolerance **31**: 253, 254
    phenotypic responses **31**: 272–274, 273
    photoadaptation **31**: 224, 259–262
    temperature acclimation/acclimatization
        anemones and **31**: 246, 247
        responses in symbiotic to changes in chlorophyll content **31**: 264–266
        to different water flow regimes **31**: 269–272
        to fluctuations in solar radiation **31**: 257–269
            behavioural responses **31**: 259
            biochemical and physiological adaptations **31**: 260–262
            daily and seasonal fluctuations in solar radiation **31**: 264–266

morphological adaptation **31**: 259, 260
photoadaptation **31**: 259–262
photoinhibition **31**: 262–264
responses to ultraviolet radiation **31**: 266–269, 274
to temperature changes **31**: 245–257
cellular mechanisms of compensation to temperature stress **31**: 247–253
coral bleaching **31**: 254–257
Reefs, coral **23**: 99–108
    algae, *see* Algae
    community structure **23**: 101–108
        carnivore exclusion effects **23**: 106, 107
        herbivore depth distribution **23**: 101
        herbivore exclusion effects **23**: 106, 107
        herbivore interactions **23**: 104, 105
        herbivore, small **23**: 107, 108
        importance of grazing **23**: 103, 104
        sessile animal interactions **23**: 108
        similarities of, between different **23**: 101
        variations in factors affecting **23**: 106
Reflectance spectra of coral reef organisms, measurement of **43**: **293–303**, 308
    see also irradiance
    fluorescence and **43**: 280, 297–302, *298, 301*
    morphology **43**: 303
Refuge behaviour, and schooling, fish larvae **28**: 123, 124
regeneration of sea cucumber **41**: 142, 143
Regions, zoogeographic (see Zoogeographic regions)
Regulation, sole fishing **29**: 323, 324, 331–333, 349
Relative condition factor **20**: 47, 48
Relative stability, principal of **29**: 332
Relativization of data **26**: 184
release strategy, enhancement **38**: 37–39
Relevés **23**: 58, 61
Relict and 'fossil' species
    bathyal zone fauna **32**: 407–413, **409**, *411*
    shallow water **32**: 407
Relocations of home range **44**: 226–228
Remipedia **29**: **131**, 186
Remote sensing *see* pigmentation in coral reef organisms
removal of **26**: 127, 128

Replacement series graphs **23**: 14, 15
  objection to use **23**: 16, 17
Reproduction **27**: 251–281; **33**: 253–293;
  34: 254–256
  see also Eggs; Mating behaviour;
    Spawning
  (see also Embryology)
  barnacle growth and **22**: 219, 220, 235
  copulation **27**: 260–262
  corals under stress **22**: 46–48
  Donax and *Bullia* **25**: 193, 194, 225, 227,
    229, 230, 231–234, **231**, *232*,
    233, 236
  fecundity **27**: 257, **258, 259**; **33**: 272, 273,
    282, 283, **284–288**
  fertilization **27**: 263–267
  gametogenesis **27**: 251–255
  mating behaviour **27**: 257–262, **261**
  of pycnogonids **24**: 21–23, 33–38
    fertilization and egg laying **24**: 36–38
    seasonality of **24**: 43, 44
    spermiogenesis **24**: 35, 36
    vitellogenesis **24**: 35
  of siphonophores **24**: 101, 102
    rates in **24**: 239–241
  predation **25**: 29, 51–53, **52**, *54*
  seasonality **33**: 253–256, 354, 363
  strategies **34**: 395
Reproduction and life cycle
  see also eggs; gametogenesis; larvae; life
    history; meiosis; vitellogenesis
  of parasites **43**: 46
    and morphology **43**: 41–44, *42–43*
    organs, copulatory **43**: *10, 12, 13, 15,*
      27–29
    rates **43**: 37, *38–40*
    strategies, latitudinal gradients
      in **43**: 51, 52
  see also fecundity; fertilization; life
    history
  in Southern Ocean sea ice **43**: 211, 214,
    215, 217–219, 222–214, 241
  of invertebrates **43**: 89, 90
    reproductive effort **43**: 91, 114, 115,
      136–140, *139*, 142
Reproduction and reproductive system
  Pleurotomarioidean gastropods **42**: *251,*
    253, 260, 261
  Protobranch bivalves **42**: 24–26
  Scaphopoda **42**: 194–196
  shelled Opisthobranchs **42**: 100–103,
    106, 107

Reproduction in groups, costs of **30**: 190,
  191
reproduction in *Pandalus* **38**: 121–166
Reproduction of bivalve embryos and
  larvae **37**: 16–25
  see *Also* fertilization; gametes; spawning
  cleavage, embryogenesis and larval
    development **37**: 21–25
  embryonal stages **37**: 21, 22
  gametogenesis **37**: 16–19
  larval stages **37**: 22, 23
  settlement and metamorphosis **37**: 23–25
  sexual maturation in field **37**: 19–21
Reproduction, length at first (L_m) **20**: 20
  multiple **20**: 11
  synchronism **20**: 11
  timing of **20**: 7–11
  total spawning **20**: 11
Reproduction, of Atlantic halibut **26**: 7–11
Reproduction, seaweed **23**: 23–27;
    see also Dispersion
  age of first **23**: 23
  fecundity
    and fertility non equivalence **23**: 25
    individual plant **23**: 24
    net rate (R_0), 25
    number of reproductive units **23**: 23–25
Reproduction, sole **29**: 266, 267
  distribution, size and age with
    maturity **29**: 275–279, *276*
  fecundity **29**: 279–290, *281*, **284**, *285*,
    289
  seasonal development and time of
    spawning **29**: 268–275, *271, 273,*
    274
  spawning behaviour **29**: 267, 268
Reproduction/spawning and juvenile fish
  adaptation
    oxygen level **36**: 39
    salinity **36**: 23, 24
    strategies **36**: 65, 66, 71, 79–81, 84
    temperature **36**: 8, 13, 14
  differentiation and variability **36**: 222,
    226
  energy metabolism **36**: 61, 65, 66, 71
  indicators of fish condition **36**: 211, *212,*
    213, 218–220
  life cycles
    abundance **36**: 123–125, 128, 129,
      133, 135, 137
    annual **36**: 104, 105, 113

ontogenesis **36**: 89, 90–96, 98, 100, 101
molecular and metabolic aspects of **36**: 90–101
mortality after **36**: 102, 103, 125, 127, 237
plastic metabolism **36**: 79–81, 84
substance and energy **36**: 141, 147, 151, 160–162, 170, 171
temperature **36**: 65, 66
reproductive behaviour, *Pandalus* **38**: 148–152
reproductive cycle of sea cucumber **41**: *158–167*
see also gametogenesis; spawning
asexual reproduction induced **41**: 167
fecundity **41**: *161,* 167
gonad morphology **41**: 159–161
sexual dimorphism and sex ratio **41**: 158, 160
size at sexual maturity **41**: 158, 159
reproductive cycles, *Pandalus* **38**: 139–148
based on external morphology **38**: 139–141
gonad development **38**: 141–147
haemolymphic proteins and **38**: 147–148
Reproductive dynamics (see Life histories)
Reproductive effort, relationship to shelter **23**: 26, 27
Reproductive facilitation **30**: 186–191
Reproductive output **44**: 8
Reproductive strategies, deep-sea
heterozygosity and **35**: 65–67, **66**, 115
fauna **35**: 6, 79, 91, 110, 116
Reproductive system
female **33**: 41–44, *42*
male **33**: 44–46, *45*
Reproductive system, external
genitalia **27**: 41–43
female **27**: 44, 45, *44*
male **27**: 43, 44, *43*
morphology **27**: 41–45
Reptiles (see Predators)
Resemblance matrix **23**: 52, 53
Resident species **39**: 48–52, *51*
Residual currents, Bristol Channel **29**: 227, *228*, 229, 236, 242
resilience of ocean system **41**: 92–95, *94*
Resin casting **43**: 182
Resource management strategies **44**: 272
Resource partitioning and Protobranch bivalves **42**: 33, 34

Resource(s), food and space
competition for **23**: 123–126
outcome **23**: 124–127
single resource **23**: 117, 118
two resources **23**: 123, 124
consumption **23**: 123
depletion, *see* Population growth and resource depletion
supply **23**: 123
Resource-limited assemblages **26**: 198
Resources
and density of population *see* deterministic
exploited *see* exploitation
of Southern Ocean sea ice *see* trophic processes
Resources *see* Ecological principles
Respiration **33**: 209–211, **212**, *213*
see also Oxygen transport
bivalve molluscs
anaerobic **22**: 146–155
heavy metal effects **22**: 164–168
in gametes **22**: 173–175
corals, stress effects **22**: 28, 29
microbial loop **29**: 104
of pycnogonids **24**: 29, 30
rates, in siphonophores **24**: 232–234
spermatophores **29**: 138, 139
stress and **24**: 357
Respiration of Protobranch bivalves **42**: 30
Respiration *see* oxygen
Respiration *see* Oxygen level
Respiration, *Donax* and *Bullia* **25**: 182, 214–224, *215*, **217**, *220*, **220**, *222*, 237
energetics **25**: 225, 229, 230, 231, **231**, **232**, 232, **233**, 234, 238
locomotion **25**: 208, 210, 211, **211**
Respiratory system, morphology **27**: 33–37, *34, 35*
Response latency, larvae **25**: 19
Response-surface analysis **20**: 162
Restricted environments **33**: 487–503
caves **33**: 488, **489**
hyperbenthic and benthopelagic **33**: 496–503, **498–501**
deeper waters **33**: 501, 502
seep sites and hydrothermal vents **33**: 502, 503
neustonic **33**: 488–490, **491**
shallow waters **33**: 497–501
under ice **33**: 490–496, **493**
Antarctic Ocean **33**: **493**, 494

Arctic Ocean **33**: **493**, 495
 ice cover at lower latitudes **33**: 496
 ice-edge zones **33**: 495, 496
Restriction fragment length
  polymorphism (RFLP) **35**: 6, 44, **45**
Restriction site analysis **43**: 6
Retention
 areas and fish spawning **26**: 257, 258
 of larvae and tides **26**: 139, 140
Retina
 cones of herring **20**: *134*
 duplex **20**: 131
 pure cone **20**: 101, 134, 138
 rod of **20**: 134, 138
Retinal pigment index **20**: *136*
Retractile filaments, peridinians **25**: 131
Reverse-phase HPLC (RPHPLC) **35**: 52
Reversibility of fishing effects **34**: 290–291
Revue des Travaux de L'Institut des Pêches
  Maritimes **29**: 328
Reynolds number **29**: 12, **13**, **14**; **33**: 147,
  148
RFLP (restriction fragment length
  polymorphism) **37**: 206, 216, 222
Rheotaxis **30**: 200, 201
Rhinomugil sp **36**: 34
Rhizopoda, parasitic **25**: 121, 122, *123*
Rhizopodial system, peridinians **25**: 131,
  134, 135
Rhodamine chloride absorption **43**: 182
Rhode River (Chesapeake Bay)**43**: 70
'Rhynchoteuthion', Cephalopoda **25**: 100
Rhythmic activity **33**: 420, 421, 452, 453
 see also Diel rhythms; Feeding
  periodicity; Vertical migration
Rhythmicity, ecological principle of **36**: 232,
  233
Ribbon worms **21**: 133
Richness *see* species richness; taxa richness
Ricker domed spawner-recruit model **25**:
  59, 61
Ridge characteristics **34**: 399
Riga, Gulf of **29**: 116
Righteye flounder **36**: 77
Ringer solution **29**: 187
RISE Project Group **34**: 386
River plume fronts **29**: 50, 52, 58
River plume fronts, biological
  accumulations **23**: 225; *see also*
  Coasts and estuaries
 red tides **23**: 221
River plumes **33**: 474, 475

Rivera Fracture Zone **34**: 398
R-mode cluster analysis in benthic
  studies **26**: 184, 220, 225, 226
RNA content **33**: 249
RNA/DNA 'critical ratio' **28**: 20
RNA/DNA ratio **33**: 256; **30**: 232–237, 245,
  246, 249, 250, 261–264, 270, 275,
  276, 277, 280, 281
Río de la Plata (Argentina) **43**: 34
Roach **36**: 190
Roach, vertical migration of **26**: 120, 121,
  126
Rock lobsters **29**: 160–165, **162**, **163**
Rock surfaces, community structure
  23: 87, 88, 96 *see also* Reefs
 emergent **23**: 89
 low intertidal **23**: 87
 upper and mid shore **23**: 88
 wave exposed **23**: 87
Rockall Trough **42**: 26
Rockfish **36**: 11, 12
Rockling **36**: 116
Rodriguez Triple Junction **35**: 17
Rose Garden hydrothermal vent **23**: 334,
  347
Rosette-glands, crustaceans **29**: 140, 141
Rosine Amine D toxicity and
  bivalves **37**: 97, **101**
Ross Sea (Antarctic) **43**: 185, 233, 243
Rothschild Osborn model **31**: 179, 180
Rotifers **29**: 77
Rotifers **37**: 103, 132
Rough endoplasmic reticulum,
  crustaceans **29**: 166
Round goby **36**: 7, 86, 159, 215, 216, 236,
  237
 life cycles **36**: *92*, 103, 127
Round herring **20**: 247, 252
Roundnose grenadier fishery **35**: 103–107
r-selection **20**: 260
R-strategic reproductive patterns,
  crustaceans **29**: 199
r-strategist species **38**: 6
R-strategist, *Donax* **25**: 194, 236
Rudd, vertical migration of **26**: 120, 121,
  126
Russell cycle **23**: 234, 235
Russia, former USSR
 MAR studies of seamounts **32**: 95
 studies of EPR seamounts
  summary **32**: 149, **157**
Ryukyu Trench, hadal fauna **32**: 369

20°S hydrothermal vents **23**: 310
  geological features **23**: 309
Sable fish **36**: 77
Saccadic eye movements **24**: 368, 371–373
"Sackhole" sampling **43**: 191
Sac-spawning copepods **44**: 10, 15, 17, 41,
    42, 51
'Safe biological limits', fishing **29**: 317
'Safe sites', predation **25:** 53
Sagami Bay (Japan) **42**: 36
Sagitta sp **36**: 36
SAHN analysis **23**: 53, 56, 57
Saldanha Bay Island **20**: 249, 257, 269
Saldanha fishery **20**: 240, 250, **251**, 252, 253
Salinity
  and bivalves **37**: 66, 68, 81
    laboratory rearing **37**: 36, 37
    spawning **37**: 30
    toxicity reaction **37**: 120, *122*, 123
  and pollock **37**: 192
  adaptation **36**: 19–28
    density and **36**: 24
    energy requirements **36**: 22–24
    hormones **36**: 24–26
    lipid compositions **36**: 26, 27
    mineral balance **36**: 19–22
    nitrogenous compounds **36**: 27, 28
  and Atlantic halibut **26**: 53, 55
  anomaly of 1970s **26**: 276, *277*, 278, 281
  barnacle shape and **22**: 211
  bivalve mollusc responses
    at sublethal level **22**: 129
    calcium reabsorption from
        shell **22**: 156, 158–160
    detection and changes **22**: 137–142
    protective **22**: 133, 134
    pumping activity **22**: 113, 116
    valve movements **22**: 110, 113
  ecological principles **36**: 239, 240, 244
  effects on plant competition **23**: 17
  front, coastal European **23**: 167, 208, 209
    major **23**: 208, 209
    minor **23**: 208
  Hiroshima Bay **21**: *9*
  Matsushima Bay **21**: *12*
  osmoregulation and **33**: 215
  plankton **29**: 6, 50, *51*
    Baltic Sea **29**: 75, 76, *75*, 114
  responses to **33**: 216
  sole **29**: 230, 232, 238
    Bristol Channel **29**: 218, 226, 227, *226*
  stratification, *see* Haline stratification

Salinity in Southern Ocean sea ice
  see also brine
  and trophic processes **43**: 199–201, 203,
    204, 210
  gradients **43**: 181
  reduction **43**: 190
Salinity tolerance **24**: 27, 28
  and siphonophore distribution **24**: 163
Salinity tolerance of Protobranch
    bivalves **42**: 30, 31
Salinity tolerances, Brachiopods **28**: 329,
    330
Salinity, and behaviour **27**: 351, 352
  juvenile emigration **27**: 306–308
Salinity, *Donax* and *Bullia* **25:** 197, 223,
    224
salinity, *Pandalus* distribution and Scotland
    plaice transplants
    (1894–1920) **38**: 8, 9
Salivary glands, octopus **21**: **183**
Salmon **37**: 217, 239
  egg quality **26**: 74, 98, 99
  spawning of **26**: 252
  vertical migration of **26**: 144, 152
Salmon and salmonids **36**: 222
  adaptation
    food and feeding **36**: 54, 58
    oxygen level **36**: 39, 42
    salinity **36**: 21–23, 25, 26
  adaptation strategies
    energy metabolism **36**: 61, 65, 66, 71,
      72
    plastic metabolism **36**: 77, 78, 79, *80*,
      81, 84, *85*
  ecological principles **36**: 232, *233*, 237,
    249
  indicators of condition **36**: 212, 213, 217,
    219
  life cycles **36**: 114, 117, 118
    abundance, dynamics of **36**: 124, *125*,
      126, 129, 133
    ontogenesis **36**: *90*, *93*, *96*, 102, 103
    substance and energy **36**: 171, 196,202
Salpa sp **36**: 36
Salpidea, parasites of **25:** 133, 134, 138,
    139, 143, 156
Salps **44**: 34, 35
Salps, role in plankton
        communities **32**: 64–68
Saltees Bank **29**: 223, *223*, 324
Saltmarshes **44**: 242
Salton Sea **43**: 65

Samoa, sea cucumber in **41**: 131, *137*, 195
Sampling **33**: 8–10, 349–352
  design and spatial distribution
      26: 193–197
  devices in benthic studies **26**: 173
  effort **26**: 175–177
  error **33**: 349, 350
  gelatinous animals, counting from
      submersibles **32**: 9
  in benthic studies **26**: 171
  macroplankton
    AseevSamyshev trawl **32**: 7, 8
    BR nets **32**: 7–9
    IsaacsKidd trawl **32**: 7, 8
  mesoplankton
    BR Juday closing net **32**: 5, 7
    JOM net **32**: 6, 7
    150-litre bottle **32**: 4, 5
  precision of **26**: 195
  preservation of samples **33**: 10, 11
  sieving **26**: 174
  size and statistical power **26**: 213, 214
  timing of in benthic studies **26**: 177, 178
Sampling analysis **23**: 47–53
  ordination, *see* Ordination
  quadrat, *see* Quadrats
  random **23**: 47
  sizes **23**: 47
  systematic **23**: 47
Sampling Southern Ocean sea ice *Plate 2*
      43: **184–196**
  in situ **43**: 186–192
    see also core sampling
  laboratory **43**: 192–196
Sampling strategies, mismatched scale
      in **44**: 234–236
Sampling systems, fish eggs and larvae **28**:
      6–23
  alternatives to nets
    acoustic backscattering
        characteristics **28**: 15, 16
    optical imaging **28**: 16, 17
  Continuous Plankton Recorder (CPR)
      Surveys **28**: 129
  flowmeters **28**: 9
  grab samplers **28**: 9
  Gulf III net **28**: 32
  Longhurst–Hardy plankton recorder **28**:
      12, 13, 14
  nets and trawls
    BIONESS **28**: 12
    Bongo **28**: 11, 12, 32
    CalVET **28**: 10, 11, **12**
    Isaac–Kidd midwater trawls **28**: 12
    LOCHNESS net **28**: 12
  visual stimuli **28**: 8, 9
Sampling systems, vertical migration
      behaviour, design of
      programme **28**: 114–117
Sand crabs **29**: 149–153, *151*, 187, 197,
      198
Sand inundation **23**: 134, 135
Sandeel
  in match/mismatch hypothesis **26**: *270*
  vertical migration of **26**: 121, 133
Sandeel populations **34**: 263, *263*
Sandeels **29**: 253
Sandy environment, invertebrates **25**:
      183–186, 234
Santa Barbara, California, USA **44**: 227
Saponins **21**: 151–168
  potency indexes **21**: **165**
Sarcophine **21**: *126*
Sardine **36**: 211
  adaptation **36**: 48, 55
  life cycles **36**: 107, 118, 129, 137
  substance and energy **36**: 190, 196
Sardine,
  binocularity in **20**: 131
  Californian **20**: 4, 5, 8, *35*, 64, 131, 144
  Indian oil **20**: 35, 36
  Japanese **20**: 4, 5, 8, 36, 140, 258
  Japanese scaled **20**: 146
  Japanese spotline **20**: 140
  Marquesan **20**: 143
  Pacific **20**: 8, 10, 20, 24, 61, **68**
  Peruvian **20**: 10, *65*, 90, 91
  scaled **20**: 43, **44**, 78
  South African **20**: 34, 36
  spawning peak in **20**: 10
Sardines, recruitment in California **26**: 260,
      261
Sargasso Sea **29**: 44; **36**: 196
Saroma Lake **20**: 310–340 *passim*
Satellite oceanographic measurements
      23: 174–179, 270, 271
  Heat Capacity Mapping Mission **23**: 166
  historical development **23**: 177
Satellites *Plate 1* **43**: 184, 185, 196
  see also remote sensing
Satiation, predators **25**: 36, 37, 53
Saturation of parasite niches **43**: 24–33
Saury and saury pike **36**: 84, **85**, 101, 152
Saxitoxin **21**: 75–88, 140

Scale
  awareness of **44**: 241
  mismatching **44**: 235
  operational measures of **44**: 241
Scaled sardine **20**: 43, **44**, 78
Scales, sole **29**: 220
Scallop culture areas, in Japan **20**: 311
Scallop culture, competition in **20**: 366
  equipment costs in **20**: 363
  Hiogi **20**: 325
  history of **20**: 313–315
  Hotate gai **20**: 310, 322, 323
  Itaya gai **20**: 325
  parasites **20**: 366, 367
  predation **20**: 365, 366
Scallop larval development **20**: 324
Scallop marketing **20**: 374–379
Scallop production **20**: 370–374
Scallop seed, transport of **20**: 338–341
Scallop veliger **20**: *328*, 329, *335*
Scallop, biology of **20**: 322–326
Scandinavia **42**: 26, 40
Scanning (electronic) **43**: 185, 196
  see also electron microscopy
Scanning Electron Microscope (SEM) **33**: 12
Scanning electron microscope **29**: 180
Scaphognathite hormonal control **27**: 201
Scaphopoda **42**: 137–236
  see also (in taxonomic index) *Antalis;*
    Cadulus; Dentalium; Pulsellum;
    Rhabdus
  associations, species-pair **42**: 210–216,
    211, 212
  bathymetric distribution **42**: 204–207, *206*
  biogeography **42**: 204–210, *206, 208,*
    209
  circulation **42**: 180–188, *183, 187*
  development **42**: 196–204, *198, 201–203*
  diversity through geologic
      time **42**: 216–219, *218*
  excretion **42**: 188–190
  feeding *see under* digestive system
  gas exchange **42**: 156, 159, 161
  latitudinal diversity
      gradients **42**: 207–210, *208, 209*
  locomotion and burrowing **42**: 161–163
  mantle cavity **42**: 153, 155–158, *157,*
    158, 160
  morphology **42**: 139–141, *140, 141*
  phylogeny **42**: 141–145, *142, 143, 144*
  reproduction **42**: 194–196

  sensation and innervation **42**: 190–194,
      195
  shells *see under* shells
SCAR (Scientific Committee on Antarctic
      Research) **43**: 174
Scavengers **34**: 283–290
  Bullia, see Feeding
  feeding preferences **34**: 288, 289
  high-energy beaches **25:** 180
Scavenging, sea ice **43**: 197, 198, 202, 203
Schiff test **29: 154, 156**
Schizogony reproduction
  peridinians **25:** 134
  Sporozoa **25:** 144
School group **20**: 105
School shape **20**: *105*
  amoeboid **20**: 103
  discoid **20**: 103
  elongated **20**: 103
  spherical **20**: 103
School, composition of **20**: 102, 103
  development of **20**: 100–102
  loss of polarization in **20**: 130
  mixed **20**: 108, *130*
  structure and density of **20**: 95–97
School, definition **30**: 158
Schooling **27**: 376, 377
Schooling, and refuge behaviour, fish
      larvae **28**: 123–124
Schools **33: 427**, 429
Sciaenidae, as predators **27: 363**
Scientific Committee on Antarctic
      Research **43**: 174
Scientific literature **33**: 5–8
  citation frequency **33: 7**
Scope for growth (SFG) **34**: 41, 42
Scophthalmidae **29**: 219
Scorpaena sp **36**: 44
Scorpion fish
  adaptation
      food and feeding **36**: 47
      oxygen level **36**: *31, 33, 35*
      pressure and buoyancy **36**: 44, 45
      temperature **36**: 7, *11*, 34
  adaptation strategies
      energy metabolism **36: 60**, 62, **64**, 66,
        67, 68, 69, 70, 72
      plastic metabolism **36**: 75, 76, **77**, 78,
        86
  ecological principles **36**: 232, 241
  life cycles **36**: 105, *115*, 116
  substance and energy **36**: 148, 151, 167

Scotia Sea (Antarctic) **43**: 216
Scotland **42**: 26
Scotland, coastal areas of **43**: 67
Scottish herring **20**: 15
Screen size in sampling methods **26**: 174
Scyllaridae **29**: 132, 160–165, **162, 163**
Scypbomedusae **44**: 29
Scyphozoa
　parasites and epibionts of **25**: 147, 152,
　　153
　predators **25**: 7, 16, 43
　symbionts **25**: 124, 125
SDF (slow-death factor) **21**: 87
SDS *see under* sodium
Sea anemones
　amino acid metabolism **22**: 71, 72
　amino acid uptake
　　localization **22**: 70, 71, 87
　　systems of **22**: 71, 72
　as prey **22**: 92, 93
　diet **22**: 93
　digestion
　　acontia **22**: 81
　　cnidae **22**: 81
　　cnidoglandular tracts **22**: 82, 87
　　endocytosis **22**: 78, 79
　　endodermal current **22**: 74, 75
　　excretion **22**: 89
　　extracellular **22**: 82–89
　　　acid polysaccharide secretion **22**: 86
　　　amylase **22**: 83, 86, 87
　　　chymotrypsin **22**: 83, 84
　　　protease **22**: 82–85
　　of bacteria **22**: 79, 80
　　of microalgae **22**: 79, 80
　　phagocytosis **22**: 74, 76–79, 88
　　pharynx, ciliary current **22**: 72–74
　　tentacles **22**: 81
　feeding behaviour
　　activation by amino acids **22**: 67, 68
　　chemoreceptors **22**: 69
　　conduction system **22**: 69
　　control of **22**: 69
　　feeding response **22**: 66, 67
　　prefeeding response **22**: 66–69, 81
　symbiosis
　　photosynthetic metabolites **22**: 91, 92
　　zoochlorellae **22**: 90, 91; **22**: 89–92
Sea anemones, toxins **21**: 104, **105,** 121
Sea bream *see* Bream
Sea cow, biotic influence **23**: 84
sea cucumber **36**: 79; **41**: **129–223**

see also anatomy; distribution of sea
　cucumber; habitat; juvenile
see also Holothuria scabra versicolor in
　taxonomic index
associations with other species **41**: 149,
　150, 151, 188, *189*
biochemistry **41**: 185–187, *186*
biotoxicity **41**: 187
burrowing, daily cycle of **41**: **176–178,**
　179
colour of **41**: *145,* 146
commercial exploitation *see*aquaculture;
　fisheries
densities **41**: 152, 153
development *see* life cycle
　diseases **41**: 188
feeding behaviour **41**: 156, 157, **181–185**
geographic range **41**: 134–139
habitat **41**: 149–151
movement and tagging **41**: 143, 155,
　156
phenotype and morphometrics **41**: *138,*
　139
physiology **41**: 185
population genetics **41**: 157, 158
predators **41**: 187
reproduction *see* reproductive cycle;
　spawning size **41**: 153, 154,
　158, 159
subspecies, possible **41**: 146–148, *147,*
　148
systematics **41**: 133, 134
Sea Fish (Conservation) Act 1967 **29**: 331,
　332
Sea Fisheries Acts **29**: 324, 332
Sea floor *see* benthos
Sea ice *see* Southern Ocean sea ice
Sea lions, Steller **37**: 198
Sea otters
　communities dominated by **23**: 78–81
　　historical development **23**: 78, 79
　foraging depth **23**: 79
　sea urchin–seaweed relationship
　　23: 78–80
Sea perch **36**: 77
Sea scorpion *see* Scorpion fish
Sea surface temperature (SST) **31**: 227, 228
　seasonality of **31**: 236, *236*
Sea urchins **29**: 253; **37**: 7, 130, 131
　disease, effects on seaweed density
　　23: 20, 21
　effects of mortality **23**: 91

effects of removal **23**: 76, 80–82
grazing **23**: 20, 21, 75, 90, 91
  food preference ranking **23**: 112, 113
  kelp destruction, *see* Kelp
  on microflora **23**: 95
  reef **23**: 102, 103, 105
  seaweed interactions in response to **23**: 84, 85
predators of **23**: 75–77, 84–86, 91–96
response to **23**: 75, 76, 94, 95
Seabird energetics **20**: 229
Seabird, Shetland **20**: *234*
Seabirds. *See* Birds
Seagrasses **44**: 242, 264
Seal **20**: 274, 275, 278
  crabeater **20**: 271, 273
  fur **20**: 252
Sealion
  Californian **20**: 246
  Steller's **20**: 246
Seamounts **30**: 305–350; **35**: 25, 26
  abundance of plankton over **30**: 315
  accumulation of fish and macroplankton **30**: 313
  allele frequencies between populations **30**: 330
  allozyme electrophoresis studies **35**: **42**, 43
  ATP concentrations **30**: 312
  benthic biology of **30**: 317–326
  benthos sampling **30**: 317, 318
  biology of hard substrata **30**: 318–320
  biology of soft substrata **30**: 320–322
  biomass of planktonic organisms **30**: 315
  commercial exploitation **30**: 331–339, *333*
  concentration of fish species around **30**: 313–315
  current–topography interactions **30**: 313
  dependent populations **30**: 329
  diurnally migrating plankton **30**: *314*, 317
  effects on ocean circulation **30**: 310
  endemism in fauna **30**: 327
  exploitation of geological and physical resources **30**: 339
  fish population feeding theory **30**: *314*
  fisheries **30**: 311–315, 331–335, *333*
  geographic affinities of organisms **30**: 326, 327

geology **30**: 308–311
hydrothermal vent
  communities **30**: 323–326
interaction with ocean currents **30**: 310
intraspecific variation in populations **30**: 330
locations **30**: *307*
micronektonic and nektonic communities over **30**: 316
morphological differentiation in populations **30**: 330
oceanography **30**: 308–311
orange roughy **30**: 336–338
overexploitation of species **30**: 339
pelagic armourhead **30**: 335, 336
pelagic communities over **30**: 315–317
pelagic ecosystems **30**: 311–317
penetrating oxygen minimum layers **30**: 323
precious corals **30**: 338
primary production **30**: 311–315
profiles **30**: 308
recruitment mechanism **30**: 329–331, *332*
reproductive and genetic isolation **30**: 327–331
review **30**: 306–308
shapes **30**: 308, *309*
speciation incidence **30**: 328
species diversity **30**: 326–331
Searching behaviour, in larva **20**: 30–32
Seashore, *see* Shoreline
Seasonal cycles **30**: 179–181
  bacterioplankton **29**: 92–97, *93*, *95*, *96*, 98
  phytoplankton **29**: 4, *5*, 22–25, *23*, *24*, 42–47, *43*, **43**, *46*, *47*, 54
    Baltic Sea **29**: 87, **89**, 90–92, **92**, **118**
Seasonal development, sole **29**: 246–249, 259, 263, 268, 269
Seasonal variations
  see also Annual cycles
  fatty acids **36**: *17*
  motor activity **36**: 65
  spawning *see* Reproduction
Seasonality
  coral reef organisms **43**: 286
  Southern Ocean sea ice **43**: 183, 185, 198, 211
    climate and weather **43**: 176, 179, 180, 286

ice formation and melting **43**: 175–180, *177*

ozone depletion **43**: 235–240

Seasons, shifts in species boundaries **23**: 132, 133

biotically affected **23**: 133

die-backs of seaweed **23**: 133

Seawater chemistry, *see* Bourrelets

Seawater quality and bivalves **37**: 41–43, **44**, 77, 136

seawater recipe **41**: **4–9**

see also concentration of elements; internal economy

chemical form and biological availability **41**: 6, 7

Seaweed dermatitis **21**: 88

Seaweeds (general features) **23**: 1–145; *see also* Algae

associations, *see* Zurich Montpelier classification

biomass/densities **23**: 3

regulatory components **23**: 4

boundaries, *see* Zonation

canopy species **23**: 75, 82–84

competition between **23**: 75

effects of removal **23**: 75, 76, 82–84

community ecology, *see* Community ecology

crustose **23**: 22

depth distribution **23**: 109

foliose **23**: 22, 96

midshore absences **23**: 98, 99

frontal accumulations **23**: 186

numbers, *see* Abundance

population ecology, *see* Population ecology

shelter provision **23**: 4

shoreline, physiological depth competence **23**: 136, 137

Sebastolobus sp **36**: 44

Secondary production and Southern Ocean sea ice **43**: **207–225**

community beneath **43**: 216–221

grazers within **43**: 207–216

Secretion of Scaphopoda **42**: 149, 189, 190

Secretions, crustaceans **29**: 149, 150, 152, 153, 175, 176, 179

Sediment habitat factors in benthic studies **26**: 227

Sedimentary environments, bioremediation **39**: 29, 30

Sedimentation

coral reef response to **22**: 8, 10, 15, 16

Sedimentation, phytoplankton **29**: 3, 22, 35–41, *39*, **40**, *41*, 42, 49, *49*, 60

Baltic Sea **29**: 77, 97, 104

Sediments and bivalve bioassays **37**: 3, 77–80, 128–127, 132, *133*

Seep sites **33**: 502, 503

Seeps **34**: 382–384, *383,* 392–394; **35**: 22–24, 115

allozyme electrophoresis studies **35**: 30, 43

Segmentation development **33**: 24–27, *25*, **27**

Seine skiff **20**: 99

Seismic surveying **35**: 99

Selection factor, harvesting **29**: 305

Selection of alleles at enzyme loci **35**: 63, 90

Selection *see* evolution

'Selective tidal transport' **29**: 248

Selectivity, predation **25**: 15

selenium **41**: *12, 13, 15, 81, 82, 121*

Self-fertilization, Chaetognatha **29**: 196

Semelparity **20**: 22

Semelparity *see* broadcast spawning

Seminal fluid

Brachyura **29**: 134–136–138, **138**, **139**, 142–144

Macrura **29**: 160, 171

Semi-permanent mounts **33**: 12

Senecioylcholine **21**: 177

Sense organs

Protobranch bivalves **42**: 16, 17

Scaphopoda **42**: 190–194

Senses **30**: 199–203

Sensillae **33**: 29, 30

Sensitivity **30**: 241–251, *242*, 285

field studies **30**: 247–251

laboratory studies **30**: 244–247

Sensitivity of bivalves **37**: 131, 132

see also bivalve embryos and larvae

Sensitivity, larvae, predation **25**: 14

Sensory mechanisms

see also Chemosensory detection, Mechano-reception, Vision

Cephalopoda **25**: 87, *95*

Donax and *Bullia* **25**: 192, 193, 236, 238

predation **25**: 10, 13, 15, 16, 25

Sensory perception

in Antarctic fishes **24**: 373–375

in pycnogonids **24**: 20, 21

of siphonophores

chemosensitivity **24**: 174
light-sensitivity **24**: 173, 174, 227
touch-sensitivity **24**: 173
"Sensory pore complex" **27**: *17*, 194
Sensory receptors of parasites **43**: 41–44,
        42–43
Sentinel species **39**: 13
Sentry effect **20**: 108
Separability of spectra of coral reef
    bleaching **43**: 306–308
    chlorophyll **43**: 306, 307, 309, 310
    fluorescence **43**: 308, 309
    organisms **43**: 279, **303–310**, *305*
Sepiids, shell **25: 87**
Sepiodea, as prey **27**: 359, 360
Sepioidea, embryonic development **25**: 87,
        88, 101
Sepiolid squids, embryonic
        development **25: 87**
Sepiolinae **44**: 171
Sequencing studies **35**: 6, 49, 50
sequestration of trace metals by
        phytoplankton **41**: 28–32, *29*
Sergestidae **27**: 115, 117, 118, 193
Serial spawning **20**: 11
Serotonin and meiosis **43**: 103, 104
Serpentine swimming, larvae **25**: 14
Serranidae, as predators **27: 364**
Sesquiterpene **21**: 100
Sestonophages (suspension-feeders), abyssal
        macrobenthos **32**: 467–470
Setae **27**: *16* ; **33**: 28–30
    filtering setae **33**: 142, 143, 148
    replacement at moulting **33**: 33–35, **35**
    spinal and setal formula **33**: *21*, 21
"Settle and stay" hypothesis **44**: 227
Settlement of bivalves **37**: 23, 25
    and metals **37**: 88, 95
    and toxicity **37**: 72, 95, 97–107
settlement of sea cucumber **41**: 174, 175
Settling velocities, phytoplankton **29**: **9**, 13,
        13, 36, 37
Seven Islands Reserve **20**: 228
Severn Estuary **29**: 223, *223*, 227
    sole **29**: 237, 243, 244, 256, 261, 349, 350
Severn, River **29**: 226, 240
Sevin toxicity and bivalves **37**: **104**, **107**
Sewage
    coral reef response to **22**: 11, 16, 19
Sewage sludge, deep-sea dumping of **35**: 6,
        93
Sewage sludge, pollution **29**: 350

sex allocation theory **38**: 57, 123–133
Sex cells *see* gametogenesis
sex change, mechanism of, in
        Pandalus **38**: 133–135
sex determination in *Pandalus* **38**: 135–139
Sex differences, sole **29**: 256, 257, **257**, 258,
        258, 259, **259**, 260, 263
    harvesting **29**: 301, 302, **301**, *302*,
        305–309, *306–309*, **308**
Sex pheromones **30**: 188, 189
Sex ratio **33**: 363–366
    sampling error **33**: 349
    seasonal variation **33**: 364, *365*
Sex ratio changes in hosts **43**: 45
sex ratio of sea cucumber **41**: 158
Sex ratios in Atlantic halibut **26**: 23–25,
        29–31
Sex ratios of fish populations **34**: 254–256
Sexual differences
    copepodids **33**: 27, 28
    development of **33**: 43, 44
    integument **33**: 30
sexual dimorphism of sea
        cucumbers **41**: 158, 160
Sexual maturation of bivalve embryos and
        larvae **37**: 19–21
Sexual maturity in Atlantic halibut **26**: 7, *8*
Shad **20**: 124–126, 133
    American **20**: 21, 22, 24, *92*, 131, 160
    gizzard **20**: 24, 35, 140
    high spawning frequency in **20**: 13
    threadfish **20**: 34, 41
'Shadow response', *Donax* **25**: 193
Shag **20**: 282, **285**, 286, 293, 294
Shannon–Weiner H′ **26**: 204, 207
Shark **36**: 36, 46, 78, 103
Sharks, as predators **27**: 366, 367
Shear rate, turbulence **29**: 15, **15**, 38, *39,40*
Shearwater **20**: 231
    Manx **20**: 282, **285**, 294
    sooty **20**: 60, *61*
Sheepshead minnow **36**: 23
Shelf break fronts, *see* Celtic Sea Shelf Break
Shelf-break front **29**: 54, 58
Shelikof Strait, pollock in **37**: 179, *180*, 183,
        184, 193
    population structure **37**: 203, *226*
        genetic **37**: *219*, 221, 222
        mechanisms **37**: 229, 231, 236
Shell size of bivalves **37**: *63*
Shelled Opisthobranchs **42**: 67–136

characters and evolutionary
      pathways **42**: 81–107
component taxa and presumed
      synapomorphies **42**: 73–81, *74,
      76–80*
   see also (in taxonomic index)
      Anaspidea; Architectibranchia;
      Cephalaspidea; Notaspidea;
      Sacoglossa
digestive system **42**: 96–100, 106
external anatomy **42**: 91–94
locomotion **42**: 105, 106
mantle cavity **42**: 94–96
modern cladists **42**: 111–113
nervous system **42**: 103–105
operculum **42**: 91
reproductive system **42**: 100–103,106,
      107
shells **42**: 82–91, *83–85*
   Odhner and **42**: 111
   Pilsbry and **42**: 68, 107–110, 129
   synthesis attempt **42**: 113–117, *114,
      116*
   taxonomic history and phylogenetics
      42: *89,* 107–117, *109, 114, 116,*
      129–136
   Thiele and **42**: 110, 111
   'typical' *see Acteon* in taxonomic index
Shells
   see also shelled Opisthobranchs
   breakage and repair **42**: 152, 281–286,
      282–284
   Pleurotomarioidean gastropods
      42: 239–241, *240,* 243–249, *244,
      247*
   Protobranch bivalves **42**: 4–9, *6*
   Scaphopoda *Plates 1 and 2* **42**: 146–154,
      148
      decollation **42**: 149, *150, 151*
      larval **42**: 200–204, *201, 202, 203*
      microstructure **42**: 152, 153, *154*
      modification **42**: 149–152, *150, 151*
Sherwood number **29**: 12, 13, **13**, 14, 15, **14,
      15**
Shetland **29**: 217; **20**: 281
   seabird colonies in **20**: 287
Shetland Islands **34**: 271
Shoal, definition **30**: 158
Shock methods of quantifying
      fecundity **43**: *94–95,* 101–103,
      102, 151
Shore rockling **36**: 116

Shorebirds
   see also Birds
   and shell-fisheries interactions **34**: *272*
   mortality or emigration **34**: *274*
Shoreline communities
   discontinuities **23**: 56
   experimental transplants **23**: 135–137
      lower to upper shore **23**: 136
      upper to mid shore **23**: 136
   gradients **23**: 68, 69
   intertidal **23**: 72–77, 86–108, 111,
      114–117, 127–142, 143–145
      animal dominated **23**: 99
      effect of uplifting **23**: 135
      foodweb **23**: 73
      numerical classification **23**: 49, 54–57
      reefs **23**: 106, 107
      relationship to sewage effluent **23**: 54
   submergence/emergence **23**: 131, 132,
      143–145
   rocky, taxonomy and associations of
      23: 59
   zonation **23**: 17, 18, 26, 59, 96
Shoreline habitats **39**: 43–70
Shoreline, *see also* Tidal levels; Tidal zone
      communities, *see* Shoreline
      communities submersion
      histories **23**: 129–132
Shrimp **36**: 36; **20**: 52
Shrimp, predator **25**: 33
Shrimps **29**: 252
   see also Penaeidae shrimps
Shrimps of hydrothermal vent
      sites **32**: 107–135
   feeding strategies **32**: 111–115
   microscale distribution **32**: 115, 116
   reproduction and life-history **32**: 116–118
   taxonomy **32**: 109, 110
Shrimps, hydrothermal vent dwelling
      23: 330, 331
Shrinkage **20**: 82, 83
Shumagin Island, pollock near **37**: 179, *180,*
      224, 238
Siamese fighting fish **36**: 99
Sicyoniidae **27**: 3, 33, 60, 115, 117, 118,251;
      29: 132, 169, 171
Sigmoidal development **33**: 319–321
Signy Island (South Orkney) **43**: 186
silicon **41**: *12,* 41, 42, *122, 124*
   biochemical function **41**: *11, 12*
   cycling ratio **41**: *84–86*
   indices of fractionation **41**: *81, 82*

interactions with trace metals **41**: *57,* 58, 63, 64, *68,* 78, *124, 125, 128*
Sillaginidae, as predators **27**: **364**
Silver carp **36**: 12, 227
Silver Pits **29**: 249
Silver salmon **36**: 133
Silver toxicity and bivalves **37**: 3, 132
    tests **37**: 87, **90–95**, 116, 123–125
Silverside **36**: 115, 183, 184, 190
Simensky–Cossins hypothesis **24**: 368
Similarity correlation coefficients **23**: 50–52
Similarity index, percent (parasites) **43**: 27
Similarity indices in benthic studies **26**: 215, 216
Simpson's rule **29**: 337
Simulation of sea ice **43**: 192–196
Singapore, sea cucumber in **41**: *189,* 190
Single habitat type studies **44**: 233, 234
'Single process', growth rates **25**: 54
Single scale approach **44**: 233
Single-locus DNA fingerprinting **35**: 46
Single-stranded conformational
    polymorphism (SSCP) **35**: 48, 49
Sinking rate
    body density and **33**: 409, 414, **415–417**
    faecal pellets **33**: 199–203, *202*
Sinking, phytoplankton **29**: 8, 9, **9**, 12, 13, **13**, 36, 40, 46
Sinuous feeding posture **20**: 27
Sinus, *Donax* and *Bullia* **25**: 200, 201, *200,* 202, *202,* 224
Siphon, *Donax* and *Bullia* **25**: *182,* 193, 196, *200,* 204, 207, 214, *215*
Siphonophora
    parasites of **25**: 138, 157
        crustacean **25**: 153, 154, 155, *155,* 156, 158
        dinoflagellate **25**: 127, 129, 131, *132*
    predators **25**: 5, 12, 30, 41, 42, 45
    symbionts **25**: 124, 125
Siphonophores **44**: 30
Siphonophores, role in plankton
    communities **32**: 64–68
Siphoxanthin and coral reef
    organisms **43**: 282, *283*
Sipunculids, parasitic **25**: 146
Site restriction of parasites **43**: 23, 24
Site-attachment **44**: 227
Size **33**: 5, *5*; **26**: 23–25, *30,* 31–33
    and egg quality in fish **26**: 74, 75
    and larvae **26**: 78, 79
    and ontogenetic variation in fish **26**: 149

body length **33**: 327–333, *329,* **331**, *332*
    Donax and *Bullia* **25**: 182, *182,* 189, 236
    egg **33**: 274–276, **275, 276,** *277,* **278,** *279,* **280,** *280*
    larvae, effect on predation **25**: 5, 9, 11, 15, 16, *17, 18,* 20, 21, *22,* 23, 25
    of embryo of Atlantic halibut **26**: 11
    reared vs wild stocks **38**: 35, 36
    size frequency **33**: 354
    sole **29**: 218, 220, 221, **222,** 254, 255
        harvesting **29**: 299, 300, 303–311, *306–309,* **308**
        spawning **29**: 272, *273, 274*
    spermatophores **29**: 134, 137
    survival and **38**: 37, 38, 39
    variation **33**: 36
    volume **33**: 235, 236
size advantage model **38**: 128
Size of eggs *see under* eggs of invertebrates
Size specificity **30**: 271, 278
Size, body *see* body size
Size, juvenile emigration,
Skagerrak **29**: *7, 33, 51,* 52, 54, *55,* 57
Skagerrak, Norwegian Sea
    cod releases **38**: 14, 19
    fisheries **38**: 221, 222
    Pandalus borealis stocks in **38**: 207, 210
Skeletal pigments in coral reef
    organisms **43**: 290–293, *291*
Skua **20**: 282
    Arctic **20**: 282, **285,** 286, 293, 294
    great **20**: 236, *238,* 282, **285,** 286, 293, 294
Slit receptors of Scaphopoda **42**: 191, 192
Slope, beach, macrofaunal distribution **25**: 183, 187, 189, 190, 236
Smalls **29**: 223, *223*
Smelt **36**: 12, 101, 103, 224
Smelt, vertical migration of **26**: 120, 139, 140, 143
Smooth hammer-head shark **36**: 78
Snailfish **36**: 62
Snails, predator **25**: 43, 47
Snails, *see* Gastropods
Snake Pit site, MAR **32**: 104, 105, **126–129**
Snake Pit vent field **35**: 88, 89
Snapper **36**: 107
Snapper, red **37**: 220
Snoek **20**: 252; **36**: 23, 24
Snow loading **43**: 183, 184
Social aggregation
    definition **30**: 158
    in pelagic invertebrates **30**: 155–216

Social systems, origin and maintenance
of **30**: 163
Sockeye salmon **36**: 133
adaptation **36**: 25, 55, 80
substance and energy **36**: 171, 196
sodium **41**: *11, 12, 120, 186*
sole **29**: 298
spermatophores **29**: 137, **139**
Sodium toxicity and bivalves **37**: 83, **100**
SDS (sodium dodecyl sulphate) **109, 110,**
115
Soft tissues of Protobranch
bivalves **42**: 9–17, *10, 12*
Sokolow's organ **24**: 21
Solbrig's components of plant population
ecology **23**: 4
Sole **36**: 28, 46, 62, 190
egg quality **26**: 74
vertical migration of **26**: 141, 142
sole rearing in UK (1957–1967) **38**: 14–16
Sole, yellowfin **37**: 218
Soleidae **29**: 219–221
Solenidae, burrowing **25**: 235
Solenoceridae **27**: 3, 59, 115, 117, 118
Soleoidei **29**: 219
Solomon Islands, sea cucumber in **41**: 132,
136
anatomy **41**: 143, *145,* 147
associations **41**: *189*
burrowing cycle **41**: 178–181
diseases **41**: 188
distribution and population structure
41: 149, *151, 152,* 153, 155,
157, 158
feeding **41**: 182
fisheries and aquaculture **41**: 191, 192,
195, 198
migration **41**: 175, 176
predators **41**: 187
reproduction and spawning **41**: 158, 163,
164, 165, 167, 169, 170
settlement **41**: 174, 175
Somatic growth, molluscs **25**: 232, 234
Sonar **44**: 263
Sonar *see* hydroacoustic
Sonar, sector-swimming **20**: 98
Sooty shearwater **20**: 60, *61*
Sorensen's coefficient **23**: 51, 52
Sound scattering **24**: 165
South Africa **20**: 247–259
beach environment **25**: 183–186

South Africa, coastal areas of:
parasites **43**: *10, 12,* 70
South African anchovy **20**: 35, 36, 240, 247
South African pilchard **20**: 5, 247
biomass of **20**: *256,* 256
eggs of **20**: 15
pelagic spawning of **20**: 24
South African sardine **20**: 34, 36
South America, coastal areas of **43**: 149
parasites **43**: *11,* 34, *50,* 51
South Georgia (Antarctic) **43**: 143, 148,
241–243
South Georgia **20**: 278, 293, 295
South Pacific **35**: 5
blue grenadier fishery **35**: 107–109
South Sandwich Trench, hadal
fauna **32**: 369, 372
South-east Asia **42**: 39, 272, *273,* 274,
284, 285
Southeast coast of England, fisheries **38**: 221
South-east Indian Ridge **35**: 17
Southern Atlantic Ocean: parasites **43**: *11,*
34, *50, 61*
Southern Bight **29**: 251, 272, 275
Southern Explorer Ridge **35**: 89
Southern North Sea herring **20**: 21
Southern Ocean **20**: 269–279
Southern Ocean and Antarctic area, trace
elements in **41**: 43
and phosphates **41**: *126*
cadmium **41**: *73, 75, 126*
copper and silicon **41**: *124*
iron **41**: 43, *45,* 50, 51, *52,* 60
Southern Ocean sea ice **43**: 172–276
see also extent and thickness; physical
environment; trophic processes;
variability in ecosystem
export of biogenic material **43**: 231–235
Southern Ocean, polar and subpolar
regions **32**: 24–29
Southern Pacific Ocean **43**: 102
parasites **43**: *10, 12,* 29, 37, *57, 58,* 69
Southwest Atlantic fisheries **39**: 271, 272,
277, 286, **266, 275**
Sowing culture, of scallops **20**: *315,* 316,
355–358, 361, 373
Space (as a consumable resource) **23**: 125,
126
supply by disturbance **23**: 125–127
Spain, coastal areas of **43**: *12,* 68, 69
Sparidae, as predators **27**: **364**
Sparrow, house **20**: 231

Spat **20**: 310, 313, 323
Spat collection **20**: 330–338
  Japanese oyster culture **21**: 13
  Miyagi **21**: *20*
  sites, Hiroshima Bay **21**: *15*
Spat collector bag **20**: 334
Spat settlement patterns, *Donax* and
      Bullia **25**: 237
Spat transport method **20**: **339**
Spatial dispersion patterns in benthic
      studies **26**: *176*
Spatial distribution **33**: 423–434
  see also Geographical distribution
  aggregations **33**: 424, 425, **426**, 428
  multispecies assemblages **33**: 429–432,
      430, 431
  schools **33**: **427**, 429
  swarms **33**: **427**, 428, 429
Spatial distribution and sampling
      design **26**: 193–197
Spatial hierarchies of structure **44**: 242
Spatial hierarchy **44**: 241–245
Spatial history **44**: 250
Spatial pattern, quantifying **44**: 266–268
Spatial statistics **44**: 260
Spatial technologies **44**: 271, 272
Spatio-temporal scale **44**: 242
Spawner-recruit relationship of
      pollock **37**: *202*, 203, 238, 239
Spawning
  bivalves **37**: 136
    see also broodstock; eggs; fertilization;
      reproduction
    ageing of gametes after **37**: 32, 33
    and bioassay methodology **37**: 46–49
    behaviour **37**: 20, 21
    counting eggs **37**: 49
    impurities eliminated **37**: 48
    induction of **37**: **29**, 30, 46, 47
    laboratory rearing **37**: 27, 30
    number of parents **37**: 47, 48
    stripping gonad **37**: 47, 135, 136
  pollock **37**: *180*, 183, 184, 203, 204
Spawning **27**: 289, 290, 291, 292, 293, 294,
      296, 297; **33**: 266–288, **268**; **20**:
      23–25
  (see also Life history)
  and food **26**: 262, 263, 284
  and upwelling **26**: 262, 263, 284
  crepuscular **20**: 22
  demersal **20**: 22, 23, 24, 25
  egg laying **33**: 266–268

  frequency of **20**: 11–13
  induced, and broodstock management of
      fish **26**: 102, 103
  migration **27**: 397, 314
  multiple **20**: 8, 11, 13, 17
  nocturnal **20**: 22
  of Atlantic halibut **26**: 9
    areas **26**: 10, 11
    distributions of species **26**: 264, *270*
    time of **26**: 9, *10*
  pelagic **20**: 24, 25
  resting eggs in sediments **33**: 268–272,
      269–271
  spread over time **26**: 282, 283
  time of
    and plankton production **26**: 258
    in temperate waters **26**: 251–255
spawning of sea cucumber **41**: **168–170**
  see also reproductive cycle
  artificially induced **41**: 170
  behaviour **41**: 168
  environmental factors and
      timing **41**: 168–170
  periodicity **41**: *164*
Spawning *see* broadcast spawning
Spawning *see* Reproduction
Spawning species **39**: 52–57, *53*, *56*
Spawning stock biomass (SSB) **29**: 315,
      316, *316*, 317, 318, 322, 349
  status of stock **29**: 334–336, 339, 340,
      343–347, *344–347*
Spawning time, population dynamics **25**: 2
Spawning, sole **29**: 218, 231–236,
      232–235, 351
  see also Fecundity
  behaviour **29**: 253, 267, 268
  ground **29**: 241, 242, 246, 248, 250, 251,
      251
  seasonal development and
      time **29**: 268–275, *271*, *273*, *274*
  weight **29**: 263–265, *264*, *265*
Spearman rank correlation **26**: 204, 231
"Specialists" (parasites) **43**: 35
Specialization
  and ecological niches **36**: 59
Speciation *see* evolution
Speciation, deep-sea fauna **35**: 71, 116
  role of depth **35**: 82
Species abundance curve **26**: 201, 202, 205
"Species alias problem" *see* synonymy
Species area curve **23**: 47, 48
Species area curves **26**: 202

Species centricity, ecological principle
    of **36**: 237
Species distribution **27**: 140–148, **142–147**
Species diversity **33**: 461–464, **462, 463**
Species diversity, deep-sea
    environment **35**: 5
Species dominance **33**: 464–466, *466, 467*
Species dominance, population
    dynamics **25:** 2
Species groups, discreteness of **26**: 188
Species identification *see* Identification
Species of bivalves **37**: 10–16
    see also taxonomic index
    listed **37**: **8–10**
    type and toxicity **37**: 118
Species pair **20**: 170, 173, 193, 295
Species replacement **34**: 276–283, *277,*
    278
Species richness and diversity **26**: 197, 198,
    200, 201, 203, *204*, 205
Species richness and parasites **43**: 33
    and latitudinal gradients **43**: 11–14, 17,
    19, 34, 47–51
    free-living organisms, problems in
        estimating **43**: 4–7
    marine, problems in estimating **43**: 8–15
    packing **43**: 31, 32
Species status of Atlantic halibut **26**: 2–4
Species, with popular names **20**: **6**
Species-pair *see* associations
Species-specificity **30**: 279–281
Specificity
    hosts of parasites **43**: *5*, 22, 23, 52, 53
    invertebrate life histories **43**: 112, *113*
Specificity indices **24**: 271–273
Specificity, ecological principles
    of **36**: 230–232
Spectra of coral reef organisms *see*
    reflectance; separability
Spectral sensitivity curve **20**: *139*, 140
Speleonectidae **29**: **131**
Sperm, crustaceans **29**: 129, 130
    see also Spermatophores
    cord **29**: 184, *185*
    plug **29**: 135, 142
Spermatheca
    Copepoda **29**: 175, *177*, 178, 180
    Decapoda **29**: 141–144, 174, 200
    euphausiids **29**: 184
Spermathecae **33**: 41, 43
Spermatocytes, parasite infection **25**: 122
Spermatogenesis **33**: 45

Spermatogenesis *see* gametogenesis
Spermatophores **29**: 129, 130, **131–133**,
    183–186, *185*, 197–200; **33**: 43,
    44–46, *45*, 254, 261, *264*
    see also Anomura, Brachyura, Copepoda,
        Macrura
    artificial insemination **29**: 189–193, **192**
    attachment **33**: *42*, 43, 261–263
        number attached **33**: 261, 262
        placement position **33**: 262, 263, *263*
    compared to other
        invertebrates **29**: 195–197
    cryopreservation **29**: 187–189, **188**
    hardening **29**: 186, 187
    pathology **29**: 193–195
    production of **33**: 264–266
    transfer **33**: *258*, 258–260
Spermatophores, Cephalopoda **25:** 90
Spermatozoa **33**: 44–46, 260
Spermatozoa, sole **29**: 269
Spermatozoids, Cephalopoda **25:** 90, 91,
    109
Spermiogenesis, pycnogonid **24**: 35, 36
Spermiotoxicity test and bivalves **37**: 50, 51
Spherule, crustaceans **29**: 178
Spinal and setal formula **33**: *21*, 21
Spines, sea urchin **21**: *149*
Spiny dogfish **36**: 46, 116
Spiny lobster **36**: 39
Spiny lobsters **29**: **132**, 160–165, **162, 163**,
    167, 168, *167*, 186, 197, 198
Spiny lobsters and plankton
    production **26**: *278*, 279
Spiralian blastulation, molluscs **25:** 108
Spitometoids, Polymastigotes **25:** 124
Splanchnotrophidae, parasitic **25:** 153
Sponge **36**: 83, 235
Sponge fisherman's disease **21**: 96
Sponge poisoning **21**: 96, **97, 101**
Sponges, symbionts of **25:** 120
Spongocoels, reproductive behaviour of
    Peracarida **39**: 153
Spongothymidine **21**: 102
Spongouridine **21**: 102
Spores and gametes, seaweed
    dispersal, *see* Dispersal
    motility **23**: 27
    production **23**: 23–27
    time spent in state **23**: 28
Sporogenesis, peridinians **25:** 126
Sporozoa, parasitic **25:** 139–144, *140, 142,*
    158

Spot, vertical migration of **26**: 121, 141
spotted seatrout (Gulf of Mexico) **38**: 27, 28
Sprat **36**: 226; **20**: 55, 280
    adaptation **36**: 76, **77**, 78
        food and feeding **36**: 48, 49, 53, *56*
        temperature **36**: 17, 18
    ecological principles **36**: 238, 239, 242,
        246, 248
    growth of **20**: 11
    indicators of condition **36**: 217, 218
        lipids **36**: 207–209, 210, *211*
    life cycles **36**: 101, 118, 129, 132
        annual **36**: 107, *108, 109*
    substance and energy **36**: 199, 200
        accumulation and production **36**: **145**,
            145–162 *passim*, *146*, *150*, **155**,
            156, *158*, **160**, **161**
        budget **36**: *197, 198*
        consumption, transformation and
            utilization **36**: **179**, 179–196
            passim, *181*, **185**, *188*, **192**, **193**
        expenditure **36**: *164*, **164**, 164–174
            passim, **166**, **168**, *169*, *173*, **174**
Spring bloom *see* Bloom conditions
Squid **36**: 103, 211, 231; **20**: 54
    adaptation **36**: 31, 32, 36, 37, 39, 44, 83
    substance and energy **36**: 170, 176
Squids, gonatid **32**: 243–324
    biomass
        distribution **32**: 299–302
        production and food
            consumption **32**: 308–310
        total in North Pacific **32**: 244
    data sources **32**: 248
    ecological divergence, size and
        distribution **32**: 289–295
    ecological role **32**: 302–310, **309**
        competitors **32**: 306
        parasites **32**: 306–308
        predators **32**: 304–306
        prey **32**: 302–304
    gelatinous degradation **32**: 295, 296
    individuality and niche
        divergence **32**: 296–299
    phylogenetic relationships **32**: *298*
    ranges, maps **32**: *246, 251*
    species accounts
        Berryteuthis anonychus **32**: 268–270
        Berryteuthis magister **32**: 249–268
        Gonatopsis japonicus **32**: 275–278
        Gonatopsis kamtschaticus **32**: 282, 283
        Gonatopsis magister **32**: 270–275

Gonatopsis onyx **32**: 280–282
Gonatus spp. **32**: 284–289
systematics **32**: 248, 249
Sri Lanka, sea cucumber in **41**: 132, 136,
    151
    fisheries and aquaculture **41**: 190, 191,
        193
    reproduction and spawning **41**: 163, *164*
S-shaped response, prey **25:** 17
St Bride's Bay **29**: 326
St George's Channel **29**: *223*, 224, 228, 324,
    325
St Gowan Light Vessel **29**: 224
St Kilda **20**: 280, 296
St Peter's fish **36**: 34
Stability in aquatic communities **26**: 199
Stability, fisheries **29**: 300, 321, 322
Stability–Time hypothesis, and diversity in
    communities **26**: 199–202
Stable ocean hypothesis **31**: 196
Stage duration hypothesis **31**: 171
Stagnant water **29**: 2
    see also Pelagic food webs
stained prawn disease (SPD) **38**: 179, 180
Stains **33**: 12
Stalk *see* Peduncle
Stalks, peridinians **25:** 134, 135
Standard dose **21**: 65
Standard safety dose **21**: 65
Standard technique, for scallop culture
    20: 310
Standardization of data **26**: 184
standing stock (*Pandalus borealis*)
    38: 208–210
Starfish
    grazing **23**: 73, 84
    removal **23**: 73–75
Start-feeding period of Atlantic
    halibut **26**: 19
Startle response **20**: *56*, 57, 94
Starvation **30**: 256–259, *258*, 283, 285;
    20: 49, 77
    in adult clupeoids **20**: 58–62
Starvation index **30**: 272
Starvation tolerance **33**: 182, **183**
Starvation, and match/mismatch
    hypothesis **26**: 283, 284
Starvation, larval mortality **25:** 2, 4
    compared predation **25:** 42, 49, 50, *49*,
        54, 55, *55*, 57, 58, 61, 67
Statistical inference in benthic
    studies **26**: 212–214

Statistical methods and bivalves **37**: 84–86
  dose-response curves **37**: 82–84
  minimum sample size **37**: 85, 86
  qualitative assessment of environmental
    samples **37**: 84, 85
Statistical power in benthic studies **26**: 213,
    214
Statocysts **27**: 11, 222, 223
Statocysts, Cephalopoda **25**: 87, *95*, 105
Statoliths **44**: 166
Status, sole *see* Stock
Ste. Anne Reserve **34**: *250*
Steam-trawl fishing **29**: 325, 327, 330,
    344–345
Steedman's fluid **33**: 10
Steelhead salmon **36**: 84, 85
Steelhead trout **36**: 222
Steinhaus' coefficient in benthic
    studies **26**: 215
Stenothermy **24**: 334, 335
Stereometry **29**: 282
Stereophotography
  coral reef assay **22**: 4
sternal spines (*Pandalus*) **38**: 63, 64
Steroidtoxins **21**: 67
Sterols, echinoderms **21**: 151
Stert Flats, Bridgwater Bay **29**: 240
Stickleback **36**: 24, 58, 196
Sting ray **36**: 116
Stochastic model of life history and
        fecundity of invertebrates **43**: 117,
        118, 140, 152
  stock delimitation in *P. borealis* **38**: 207,
    208
Stock, sole **29**: 249–251, *251*, 265, 266
  biomass **29**: 265, 266, 298
  future **29**: 333–335
  harvesting **29**: 299, *300*, 316–318, *318*,
    322
  reproduction **29**: 280
  status **29**: 217, 218, 334, 335, 347, 348,
        351, 352
    comparison of assessment
        methods **29**: 339–342, *341*
    egg-production based
        estimates **29**: 336–339, *339*
stock: recruitment relationship **38**: 7
stocking, definition **38**: 5
Stocks of fish, protecting **36**: 246, 247
Stocks of pollock **37**: 179, 181, **185**, *186*,
    192
Stoke's law **29**: 8, 9, **9**, 36

Stomach contents analysis **33**: 149–151,
    151, 152
Stomach contents, predation **25**: 39, 40, 41
Stomach, sole **29**: 252
Stomato-pharyngeal system, peridinians
    25: 126
Stomatopoda, spermatophores **29**: **131**, 184,
    185
"Stomodaeal apparatus" **27**: 22, 23
Storage of fish eggs, *see* Overripening and
    storage
Storage, spermatophore **29**: 130, 198–200
  Anomura **29**: 146
  Brachyura **29**: 135, 137, 139, 141–145,
    198, 200
  Copepoda **29**: 180–183
  Macrura **29**: 168, 169
Storm, phytoplankton **29**: 27–29, *28*, 50
Strategies of adaptation **36**: 59–87
Stratification
  see also Pelagic food webs
  Baltic Sea **29**: 97
  Bristol Channel **29**: 231, 232, *234*
  critical values **23**: 180–182
  h/u³ criterion **23**: 180
  parameters **23**: 180, 181, 182
  phytoplankton **29**: 2, 3
  shear in areas of **23**: 189
  spring tide effects **23**: 182, 183
  thermal, *see* Thermal fronts
Stress **36**: 4, 25, 80, 118, 120, 234, 235,
    250
Stress and Atlantic halibut **26**: 53, 54,57,
    58
Stress, *Donax* and *Bullia* **25**: 197, 207, 235
Stress, environmental
  corals, *see* Corals, stress
  marine bivalve molluscs, *see* Bivalve
    molluscs, stress
striped bass (Chesapeake Bay) **38**: 28, 29
Striped bass **36**: 26, *92*
Striped greenling **36**: 62
Stripping
  of Atlantic halibut in aquaculture
    26: 54
  overripening and storage of fish
    eggs **26**: *91*, 92, 93
"stripping", gonad **37**: 47, 135, 136
strontium **41**: *11, 12, 122*; **44**: 258
Structural functional homeostasis
    and energy optimum **36**: 243–245
Structure *see* morphology

Sturgeon **36**: 214; **38**: 5
  adaptation **36**: 37, 39, 71
  ecological principles **36**: 247, 249
  life cycles **36**: 90, 103, 129
Stylet, peridinians **25:** 131
Styliform prolongations, Amphipoda **25:** 156
Sub-carangiform swimming **20**: 88
Subertine **21**: 99
Sub-genital segment, crustaceans **29**: 175
Subintegumental glands **33**: 29, 30
Submarine ridges **32**: 145–241
Submersibles
  Argus **32**: 8
  Mir 1 and 2 **32**: *8*
Sub-populations of pollock **37**: 205, 238
Subradular organ, *Bullia* **25**: 193
Substance and energy (mainly Black Sea)
      36: 2, 139–203, 238
  accumulation and production **36**: 142–162
  budget **36**: 196–199
  consumption, transformation and
      utilization **36**: 174–196
  defined **36**: 139–142
  ecological metabolism **36**: 201–203
  expenditure **36**: 162–174
  trophic significance **36**: 199, 201
substance and energy **36**: *197, 198*, 199
  accumulation and production **36**: 143,
      147, 148, 150, 154–162
  accumulation and production **36**: 143,
      147–162 *passim*, **155, 156, 160,
      161**
  consumption, transformation and
      utilization **36**: **179**, 179–196
      passim, *181*, **185**, *188*, **192, 193**
  expenditure **36**: *164*, **164**, 164–174
      passim, **166**, *169, 173*, **174**
    trophic significance **36**: *199*, 200, 201
  passim, **155, 156**, *158*, **160–161**
  passim, **166**, *169, 173*, **174**
  passim, *181*, **185**, *188*, **192, 193**
Substratum
  barnacle shape and **22**: 209–211
Substratum, and behaviour **27**: 341–349
  characteristics of **27**: 341, 342
  distribution in mud **27**: *344*
  experimental studies on **27**: 345–347
  organic content **27**: 345
  particle size **27**: 341, 342, **342**
    determination **27**: 342
  penaeid distribution **27**: 342–345
  preference for sediments **27**: 347–349

Subtidal zone **39**: 34–43
Subtidal zones, *see* Tidal zone
Subtropical life history strategies **33**: 382,
      383
Succession
  algal, *see* Algae
  models **23**: 114, 120, 121
  plankton **23**: 212
Succession, phytoplankton **29**: 87
Succinate dehydrogenase,
      crustaceans **29**: 137, 143
Sucking device
  Ciliophora **25**: 137, 139
  Ellobiopsidea **25**: 135, 137
  peridinians **25:** 135
Sucrose proline, crustaceans **29**: **188**
Sucrose, picoplankton **29**: *108*, 109, 110,
      110, 116, 117
Suez Canal **43**: *58*
Sugars, crustaceans **29**: 137, **138**
Sulfur oxidizing bacteria **23**: 316, 317, 325,
      340
Sulphated acid mucopolysaccharide
      (sAMPS), crustaceans **29**: 136
sulphur **41**: *11, 12, 13, 15, 120*
Summer heating **23**: 169
Sunda Trench **42**: 205
Sunda-Arafura Shelves, fisheries **39**: 268,
      269, 276, **266, 275**
Sunfish **36**: 55
Superblooms **31**: 353, *354*
Surface area, excretion **29**: 32
Surface convergence **23**: 175, 184, 185,
      277
  ecosystem maturity at sites of **23**: 230,
      231
  evidence **23**: 185–189
  biological **23**: 185, 186, 256
Surface nutrient depletion **23**: 192–194
Surfactants, *see* detergents and oil
Surfing species, *Donax* and *Bullia* **25:** *181*,
      182, 189, 191, 197, 198–200, 207,
      208, 235
  energetics **25:** 212, **213**, 218
  tidal migration **25:** 187, *188*
Surimi **37**: 182, 183
Surugatoxin **21**: 178
survival
  after release **38**: 36, 37
  of *Pandalus* larvae **38**: 164–166
Survival curves **33**: 367, 367, 368, *368*
  see also Mortality

Survival rates
and tests of match/mismatch
hypothesis **26**: 264, 265
of larvae of Atlantic halibut **26**: *60*
Survival time, marine venoms **21**: 65
Survivorship, seaweed *see also* Mortality
rates
age specific **23**: 23, 25
curves **23**: 5
drought **23**: 139, 140
juvenile **23**: 12
lateral branch of marked shoots **23**: 8, 9
life tables calculated from **23**: 7
long life span problems in studying **23**: 6
primary shoot **23**: 8–9
Susceptibility, predation **25**: 16–22, *17*, *18*,
20, 22
Suspended particulate matter (SPM) **24**: 414,
419
Suspension feeding, *see* Feeding
Swansea Bay, sole **29**: *223*, 227, 237, 245,
295, 330
early fisheries **29**: 323, 326, 328
future **29**: 349, 350
Swarm, definition **30**: 158
Swarms **33**: **427**, 428, 429
multispecies swarms **33**: 429–432, **430,**
431
Swim bladder, sole **29**: 230
Swimbladder **20**: 113, 115, 118, 155–160
development of **20**: 158–160
gases in **20**: 157
Swimming **27**: 17, 179, 222
bacterial infections **27**: 382,
in Antarctic fishes **24**: 368–371
labriform **24**: 368, 370
subcarangiform **24**: 368–370
in pycnogonids **24**: 31, 32
in siphonophores **24**: 164, 177–182
and feeding **24**: 223–225, 229
speeds **27**: 313
temperature **27**: 349–351
tides and water currents **27**: 339, 340
Swimming activity **33**: 401–423
energetic cost **33**: 421–423
escape reaction **33**: 414, **418**, 419
hydrodynamic features **33**: *422*, 422, 423
rhythmic activity **33**: 420, 421
sinking rate, body density and **33**: 409,
414, **415–417**
swimming pattern **33**: 402–407, *404*
copepodid **33**: 403–407

nauplius **33**: 402, 403
swimming speed **33**: 407–409, **408,**
410–413
Swimming activity of bivalves **37**: 68, *69*
Swimming appendages **33**: *18*, 21,
402–406, *403*
development **33**: **27**, 27, 55
Swimming *see* Motor activity
Swimming velocity, phytoplankton **29**: 13,
14, 26
Swimming, predation
direction **25**: 12–13, 15
speed **25**: 5, 10, 11–12, *13*,14, 15, 19–20,
20, 21, 25
Swordfish, vertical migration of **26**: 124,
125, 135
Sylt island **43**: 7
Symbionts, zooplankton **25**: 120, 122, 123,
124, 125, 126, 129
Symbioses, Scaphopoda **42**: 211–214,
211, 212
Symbiosis
microalgae in sea anemones
content **22**: 90, 91
localization **22**: 89, 90
photosynthetic products **22**: 91, 92
Sympagic organisms **43**: 214, 215
see also trophic processes
Sympatric speciation **43**: 46, 143
Synchronism reproduction **20**: 11
Synchronous oogenesis **43**: 89
Synecology, pycnogonid **24**: 54–66
fouling communities **24**: 64–66
marine vegetal communities **24**: 55–60
sessile invertebrate communities
24: 60–63
soft sediment communities **24**: 63, 64
Synodontidae, as predators **27**: **365**
Synonymy of parasites **43**: 5, 8, 9
Synxenic speciation **43**: 46
Syrman goby **36**: **42**, 78, 236
Systematics, *see* Classification

TAC (Total Allowable Catch) **29**: 299, 332,
348
Tactile sense **30**: 202, 203
Tactile systems, larvae **25**: 16, 19
TAG site, Mid-Atlantic Ridge **32**: 79–81,
97, 98–100, **126–129**
TAG vent fields **35**: 88, 89
Tagging **27**: 312; **20**: 119, 185
external **20**: 123, 125

internal **20**: 120
of herring **20**: 126–128
of shad and alewife **20**: 124–126
ultrasonic **20**: 120, 125
tagging of sea cucumber **41**: 143, 155, 156
Tagging pollock **37**: 206, *214, 215*, 240
tagging techniques **38**: 40
Tagging, sole **29**: 244–246, *245, 247*, 248,
        250, 251, *251*, 296
Tail beating, larvae **25:** 14
"tail fan" **38**: 63, 64
Taiwan, Kuroshio Current **39**: 265–267,
        275, 276
Tamar Estuary **29**: 244
Tamar Estuary, England **44**: 226
Tanning, crustaceans **29**: 140, 168
Tapeworm **36**: 57
Targets, fishery **29**: 321, 322
Tarpon **20**: 34
Tasmania **35**: 84–86, 109; **43**: *10*
Tasmania, spiny lobsters in **26**: 278, 279
Taste organs **27**: 194
Taxa richness of parasites of marine
        fish **43**: 16, 17
Taxonomic history **27**: 1, 2
Taxonomic history and phylogenetics of
        shelled Opisthobranchs **42**: *89,*
        107–117, *109, 114, 116*
Taxonomic specificity, ecological principle
        of **36**: 230, 231
Taxonomy **27**: 55–126
    adults **27**: 58–106
        key to families, Penaeoidea **27**: 58–60
        key to genera, Penaeidae **27**: 61–63
        Artemesia **27**: 61, 63, 64, *64*
        Atypopenaeus **27**: 61, 64–66, *65*
        diagnostic larval features **27**: **116**
        divergence **27**: 150, 152
        Funchalia **27**: 61, 66–68, *67*
        Heteropenaeus **27**: 61, 68, 69, *69*
        juveniles, species identification **27**: 126
        larvae and postlarvae,
            key to families, Penaeoidea
                27: 115–118
            key to genera, Penaeidae **27**: 119–123
            species descriptions **27**: **108–112, 113**
            species identification **27**: 123–126
        Macropetasma **27**: 61, 70, *70*
        Metapenaeopsis **27**: 61, 70–78, *71*
        Metapenaeus **27**: 61, 79–83, *80*
        Parapenaeopsis **27**: 61, 83–87
        Parapenaeus **27**: 61, 87–90, *88*

Pelagopenaeus **27**: 61, 91
Penaeopsis **27**: 61, 91–93, *92*
Penaeus **27**: 16, 93–98, *94*
    subgenera **27**: 94, 95
Protrachypene **27**: 61, 99
Tanypenaeus **27**: 61, 99, 100, *100*
Trachypenaeopsis **27**: 61, 100–102, *101*
Trachypenaeus **27**: 61, 102–105, *103*
tribes **27**: 150
Xiphopenaeus **27**: 61, 106, *106*
zoogeographic regions **27**: 55–57
taxonomy of *Pandalus* **38**: 58–60
Taylor column **32**: 155
Taylor column circulation **28**: 46
Taylor columns **35**: 25
Taylor's power law **26**: 184
    and variance **26**: 194, 195
TBT (tributyl-tin) toxicity and bivalves
        37: 3, 4, 13, 116, 132, *133*
    salinity and **37**: 123
    tests **37**: 96, **98–100**
        larval physiology **37**: 66, 70, 72, 73
TCA (Tricarboxylic acid) cycle,
        crustaceans **29**: 143, 144
TCC (trichlorocarbanilide) toxicity and
        bivalves **37**: 97, **100**
TCP (tetrachlorophenol) toxicity and
        bivalves **37**: 97, **105**
Technology for using marine fish
        resources **36**: 251–253
Tegumental glands, crustaceans **29**: 140
Telemetry **44**: 217, 222, 223, 251, 252, 258,
        259, 272
Teleost, diel vertical migrations of
        26: 115–168
Teleplanic reproductive strategy **35**: 65–67,
        79, 115
Tellinacea **25:** 186
Tellinacea, growth performance **35**:
        167–171, **173**, *174*
Temperate life history strategies **33**: 383,
        384
Temperature
    and bivalves
        incubation **37**: 52, 53
        laboratory rearing **37**: 33, 34, 36
        spawning **37**: 13, 16, 20, 29, 31, 46
        toxicity reactions **37**: 118, 119, 120,
            122
    and pollock **37**: 190, 192, 203, 204
Temperature **36**: 210, 227
    see also Annual cycle

see also climate; latitude
adaptation **36**: 8–19
  collagen **36**: 15
  contractile proteins **36**: 14, 15
  enzymes and metabolism **36**: 8–14
  glycoproteins and other
     substances **36**: 18, 19
  lipids **36**: 15–18
  oxygen level **36**: 8, 10–12
and diel vertical migration **26**: 125
and fecundity of invertebrates **43**: 99,
    101, 102
  thermal shock method of
     quantifying **43**: *94*, 101, 102
and food density, effects on growth rates,
    fish larvae **28**: 106–111
and incubation of Atlantic halibut **26**: 12
and parasites **43**: 46, 48–51, *49, 50*
  host ranges **43**: *53*
as environmental factor in vertical
    migration **26**: 145–147
barnacle growth and **22**: 212, 215, 235
bivalve mollusc heart rate and
    22: 130–132
burrowing **25**: 208
coral reefs and **22**: 11, 14–16, 26
deep-sea environments **35**: 8–11
development time and **33**: 298, 299, 344
distribution of *Paridalus* and **38**: 118–121
ecological principles **36**: 229, 230, 232,
    233, 236, 238, 239, 241, 242, 244
effects on growth rates, fish larvae **28**:
    104, 105
egg production and **33**: 289, 290
enzyme thermal adaptations **35**: 28
food and feeding **36**: 107, 108, 126
general effects **36**: 7, 8
growth performance and **35**: 165, 166,
    168, 169
growth, in *Pandalus* **38**: 201–203
hatching behaviour in *Pandalus*
    and **38**: 153–156
hatching Cephalopoda **25**: 93, 105, *106*
hydrothermal vents **35**: 8, 10, 17, 18, 28
life cycles **36**: 106, 107, 118, 119, 126,
    128, 130, *131*
mollusc distribution **25**: 184, 185, 189,
    196, 197, 237
osmoregulation and **33**: 215
plankton **29**: 4, 6, *6*, 25, *23*, *24*, 45
  Baltic Sea **29**: *75*, 76, 91, 97, 98,
    102–104, 106, 110

predation **25**: 11, 20, 27, 30
preference **36**: 19
respiration **25**: 216, 219–221, *220*, **220**,
    237
responses to **33**: 216
size relationship **33**: 329, 330, **331**
sole **29**: 217, 218, 221
  adults **29**: 248, 249
  Bristol Channel **29**: 224, 225, *225*
  eggs and larvae **29**: 230–232, *234*
  fecundity **29**: 290
  feeding **29**: 254
  global warming **29**: 350
  growth **29**: 255, 256
  juvenile **29**: 244, 246
  mortality **29**: 298, 299
  spawning **29**: 269–272, *271*, 336, 338
Southern Ocean sea ice **43**: *177*, 178, 179,
    181, 182, 184, 199–201
weight relationship **33**: 335, **337**
Temperature adaptation **36**: 7–19
Temperature effects and plant
    competition **23**: 15, 17
Temperature influence on growth rates in
    copepods **44**: 7
Temperature tolerance **24**: 376
  see also Antifreeze, biological
  and metabolic rate **24**: 340, 341, 344,
    345
  and oxygen affinity in blood **24**: 356
  enzymes and **24**: 357–360
  neural function and **24**: 361–368
  proteins and **24**: 359, 360
Temperature, behaviour **27**: 349–351
  juvenile emigration **27**: 308
Temperature, in growth of larva **20**: 77
Temperature-dependent
    development **44**: 251
Temperature-independence, of *zeitgeber* in
    fish migration **26**: 118
Temperatures
  Hiroshima Bay **21**: *4*
  Matsushima Bay **21**: 12
Temporal sampling design **26**: 177, 178
Tenby **29**: 324
Ten-pounder **20**: 34
Tentacle extracts, *Chironex* **21**: **118**
Tentacles, Cephalopoda **25**: 99, 100, *100*
TEP (transparent exopolymer
    particles) **43**: 234, 235
Teraponidae, as predators **27**: **365**

Terebra cinerea
  burrowing performance
    burial mechanism **28**: 410, 411
    comparisons **28**: *417*
  digging cycles **28**: 405
Terebratalia transversa
  development **28**: 275, **280**, 281
  ecological tolerance **28**: 337, 340, 342,
      345
  feeding **28**: 300, 306, 308, 309, 312
  metabolic pathways **28**: 325, 327, 328
  reproduction **28**: 254, 258, 265–267, 272
  structure
    digestive system **28**: 231
    lophophore **28**: 218, 223, 225
    mantle **28**: 181, 185, **186**, **187**, 188, 200
    muscular system **28**: 237, **239**, 240,
        241, 242, 243
    nervous system **28**: 245–247
    pedicle **28**: 203, 205
    respiratory system **28**: 316, 318, *319*
    sensory system **28**: 283, 285–288
Terebratulina retusa
  amoebomocytes **28**: 215
  caeca **28**: 199
  coelomocytes **28**: *213*
  development **28**: 283
  digestive system **28**: 227, **228**, 231
  epithelia **28**: 205
  genetics **28**: 347–349, 351
  life history **28**: 333, 341, 344–346
  lophopore **28**: 217, 223
  mantle
    contents, variation **28**: 198
    dorsal and ventral views **28**: **184**
    membrane **28**: **191**
    outer **28**: 193
    seasonal changes **28**: 193
    section **28**: **190**
    volume data **28**: *357*
  muscular system **28**: 236, 237, 243
  nerves **28**: 246, 247
  pedicle **28**: 200, 203, 211, 335
  physiology and nutrition **28**: 286–288,
      299, 309, 310, 313–316, 318, *319*,
      321, *324*, 326
  principal organs **28**: **181**
  reproduction **28**: 248, 249, 251–273, **252**,
      275–279, 282
  rootlets **28**: 208
  sensory organs **28**: 248

Terminal/posterior globules of
    parasites **43**: 41–44, *42, 43*
Tern **20**: 360
  Arctic **20**: 236, *238*, **285**, 286, 293, 294
Terra Nova Bay **43**: 180
Territoriality and distribution of
    species **26**: 193, 194
Tertiary **42**: 217
Testes, sole **29**: 267, 269
Testis **33**: 44, 46
Tethys Sea **43**: 53
Tetrachlorophenol, *see* TCP
Tetrapyrrole **43**: *291*
Tetrodotoxin **21**: 82, 88, 141, 186
Teuthoidea, embryonic development **25**: 88,
    101
Thailand, sea cucumber in **41**: 136, *137*
Thalia sp **36**: 36
Thaliacea **44**: 3, 10, 21–24, 28, 30, 32–35,
    37, 42, 132–136
  numbers of species, data points and
      ranges **44**: 11
Thames Estuary **29**: 270
Thelycum
  artificial insemination **29**: 190, 191, **192**
Theory of multispecies harvesting
    regime **20**: 280
Thermal fronts (transects) **23**: 169, 170, 175
  annual variation, models
      explaining **23**: 183, 184
  measurement
    airborne **23**: 176
    improvements **23**: 170
  mixed with haline fronts **23**: 172, 174, 175
  stratification and homogeneity at **23**: 169,
      170, 172–174
  annual cycle **23**: 170
Thermal fronts in tidal seas **26**: 254, 257, 258
Thermal shock method of quantifying
    fecundity **43**: *94*, 101, 102
Thermal stratification, Bristol Channel
    29: 224, 225, 231, 232, *234*
Thermal tolerance **33**: 216
Thermal tolerance, *Donax* and *Bullia* **25**:
    196, 197
Thermocline responses **33**: 443, 444
Thermocline(s) 187; *see also* Stratification
  depth **23**: 180, 203
  levels
    chlorophyll maximums **23**: 236, 251,
        257, 258
    oscillations **23**: 264–267

model, one-dimensional **23**: 180
plankton cycles and summer **23**: 214, 215
  establishment/breakdown **23**: 215
shelf break **23**: 199, 203
time–space pattern **23**: 200, 203
Thermocline, Baltic Sea **29**: 76, 90
[H^3]Thymidine incorporation into cold
  TCA precipitate (TTI) **29**: 86, 105,
  106, *107*, *108*, *110*, 111, 113–115,
  115, 116
Thermoclines, and vertical
  migration **26**: 146, 147, 148
Thermohaline fronts **23**: 172, 174, 175
Thickness of ice *see* extent and thickness
Thiele, Johannes **42**: 110, 111
Thinning
  alien- (interspecific) **23**: 10, 12–23
  self- (intraspecific) **23**: 10–22
  law **23**: 10
thorium **41**: *12*, 44, 45
threadfin (Hawaii) **38**: 29, 30
Three-spined stickleback **36**: 196
Thresher shark **36**: 78
Thyroid activity **36**: 9, 86
Tidal
  barrage, Severn Estuary **29**: 349, 350
  fronts, plankton **29**: 4, *6*, 50, 52–54, *53*,
    58
Tidal circulation, dispersal processes **28**:
    46–51
Tidal currents **23**: 169
Tidal energy, mollusc distribution **25**: 180,
    183, 184
Tidal fronts **23**: 167–185
  biological characteristics **23**: 210–276
  bottom water near **23**: 189–197
  cold masses, *see* Bourrelets
  circulation patterns, *see* Circulation
    patterns
  eddies, *see* Eddies
  European shelf, *see* European shelf
  general features **23**: 167–179
  model construction **23**: 179–185
    one-dimensional **23**: 179–184, 278
    significant variables **23**: 180
    summer establishment **23**: 179–182
    three-dimensional **23**: 185, 189, 190
    time changes **23**: 182
    two-dimensional **23**: 184, 185
  particle exchange **23**: 179
  stratification, *see* Stratification
  thermal transects, see Thermal fronts

water exchange **23**: 179–184, 243, 277
  importance **23**: 267
Tidal fronts, dispersal processes **28**: 46–51
Tidal levels
  barnacle growth and **22**: 212, 213, 216,
    234–236
Tidal migrations, *Donax* and *Bullia*
    25: 187–189, *188*, 196, 207, 208,
    212, 213, **213**, 235, 236
Tidal movements **44**: 219–222
Tidal stream transport, selective **28**: 61–67
Tidal streaming **23**: 199, 200
  velocity (u) **23**: 180
Tidal zone
  communities, *see* Shoreline communities
    in pools, interactions **23**: 88–99
  sub- **23**: 46, 54, 77–86, 96–99
    numerical classification **23**: 49–51, 54
    physiochemical determinants **23**:
      141–143
    reefs **23**: 106
  Swinbanks classification **23**: 129, 130
Tide levels
  actual and predicted **23**: 129–131
  critical, hypothesis **23**: 127–132
    biological zones coinciding with
      23: 129
    Colman's **23**: 129
    Doty's **23**: 129
    refutation **23**: 128, 129
  emersion
    and submersion **23**: 131, 132, 143–145
    curves **23**: 128
    physiological effects **23**: 141
  neap- **23**: 137, 138
Tide(s)
  coefficient **23**: 183, 189, 194, 196
  internal, *see* Internal waves
  levels, *see* Tide levels
  neap **23**: 137, 138
    -spring frontal cycle **23**: 182, 183, 241,
      267, 268, 280, 281
  pools, community interactions **23**: 88–99
  red, *see* Red tides
Tides
  and offshore activity **26**: *139–141*
  and retention of fish larvae **26**: 139, 140
  as *zeitgeber* **26**: 137–143
  intensity of **26**: 138, 139
  stream transport and vertical
    migration **26**: 142

Tides, and behaviour **27**: 338, 339
Tides, as cue for diel vertical migration
    33: 451, 452
Tides, Bristol Channel **29**: 227
    sole migration **29**: 236, 242, 248, 351
Tiger prawn **29**: 191
Tilapia **36**: 20
Tilman's theory **23**: 89, 117, 118, 122–125,
    126
Time
    and diversity in communities **26**: 198, 199
    of plankton production **26**: 255, 258, 259
    of spawning
        in temperate waters **26**: 251–255
        of Atlantic halibut **26**: 9, 10
Time response **30**: 251, *253, 255, 258*, 285
Time series analysis in benthic
        studies **26**: 231, 232
Time, 'macroscale' and 'microscale'
    36: 232–234
Time-energy budgets, *Donax* and *Bullia*
    25: 211–214, *213*, 236
Time-space diagram **44**: 214
Timing, spawning, sole **29**: 269–275, *271,
    273, 274*
Tintinnids **20**: 27, *46*
Tintinnids, *see* Ciliophora
Tissue pigments in coral reef
        organisms **43**: 287–290
Toad goby **36**: 78, 86
tolerance measurement **23**: 138
    photosynthetic recovery, *see*
        Photosynthesis
Toluidine blue test **29**: 155, **156**
Tonga Trench, hadal fauna **32**: 369
Toothfish **36**: 78
"top-down" approach to density-dependent
        effects **38**: 40, 45
'Torbay' sole **29**: 261
Tortonian **42**: 217
Total Allowable Catch (TAC) **38**: 4
Total allowable catches (TAC) **29**: 299, 332,
    348
Total particulate carbon, phytoplankton
    29: 45, *46*
Total spawning reproduction **20**: 11
Toxaphene toxicity and bivalves **37**: 97, **103**
Toxic
    bloom **29**: 79
    flagellates **29**: 21
Toxic metals

bivalve mollusc stress and **22**: 104, 129,
    130, 132, 133
    copper, *see* Copper
Toxicity **21**: 63
Toxicity *see* Pollution and toxicity
Toxicity testing **20**: 187
Toxicity tests and bivalves **37**: 88–125
    see also bioassay; pollutants
    assessment **37**: 132, 134
    exposure and observation of
        response **37**: 136
    factors affecting, *see* extrinsic; intrinsic
    interaction between toxicants
        37: 123–125
    spermiotoxicity **37**: 50, 51
    toxicity defined **37**: 123, 124
    with pure chemicals **37**: 74–76
Toxicology
    Cnidaria **21**: 112
    echinoderms **21**: 151
    Holothurioideae **21**: **159**
Toxinology **21**: 59–217
Toxins **21**: 59–217
    concentrating **21**: *86*
    crabs **21**: **141**
    isolation technique **21**: *65*
    purification, *Goniopora* **21**: *128*
    separation from *Sphaerotrichia* **21**: *93*
Trace element composition **33**: 243–245,
    244
trace metals *see* micronutrients;
        phytoplankton and trace metals
        tungsten **41**: *12, 13, 122*
Trachypeneini **27**: 150
Tracking devices **44**: 272
Tracking technology **44**: 251, 256, 257, 259,
    271
Trade-offs between life-history traits of
        invertebrates **43**: 112, *113*, 124,
        125
transfers, definition **38**: 5
Transformation of substance and energy
        see Consumption
Transmitters **44**: 252
Transparency, predation **25**: 15
Transparent exopolymer particles
        (TEP) **43**: 234, 235
Trawl fisheries **29**: 324–326
Trawls **34**: 209–213, 287
    area trawled by English North
        Seafleet **34**: *216*
    effects on epifauna **34**: 219–222, **220**

effects on infauna **34**: 213–219
sediment suspension **34**: 228, *228*
Treaty of Rome **29**: 331
Tree diagram **23**: 70, 71
   minimum-spanning **23**: 71
Tree duck, black-bellied **20**: 231
Trehalose **29**: **188**
Trellis diagram **23**: 52, 53
Trematoda, parasitic **25**: 146, 147–150, 158
trematode parasites of *Pandalus* **38**: 180
Trematodes **36**: 57
Trematodes and Atlantic halibut **26**: 38, 41
Trematomus spp **36**: 233
Trevose, sole **29**: 253, 256, 263, 342
   adults **29**: 246–248
   juveniles **29**: 240, 245
   reproduction **29**: 268, 276, 277, *276*, 278
Triacylglycerol (TAG) **30**: 275
Triacylglycerol, marker of larval
      nutrition **28**: 19
Triacylglycerols **33**: 246–248
Triassic, Pleurotomarioidean gastropods
      42: 239, *242*, 243, 266, *268*
Tributyl-tin, *see* TBT
Tricarboxylic acid cycle,
      crustaceans **29**: 143, 144
Trichiuridae, as predators **27**: **365**
Trichloroacetic acid (TCA) **29**: 86
Trichlorocarbanilide, *see* TCC
Trichocystoid filaments, Copepoda **25**: 11
Triglyceride **20**: 48
Triglycerides (TAG) **30**: 260, 280, 285
Tri-phasic life cycle **44**: 211–213
Trondheim, sole **29**: 217
Trondheimsfjord, Norway, hatchery **38**: 9
Trophic interaction **44**: 225; **34**: 257–292,
      306–308
Trophic Level Formalism **26**: 209
Trophic processes in Southern Ocean sea
      ice **43**: **196–231**
   see also secondary production
   exploitation **43**: 229–231
   microbial activity **43**: 202–207
   predation **43**: 225–229, 241, 274
   primary production **43**: 196–202
Trophic relay **44**: 225
Trophic significance of substance and
      energy **36**: 199–201
Trophomere, Ellobiopsidea **25**: 135
Trophozoites
   peridinians **25**: 129, *130*, 131
   Sporozoa **25**: 139, 143

Tropical areas
   see also latitude
   land and sea species differences **43**: 7
   species of **43**: 4–5, *47–50*
Tropical fishes, predators **25**: 4, 51, **52**, 53,
      55, 57, 64, 66
Tropical life history strategies **33**: 382, 383
Tropical oceans **32**: 30–37
   divergences and upwellings **32**: 34–37
   oligotrophic regions **32**: 30–34
   oxygen minimum layer **32**: 41–48
   tropical intermediate water (TIW)**32**: 154
Tropical Prawn Research Project **27**: 5
Tropical Shelf Fauna, Ekman's
      classification **27**: 127
Tropism, pycnogonid **24**: 32
Trout **36**: 222, 249, 250
   adaptation **36**: 9, 12, 54
      salinity **36**: 22, 23, 24, 25, 27
   egg quality **26**: 74, 81, 82, 98, 99
   egg viability **26**: 92–95, *93, 94*
   life cycles **36**: 126, 136
   substance and energy **36**: 152, 170, 190
Trypsin, crustaceans **29**: 191
Tryptophane, crustaceans **29**: 153
Tryptophanyl, crustaceans **29**: 136
Tsushima Current **39**: 267
't'-test **26**: 213
TTI *see* [^{3}H]Thymidine incorporation
Utilization of substance and energy *see*
      Consumption
Tubule, venom apparatus, *Physalia*
      21: *110, 111*
Tumours
   corals under stress **22**: 49
Tuna **36**: 176
   adaptation **36**: 17, 36, 55, 61, 71, 72, 77, 86
Tuna, blue fin **20**: 62
Tuna, spawning **26**: 262, *263*
Tunicates
   parasites of **25**: 125, 143, 154, 155, *155*,
      156, 158
   symbionts **25**: 120
Turbellaria, parasitic **25**: 147
Turbellarians **21**: 133
Turbidity and turbulence and
      bivalves **37**: 37, 80
Turbidity maximum, estuarine **24**: 413
Turbidity, and predation **27**: 376, 377
Turbidity, predation **25**: 15, 27
Turbidity, tolerance, Brachiopods **28**:
      331–332

turbot (Denmark) **38**: 23
Turbot
    adaptation **36**: 14, 15, 23, 41, 54, **77**, 78
    egg quality **26**: 74
    egg viability **26**: 92
    life cycles **36**: *99*
Turbulence **29**: 2, 3
    see also Pelagic food webs
Turbulence in feeding ecology of larval fish
    effect on encounter rates between predator
        and prey **31**: 199, 200
    effect on growth of larval fish **31**: 201
    effect on ingestion rates of
        predators **31**: 200
    effect on recruitment of larval fish **31**: 201,
        202
    field studies of effect of tubulence on
        feeding **31**: 195–199, 208, 209
    future research **31**: 210, 211
    laboratory studies of effect of tubulence on
        feeding ecology **31**: 192–195,
        207, 208
        larval fish **31**: 192–4
        other zooplankton **31**: 194, 195
    modeling turbulence-dependent feeding
        rates **31**: 182–192, 203–207
        components of the feeding
            cycle **31**: 183, 184
            predator search behavior **31**: 184–188
            predictive models **31**: 188–192
        recruitment and **31**: 196, *197*
        size-dependency and **31**: 204, *205*
        turbulence in the ocean **31**: 173–175
Turbulence theory **31**: 172, 198, 199,
        208, 210
Turbulence, predation **25**: 27
Turbulent dissipation rate **31**: 173
Turbulent kinetic energy **31**: 173
Tvarminne **29**: 79, 89, 91, **92**, 98, **99**, **100**,
        102, 104, 111, **113**
Type 2 functional response **20**: 30
Typhlosole, crustaceans **29**: 165, 167, 168,
        167, 178
Tyrosine, crustaceans **29**: 187
Tyrosyl, crustaceans **29**: 136

UK, rearing and release of plaice and sole
        (1957–1967) **38**: 14–16
Ultrafiltration, Scaphopoda **42**: 188, 189
Ultrasonic tag data **20**: 93
Ultrasonic telemetry **44**: 217, 223
Ultrasonic tracking technique **20**: 161

Ultraviolet radiation *see* light
Umea station, Baltic Sea **29**: **89**, **92**, 104
United Kingdom, halibut fisheries in **26**: 48
United States, coastal areas of
    invertebrates **43**: 125, 145
    parasites **43**: *11*, 65, 69, 70
Univariate community analysis **26**: 171, 172
uptake of trace metals *see* biological
        availability
Upwelling **33**: 482, **484, 485**; **20**: 8, 10, 36,
        45
    and fish recruitment **26**: 250, 259–261,
        284
    and Lasker events **26**: 261
    and spawning **26**: 262, 263, 284
    and survival rates **26**: 264, 265
Upwelling areas **44**: 40
Upwelling regions, phytoplankton **29**: 4, 60
Upwellings, tropical oceans **32**: 34–37
Urban and industrial effluents **37**: 127–129,
        132, *133*
    see also biocide; detergents and oil; metal;
        toxicity
Urchins, sea, *see* Sea urchins
Urea excretion, molluscs **25**: 229
Uric acid excretion, molluscs **25**: 229
Urocanylcholine **21**: 176
uropods **38**: 64
Uropods, morphology **27**: 18, *18*
    larvae **27**: **47**, 51, **53**
Urosome **33**: *15*, 16, 17, 24, 25
    musculature **33**: 38
USA, enhancement in (1885–1930) **38**: 8
use in hybridization **31**: 7
Use of GIS **39**: 293, 294
Uses and importance of parasites **43**: **62–70**
    see also under mariculture; pollution
    biological control of introduced
        pests **43**: 70
    human disease caused by **43**: 68, 69
    introduction into new habitats **43**: 69, 70
Ushant front **29**: 34
Ushant fronts **23**: 169–171, 174, 278
    chlorophyll surface content **23**: 236–238,
        257
        vertical distribution **23**: 238
    circulation patterns **23**: 174, 175,
        187–189
    cold bottom waters **23**: 190, 191
    eddies **23**: 178, 179, 244
    foam accumulation **23**: 227
    food chain structure **23**: 263

internal waves off the **23**: 207
mixing time **23**: 249
nitrate transect **23**: 193
red tides **23**: 232, 233, 241
temperature transect **23**: 193
zooplankton distribution **23**: 264, 265
Uskmouth power station **29**: 237
USSR, halibut fisheries in **26**: 48, 49
Utermöhl method **29**: 78, 81, 89
UV irradiation **33**: 217, 218

Vagina, crustaceans **29**: 141, 142
Valliculture *sensu lato* **34**: 82
Valve closure, *Donax* **25**: 235
Vampyromorpha, eggs **25**: 88
vanadium **41**: *12, 13, 122*
Variability in sea ice ecosystem, responses
    to **43**: **235–244**
  ozone depletion **43**: 235–240
Variability *see* Differentiation
Variance, Analysis of in benthic
        studies **26**: 182, 185, 213, 220
  and Taylor's power law **26**: 194, 195
  F-test **26**: 228
Variation
  see also evolution
  genetic **33**: 63, 64, 371
  geographical **43**: 9
  growth rate **33**: 316
  morphological **33**: 36, 37
Vas deferens **33**: 44
Vas deferens **29**: 193, 198
  see also Origin-spermatophores
  Decapoda **29**: 139, 150, *151*, 174
  Stomatopoda **29**: 184, *185*
vas deferens gland system **38**: 133
VDF (very fast death factor) **21**: 88
Vegetation classification **23**: 43–57
  dominance types, *see* Dominance types
  fioristic, *see* Floristic classification
  numerical **23**: 46–57
  total species composition **23**: 46
  Zurich-Montpelier principles, *see* Zurich-
        Montpelier classification
Vegetation structure **23**: 43–57
  causal factors **23**: 43
  classification, *see* Vegetation
        classification
  determinants, *see* Biotic interactions;
        Physiochemical determinants
Veil line **26**: 205, *206*
Venerupin **21**: 179

Venice **43**: 66
Venom apparatus
  animals **21**: 63
  cnidarians **21**: 109
  Conus **21**: *171*
  echinoderms **21**: 148
  molluscs **21**: 170
  octopus **21**: *172*
Venomous marine animals **21**: 59–217
Venomous marine plants **21**: 59–217
Ventral nerve cord **33**: 38, 39
Vents, hydrothermal, *see* Hydrothermal vents
Verkhova **36**: 183
Vermilion rockfish **36**: 11, 12
Vertebrae, sole **29**: 221, **222**, 230
Vertebrates
  genetic identity related to taxonomic
        divergence **35**: 69
  heterozygosity **35**: 59, 61, 62, 115
  See also Fish
  spatial patterns of genetic variation **35**:
        79, 83–87
Vertebrates, predation **25**: 6, 9, 10, 14–15,
        36, 37, 45, 47
vertical concentration profiles of
        elements **41**: 7–9, *8*, 9, 43, *44*
Vertical distribution **33**: 458, 459, *459, 460*,
        486, 487, 504, 505
Vertical distribution, predation **25**: 27
Vertical migration **33**: *489*, 439–452, *441*,
        454, 455
  cues for **33**: 446–452
    food availability **33**: 450, 451
    light **33**: 446–449, *448*
    predators **33**: 449, 450
    tides **33**: 451, 452
  diel feeding periodicities and **33**: 170,
        171, *442*
  effects of abrupt bottom topography
        33: 444
  halocline responses **33**: 443
  multispecies studies **33**: 442, 443
  ontogenetic migrations **33**: 445, **446**
  oxycline responses **33**: 444
  thermocline responses **33**: 443, 444
Vertical migration behaviour, fish eggs and
        larvae **28**: 114–123
Vertically-migrating drifter (VMD) **44**: 251
Vestibulo-ocular reflex (VOR) **24**: 373, 374
Vestimentiferans
  development and dispersal **34**: 407
  distribution **34**: 404–410, *406*

gene flow **34**: 407
hydrothermal vents **35**: 18, 19, 27
phylogenetic relationships **34**: 408–410, *409*
seeps **35**: 23
Vestimentiferans, hydrothermal vent dwelling **23**: 324–326, 336
copepod associations **23**: 329
physiology **23**: 325
prokaryotic symbionts **23**: 325
viability of reared fish **38**: 33–36
Vicariance **34**: 399–401
Videos **43**: 188
Vietnam **41**: 136, 201
Vilaine, Bay of **29**: 238, 240, 242, 244, 261, 294, 299
Violaxanthin **43**: *283*
Virtual population analysis (VPA) **29**: 218, 250, 297, 334–336, 339–342, *341*, 346, 347, *346*, 347, 352
Virtual population analysis (VPA) **34**: 305; **38**: 203; **28**: 33; **20**: 253
Viruses (see Parasites)
Viruses **29**: 77
Viruses, plankton **25**: 118, 119
Viscosity, effect on swimming speed **25**: 20
Visibility and sampling, eggs and larvae **28**: 8, **9**
'Visibility', of prey **25**: 10, 15, 25
Vision **30**: 199, 200
Vision, predation **25**: *5, 6,*7, 8
fish **25**: 9, 10, 15, 16, 25
invertebrates **25**: 13, 14
Vitamin C, *see* Ascorbic acid
Vitamin content **33**: 250, 251
Vitamins
and feed quality in fish eggs **26**: 98, 99, 101, 102
in Atlantic halibut **26**: 36, 37
vitellin **38**: 147
Vitellogenesis **24**: 35; **33**: 43
crustaceans **29**: 143, 184
sole **29**: 279–282, *281*
Vitellogenesis of invertebrates **43**: 89, 90, 129, 141, 150
Vitellogenesis, Brachiopods **28**: 259–270
Vitellogenin **29**: 269; **38**: 147
Viviparity/ovoviviparity **43**: 51, 111
Viviparous blenny **36**: 17
Vlymen's searching model **20**: 31, 32
Volent nematocysts **21**: 107
Volume **33**: 235, 236

von Bertalanffy growth equation (VBG) **38**: 193, 194
modifications **38**: 193–196
von Bertalanffy growth function (VBGF) **35**: 159

Von Bertalanffy's equation **20**: 20
von Bertalanfy, growth curves **27**: *244*
growth equation **27**: 234
applications of **27**: 241–246, *244, 245, 247*, **248–250**
Vostok, Lake (Antarctic) **43**: 246
VPA (virtual population analysis) **37**: 194
VPA *see* Virtual population analysis
Vulnerability
eggs and larvae compared **25**: 23–27, *24*, **26**
population **25**: 27–37, *28*, **31**, *32*, **34**, **35**
Vulvae, crustaceans **29**: 144, 148, 182, 183

Wadden Sea **29**: 242–244, 252, 261; **43**: 34
Wakame **20**: 320
Walking in pycnogonids **24**: 31
Walking legs, pycnogonid **24**: 5,12, 13
Walleye pollock **36**: 132; **37**: 177–253
see also fisheries; population ecology; population structure
background **37**: 182–187
life history **37**: 183–185
predator, prey interactions and ecosystem **37**: 185–187, 190, 202, 203
Walvis Bay **20**: 254, 255
Wastes, *see* urban and industrial
Water
see also Thermal fronts; Thermoclines column
capsizing **23**: 198
height **23**: 180
mixing **23**: 241, 246; *see also* Wind mixing
content in Atlantic halibut **26**: *35*
stratification as ecosystem **26**: 254, 259
surface nutrients **23**: 192–194
temperatures, hydrothermal vent **23**: 302, 303, 318;
transparency and diel vertical migration **26**: 127
upwellings **23**: 217–219
Water column
and migratory plankton species **32**: 3

vertical zonation of plankton **32**: 68–74
    schema **32**: *70*
Water column structure **29**: 2, 3
    see also Pelagic food webs
Water content **33**: 226, 227, *227*
Water currents, macrofauna distribution
        25: 185, 187, 189, 190, 236
Water density, Skagerrak **29**: 55
Water movement, response to **30**: 200, 201
Water, currents, depths and
        behaviour **27**: 339–340
Waterford Bank **29**: 325
Wave
    action
        competition in areas of **23**: 110
        effects **23**: 89
    internal, *see* Internal waves
Wave energy, mollusc distribution
        25: 180, 183, 184, 187, 189, 190,
        234
    Donax and *Bullia*, burrowing **25**: 207,
        208, 236
Wave method of quantifying
        fecundity **43**: *94*, 103
Wax esters **33**: 246, 248; **36**: 45, 46, 55
Weather and vertical migration **26**: 123
Weddell Sea (Antarctic) **43**: 233
    invertebrates **43**: 135, 143, 148, 149
    ozone depletion **43**: 236, 242, 243
    predation **43**: 225–229
    sampling **43**: 185, 187, 188
    seasonality **43**: 176, 189
    trophic processes **43**: 208, 209, 211,
        216–222
Weibull function **23**: 8, 9
Weight *see* Body weight
Weight, mortality rates **25**: 53, 54
Weight, plant
    –density relationships **23**: 10, 11
    of reproductive bodies **23**: 25–27
Weight, sole **29**: 218, *262–266*, *264*, **265**
    fecundity **29**: 283–286, *284*, *285*
    spawning **29**: 269
    yield per recruit **29**: 301–303
Weight-specific growth rates **44**: 51
    epipelagic metazoans other than
        copepods **44**: 119
    epipelagic metazoan zooplankton **44**: 8
Weih's model **20**: 97
Weirs **29**: 323
Welsh Port Books **29**: 323
Wentworth scale, particle size **27**: 341

West Bank **29**: 247
Western Atlantic
    see also Caribbean
    Pleurotomarioidean gastropods **42**: 238,
        243, 250
    ecology **42**: 272–279, *273*, *275*, *276*,
        281, *283*
    Protobranch bivalves **42**: 31, 32, 40
Western Atlantic Ocean **43**: 233
    invertebrates **43**: 145
    ozone depletion **43**: 236, 242, 243
    parasites **43**: *10, 11, 18, 19, 50, 53, 59, 60,
        61*, 69, 70
    predation **43**: 225–229
    sampling **43**: 185, 187, 188
    seasonality **43**: 176, 180
    secondary production **43**: 208, 209, 211,
        216–222
Western Pacific **42**: 26
    see also Japan; Philippines
    Pleurotomarioidean gastropods **42**: 239,
        267
    ecology **42**: 272–275, *273*, *275*,
        277–279, 281, *283*
Western Pacific Ocean, parasites **43**: *10 12,
        13, 18, 37, 38*, 53, 59, *60, 61*
Weston-super-Mare **29**: 252
Whale **36**: 103; **20**: 271–275
    baleen **20**: 271, 273, 274, 278
    biomass of **20**: 274, 275
    blue **20**: 273
    fin **20**: 273
    sei **20**: 273
    sperm **20**: 271, 278
Whale carcasses **35**: 22, 24
Whales **34**: 264, 392–394
Whaling *see under* exploitation
White band disease
    corals under stress **22**: 48
White muscle fibres **20**: 75
White perch **36**: 196
White Sea **36**: 101, 223, 226; **43**: *11, 14*
White smokers **35**: 17
Whitefish **36**: 101, 103, 231; **20**: 279
Whiting **20**: 291
    adaptation strategies
        energy metabolism **36**: **60**, 61, 64, 70,
        72
    chromosomal aberrations in **26**: 86
    in match/mismatch hypothesis **26**: *270*
    plastic metabolism **36**: **75**, **77**
Wild fish **20**: 84–86

Wind **43**: 180
  phytoplankton **29**: 4, 27, *27*, 29, 47–50,
      48, *49*, 60
  sole **29**: 227, 236, 351
Wind mixing **23**: 276
  and phytoplankton growth **23**: 221, 240,
      241
    laboratory experiments **23**: 247, 248
    interactions with internal tides **23**: 198,
        199, 200, 202–204
    surface layer **23**: 180, 181
Winemiller-Rose demographic model of life
      history and fecundity of inverte-
      brates **43**: 120–123, *121*, 152
Wolf herring **20**: 34
Wolf-fish **36**: 176
Wolfish predation on sea urchins **23**: 93–94
Woodhead drifters **29**: 229
Wood-lice **29**: 130
Woods Hole, MA, hatchery **38**: 8
World Ocean
  abyssal and hadal zones **32**: 327, 328, *348*
  zoogeographic regions **32**: *353*
  articulate and inarticulate
      brachiopods **32**: *403*
  bathyal zone proportion **32**: 390
  Exclusive Economic Zones (EEZs)**32**: 413
  main gyres **32**: *402, 534*
  phytogeographical divisions **32**: *559*
  phytoplankton **32**: *534*, 545–547
  phytoplankton distribution
      maps **32**: *530–543*
  regions, faunal similarities **32**: 336–347
  sub-littoral and bathyal regions, warm-
      water species **32**: *405*
Worms **36**: 35
Wounding, predation **25:** 21, 22
Wrasse **36**: 77

'Xanthellae', Dinoflagellata **25:** 125
Xanthidae **29**: 133
Xanthophylls and coral reef organisms
      43: *283*, 285, 286, 292
Xenophyophores **30**: 321, 322
X-organ sinus gland **38**: 133, 135
X-radiography
  coral growth **22**: 21, 22, 25

Year class abundance of pollock **37**: *199,
      200*, 201
Yeasts, plankton **25:** 121
Yellow Sea fisheries **39**: 267, 268, 276

Yellowfin and match/mismatch
      hypothesis **26**: 263
Yellowfin sole **36**: 62
Yellowfin tuna **36**: 176
Yellowfish **36**: 132
Yemen, sea cucumber in **41**: 136, *137, 151*
  Oygarden archipelago, cod release **38**: 23
Yield, sole **29**: 218, 336, 348
  see also Harvesting
Yield. See Coastal lagoon fisheries
Yolk
  crustaceans **29**: 184
  sole **29**: 229, 230, 279, 282
Yolk-sac
  Cephalopoda **25:** 93, 94, 95, *100*, 102, 103
  larvae, prey **25:** *3*, 20, 23, 25, **26**
Yolk-sac larvae of Atlantic halibut
      26: 14–18, 57–60
Yoshuku ren **21**: *22*

Zeaxanthin and coral reef organisms **43**: *283,
      291*, 292
Zeitgeber
  cyclic events as **26**: 122–145
    bioenergetic advantage **26**: 136, 137
    food **26**: 129–136
    light **26**: 122–129
    predator avoidance **26**: 143, 144
    tides **26**: 137–143
Zero net growth isocline(s) **23**: 122–125
  crossing at equilibrium points **23**: 124, 125
zinc **41**: 22, **67–71**, 88
  impact on primary production **41**: *13, 23,
      69–71, 70*
  biochemical function **41**: *11, 12*
  biological availability and uptake **41**: 26,
      27–30, 31, 32, 48
  cycling ratio **41**: *84, 85, 86*
  distribution in oceans **41**: 67–69, *68, 125*
  functions and properties **41**: *15–17,
      21–22*
  indices of fractionation **41**: *81, 82*
  inputs to system **41**: *123*
  interactions **41**: 33, 34–39, *34–36*, 61,
      64, 74–77, *78, 125*
Zinc toxicity and bivalves **37**: 45
  tests **37**: 88, 89, **90–95**, 95, 116, 123, 124
    larval physiology **37**: 66, 67, 68, 70
Zinc, bivalve molluscs
  accumulation in tissues **22**: 163
  behavioural responses to **22**: 117–119,
      165–168

detection and responses to **22**: 143–145
effects on
   gametogenesis **22**: 166, 168, 169, 172,
     173
   mitochondrial respiration **22**: 176, 177,
     180, 181
   physiology **22**: 165–168
   unfertilized eggs **22**: 173–175
mortality **22**: 122, 123T
sublethal levels **22**: 130
Zinc, crustaceans **29**: **139**
Zonation **23**: 59
lower boundary controlling factors
   competition **23**: 108, 109
   herbivory **23**: 111, 112
physiological determinants **23**: 137, 138,
   141
role of competition in structuring **23**: 108,
   109
shoreline **23**: 17, 18, 108, 109, 116, 117
upper boundary controlling factors
   competition **23**: 109
   herbivory **23**: 111, 112
Zoochlorellae
in sea anemones
   content regulation **22**: 90
   localization **22**: 90
   photosynthetic products **22**: 91
'Zoochlorellae', symbionts **25**: 124
Zoogeographic regions **27**: 55–57, **130, 131**
Eastern Atlantic **27**: **131**, *137*, 139, 140,
   146, 147
Eastern Pacific **27**: **130**, 136, *137*,**145**
Indo-West Pacific **27**: **130**, 131, 133, **142,**
   145, *156*
Western Atlantic **27**: **130, 131**, *137*, 138,
   145, 146
Zoogeographical regions, parasites as
   indicators of **43**: 55–59
Zoogeography **27**: 127–157
fossil record **27**: 147–157
isotherms, global **27**: *128*
of herring parasites **24**: 285–292
   Arctic and sub-Arctic **24**: 291
   Atlantic only **24**: 290, 291
   fresh and brackish-water **24**: 292
   Pacific only **24**: 289, 290
   temperate and tropical **24**: 292
   widespread occurrence **24**: 287, 289
of pycnogonids **24**: 6, 67–76
   deep-sea **24**: 72–75
   in plankton **24**: 75, 76

regional faunas **24**: 68–72
of siphonophores **24**: 139–151
regional distribution **27**: 127–140, 147,
   148
taxonomic literature **27**: 55–57
Zoogeography *see* Geographical distribution
Zoogeography, Protobranch
   bivalves **42**: 36–40, *37, 38*
Zooplankton **23**: 210; **36**: 31, *48, 49, 56, 57*,
   107, 108, 200
accumulation **23**: 226
   phytoplankton dependent **23**: 210, 259
and commensal species **26**: 145
and vertical migration **26**: 134–136
distribution **23**: 263–266
   Celtic Sea Shelf Break **23**: 271, 272
   frontal concentration **23**: 225, 264, 265
   vertical **23**: 255, 256, 269
in rearing experiments of Atlantic
   halibut **26**: 58, 62
micro- **23**: 229
optimal food **23**: 269
seasonal distribution of **26**: 265–269
Zooplankton *see* Mesozooplankton
Zooplankton, *see* Ectoparasitism, Endoparasitism, Parasitology, Symbionts
Zoospore density estimations **23**: 23, 24
Zooxanthellae
in corals
   loss as stress indicator **22**: 29–35, 39
     by various species **22**: 30, 31T, 32
     polyp size and **22**: 32–34
in sea anemones
   content regulation **22**: 90, 91
   localization **22**: 89, 90
   photosynthetic products **22**: 91, 92
'Zooxanthellae', symbionts **25**: 124
Zooxanthellar pigments in coral reef
   organisms **43**: 284–287, 289, 292,
   293, 303, 307, 310
Zurich–Montpelier classification of
   associations **23**: 43, 44, 57–65
departures from **23**: 60–63
Dutch coast **23**: 62, 64
Japanese islands **23**: 62–64
Mediterranean areas **23**: 57–62
non-equivalent measurements **23**: 65
subarctic zone **23**: 63
subtropical zone **23**: 63
temperate zone **23**: 63
Zurich-Montpellier index in benthic
   studies **26**: 215, 216

CUMULATIVE TAXONOMIC INDEXES FOR VOLUMES 20-44

Abalone (*Haliotis tuberculata*) **43**: 70, *94*, *95*, 101, 103, 104
Abothrium **24**: 289
  gadi **24**: 274
Abothrium gadi **40**: 7, 13, 14, 28, 33, 36, 37
Abra **32**: 459; **35**: **169**
  alba **35**: 163, 167, **168**, **173**, *174*
  nitida **35**: **173**, *174*
  ovata **35**: 163, 167, **173**, *174*
  prismatica **35**: **173**, *174*
Abra alba **40**: 25
Abralia vernayi **32**: 63
Abramis brama **26**: 83, 84, **120**, 126
Abudefduf saxatilis **21**: 99
Abyla **24**: 200
  haeckeli **24**: 132
  trigona **24**: 132
Abylopsis **24**: 101, 107, 180, 207, 226; **25**: 131, 138
  eschscholtzi **24**: 142, 167
  pentagona **25**: 124
  tetragona **24**: 107, 121, 132, 163, 169, 179; **25**: 156
Abyssoniscus sp. **32**: **161**
Abyssorchomene **35**: **32**, 50, 80
Abyssorchomene chevreuxi **32**: **77**
Abyssorchomene rossi **43**: 220
Acanella **30**: 338
Acanella arbuscula **35**: 110
Acania grani **44**: 53, 310
Acania spp. **44**: 61, 297
Acantharians **25**: 129
Acanthaster planci **21**: 144, 154, 167, 168; **34**: 230
Acanthastrea echinata **22**: 11
Acanthephyra **25**: 119, 136; **32**: 60
Acanthephyra eximia **32**: **162**
Acanthephyra kingsleyi and *A. purpurea* **43**: 97

Acanthephyra microphthalma **32**: **77**
Acanthephyra purpurea **34**: **378**
Acanthina angelica **34**: 49
Acanthina spirata **21**: 177
Acanthocephala **40**: 8, 9, 26, 28; **43**: *21*, *32*, 54-66
Acanthocephalus clavula **40**: 8, 13, 21
Acanthocephalus lucii **40**: 8, 19, 28, 34
Acanthochirana,
  Zoogeography/Evolution **27**: 150
  cordata, Zoogeography/Evolution **27**: *151*
Acanthochiton communis **42**: 200
Acanthochromis polyacanthus **44**: 210
Acanthocottus bubalis **26**: 138
Acanthocybium solandri **30**: 334
Acanthocyclops vernalis **25**: 5
Acanthoglobius flavimanus **26**: 102
Acanthogorgia **32**: **160**
Acanthohaustorius millsi **39**: 195
Acantholaimus **30**: 48, **50**, 50
Acantholatris **32**: 186, 218
Acantholatris gayi **32**: **173**, 188, 218, 224
Acantholatris monodactylis **30**: 329
Acanthometra **25**: 125, 129
Acanthonotozoma **39**: 184
Acanthopagrus **43**: 55
  A. australis **43**: *32*, *56*
  A. berda, A. butcheri and *A. latus* **43**: *56*
Acanthopagrus australis **44**: 212; **28**: 66
Acanthopharynx sp. **32**: **126**
Acanthophthalmus kuhlii **40**: 264
Acanthurus achilles **31**: 9, 16
Acanthurus glaucopareius **31**: 9, 16
Acanthurus nigrofuscus **44**: 223, 231
Acanthurus olivaceus **30**: 334
Acanthurus rackliffei **31**: 9
Acanthus ilicifolius **40**: 97, 103, 107, 108, 122, 167
Acar **32**: 493

Acartia **20**: 30, 40, 41; **25**: 122, 133, 137,
        138, 147, 148, 151; **26**: 135;
        28: 100, **116**; **33**: *4*, 37, 218, 236,
        298; **38**: 18, 33; **40**: 129, 130
    behaviour **33**: 402, **427**, 428, 429, 445, 451
    clausi **29**: *23*, 48, *55*; **30**: 245
    ecology **33**: 469, 473, 475, 480, 481, 483,
        490, 497, 524, 527, 528
    growth/development **33**: 298, 321, 329,
        341
    life history **33**: 347, 384, 389, 390, **395**
    nutrition **33**: **152**, 164, 166, 169, 177, 195,
        208
    reproduction **33**: 257, 265, **268**, 268, **269**,
        272, **285**, 290
    size/weight **33**: 328, *332*, 333
    taxonomy **33**: **51**, 60, *61*, 61, 64, 95, 110,
        111
        clausi **25**: 132
    tonsa **29**: 20, 27, *28, 52, 53*, **131**, 180,
        181; **30**: 245–247; **20**: 40
Acartia adriatica **33**: 110
    africana **33**: 110
    amboinensis **33**: 110
    australis **33**: 110, 111, **427**, 429, **430**
    bacorehuisensis **33**: 110
    baylyi **33**: 110
    bermudensis **33**: 110
    bifilosa **33**: 7, **26**, 110
        biochemistry **33**: 237, **238**
        ecology **33**: **375**, **395**, 398
        growth/development **33**: **301**
        nutrition **33**: 143, 155, **186**
        reproduction **33**: **269**
    bilobata **33**: 110
    bispinosa **33**: 110, **427**
    bowmani **33**: 110
    brevicornis **33**: 110
    californiensis **33**: **26**, 110
        ecology **33**: **395**
        growth **33**: **300**, **320**
        reproduction **33**: **269**, **285**
    centrura **33**: **47**, 110, 506, **507**
    chilkaensis **33**: 110
    clausi **33**: 7, **26**, 37, 60, 64, 110
        behaviour **33**: 405, 406, **408**, **410**, **415**,
            418, **426**, **427**, **446**
        biochemistry **33**: 237, **238**, 240, **241**,
            242, 250, 251
        chromosomes **33**: 47
        distribution **33**: 508, **512**

    ecology **33**: 351, **355**, 363, 366, 367,
        367, 369, **375**, 380, **395**, 398, 469
    growth/development **33**: 289, **300**, **310**,
        311, **312**, *316*, 316, **320**, 339, **341**
    life history **33**: 340, **345**, **386**
    nutrition **33**: 150, 156, 159, 161, 164,
        167, **184**, **186**, **187**, **189**, 197, **198**
    physiology **33**: 215, 216, 217
    reproduction **33**: 260, 267, **269**, 281,
        282, **285**, **288**, 290
    size/weight **33**: **228**, **331**, 332, 336, **337**
danae **33**: 7, **26**, 110
denticornis **33**: 110
discaudata **33**: 7, 110, 216, **301**
dubia **33**: 110
dweepi **33**: 110
ensifera **33**: 110
enzoi **33**: 110
erythraea **33**: 110, **269**, **427**, **431**
fancetti **33**: 110
floridana **33**: 110
forcipata **33**: 110
fossae **33**: 110, 149, **418**, **507**
giesbrechti **33**: 110
grani **33**: **26**, 110
    growth/development, **301**, **310**
    nutrition **33**: **186**
    reproduction **33**: 268, 282, **285**
gravelyi **33**: **47**, 110
hamata, 110, **427**
hudsonica **33**: 110, 228
    behaviour **33**: 406, **410**, 414, **415**, **418**,
        419, 420, **433**, 439, **446**, 449
    ecology **33**: 373, 374, **375**, 469, 475,
        525
    growth/development **33**: 339
    nutrition **33**: 143, 157, **187**, **189**
    physiology **33**: 217
    reproduction **33**: 258, **269**, 282, **285**,
        291, 294
iseana **33**: 110
italica **33**: 110
japonica **33**: 110, 111, **427**
jilletti **33**: 110
josephinae **33**: 110, **269**
kempi **33**: 110
keralensis **33**: **47**, 110
latisetosa **33**: 110, **269**
laxa **33**: 110
lefevreae **33**: 110, 508
levequei **33**: 110
lilljeborgi **33**: **26**, 110

longipatella **33**: 110, 451
longiremis **33**: **7**, **26**, 110
  behaviour **33**: 444
  biochemistry **33**: 247
  ecology **33**: **355**, 364, **375**, 481, **493**,
    495, 496
  growth/development **33**: **310**
  life history **33**: **386**
  nutrition **33**: 143, 150, **189**
  reproduction **33**: 43, **255**, **269**
longisetosa **33**: 110
macropus **33**: 110
major **33**: 110
margalefi **33**: **47**, 64, 110, 507, 508
minor **33**: 110
mossi **33**: 110
natalensis **33**: 110, 451
negligens **33**: **7**, **26**, **47**, 110, **355**, **415**
nicolae **33**: 110
omorii **33**: 110
  behaviour **33**: **427**, 429, **431**, **446**
  biochemistry **33**: **238**
  distribution **33**: **507**
  ecology **33**: **355**
  nutrition **33**: 157, 164, **200**, 203
  reproduction **33**: 258, **285**
  size **33**: 330
pacifica **33**: 110, **269**
pietschmanni **33**: 110
plumosa **33**: **47**, 110
ransoni **33**: 110
remivagantis **33**: 110
sarojus **33**: 110
sewelli **33**: 110
simplex **33**: 110
sinensis **33**: 110
sinijensis **33**: 110, **269**, **301**, **427**, **430**
southwelli **33**: **47**, 110
spinata **33**: 110, **427**, 428
spinicauda **33**: 22, **47**, 110, **269**
steueri **33**: 110
  behaviour **33**: **427**, 429, **431**
  growth/development **33**: 298, **301**, **310**,
    312, **320**
  life history **33**: **386**
  reproduction **33**: **269**, **285**
teclae **33**: 110, **269**
tonsa **33**: **7**, 60, 110, 297
  behaviour **33**: 406, **408**, **410**, **415**, **418**,
    419, **427**, **433**, 447, 452
  biochemistry **33**: 237, **238**, 240, **241**,
    242, 243

  distribution **33**: 506, **507**
  ecology **33**: 351, **355**, 365, 369, 372,
    374, **375**, 474, 475, 476, 520
  growth/development **33**: **301**, **310**, **320**,
    329, **341**
  longevity **33**: 340, **345**
  morphology **33**: **26**, 36, 37, 38
  nutrition **33**: 143, 146, 151, 153, 154,
    157, 159, 162, 164, 167, 171, 177,
    178, **181**, **183**, **184**, **186**, **187**, **189**,
    192, **194**, **197**, **198**, **200**, 203, 205,
    206, **209**
  physiology **33**: 217
  reproduction **33**: **255**, 260, 262, 265,
    269, **272**, 281, 282, **285**, **288**, 290,
    291, 520
  size/weight **33**: **229**, 328, 330, **331**,
    333, **334**, *335*, 336, **337**
tortaniformis **33**: 110
tranteri **33**: 110, **229**
  behaviour **33**: 420, 447, 450, 452
  ecology **33**: 369, **375**
  reproduction **33**: **285**
  size/weight **33**: **229**, **337**, 340, **341**
tropica **33**: 110
tsuensis **33**: 110
  biochemistry **33**: **238**, **241**
  growth/development **33**: **301**, **320**
  reproduction **33**: **269**
  size/weight **33**: **229**, 332
tumida **33**: 110
Acartia bifilosa **44**: **7**, 51
Acartia clausi **44**: 31, 51–53
Acartia erythraea **44**: 53
Acartia fossae **44**: 53
Acartia grani **31**: 195
Acartia hudsonica **44**: 31, 53, 54
Acartia lilljeborgi **44**: 54, 55
Acartia longiremis **44**: 55–57
Acartia negligens **44**: 58
Acartia omori **44**: 31, 58
Acartia pacifica **44**: 58
Acartia spinata **44**: 54, 55
Acartia thonsa **44**: 298
Acartia tonsa **44**: 21, 55, 58–60, 307;
    31: 200, 316
Acartia tranteri **40**: 130
Acartia tumida **44**: 60, 61
Acartiidae **33**: **51**, 110, 111, 268, **269**
Acasta **22**: 204, 233
Aceridae **42**: 78
Acesta **32**: **159**

Acesta bullisi **23**: 339
Acesta excavata **32**: 119
Acesta lopezi **26**: **179**
Acesta neosuecica **26**: **179**
Acetes **27**: 1
   australis, Predation **27**: 361
     Predation **27**: 369
   japonicus **30**: **164**
   sibogae australis **30**: 179, 181
Acetes sibouc australis **25**: 8
Achaea serva **40**: 178
Acharacidae **42**: *5, 6*, 7
Acharax **42**: 6
   A. johnsoni **42**: 36
   A. tokunagai **42**: 8
Acharax alinae **34**: **374**
Achatina fulica **34**: 33, 34
Achelia **24**: 11, 22, 38, 43, 63, 69, 70
   alaskensis **24**: 43
   chelata **24**: 46, 52
   communis **24**: 60
   echinata **24**: 13, 14, 15, 28, 35, 44–47,
     57, 63, 66, 67, 72
   langi **24**: 57, 66
   longipes **24**: 57
   shepherdi **24**: 4, 12
   vulgaris **24**: 57
Acheliana tropicalis **24**: 17
Acidostoma **22**: 92
Acila **42**: *18*, 44
   A. castrensis **42**: 18, 26, *27*
Acineta **25**: 138
Acipenser **40**: 26
Acipenser queldenstaedti **26**: **120**
Acipenser ruthenus **40**: 309
Acmaea
   (Macclintockia) scabra **34**: 32
   (Tectura) digitalis **34**: 32
   (Tectura) tessulata **34**: 12
   persona **34**: *32*
Acmaea testudinalis **23**: 22
Acochlidiacea **42**: 74, 89, 111
Acoela **42**: *89*, 110; **43**: *22*
Acontiostoma **39**: 114
Acrocalanus **33**: **52**, 77, 120, 483; **40**: 129
   andersoni **33**: 120
   gibber **33**: 120, **285**, **331**
   gracilis **33**: 120, **285**
   indicus **33**: 120
   inermis **33**: 37, **375**
   longicornis **33**: 120
   monachus **33**: 120

Acrocalanus gibber **44**: 86, 87
Acrocalanus gracilis **44**: 87
Acrocalanus spp. **44**: 87
Acrocystes nana **23**: 138
Acromitoides purpurus **21**: 105
Acropora **21**: 127, 132; **22**: 7, 10, 11, 17, 42,
     49; **24**: 60; **31**: 54, 256, 260, 264,
     265; **43**: 302, *305*
   A. aspera **43**: 290
   A. formosa **43**: 285
   A. hyacinthus, A. intermediata and
     A.'millepora **43**: 105
   acuminata **22**: 42
   cervicornis **22**: 9, 11, 12, 15, 17, 21, 23,
     26, 30, 31, 33, 38, 40–42, 45,
     48, 50
   formosa **22**: 14, 23, 26, 31, 33, 41, 42
   palmata **21**: 106; **22**: 7, 9, 12, 18, 23, 30,
     33, 40, 41, 45, 48
   prolifera **22**: 48
Acropora cervicornis **31**: *224, 241*, 243, 262
Acropora cythera **31**: 243
Acropora granulosa **31**: 264
Acropora hyacinthus **31**: 243
Acropora palifera **31**: 270
Acropora palmata **31**: 241, 270
Acroporidae **43**: 287, 288
Acrostichum **40**: 92, 157, 177
Acrothoracica **22**: 200
Acteocina **42**: 82, 83, 87, 90, 92, 97, *116*,
     132, 135
   A. wetherilli **42**: 87
Acteon **28**: *393*, 412, 419; **42**: 68, 106,
     117–119, 135
   A. tornatilis **42**: *72*, 73, 117, 118
   characters and evolutionary pathways
     42: 82, *83, 84, 85, 89*, 90
   digestive system **42**: 96, 98, 99
   external anatomy **42**: 91, 93
   mantle cavity **42**: 94
   nervous system **42**: 103, 105
   reproductive system **42**: 101, 102
   taxonomic history **42**: 112, 115, *116*
Acteonellidae **42**: 87, 111
Acteonidae **42**: 75, 87, 90, 91, 110, 118,
     119, 129
Actinia **24**: 49
   equina **24**: 45
   tenebrosa **24**: 45
Actinia equina **21**: 106, 121; **22**: 36, 46, 47,
     68, 70, 72, 76, 79–84, 86, 88, 93
   sulcata **22**: 93

Actiniaria sp. **32**: *97*, 98, *104*, 108, 357, **438**, 445, **446**, **448**, **483**, **486**, **492**
Actinobranchium **25**: 138
  salparum **25**: 137, 139
Actinocyclus (*Charcotia*) *actinochilus* **32**: 542, *552*
Actinodendron plumosum **21**: 106
Actinodinium **24**: 207
Actinodinium apsteini **25**: *132*, 133
Actinonema **30**: **50**
Actinopyga **41**: 190
Actinopyga agassizi **21**: 146, 150, 157–159, 166, 167
  tecanora **21**: 146
Actinopyga mauritana **43**: 105
Actinostola **22**: 74
  callosa **22**: 92
Actinostola callosa **34**: **365**
Actinostolidae **32**: 474
Adalaria proxima **43**: 140
Adamsia carciniopados **24**: 45
Adamsia palliata **21**: 106; **22**: 83
  sulcata **22**: 83
Adamussium colbecki **35**: *175*
Adelogorgia phyllosclera **43**: *291*
Adipicola **34**: 384
  n. sp. **34**: **374**
Adocia **21**: 97
Adrana **42**: 7, 44
  A. superba **42**: *6*
Aechmophorus occidentalis **20**: 241
Aedes aegypti **40**: 106, 116
Aedes taeniorhynchus **40**: 146, 147
Aega **32**: **161**
Aega monophthalma **40**: 9, 19
Aega psora **40**: 9, 19
Aega stromii **40**: 9, 19
Aega tridens **40**: 9, 19
Aeger, Zoogeography/Evolution **27**: 150
  tipularis,
        Zoogeography/Evolution **27**: *151*
Aegialitidaceae **40**: 93
Aegialitis **40**: 102
Aegialitis annulata **40**: 95, 104, 105
Aegialitis rotundifolia **40**: 95, 101
Aegialolaimus **30**: 50
Aegiceras **40**: 101–103
Aegiceras corniculatum **40**: 95, 102, 104, 107, 112–114, 158, 184
Aegiceras floridum **40**: 95
Aegina sp. **32**: *72*

Aeginina longicornis **39**: 206
Aelolidia papillosa **22**: 92
Aequipecten **22**: 136
Aequipecten opercularis **34**: 287
Aequorea **24**: 203; **25**: 32, 33, 45
  victoria **24**: 208; **25**: 20, *28*, **31**
Aequorea Victoria **44**: 29, 121; **43**: 287; **28**: *87*, 91, 92, **93**
Aeromonas **25**: 120
Aeromonas, Parasites **27**: 382
Aeropsidae **32**: 466
Aeropsis fulva **32**: **462**
Aetedidae **32**: 55
Aetideidae **32**: 193; **33**: 3, 434
  ecology **33**: 470, **499**, 502
  morphology **33**: 21, 27, 28
  taxonomy **33**: **53**, 124–129
Aetideopsis
  ecology **33**: 528
  nutrition **33**: **152**
  taxonomy **33**: **53**, 88, 90, 92, 124
Aetideopsis albatrossae **33**: 124
  antarctica **33**: 124
  armata **33**: 124
  carinata **33**: 124
  cristata **33**: 124
  divaricata **33**: 124
  inflata **33**: 124
  magna **33**: **499**
  minor **33**: 124
  modesta **33**: 124
  multiserrata **33**: *4*, 124, 522
  retusa **33**: 124
  rostrata **33**: 124
  trichechus **33**: 124
  tumerosa **33**: 124
Aetideus **25**: 138
  ecology **33**: 480, 528
  nutrition **33**: **152**, 166
  taxonomy **33**: **53**, 92, 124, 125
Aetideus acutus **33**: 124
  arcuatus **33**: 124
  armatus **33**: 7, **26**, 124, **223**, 268, **512**
  australis **33**: 124
  bradyi **33**: 124
  divergens **33**: 124, 165, **188**, **189**
  giesbrechti **33**: 124
  mexicanus **33**: 124
  pseudarmatus **33**: 124
  truncatus **33**: 124
Afrenulata **32**: 408
Afrocucumis africana **21**: 147

Afrocypholaelaps africana **40**: 179
Agalma **24**: 126, 196, 215
  clausi **24**: 116
  elegans **24**: 103, 178, 193, 195, 198, 217,
      225, 232
  haeckeli **24**: 116
  okeni **24**: 174, 178, 198, 200, 202, 207,
      218, 219, 225, 227, 230, 232, 234,
      239
Agamonema **40**: 7
Agaricia **31**: 241
Agaricia agaricites **22**: 10–12, 18, 27, 30,
      33, 37–39; **31**: 271
Agaricia Plate 4 **43**: 288
  A. *agaricities* **43**: 285
  A. *tenifolia* **43**: 303
Agaronia testacea **28**: *393*, 411
Agarum **23**: 46; **38**: 116
  cribosum **23**: 79, 82, 84, 90, 95, 122
Agarum clathratum **39**: 40
Agarum cribrosum **38**: 116
Agarum fimbriatum **38**: 116
Agauopsis auzendei **32**: **127**
Agelaius phoeniceus (red-winged
      blackbird) **39**: 45
Agelas **21**: 104
  dispar **21**: 97, 100
  oroides **21**: 104
Aggregata **25**: 144
Agkistrodon contortrix **21**: 64
Aglaia cucullata **40**: 95
Aglajidae **42**: 97, 109, 110, 129
Aglantha **24**: 102; **25**: 32, 33
  digitale **24**: 167
Aglantha digitale **23**: 234; **32**: *13*, 305
Aglaophamus malmgreni **26**: **179**
Aglaophenia cupressina **21**: 105
Agmasoma, Parasites **27**: 384
  duorara, Parasites **27**: 384
  penaei, Parasites **27**: 384
Agnesiopsis translucida **32**: 361
Agonus cataphractus **26**: **121**
Aigialus parvus **40**: 120
Aigialus striatispora **40**: 119
Aiptasia **22**: 92
  pallida **22**: 91
  sulcata **22**: 91
Aiptasia pallida **21**: 106, 125; **31**: 250, 260
Aipysurus laevis, Predation **27**: 371
Ajala ajala **40**: 155
Akera **42**: 71, *78*, 132, 135
  A. *soluta* **42**: 72, 73

  external anatomy **42**: 92, 93
  mantle cavity **42**: 95
  shell **42**: *84, 85, 89*
  taxonomic history **42**: 105, 106, 115, *116*
Akeridae **42**: 78, 87, 88, 109, 129
Akerogammarus knipowitschi **39**: 195
Alaria **39**: 24, 26
  esculenta **23**: 111
  fistulosa **23**: 82, 83
  nana **23**: 131, 132
Alaska (walleye) pollock *see Theragra
      chalcogramma* **38**: 53
Alaska pollock *see Theragra chalcogramma*
Alaysia spiralis **34**: **366**
Albula **30**: 278; **20**: 34
Albula vulpes **44**: 231
Albunea symnista **29**: **133**, *151*, 152, 153,
      154, **155**, **158**, 186, 187, 197,
      198
Alburnus alburnus **26**: **120**, 126
Alca torda **20**: 282
Alciopa **25**: 139, 150
Alcyonaria **32**: *333*
Alcyonidium gelatinosum **21**: 142, 143
  hirsutum **21**: 142
Alderia modesta **43**: 125, 146
Aldrichetta forsteri **24**: 329
Aldrovandia **32**: 167, 186
Aldrovandia affinis **32**: **168**, 206
Aldrovandia oleosa **32**: **168**
Aldrovandia phalacra **32**: **168**, 206
Alepes djedaba, Predation **27**: **362**
Alepisaurus ferox **32**: 270, 281
Alepocephalidae **32**: 167, **169**
Alepocephalus sp. **32**: **169**, 186
Alexandrella carinana **32**: 372
Alexandrium **31**: 309, 348, 356
Alexandrium excavatum **33**: 157
Alexandrium fundyense **31**: 313, 321, **321**
Alexandrium minutum **31**: 337
Alexandrium ostenfeldii **31**: 313
Alexandrium tamarense (formerly
      Gonyaulax tamarensis) **31**: 304,
      313, 321, 350, *351*, 363
Algae **40**: 122–126
  Chaetoceros **41**: 36, 182, 199
  Isochrysis galbana **41**: 182, 199
  pigmentation in coral reef organisms
      43: 281, *305*, 308, 309
  macroalgae **43**: 281–284, 306
  Schizochytrium **41**: 170

Southern Ocean pack ice **43**: 174, 184, 187, 191, 192, 208, 209, 218, 224
  and bacteria **43**: 182, 193, 205
  blooms **43**: 232, 308
  excessive lipid formation **43**: 201, 202
  micro- **43**: 213, 221–223
  salinity, effects of **43**: 199, 200
  summer ice **43**: 210
  uni-algal isolates **43**: 196
  UV radiation **43**: 236–238
  winter survival **43**: 201
Algae *see under subject index*
Alicella gigantea **39**: 195
Alicia costae **21**: 106
Allia quadrilobata **26**: **179**
Allocentrotus fragilis **35**: **185**
Allocyttus
  niger **35**: **37**
  verrucosus **35**: **37**
Allogastropoda *see* Lower Heterobranchia
Alloiopodus **33**: **52**, 121
  pinguis **33**: 121, **498**
Allonautilus **44**: 146
Alloniscus perconvexus **30**: 106, 118
Allopora californica **43**: *291*, 292
Almyracuma proximoculi **39**: 144, 217
Alona taraporevalae **31**: 87, 149
Alosa **24**: 275; **20**: 22, 160
  aestivalis **25**: 15
  aestivalis **20**: 39, 174
  alosa **24**: 276
  fallax **20**: 86
  kessleri **20**: 133
  pseudoharengus **25**: 5, 58, 59; **20**: 6, 11, 39
  sapidissima **20**: 6, 39, 131
Alosa pseudoharengus **26**: **120**
Alosa sapidissima (shad) **38**: 7, 54
Alpheus romensky **32**: **162**
Alrhabdus **33**: **51**, 70, 106
  johrdeae **33**: 106
Alternaria alternata **40**: 121
Alteromonas, Parasites **27**: 382
Altiverruca mollae **32**: **161**
Alutera scripta **21**: 126
Alvinella **34**: 385, 410, 411, 412; **35**: 22, 30, 88
  caudata **34**: **368**, 411, *413*; **35**: **41**, 87, 89
  hirsuta **35**: **41**
  pompejana **34**: **368**, 411, 413; **35**: 18, 44, 87, 89

Alvinella pompejana **23**: 326, 327, 335,343, 344
Alvinellidae **32**: 120
Alviniconcha **32**: 120; **35**: **40**
Alviniconcha hessleri **34**: **372**
Alvinocarididae **32**: 99, *100*, 105, 110, 120, **129**, 130
Alvinocaris **32**: 107, 118, 120, 124, 130, 131; **34**: **377**
  longirostris **34**: **377**
  lusca **34**: **377**
  markensis **34**: **377**
Alvinocaris longirostris **32**: 110, 111, **129**
Alvinocaris lusca **23**: 330, 331, 335; **32**: 110
Alvinocaris markensis **32**: 99, 102, 105, 107, 110, 111, 114, **129**, 130
Alvinocaris muricola **32**: 111, **129**, 130, 131
Alvinocaris sp. **32**: 114, 116
Alvinocaris stactophila **32**: 110
Alvinoconcha **34**: 397
Amallocystis **25**: 137
Amallophora **33**: **53**, 134
  elegans **33**: 124, 135
  impar **33**: 124, 136
  macilenta **33**: **500**
  obtusifrons **33**: 134
  rotunda **33**: **500**
Amallothrix **25**: 138
  ecology **33**: 486, **500**
  taxonomy **33**: **53**, 57, 81, 82, 134
Amallothrix arcuata **33**: 134
  dentipes **33**: 134
  emarginata **33**: 134
  falcifer **33**: 134
  farrani **33**: 136
  gracilis **33**: 134
  hadrosoma **33**: 134
  indica **33**: 134
  invenusta **33**: 134
  longispina **33**: 134
  parafalcifer **33**: 134
  profunda **33**: 136
  pseudoarcuata **33**: 134
  pseudopropinqua **33**: 134
  robusta **33**: 134
  sarsi **33**: 136
Amathillina cristata **39**: 195
Amathillina spinosa **39**: 195
Amathimysis cherados **39**: 213
Amathimysis polita **39**: 213
Amathinidae *see* Lower Heterobranchia

Amathys **32**: 130
Amathys lutzi **34**: **369**
Ambassis commersoni, Predation **27**: **362**, 367, 368
Amblyomma **43**: 35
Amblyops **25**: 154
Americamysis almyra **39**: 213, 215
Americamysis bahia **39**: 192, 213
Americamysis bigelowi **39**: 184, 213
American oyster *see Crassostrea virginica*
Americonuphis reesei, Physiology **27**: 198
Ameson, Parasites **27**: 384
Amia calva **40**: 269
Ammallothrix **25**: 138
Ammodramus maritimus (sea sparrow) **39**: 45
Ammodytes **25**: 6; **28**: 28, 87, 89, 100, 101, 127; **30**: 233, 246, 270; **34**: 257; **20**: 53
  hexapterus **30**: 250; **34**: 270
  marinus **25**: 12, 14, 15, 30, **31**, **35**, 37, 40, 42; **20**: 280
  personatus **25**: 8, 21, 23, **31**, **34**, 47, 48
  tobianus **25**: **35**
Ammodytes americanus **26**: 133
Ammodytes hexapterus **39**: 52, 56, 77, 70, 72, 80
Ammodytes marinus **26**: **121**, 133
Ammodytes personatus **26**: **121**, 149
Ammodytes sp. **26**: 27, **120**, 133, 149, 270
Ammothea sp. **32**: **127**
Ammothea **24**: 11
  allopodes **24**: 71
  caroliensis **24**: 71
  hilgendorfi **24**: 16, 52, 57, 66, 69
  magniceps **24**: 32
  striata **24**: 20, 29
Ammothea verenae **34**: **374**
Ammothella **24**: 8, 9, 22, 64, 66, 69
  appendiculata **24**: 64, 66
  biunguiculata **24**: 13, 44, 72
  indica **24**: 66
  longipes **24**: 38, 41, 44, 47
  setacea **24**: 71
  stauromata **24**: 66
  uniunguiculata **24**: 44
Amoeba proteus **21**: 164; **31**: 269
Amoebae **43**: 208, 213
Amoebophrya **25**: 126, 127, 129
  acanthometrae **25**: 129
  grassei **25**: 129

  rosei **25**: 129
  sticholonchae **25**: 128
Amonardia **44**: 309
Ampelisca **35**: **169**, 174, **177**, *178, 179*
  agassizi **35**: 175, *179*
  armoricana **35**: 174, *179*
  auracana **35**: 175, *179*
  brevicornis **35**: **163**, **168**, 174, 175, *179*
  sarsi **35**: 174, 175, *179*
  tenuicornis **35**: 174, *179*
  typica **35**: 174, *179*
Ampelisca abdita **39**: 195
Ampelisca araucana **39**: 195
Ampelisca brevicornis **39**: 195
Ampelisca diadema **39**: 195
Ampelisca macrocephala **39**: 195
Ampelisca sarsi **39**: 195
Ampelisca spinipes **34**: 287
Ampelisca tenuicornis **39**: 195
Ampelisca typica **39**: 195
Ampelisca vadorum **39**: 195
Ampeliscidae **32**: 468
Ampharete **29**: 252; **35**: **163**, **169**
  acutifrons **35**: **168**
Ampharete acutifrons **26**: **179**
Ampharete finmarchica **26**: **179**
Ampharetidae **26**: **179**
Ampharetidae **32**: **126**, 463, 466; **39**: 36
Amphascandria **33**: 17
Amphianthus
  inornata **35**: **33**, 110
  radiatus **35**: 52
Amphiascoides **44**: 301
Amphiascoides atopus **44**: 305, 306
Amphiascoides subdebilis **44**: 308
Amphibia **40**: 154, 155
Amphicaryon **24**: 139
  intermedia **24**: 119
Amphicrossus altalis **34**: **375**
Amphicteis mederi **32**: **460**
Amphictenidae **39**: 36
Amphicypellus elegans **25**: 122
Amphidinium **21**: 72
  caterae **21**: 72
  rhynchocephalum **21**: 72
Amphiduros axialensis **34**: **366**
Amphilochidae **32**: **128**
Amphinema brasiliensis **21**: 138
Amphineura **32**: 335
Amphinomidae **32**: **126**
Amphiophiura sp. **32**: **166**

Amphiophiura bullata pacifica **32**: **447**, **448**, **452**
Amphiophiura concava **32**: **447**, **448**, **452**; **32**: **447**, **448**, **452**
Amphiophiura pachyplax **32**: **447**, **452**
Amphiophiura sculptilis **32**: **462**
Amphiophiura vitjiazi **32**: **447**, **448**, **452**
Amphioporus lactifloreus **21**: 134
Amphiplica gordensis **34**: **370**
Amphipoda **32**: **56**, 57, **162**, 328, **335**, 355, 357, 445, 447, **452**, **453**, **454**, 463, 468; **39**: 107, 157, 195–205; **40**: 9, 26–28; **43**: 93, 140, 142, 208, 210, 215, 220
  caprellids **39**: 157, 186, 187, 189, 206
  gammarids **39**: 157, 186, 187, 189, 193, 195–205
  hyperiids **39**: 157, 186, 187, 189, 207
Amphiporeia lawrenciana **39**: 196
Amphiporeia virginiana **39**: 152, 196
Amphiprion clarkii **21**: 107
Amphiprion percula **25**: 15, 16, *17*, 19, *22*; **20**: 57
Amphiprion perideraion **31**: 186
Amphiprora paludensis var. *hyperborea* **32**: *547*
Amphiprora paludosa var. *hyperborea* **32**: 537, 557
Amphiroa **23**: 100
Amphisamytha
  fauchaldi **34**: **369**, 414
  galapagensis **34**: **369**, 402, 414
Amphisamytha galapagensis **23**: 327, 328, 346
Amphisamytha galapagensis **35**: 20
Amphisolenia **32**: 560
Amphisolenia globifera **32**: 545
Amphithoe **39**: 151
Amphithoe lacertosa **39**: 196
Amphithoe longimana **39**: 196; **43**: 142
Amphithoe ramondi **23**: 101
Amphithoe rubricata **39**: 196
Amphithoe vaillante **39**: 196
Amphitretidae **32**: 62
Amphitritinae **26**: **179**
Amphiura **34**: 287
  filiformis **34**: 235
Amphiura filiformis **26**: 187
Amphiura sculptilis **32**: 464
Amphora coffeaeformis **31**: 311
Amphorme horrida **32**: **160**, 216
Amylodinium **25**: 134

Amyloodinium **25**: 127
Amyloodinium ocellatum **43**: 65
Anabaena flos-aquae **21**: 71, 87
Anacystis **20**: 183
Anacystis nidulans **31**: 269
Anadara senilis **22**: 103, 165
Anaitides **32**: **161**
Anakinetica cumingi **28**: 178, 201, **204**, **206**
Anapagurus hyndmani **29**: **132**, 148
Anaplopoma fimbria **39**: 279
Anarhichas lupus **40**: 23, 24
Anaspidea **42**: 68, 71, 118
  see also Akera; Aplysia
  digestive system **42**: 96, 98, 99
  external anatomy **42**: 93
  morphology **42**: 77, 78, *78*
  nervous system **42**: 103, 104, 105
  reproductive system **42**: 100
  shell **42**: 88, 90, 91
  taxonomic history **42**: 73, *89*, 108, 110, 112, 114, 117
Anatanais **39**: 160
Anatolanthias **32**: 186, 211
Anatolanthias apiomycter **32**: **172**
Anatoma **34**: **371**
Anaulus **25**: 191
  birostratus **25**: 185
Anawekia **33**: **53**, 78, 81, 130
  bilobata **33**: 130
  robusta **33**: 130
  spinosa **33**: 130
Anchialina typica **39**: 175
Anchoa **20**: 161
  compressa **20**: 131
  hepsetus **20**: 180
  lamprotaenia **20**: 6, 28
  marinii **20**: 180
  mitchilli **20**: 6, 26, 102
  tricolor **20**: 180
Anchoa mitchilli **25**: 3, 14, 15, 17, *20*, 21, **31**, 33, 34, **34**, **35**, 36, 37, 42, 50, 57, 58
Anchoviella choerostoma **20**: 98,100
Ancistrosyllis groenlandica **26**: **179**
Ancyananiscus **39**: 184
Ancylus fluviatilis **34**: 18, 22, 37, **41**
Andaniexis subabyssi **32**: 74
Anemonia sulcata **24**: 45; **21**: 106, 121, 122, 125, 132
Anemonia sulcata **22**: 68, 71
Anemonia viridis **31**: 250

Angatea **21**: 139
Angelopsis dilata **24**: 118
  euryale **24**: 118
Angliera phreaticola **39**: 208
Anguilla **28**: 56, 66
Anguilla anguilla **26**: 141; **38**: 33, 53;
    40: 286, 291; **43**: 66
Anguilla japonica **40**: 267
Anguilla rostrata **26**: **121**, 130, 142; **30**: 226
Anguilla sp. **26**: 149
Anguilla spp. **44**: 210
Anidolyta **42**: 113, 114, 132
Anilocra physodes **39**: 129
Aniptodera longispora **40**: 119
Aniptodera salsuginosa **40**: 119, 122
Anisakids **43**: 68
Anisakis **24**: 296, 303; **25**: 151; **32**: 308;
    40: 7, 23, 52
  simplex **24**: 273, 277, 288, 293, 294, 297,
    298, 302, 305–307; **25**: 151
Anisakis simplex **32**: 307, 308; **40**: 7, 18, 26,
    33, 35, 36, 39–41, 43, 44, 57
Anisakis spp. **26**: 38
Anisodoris nobilis **34**: 33
Anisogammarus anandalei **39**: 110
Anisomysis pelewensis **33**: **430**
Annelida **31**: 39
  see also Polychaeta
  fecundity **43**: 103, 104, 106, 136, 137,
    143–145
  parasitism **43**: *21*
Annulipulsellum **42**: *144*
Anobothrus **23**: 340
Anobothrus gracilis **26**: **179**
Anodonta **22**: 111; **35**: **169**
  anatina **22**: 103, 111, 148
  cygnea **22**: 103–105, 131, 146, 148, 152,
    156, 157
Anomalocera **23**: 241; **25**: **34**; **33**: **52**, 66, 95,
    113
  opalus **33**: 113
  ornata **25**: 30; **33**: 39, 113, **270**, 291
  patersoni **23**: 185, 225, 226, 256; **33**: **7**,
    49, 113, 218
    behaviour **33**: 406, 420
    biochemistry **33**: 243, 251
    distribution **33**: 505, **512**
    ecology **33**: 370, **491**
    morphology **33**: 30, **32**, *492*
    nutrition **33**: 150, 202
    reproduction **33**: 268, **270**, **288**
Anomalodesmata **42**: *37*

Anomotaenia **25**: 151
Anomura **32**: **163**, **210**
Anonyx nugax **39**: 196
Anonyx sarsi **39**: 196
Anopheles maculatus **40**: 116
Anopheles stephensi **40**: 106
Anoplarchrus purpurescens (high
    cockscomb) **39**: 50
Anoplodactylus **24**: 10, 11, 13, 15, 22, 26,
    32, 60, 63, 64, 66, 69
  angulatus **24**: 44, 45
  antillianus **24**: 38
  arescus **24**: 17, 55, 64
  australis **24**: 60
  cribellatus **24**: 64
  digitatus **24**: 8
  evansi **24**: 16
  gestiens **24**: 23, 24
  glandulifer **24**: 66
  insigniformis **24**: 13
  lentus **24**: 19, 21, 32, 37
  marshallensis **24**: 66
  massiliensis **24**: 64
  multiclavus **24**: 13
  ophiurophilus **24**: 46, 52
  perforatus **24**: 13, 44
  petiolatus **24**: 28, 43, 44, 46, 57, 64,
    66, 75
  portus **24**: 14, 23, 64, 69
  pulcher **24**: 13
  pygmaeus **24**: 21, 28, 45, 57, 64, 66, 67
  saxatilis **24**: 69
  stellatus **24**: 4, 5, 8, 10, 12, 13
  tubiferus **24**: 66
  typhlops **24**: 75
  virescens **24**: 13, 43, 67
Anoplogaster **32**: 61
Anoplogaster cornuta **32**: 64
Anoplopoma fibriata (sablefish) **38**: 178
Anoplopoma fimbria **30**: 331
Antalis **42**: *143*, 191, 192
  A. dentalis **42**: 197
  A. entalis **42**: 162, 165, 176, 189, 195, 214
    development **42**: 196, 197, 199, 200,
    201, *202*, 204
  A. entalis stimpsoni **42**: 163, 164
  A. pretiosum **42**: 152, *166*, 215, 216
  A. pseudohexagonum **42**: 162, 169
  A. vulgaris **42**: 140, 161, 197
Antarctomysis maxima **39**: 213
Antarctomysis ohlini **39**: 143, 144, 213
Antarcturus polaris **39**: 208

Antestiopsis **40**: 178
Antharcturus sp. **32**: **460**; **32**: 463
Antharcturus hirsutus **32**: **460**
Anthomastus **30**: 325
Anthopleura **22**: 90, 91
  artemisia **22**: 92
  ballii **22**: 83
  elegantissima **22**: 68, 90, 92, 93
  midorii **22**: 68
  palliata **22**: 93
  xanthogrammica **22**: 90, 93
Anthopleura elangitissima **31**: 250
Anthopleura xanthogrammica **21**: 106, 125;
  23: 75
Anthosactis nomados Plate 3 **42**: 213,214
Anthozoa **43**: 287
  octocorallia **43**: 290–292, *291*
Antias unioramea **39**: 208; **39**: 159
Anticoma sp. **32**: **126**
Antigonia **32**: 186
Antigonia aurorosea **32**: **172**, 188
Antigonia capros **32**: **172**, 188, **195**
Antimora **32**: 186
Antimora rostrata **32**: **170**; **35**: **36**
Antipatharia **32**: *333*
Antipatharia **32**: **160**, 468, 470, 500, **504**
Antipathes **30**: 319, 338
  dichotoma **30**: 338
Antipathes fiordensis **43**: 105, 133
Antipathes lentipinna **23**: 338
Antrimpos kiliani,
          Zoogeography/Evolution **27**: *151*
  speciosus,
          Zoogeography/Evolution **27**: 150,
          150
Antrisocopia **33**: **51**, 76, 99
  prehensilis **33**: 99, **489**
Anulidentaliidae **42**: *143*
Anulidentalium **42**: *143*
Anuropidae **39**: 157
Apagis **31**: 86
Aphanizomenon **21**: 88
  flos-aquae **21**: 71, 87, 88
Aphanopus **32**: 206
Aphanopus capricornis **32**: **174**, 218
Aphelasterias japonica **21**: 144
Aphotopontius
  acanthinus **34**: 375
  arcuatus **34**: 375
  atlanteus **34**: 375
  baculigerus **34**: 375
  flexispina **34**: 375

forcipatus **34**: 375, 419
  hydronauticus **34**: 375
  limatulus **34**: 375
  mammillatus **34**: 375
  probolus **34**: 375
  temperatus **34**: 375
Aphotopontius atlanteus **32**: **128**
Aphotopontius forcipatus **32**: **128**, 131
Aphriza virgata (surf bird) **39**: 15, 64, 66,
          68, 80
Aphrodite australis **21**: 138
Aphroditidae **32**: 455
Apicomplexa **40**: 4; **43**: 20
Apis dorsata **40**: 145
Apis mellifera **29**: **188**; **40**: 179
Aplacophora **31**: 39; **32**: 369; **42**: 42, 97
Aplustridae **42**: 75, 86, 90
Aplysia **21**: 170, 179, 189; **30**: 179, 190;
          42: 71, 105, 106, 132, 135; **43**: 104
  brasiliana **21**: 182
  californica **21**: 121, 177, 180, 182
  characters and evolutionary
          pathways **42**: 86, *89*, 90, 92, 98
  delipans **21**: 179
  kurodai **21**: 180
  limacina **21**: 179
  pulmonica **21**: 181
  taxonomic history **42**: 107, 115, *116*,117
Aplysia benedicti **24**: 53
Aplysiacea *see* Anaspidea
Aplysiidae **42**: 78, 109, 129
Apoda **32**: **462**
Apodida **32**: **438**, 466
Apodinium **25**: 125, 127, 134, 135
  chattoni **25**: 134
  mycetoides **25**: 134
  rhizophorum **25**: 134
  zygorhizum **25**: 134
Apolemia **24**: 145, 191, 199
  uvaria **24**: 208, 212, 217, 221, 223,
          224, 229
Aporrhais **28**: *392*, 411
Apotectonia heterostegos **34**: **377**
Appendicularia **44**: 34
Appendicularia sicula **44**: 129, 130
Aprion virescens **30**: 334
Apristurus maderensis **34**: **380**
Apseudes **39**: 159
Apseudes chilkensis, Food/Feeding **27**: 327
  gymnophobia, Food/Feeding **27**: 327
Apseudes diversus **32**: **161**
Apseudes hermaphroditicus **39**: 114, 122

Apseudes latreillei **39**: 125, 219
Apseudomorpha **39**: 159
Aptenodytes forsteri **43**: 228, 243
Aptenodytes patagonica **20**: 273
Apteronotus leptorhynchus **40**: 307, 308
Araceae **40**: 128
Arachis hypogea **40**: 259
Arachnomysis **39**: 161
Araeolaimus sp. **32**: **126**
Araeosoma thetidis **21**: 145, 150, 169
Araeosomoa **32**: *214*
Aratus pisonii **40**: 141
Arbacia **37**: 131
Arbacia lixula **21**: 145; **43**: 147
  punctulata **21**: 164
Arca (Bentharca) asperula **32**: 503
Arca orbiculata **32**: 503
Archachatina marginata **34**: 33
Archaea **43**: 202
Archaeogastropoda **32**: **127**, 409, 463
Archeolithothamnium **32**: **160**, 176
Archeomysis **39**: 131, 133
Archeomysis grebnitzki **39**: 123, 128
Archescolecithrix **33**: **53**, 81, 84, 134
  auropecten **33**: 134
Archidiaptomus **33**: **52**, 67, 115
  aroorus **33**: 115
Archidoris
  montereyensis **34**: 33, 49
  pseudoargus **34**: 33
Archinome (Euphrosine) rosacea **34**: **366**
Archinome rosacea **32**: **126**, 130, 131
Archinomidae **32**: **126**
Architectibranchia **42**: 68, 106, 118
  see also Hydatina
  characters and evolutionary
    pathways **42**: 90, 93, 100, 103, 104
  morphology **42**: *74*, 75
  taxonomic history **42**: 73, *89*, 112,
    114, *116*
Architectonicidae *see* Lower Heterobranchia
Archosargus probatocephalus **40**: 266
Arcidae **32**: 470, 495
Arcopsis **32**: **159**
Arcovestia ivanovi **34**: **366**
Arctic shanny *see Stichaeus punctatus*
Arctica islandica **22**: 103, 107, 146–148;
    34: 213, 214
Arctocephalus australis australis **34**: 274
Arctocephalus forsteri **39**: 279
Arctocephalus gazella **39**: 281; **43**: 227;
    20: 273

  pusillus **20**: 252
Arctocephalus pusillus **39**: 282
Arctocephalus tropicalis **39**: 281
Arctoscopus japonicus **24**: 275
Arcturella sawayae **39**: 159, 208
Arcturidae **32**: 372; **39**: 159
Arcturus **39**: 159
Arcturus longicornis **39**: 208
Ardea cinerea, Predation **27**: 372
  goliath, Predation **27**: 372
  grayii, Predation **27**: 372
  purpurea, Predation **27**: 372
Arenaria interpres (ruddy turnstone)
    39: 66–7
Arenaria melanocephala (black
    turnstone) **39**: 15, 64, 66
Arenicola marina **34**: 211
Argentinidae **32**: **169**, 211
Argis sp. **32**: 266
Argonauta **25**: 88–90, 97
  argo **25**: 104
Argonauta spp. **32**: 64; **44**: 168
Argopecten irradians **37**: **10**
Argopecten purpuratus **35**: 166, **168**, 173,
    175, **184**
Argopecten ventricosus **43**: 101
Argulus alosae **24**: 276, 291
  coregoni **24**: 273
Argyripnus **32**: 187, 198, 217
Argyripnus electronus **32**: **169**, 188, **195**,
    217
Argyropelecus hemigymnus **40**: 295
Argyrosomus argentatus, Predation **27**: 369
  hololepidotus, Predation **27**: **363**, 367
Argyrotheca **28**: 249–251, 265, 271, 275
Argyrotheca cistellula **28**: 211, 341
Argyrotheca cuneata **28**: 236
Aricidea **26**: **179**
Arietellidae **33**: **51**, 101–103, 470, **489**, **498**
Arietelloidea
  ecology **33**: 502
  morphology **33**: 17
  taxonomy **33**: 50, **51**, *54*, *55*, 101–109
Arietellus **33**: **51**, 71, 74, 101, 486
  aculeatus **33**: 101
  armatus **33**: 101, 422
  giesbrechti **33**: 101
  minor **33**: 101
  mohri **33**: 101
  pacificus **33**: 101
  pavoninus **33**: 101
  plumifer **33**: 101, 444

setosus **33**: 101, 142
simplex **33**: 101
tripartitus **33**: 101
Ariolimax columbianus **34**: 9, 13, 22, 31,33
Ariomma **32**: 186
Ariomma lurida **30**: 334
Ariomma luridum **32**: **174**
Ariommidae **32**: **174**
Arion
ater **34**: 7
empiricorium **34**: 33
Aristaeomorpha, Taxonomy **27**: 115
Aristeomorpha foliacea **32**: **162**
Aristeus, Taxonomy **27**: 64
Arius felis **26**: 139
Arius felis, Predation **27**: 361, **362**
proximus, Predation **27**: **362**
thalassinus, Predation **27**: **362**, 367
Arius thalassinus **40**: 269, 284
Armadillidium **30**: 90; **39**: 135
vulgare **30**: 114, 124
Armadillidium vulgare **29**: 185, 186;
                          39: 109, 137, 139
Armena-Tritonia **42**: 117
Armina **42**: *116*, 133, 135
Arnoglossus **32**: 186
Arnoglossus multirastris **32**: **174**, 190, 193,
                              194
Arrhis phyllonyx **39**: 196
Arrowtooth flounder *see Atheresthes stomias*
Artemesia, Taxonomy **27**: 61, 62, 63, 64,
                          108, **113**
longinaris, Taxonomy **27**: 64, *64*, **108**
Life histories **27**: 296, 312, **313**
Zoogeography/Evolution **27**: 139, **145**,
                          147, **14**
Life histories **27**: 285, **313**
Physiology **27**: 203
Zoogeography/Evolution **27**: 148, 149,
                          150, 154, 156, 157
Artemia **22**: 67, 88; **24**: 41, 42; **25**: 25, 36;
          30: 244, 245, 254, 271; **33**: **188**,
          297; **20**: 32, 39, 40, 43, 52, 67, 183
salina **20**: 182
Artemia salina **37**: 131, 132; **38**: 162
Artemia sp. (brine shrimp) **38**: 14, 15, 36,
                159, 161, 162, 163, 226
Artemia spp. **44**: 183, 296, 297, 310, 311
Artemia, Physiology **27**: 180, 210
Food/Feeding **27**: 327
Arthessidae **42**: 80
Arthrophryxus **25**: 154

Arthropoda **43**: *21*
see also Crustacea
Asbestopluma sp. **32**: 98, **126**, 495
Asbestopluma biseralis **32**: 503
Ascarophis **25**: 151; **40**: 11, 28, 36, 37
Ascarophis arctica **40**: 8, 18, 31
Ascarophis crassicolis **40**: 8, 13, 14, 31
Ascarophis extalicola **40**: 8, 18, 31
Ascarophis filiformis **40**: 8, 13, 14, 31
Ascarophis japonicus **40**: 7, 10, 11
Ascarophis morrhae **40**: 8, 18, 31
Ascarophis pacificus **24**: 277, 290
Ascidia **40**: 131–134
Ascidiacea **32**: 468, **469**, **483**, **504**
Ascidiae **32**: **335**, **438**
Ascidians **43**: 111
Ascobulla **42**: 71, 83, 90, 95, 97
Ascobullidae **42**: 80
Ascoglossa **42**: 78, 108
Asconiscus **25**: 153
Ascophyllum nodosum **23**: 6, 8, 9, 11, 26, 27,
                        29, 30, 108
Ascorhynchus **24**: 8, 9, 16, 53, 64, 69, 75
abyssi **24**: 22
arenicola **24**: 17, 63
armatus **24**: 14
birsteini **24**: 22, 75
breviscapus **24**: 12
castelli **24**: 44
corderoi **24**: 22
endoparasiticus **24**: 53
ovicoxa **24**: 22
pararmatus **24**: 14
prosum **24**: 5
ramipes **24**: 44
simile **24**: 10, 11, 17, 32, 55, 63
utinomii **24**: 33, 44
Asellota **32**: 335; **39**: 118, 119
Asellus **39**: 118
Asellus aquaticus **35**: 166, **168**
Asellus aquaticus **39**: 133, 138; **40**: 28
Asparagopsis **21**: 93
toxiformis **21**: 93
Aspergillus **40**: 121
Aspidiotus destructor **40**: 178
Aspidochirotida **32**: 466; **41**: 133
Aspidodiadema sp. **32**: **165**, 176, 179
Aspidoecia **25**: 153
Aspidogastrea **43**: 22, 41
Aspidophryxus **25**: 154
Aspirochirotida **32**: **438**
Assurger anzac **32**: **174**, 188

Astacilla **39**: 159
Astacilla ceaca **39**: 159
Astacus leptodactylus **27**: 179
Astarte **34**: 221
Asterias **23**: 87
Asterias amurensis **21**: 144, 152, 166;
    20: 365
  rubens **21**: 144, 153
Asterias epichlora **31**: 20
Asterias hexactis **31**: 20
Asterias rubens **34**: 288
Asterina
  A. *gibbosa* and A. *minor* **43**: 100
  A. *phylactia* **43**: 130, 131
Asterina petinifera **21**: 144, 152
Asteroidea **31**: 39; **32**: 97, **165**, 329, 333,
    335, **335**, 344, 429, **438**, 445, **446**,
    447, **453**, **462**, **483**, **486**, **492**
Asteroidea and fecundity
  environmental factors **43**: 147
  life-history stategies **43**: 126, 130, 131,
    136
  quantification methodology **43**: 93, *94*,
    95, 100, 103, 105, 107
Asterolampra **32**: 560
Asteroschema sp. **32**: **166**
Asthenosoma periculosum **21**: 168
  varium **21**: 145, 150, 169
Astrangia danae **22**: 50
Astreopora **21**: 106
Astropecten sp. **32**: **165**, 177, 455
Astropecten sp.n.aff. *ornatissimus* **32**: **165**
Astropecten **22**: 109; **42**: 35
Astropecten irregularis **34**: 235
Astropecten scoparius **21**: 144
Astropectinidae **32**: 445, 446, **453**, 455
Astrophiura sp. **32**: **166**
Asychis **26**: **179**
Atapozoa **34**: 33
Ateleopodidae **32**: **171**
Ateleopus sp. **32**: **171**
Atelodinium **33**: 373, 524
Atergatis floridus **21**: 140, 141
Atheresthes stomias (arrowtooth
    flounder) **38**: 178
Athorybia **24**: 232
  lucida **24**: 118
  rosacea **24**: 126, 177, 178, 200, 218, 219
Atilax paludinosus, Physiology **27**: 373
Atlanta sp. **32**: **452**
Atlantic cod *see Gadhus morhua*
Atlantic herring *see Clupea harengus*

Atlantic salmon *see Salmo salar* **40**: 18, 23
Atlanticellodinium **25**: 127
  tregouboffi **25**: 128
Atlantopandalus **38**: 66
Atractoscion nobilis (white sea-bass)
    38: 54
  stock enhancement studies **38**: 31
  survival from hatchery releases **38**: 41
Atyidae **42**: 75
Atyloella magellanica **39**: 196
Atypopenaeus, Taxonomy **27**: 61, 63,
    64–66, **108**, **113**, 119, 120, *120*,
    121, 123
  bicornis, Taxonomy **27**: 66,
    Zoogeography/Evolution **27**: **142**
  dearmatus, Taxonomy **27**: 66
    Zoogeography/Evolution **27**: **142**
  formosus, Taxonomy **27**: 65, 66, 108, **123**
    Zoogeography/Evolution **27**: **142**
    Life histories **27**: 285
    Physiology **27**: 203
    Zoogeography/Evolution **27**: 147, 150,
    154
  stenodactylus **27**: 66, **108**, **142**
Augaptilidae **33**: 29, **51**, 103–106, **435**, 436,
    498
Augaptilina **33**: **51**, 66, 103
  scopifera **33**: 103
Augaptilus **25**: 138
  morphology **33**: 32
  taxonomy **33**: **51**, 55, 71, 74, 103
Augaptilus anceps **33**: 103
  cornutus **33**: 103
  glacialis **33**: 103
  lamellifer **33**: 103
  longicaudatus **33**: 103
  megalurus **33**: 103
  spinifrons **33**: 103
Aulacantha **25**: 127
Aulacodiscus kittoni **25**: 185
Aulacothyropsidae **32**: **409**
Aulopidae **32**: **169**
Auloressella **32**: **165**
Aurantiactinomyxon **25**: 146; **25**: 32
Aurelia **28**: 89; **20**: 53
Aurelia aurita **33**: 527; **44**: 29, 40, 121;
    21: 106
Aurelia sp. **26**: 145
Aureococcus anophagefferens **31**: 309, 310,
    361, 362
aurita **25**: 6, 7, 12, 17–19, 21–23, **31**, 33, **34**,
    36, 41, 44, *44*, 45–47, 46, 47

victoria **25**: 23, 33, 36, 45
Austinograea
  alaysae **34**: **378**
  williamsi **34**: **378**
Austrodecus **24**: 8, 9, 14, 15, 26, 45, 54,63
  breviceps **24**: 11, 60
  fagei **24**: 60
  glaciale **24**: 18, 46
Austroglossus **29**: 217
Austroniscus sp. **32**: **161**
Austropallene tcherniae **24**: 60
Austrosignum **32**: **161**
Austrosignum maltinii **39**: 208
Austrovenus stutchburyi **43**: 34, 35
Autobranchia *see* Bivalvia
Autolytus **26**: **179**
Auxis **30**: 228, 279
Aves **40**: 25, 155–157
Avicennia **24**: 60; **40**: 88, 91–94, 98–100,
    103, 108–115, 126, 141, 146, 152,
    160, 163, 167, 169, 170, 175, 178
Avicennia africana **40**: 94, 114
Avicennia alba **40**: 94, 104, 111, 120, 121,
    167
Avicennia balanophora **40**: 94
Avicennia bicolor **40**: 94, 166, 169
Avicennia eucalyptifolia **40**: 94
Avicennia germinans **40**: 94, 97, 104–106,
    110, 112–114, 156–161, 163, 167,
    168, 178
Avicennia lanata **40**: 94, 99
Avicennia marina **40**: 92, 94, 98–100, 104,
    105, 107, 108, 110–114, 116
Avicennia marina var. *australasica* **40**: 164
Avicennia marina var. *resinifera* **40**: 92
Avicennia officinalis **40**: 94, 106, 107, 111,
    114, 121, 158
Avicennia schaueriana **40**: 94, 174
Avicennia, Food/Feeding **27**: 318
Avicenniaceae **40**: 93, 94
Avrainvillea obscura **40**: 126
Axinella **21**: 104
Axiothella millsi **34**: **366**
Axis axis **40**: 158
Axis porcinus **40**: 157
Azospirillum brasilense **40**: 115
Azotobacter **40**: 115
Azygokeras **33**: **53**, 80, 81, 125
  columbiae **33**: 125

Babylonia japonica **21**: 179
Bacillariophyceae **40**: 124

Bacillus alvei and *B. thuringiensis* **37**: 107
Bacillus thuringiensis **40**: 116
Bacteria **41**: 4, 35, 45, 49, 55, *60*, 61;
    43: 64
  see also Algae; Diatoms
    activity within **43**: 202–208
    and Southern Ocean pack ice **43**: 174,
      182, 192, 193, 213, 232
  Vibrio **41**: 188, *189*, 198
Bacteriastrum **32**: 560
Bacterioplankton **43**: 238, 239
Bacteriosira **32**: 560
Bacteriosira bathyomphala **32**: 537, *546*,
    557
Bacteriosira fragilis **32**: 557
Baenectes **32**: **161**
Baetis **35**: 166, **169**
Bagarius bagarius **40**: 274, 275, 276, 277
Bagatus algicola **39**: 208
Balaena glacialis **34**: 275
Balaena mysticetus **38**: 178
Balaenoptera
  acutorostrata **34**: 270
  B. acutorostrata **43**: 227, 229
  B. musculus **43**: 229, 230
  B. physalus **43**: 213, 229
  borealis **34**: 264
  musculus **34**: 259
  physalus **34**: 275
Balaenoptera borealis **20**: 273
  musculus **20**: 273
  physalus **20**: 273
Balaenoptera musculus **31**: 20
Balaenoptera musculus × *Balaenaoptera*
    physalus **31**: 41
Balaenoptera physalis **32**: 270
Balaenoptera physalus **31**: 20
Balanion (ciliates) **41**: *34*, 38
Balanomorpha **32**: **127**, **409**
Balanus (*Solidobalanus*) *nascanus* **32**: **159**,
    161
Balanus **22**: 201, 228, 232; **23**: 87; **37**: 132;
    40: 26
  amaryllis **22**: 234
  amphitrite **22**: 208, 214, 229, 232
  amphitrite amphitrite **22**: 236
  balanoides (*Semibalanus balanoides*)
    22: 203, 208, 209–211, 215–217,
      209, 220, 222, 225, 229, 234–238
  balanus (*B. porcatus*) **22**: 214, 234, 237
  crenatus **22**: 214, 219
  glandula **22**: 213, 215

hameri **22**: 214
improvisus **22**: 208, 219
perforatus **22**: 204, 214, 229, 230,
    232, 234
psittacus **22**: 234
Balanus amphitrite **40**: 138, 139
Balanus balanoides **30**: 245
Balanus balanus **29**: 139
Balanus glandula **34**: 48, 49; **39**: 26, 29
Balanus rostratus **20**: 366
Balanus tintinnabulum **24**: 66
Balanus–Semibalanus spp. **39**: 18
Bald eagle *see Haliaetus leucocephalus*
Balistapus undulatus **34**: 230
Baltodentaliidae **42**: *143*
Baltodentalium **42**: *143*
Bankia campanellata **40**: 149
Bankia carinata **40**: 149
Bankia fimbriatula **40**: 149
Bankia gouldi **40**: 149
Bankia rochi **40**: 149
Bargmannia **24**: 117, 118
    elongata **24**: 113, 114
Barleeia **34**: 24
Barnacle *see Balanus; Chthalamus*
Barnacles **43**: 35, 70
Barramundi *see Lates calcarifer*
Barrow's goldeneye *see Bucephala islandica*
Bartholomea annulata **33**: 503
Bassanago **32**: 186, 218
Bassanago nielseni **32**: **169**, 189, **194**,
    206, 218
Bassia bassensis **24**: 132, 142, 165, 166,
    170, 208
Bassogigas profundissimus **32**: 359
Bassozetus **23**: 328
Batheuchaeta **33**: **53**, 88, 90, 125, 486
    anomala **33**: 125
    antarctica **33**: 125
    brodskyi **33**: 125
    enormis **33**: 125
    gurjanovae **33**: 125
    heptneri **33**: 125
    lamellata **33**: 125
    peculiaris **33**: 125
    pubescens **33**: 125
    tuberculata **33**: 125
Bathochordaeus **33**: 160
    charon **33**: 503
Bathoxiphinae **42**: *144*
Bathoxiphus **42**: *144*, 164
    B. ensiculus **42**: 179

Bathyacmaea
    jonassoni **34**: **371**
    secunda **34**: **371**
Bathybdella sawyeri **23**: 328; **34**: **369**
Bathybembix bairdii **35**: **32**, 80
Bathyberthella **42**: 89, 96, 132, 133, 135
    taxonomic history **42**: *114*, 115, *116*
Bathycadulus **42**: *144*
Bathycalanus **33**: **52**, 71, 119, 486
    bradyi **33**: 119
    eltaninae **33**: 119
    eximius **33**: 119
    inflatus **33**: 119
    princeps **33**: *4*, 119
    richardi **33**: 119
    sverdrupi **33**: 5, 119
    unicornis **33**: 119
Bathycatalina filamentosa **34**: **367**
Bathyceradocus stephenseni **32**: 463
Bathychaunax roseus **34**: **380**
Bathycheles incisus **42**: 216
Bathycopea **39**: 176
Bathydoris **42**: *116*, 133, 135
Bathyeliasonia kirkegaardi **32**: 349
Bathygadidae **32**: 211
Bathygnathia **32**: **161**
Bathyhalacarus **34**: **374**
Bathykermadeka hadalis **32**: 369
Bathykurila guaymasensis **34**: **367**
Bathylagus **32**: 307
Bathylagus nigrigenys **32**: 61
Bathylagus pacificus **32**: 307
Bathymargarites **34**: 415, 416
    symplector **34**: **371**
Bathymodiolid n. sp. **34**: **374**
Bathymodiolus sp. **32**: 102, 103, *104*, 107,
    108, 120, **127**, 130, 131
Bathymodiolus **23**: 331; **34**: 397, **419**; **35**: **40**,
    90
    aduloides **34**: **373**
    brevior **34**: **373**
    elongatus **34**: **373**
    japonicus **34**: **373**
    n. sp.1 **34**: **374**
    n. sp.2 **34**: **374**
    n. sp.3 **34**: **374**
    platifrons **34**: **373**
    puteoserpentis **34**: **373**
    septemdierum **34**: **373**
    thermophilus **23**: 303, 320, 321, 333, 334,
        335, 338, 344–346; **34**: **373**, 418,
        419; **35**: 19, **40**, 87, 88, 90,91

Bathymodiolus puteoserpentis **32**: 105, 120, **127**, 130
Bathynomus **30**: 91; **39**: 157
Bathynomus giganteus **32**: **409**, 410; **39**: 127, 130, 185, 188, 208
Bathypathes **32**: **160**
Bathypathes lyra **32**: 362, 503
Bathypathes patula **32**: 362
Bathypecten vulcani **32**: 408; **34**: **374**
Bathypectinura heros **32**: **447**
Bathyphysa **24**: 116, 177, 185
  conifera **24**: 113, 116, 223
  japonica **24**: 112
  sibogae **24**: 113, 116, 205, 215, 223, 229
Bathypolypus arcticus **44**: 150; **25**: 108
Bathypontia **33**: **52**, 93, 121, 486, **498**
  elegans **33**: 121
  elongata **33**: 121
  inispina **33**: 121
  intermedia **33**: 121
  kanaevae **33**: 121
  longicornis **33**: 121
  longiseta **33**: 121
  major **33**: 121
  michelae **33**: 121
  minor **33**: 121
  regalis **33**: 121
  sarsi **33**: 121
  similis **33**: 121
  spinifera **33**: 121
Bathypontiidae **33**: **52**, 121, 122, **152**, **498**
Bathypontioidea **33**: **52**, 54, 55, 121, 122, 502
Bathyporeia guilliamsoniana **39**: 196
Bathyporeia pelagica **39**: 170, 196
Bathyporeia pilosa **39**: 170, 196
Bathyporeia sarsi **39**: 196
Bathypterois **32**: 186
Bathypterois atricolor **32**: **169**
Bathysolea profundicula **29**: 220
Bathyspinula **42**: 6
  B. calcar **42**: 39
Bathyteuthidae **32**: 63
Bathytroctes oligolepsis **32**: **169**
Bathytyphlops **32**: 186
Bathytyphlops marionae **32**: **169**
Batysaurus ferox **32**: **169**
Bdellocephala punctata **21**: 133
Bdelloura candida **21**: 133
Bearded seal *see Erignathus barbatus*
Bears *see Ursus* spp.
Beggiatoa **23**: 317, 338
Bembicium auratum **40**: 147

Bembradium **32**: 186
Bembradium roseum **32**: **172**, 206, **218**
Beneckea, Parasites **27**: 382
Benthesicymus **25**: 119
Benthesicymus investigatoris **32**: **162**
Benthesicymus iridescens **32**: **77**
Bentheuphausiidae **32**: 63
Benthoctopus **34**: **380**
Benthodesmus **32**: 186, 198
Benthodesmus elongatus **32**: **174**, **195**, 218
Benthodytes typica **35**: **31**
Benthogone rosea **35**: **31**
Benthomangelia brevis **32**: **163**
Benthomisophria palliata **32**: **77**
Benthopecten armatus/spinosus **35**: **31**
Benthopectinidae **32**: 445, 455
Benthosaurus **32**: 104, 186
Benthosema glaciale **23**: 231; **26**: **120**, 130
Benthoxynus
  spiculifer **34**: **375**
  tumidiseta **34**: **375**
Benthoxynus spiculifer **23**: 329
Berardius bairdi **32**: 267
Beroë **25**: 147, 150
  abyssicola **25**: 124
Berryteuthis **32**: 245, 295, 296, 298, 299
Berryteuthis anonychus **32**: 244, 248, *250*, 268–270, 290, **290**, 291, 293, 295, 296, **297**, *298*, 299, 302, 303, 304, 310
Berryteuthis magister **32**: 244, 245, 247, 248, 249–268, *250*, 271, 275, 277, 278, 290, **290**, 291, 292, 294, 295, 296, **297**, *298*, 299, 301–304, **306**, 307–309; **35**: **32**, 103; **39**: 273
Berryteuthis magister nipponensis **32**: 249
Berthella **42**: *114*, 132
Berthellina **42**: 88, *114*, 132
Berycidae **32**: **171**
Beryx **32**: 186, 187, 193
  decadactylus **30**: 334
  splendens **30**: 331, *333*
Beryx splendens **32**: **171**, 189–191, **196**; **34**: **380**; **35**: **34**
Bestiola **33**: 120
Bestiolina **33**: **52**, 77, 120, 483
  amoyensis **33**: 120
  inermis **33**: 120
  similis **33**: 120
  sinicus **33**: 120
  zeylonica **33**: 120
Bestiolina similis **44**: 87

Betaeus emarginatus **43**: 148, 149
Betamorpha **32**: **161**
Bichir *see Polypterus senegalus*
Biddulphia **40**: 124
Biddulphia sinensis **32**: 555
Bignoniaceae **40**: 94
Billia **22**: 150
Biolitaenidae **32**: 63
Birgus latro **21**: 140; **29**: **132**, 146
Birsteiniamysis **32**: 372
Bithynia tentaculata **34**: 31
Bivalvia **31**: 39; **32**: 210, **335**, 357, 429, **438**,
      467, 468, **483**, **486**, **492**, **504**;
      40: 25, 28; **42**: 145, 204
  see also Protobranch bivalves *in subject*
    index
Bivalvia and fecundity
  see also Crassostrea; Mytilus
  environmental factors **43**: *145*, 149
  life-history strategies **43**: 115, 132, 133
  quantification methodologies **43**: *94*, 101,
    103, 104, 105
Bivalvopagurus sinensis **32**: **163**, 216
Black bear *see Ursus americanus*
Black oystercatcher *see Haematopus*
    bachmani
Black turnstone *see Arenaria melanocephala*
Black-legged kittiwake *see Rissa tridactyla*
Blastodinium **25**: 119, 125–127, 131, 133;
    33: 524
  contortum **33**: 524
Blennius pholis **26**: 137, 142; **40**: 287
Blennius sanguinolentus **40**: 259–261, 289
Blenny *see Blennius sanguinolentus*
Blue whiting *see Micromesistius poutassou*
Boccardia ligerica **29**: 253
Bodotria scorpioides **39**: 217
Boeckella **33**: **52**, 111, 113
Boehmia chelata **24**: 14, 16, 63, 71
  longirostris **24**: 71
  tuberosa **24**: 71
Boergesenia forbesii **23**: 138
Bohadschia argus **21**: 147
  bivittata **21**: 147
  koellikeri **21**: 159
Boholina **33**: **51**, 70, 73, 100, **489**
  crassicephala **33**: 100
  purgata **33**: 100
Boholiniidae **33**: **51**, 100, 489
Bolbosome caenoforme **25**: 152
Bolionopsis infundibulum **44**: 128, 129
Bolocera **22**: 74, 92

  tuediae **22**: 84, 85, 92
Bolocera tuediae **38**: 167
Boloceroides **22**: 67, 68
Bombacaceae **40**: 94
Bomolochus cuneatus **24**: 276, 290, 294
Bonelliidae **32**: 330
Bonus petrochenkoi **32**: 369
Boodleopsis carolinensis **40**: 125
Boodleopsis pusilla **40**: 126
Bopyridae **33**: 524
Bopyroides hippolytes **38**: 182
Borenymphon **24**: 26
  abyssorum **24**: 18, 42
Boreogadus saida **24**: 341, 345; **29**: 317;
    40: 22, 23
Boreomysinae **39**: 160
Boreomysis **25**: 153, 154; **32**: 57
Boreomysis arctica **39**: 213
Boreomysis incisa **32**: 57
Boreomysis nobilis **39**: 213
Boreoteuthis **32**: 245
Boreoteuthis borealis **32**: 248
Boschmaella **22**: 219
Bosmina (Eubosmina) longispina **31**: 87,
    132
Bosmina **30**: 183; **31**: 102
Bostrychia **40**: 125, 126
Bostrychia calliptera **40**: 125
Bostrychia moritziana **40**: 125
Bostrychia pinnata **40**: 125
Bostrychia radicans **40**: 125, 126
Bostrychia simpliciuscula **40**: 125, 126
Bostrychia tenella **40**: 125
Bothidae **32**: 167, 211; **40**: 25
Bothriocephalus **40**: 6
Bothriocephalus crenatus **24**: 273, 276, 285,
    287, 293, 302, 304
Bothriocephalus gregarius **40**: 17
Bothriocephalus scorpii **40**: 6, 17, 25, 34, 46
Bothriotaenia **24**: 289
Bothrocara hollandi **32**: 266
Bothrocarichthys microcephalus **32**: 307
Botrynema brucei **32**: 72
Botryoglossum farlowianum **23**: 80
Bougisia **25**: 154, 156
Bouvierella curtirama **32**: **128**; **34**: **376**
Bovallia **35**: 173, **177**, *178*
  gigantica **35**: 174
Bovallia gigantea **39**: 196
Bowfin *see Amia calva*
Bowmaniella floridana **39**: 213

Brachionus plicatilis **30**: 225, 232, 246;
    31: 92; **37**: 131, 132; **20**: 30, 182
Brachiopoda **32**: **165, 210**, 329, **335**, 357,
    389, 404, **409**, 467, 468, **483, 504**
Brachycalanus **33**: **53**, 84, 132, **500**
  atlanticus **33**: 132
  bjornbergae **33**: 132
  minutus **33**: 132
  ordinarius **33**: 132
  rothlisbergi **33**: 132, **500**
Brachydanio rerio **25**: 19, 20; **31**: 194;
    40: 264
Brachydiastylis resima **39**: 217
Brachyosoma scottocalani **25**: 139
Brachyphallus crenatus **40**: 6, 16, 27, 31,
    36, 43, 46
Brachyramphus marmoratus **34**: 270
Brachyscelus **25**: 156
Brachyscelus crusculum **39**: 207
Brachyura **32**: **163, 210**
Brada sp. **32**: **460**
Brada villosa **26**: **179**
Bradycalanus **33**: **52**, 71, 119, 486
  gigas **33**: 119
  pseudotypicus **33**: 119
  pseudotypicus enormis **33**: 5, 119
  sarsi **33**: 119
  typicus **33**: 119
Bradyetes **33**: **53**, 87, 90, 125
  brevis **33**: 125
  florens **33**: 125
  inermis **33**: 125
  matthei **33**: 125
Bradyidius
  nutrition **33**: **152**, 166
  taxonomy **33**: **53**, 78, 80, 88, 90, 92,
    93,'125
Bradyidius angustus **33**: 125
  armatus **33**: 125
  arnoldi **33**: 125
  bradyi **33**: 125, 268, **499**
  brevispinus **33**: **499**
  curtus **33**: 125
  dentatus **33**: **499**
  hirsutus **33**: 125
  luluae **33**: 125
  pacificus **33**: 125
  plinioi **33**: 125, **499**
  rakuma **33**: 125
  robustus **33**: **499**
  saanichi **33**: 125
  similis **33**: 125, **499**

spinibasis **33**: 499
spinifer **33**: 125
styliformis **33**: 125
subarmatus **33**: 125
tropicus **33**: 125
Brama japonica **32**: 267, 269, 281, 283, 285,
    286, 305
Branchinotogluma
  burkensis **34**: **367**
  grasslei **34**: **367**
  hessleri **34**: **367**
  sandersi **34**: **367**
Branchiophryxus **25**: 154
  caulleryi **25**: 154
  koehleri **25**: 154
  nyctiphanae **25**: 154
Branchiopoda **43**: *21*
Branchiplicatus cupreus **34**: **367**
Branchipolynoë
  pettiboneae **34**: **367**
  seepensis **34**: **367**, 394
  symmytilida **34**: **367**
Branchipolynoe **32**: 98, 130
Branchipolynoe seepensis **32**: **126**, 130, 131
Branchipolynoe symmytilida **23**: 326, 327
Branchiura **43**: 55
brasiliensis, Taxonomy **27**: 98, **110**
  (*see also Penaeus aztecus*)
  Behaviour **27**: 336
  Food/feeding **27**: **317**, 319, 331
  Moulting/growth **27**: 226, **240**
  Parasites **27**: 386
  Reproduction **27**: 257
  Zoogeography/Evolution **27**: **146**
Brasilomysis casyroi **39**: 214
Brassavola nodosa **40**: 129
Brattstromia **33**: **51**, 70, 73, 100
  longicaudata **33**: 100, **489**
Bresiliidae **32**: 110, 120, **129**
Brevinucula **42**: 6
brevirostris, Taxonomy **27**: 97, **110**
  Life histories **27**: **304**
  Zoogeography/Evolution **27**: **145**
Brevitricolpites variabilis **40**: 92
Brevoortia **25**: 6; **20**: 14, 21, 161, 177
  maculata chilcae **20**: 167
  patronus **20**: 6, 39
  pectinata **20**: 180
  tyrannus **25**: 9, 21, 22, 25, 30, **34**; **20**: 6,
    11, 29, 37, 39, 102
Brevoortia patronus **26**: **121**, 137
Brevoortia smithi **31**: 50

Brevoortia tyrannus **44**: 215, 249; **26**: 121
    28: 63, **66**, 120; **31**: 50, 201; **34**: 81
Brine shrimp *see Artemia* sp.
Brisaster latifrons **32**: **462**
Brisinga eucoryne **32**: **165**, 180, 184
Brisingella sp. **32**: **165**
Brisingidae **32**: *97*, 445
Brissidae **32**: 466
Brittle star (*Ophiura ljungmani*) **43**: 107
Brnnich's guillemot *see Uria lomvia*
Bromeliaceae **40**: 128
Brosme brosme **26**: 33, 34; **40**: 14
Brotulidae **32**: 349
Brown algae
    Padina, Sargassum and Turbinaria
      43: 308
    Phaeophyta *Plate 4* **43**: 281
Brown algae *see Fucus gardneri*
Brown bear *see Ursus arctus*
Brown trout *see Salmo trutta*
Bruceiella globulus **34**: **371**
Bruguiera **40**: 99, 103, 104, 107
Bruguiera cylindrica **40**: 95, 104, 121
Bruguiera exaristata **40**: 95
Bruguiera gymnorhiza **40**: 95, 99, 102, 112,
    114, 141, 142, 165, 170, 177
Bruguiera hainesii **40**: 95, 107
Bruguiera parviflora **40**: 95
Bruguiera sexangula **40**: 95, 158, 184
Bryopsis hypnoides **23**: 108
Bryothamnion seaforthii **23**: 100
Bryozoa **32**: **335**, 467, 468, 470, **483**; **43**: *21*,
    102, 110, 111, 150
Bubalus bubalis **40**: 157
Buccinidae **32**: 371
Buccinum **21**: 139, 153, 154; **25**: 186;
    34: 37335: 32
    undatum **34**: **6**, 18, 22, 33, 37, **41**, 288
    undulatum **21**: 153, 154, 177, 178
    viridum **34**: **372**
Buccinum undatum **40**: 27; **43**: 67
Bucephala **39**: 15, 17
Bucephala clangula (common
    goldeneye) **39**: 68
Bucephala clangula **34**: 270
Bucephala islandica (Barrow's golden
    eye) **39**: 64–5, 68, 78, 80
Bucephaloides arcuatus **40**: 6, 13, 20
Bufo **40**: 154
Bufo marinus **24**: 366
Bufo melanostictus **40**: 154, 155

Buglossidium luteum **29**: 220, 229, 230, 237
    253
Bugula neritina **43**: 102
Bugula turbinata **24**: 45
Bulla **42**: 105, *116*, 132, 135
    B. ampulla **42**: *72*, 73
    external anatomy **42**: 91, 92, 93
    morphology **42**: 75, *76*
    shell **42**: *84, 85*, 89
Bullariidae **42**: 75
Bullhead *see Ictalurus nebulosus*
Bullia **30**: 114, 117
    digitalis **25**: 180–238, *181*, *188*, *200*, *202*,
      204, *205*, **209**, **211**, **213**, *215*, **217**,
      220, **220**, 222, **226**, **231**, *232*, *233*
    digitalis **30**: 103
    laevissima **25**: 196
    pura **25**: 180
    rhodostoma **25**: 180, 189, 195, 196, 208
    tenuis **25**: 194
Bullia digitalis **28**: 390–426
Bullia laevissima **28**: *392*, 402, **403**, 407,
    414, *417*
Bullia melanoides **28**: 406
Bullia rhodostoma **28**: *392*, **403**, 404, 418
Bullidae **42**: 75, 86, 87, 90, 97, 109, 110,
    129
Bullinidae **42**: 75
Bulloidea **42**: 75, *76*, *84, 85*, 98, 106, 115
Bullomorpha **42**: 75
Bumeralius buchalius **39**: 196
Bunodactis elegantissima **21**: 106
Bunodactis verrucosa **22**: 72
Bunodeopsis **22**: 90
Burbot *see Lota lota*
Burnupena **28**: 425
Bursatella **42**: 71, 88
Busycon **25**: 200; **28**: *392*
    canaliculatum **34**: 4, 5, 6, 7
    undatum **34**: 5
Busycon carica **28**: 397–400, 411, 426
Busycon contrarium **28**: 412
Butoroides striatus, Predation **27**: 372
Butter clam *see Saxidomus giganteus*
Byblis japonicus **39**: 196
Bythitadae **32**: **129**
Bythites hollisi **34**: **381**
Bythograea **23**: 322
    intermedia **34**: 378
    laubieri **34**: 378
    microps **34**: 378
    thermydron **34**: 378

Bythograeidae **32**: **129**
Bythotrephes **31**: 93, 95, 97, 140
Bythotrephes cederstroemi **31**: 86
Bythotrephes longimanus **31**: 86, 94, 95,
      105, 106

Cachonella **25**: 127
    paradoxa **25**: 131, *132*
Cactaceae **40**: 128
Cactospiza heliobates **40**: 157
Cadulus **42**: *144, 177*
    C. austinclarki **42**: *205*
    C. californicus **42**: 196
    C. quadridentatus **42**: 163, 171
    C. subfusiformis **42**: 191, 196, *203*
    C. tolmiei **42**: 149
       feeding and digestion **42**: 163, 164,
          166, 169, *171*, 172
    C. vincentianus **42**: 204
Caecidae **39**: 36; **42**: 149
Caelorinchus **32**: 167, 186, 191
Caelorinchus immaculatus **32**: 146, **170**,
      188, **194**, 206, 208, 211, 220, 227
Caelorinchus multifasciatus **32**: **170**
Caelorinchus nazcaensis **32**: **170**, 191, **194**
Caelorinchus spilonotus **32**: **170**, 186, 188
Caenogastropoda **43**: 93
Caenopedina sp. **32**: **165**
Caesalpiniaceae **40**: 94
Cajanus **40**: 121
Calamoecia **33**: 52, 111
Calanidae **32**: 11, *35*; **33**: 222, 347, 365,
      434, 528
    morphology **33**: 21, 30, 37
    taxonomy **33**: **52**, 118, 119
Calanipeda **33**: **52**, 115
    aquae-dulcis **33**: 115, **375**
Calanoida **33**: 1
    phylogeny **33**: 2-3, *3*
    taxonomy **33**: 2, 50, **51-53**, 65-97,
      99-138
Calanoida **26**: *131*
Calanoides
    ecology **33**: 480, 482, 487, 528
    growth **33**: 299
    nutrition **33**: **152**
    taxonomy **33**: **52**, 68, 118
Calanoides spp. **32**: 29
Calanoides acutus **33**: **7**, 118
    carinatus **33**: **7**, **26**, 118, 298
       behaviour **33**: **415**, **446**
       biochemistry **33**: 245

distribution **33**: **512**
    ecology **33**: **356**, 365, **375**, 481, 482,
       484, 485
    growth/development **33**: **301**, **320**
    life history **33**: 340, 383, 384, 385, **386**
    nutrition **33**: **183**, **186**, **189**, **198**
    reproduction **33**: **255**, **268**, 282, **285**
    size/weight **33**: **223**, **229**, **331**
macrocarinatus **33**: 118
natalis **33**: 118
patagoniensis **33**: 118
philippinensis **33**: 118, 482, **485**
    biochemistry **33**: 245, 247, 248
    ecology **33**: **355**, 492, 494
    growth/development **33**: 344
    life history **33**: 385, **386**, 390
    nutrition **33**: 159, 170, **186**, **189**, 196,
       208
    physiology **33**: 210
    reproduction **33**: 266, **285**, 292
    size/weight **33**: **223**, **229**, **234**
Calanoides acutus **44**: 61; **32**: 26, 27, *28*;
      43: 217-219
Calanoides carinatus **44**: 61
Calanopia **33**: **52**, 93, 113, 518
    americana **33**: 113, **270**, **489**
    aurivilli **33**: **47**, 113
    australica **33**: 113
    biloba **33**: 113
    elliptica **33**: **47**, 113, **430**, **507**
    herdmani **33**: 113
    media **33**: 113, **507**
    minor **33**: **47**, 113, **491**
    parathompsoni **33**: 113
    sarsi **33**: 113
    sewelli **33**: 113
    seymouri **33**: 113
    thompsoni **33**: **26**, 113, **270**
Calanozoa **43**: 209, 210, 217
Calantica moskalevi **32**: **409**
Calanus spp. **32**: 29, 34
Calanus **23**: 230, 234; **25**: 137, 142; **26**: 272,
      274-276, 279, 281; **28**: 101, **103**,
      130; **29**: 45, 175; **32**: 303; **33**: 8,
      41, 218, 236, 298, 346, 518;
      20: 168
    behaviour **33**: 402, 405, 425, **426**, 428,
      429, 451
    biochemistry **33**: 237, 240, 243, 245
    distribution **33**: 508
    ecology **33**: 366, 370, 374, 480, 483, 487,
      490, 495, 497, 528

finmarchicus **25**: 122, 138, 148, 150;
    29: *27, 34*, **131**, 178, 179; **30**: **164**
growth/development **33**: 299, 330
helgolandicus **25**: 150
nutrition **33**: 140, 150, **152**, 168, 169, 172,
    208
pacificus **30**: **164**
pacificus californicus **30**: 172
plumchrus **25**: 121
reproduction **33**: 265, 283
sinicus **25**: 151
taxonomy **33**: 52, 60, 62, 63, 64, 67, 118
tonsus **30**: **164**
Calanus agulhensis **33**: 60, 118, 516
  australis **33**: 118, 516
    distribution **33**: **512**
    growth **33**: **302, 320**
    reproduction **33**: **286**, 291
  chilensis **33**: 118, **356, 485**, 516
  euxinus **33**: 48, 60, 118, 516
  finmarchicus **33**: 7, 10, **26**, 516
    behaviour **33**: 405, **408, 410, 415, 418**,
      419, 425, **426, 427**, 428, **433**, 439,
      441, **442**, 443, **446**, 450
    biochemistry **33**: 237, 240, 242, 247,
      250
    chromosomes **33**: **47**, 48
    distribution **33**: 506, 508, **512**, 516, 517
    ecology **33**: 351, **356**, 365, 369, 370,
      376, 380, 394, **395**, 398, 468, 470,
      481, **493**, 495, 523, 528
    growth/development **33**: **302, 310, 311**,
      312, *318*, **320**, 323, 338, 339, **341**
    life history **33**: 385, **386**
    morphology **33**: *4, 22*, 28, 31, **32**, 32,
      38, 39, 46, 47
    nutrition **33**: 150, **151**, 157, 163, 165,
      172, **183, 186, 187, 189, 192**, 197,
      200, 203, 204, **206, 208**
    physiology **33**: 215, 216, 217
    reproduction **33**: 43, 254, **255**, 260,
      261, 265, 267, **268**, 273, 282, **286**,
      288, 289
    size/weight **33**: **229, 234**
    taxonomy **33**: 49, 63, 64, 118
  fonsecai **33**: 118
  glacialis **33**: 7, **32**, 47, 118, 516
    behaviour **33**: 405, **426, 446**, 451
    biochemistry **33**: 247
    distribution **33**: **512**
    ecology **33**: **357**, 367, **376, 493**, 495,
      528

    growth/development **33**: **310, 311**
    life history **33**: 385, **386**
    nutrition **33**: **186, 190**
    physiology **33**: 216
    reproduction **33**: 281, 283, **286, 288**,
      292
    size/weight **33**: **234**, 334
  helgolandicus **33**: 7, 9, **26**, 63, 118, 297,
    516
    behaviour **33**: 402, **410, 418**, 419, 420,
      422, 428, 443, 452
    biochemistry **33**: 237, **239**, 240, 251
    chromosomes **33**: **47**
    distribution **33**: **512**, 517
    ecology **33**: **357**, 365, **376, 386, 396**,
      398, 522, 528
    growth/development **33**: **310, 311**, 319,
      320, 321, 326, *327*, **341**
    morphology **33**: *40*, 41
    nutrition **33**: 145, 150, **151**, 166, 171,
      187, 192, 197, **198, 200**, 203, 205,
      206, 208
    reproduction **33**: 43, **268**, 282, **286**,
      288, 290
    size/weight **33**: 221, **223, 229**, 233
  hyperboreus **33**: 7, **26**, 49, 63, 118, 516
    behaviour **33**: 406, **410**, 419, **426**, 445
    biochemistry **33**: 247
    distribution **33**: **512**
    ecology **33**: **357**, 369, **376**, 399, **493**,
      495
    growth/development **33**: **302, 310, 312**,
      321
    life history **33**: 385, **386**
    morphology **33**: **32**, 37, 46
    nutrition **33**: 156, 166, **190, 206**
    physiology **33**: 216
    reproduction **33**: 260, **267**, 272, 277,
      283, **286, 288**, 292
    size/weight **33**: **229, 234**, *332*
  jaschnovi **33**: 36, 60, 118, 516
  magnellanicus **33**: 118
  marshallae **33**: **32**, 118, 516
    distribution **33**: **512**
    ecology **33**: **357**, 365, **485, 493**, 495
    growth/development **33**: **302, 310, 312**,
      314, 316, **320**, 321, 323, 329, **341**
    life history **33**: **386**
    nutrition **33**: 196, **208**
    reproduction **33**: 281, 282, **286**
    size/weight **33**: **230**, 328, 333, *334, 335*
  minor **33**: 7, **26, 40**, 64, 118, 516

distribution **33**: **512**
ecology **33**: **357**, 503
growth/development **33**: **302**
nutrition **33**: **194**
size/weight **33**: **223**, **331**
pacificus **33**: **7**, 118, 297, 516
  behaviour **33**: 406, **408**, **410**, **415**, **426**,
    427, 428, **446**, 449, 451, 453, 455
  biochemistry **33**: 245, 249
  distribution **33**: **512**
  ecology **33**: **357**, 365, 366, 371, **376**
  growth/development **33**: **302**, **310**, **312**,
    315, 316, 316, **320**, 326, 338, 340,
    341
  life history **33**: **386**
  nutrition **33**: 140, 149, 154, 159, 163,
    164, 171, 178, 182, **183**, **186**, **187**,
    188, **190**, **192**, **198**, 203, 204
  reproduction **33**: **255**, 257, 259, **268**,
    281, **286**, **288**, 289, 290
plumchrus **33**: 385
ponticus **33**: 516
propinquus **33**: **7**, 118, 516
  behaviour **33**: **415**
  biochemistry **33**: 226, 237, 240, 242,
    243, 245, 246, 247, 248
  distribution **33**: **512**
  ecology **33**: **357**, 492, 494, 528
  life history **33**: **386**
  nutrition **33**: 163, 166, **190**, **208**
  physiology **33**: 218
  reproduction **33**: **268**, **286**
  size/weight **33**: **223**, **230**, **234**
simillimus **33**: 118, 516
  biochemistry **33**: 247
  distribution **33**: **512**
  ecology **33**: **357**, 528
  life history **33**: **387**, 390
  nutrition **33**: **190**
  reproduction **33**: **286**
sinicus **33**: 22, 36, 118, 516
  behaviour **33**: **446**
  biochemistry **33**: **239**, **241**
  distribution **33**: 506, **512**
  ecology **33**: **357**, **376**
  growth/development **33**: **310**, **312**, **320**,
    344
  nutrition **33**: **200**
  reproduction **33**: **268**
  size/weight **33**: **230**, **331**, 333
  ecology **33**: **375**, **484**
  size **33**: 338, *340*

tenuicornis **33**: **223**
Calanus agulhensis **44**: 61–63
Calanus australis **32**: 34, *35*
Calanus carinatus **32**: 36
Calanus chilensis **44**: 63, 64
Calanus cristatus **32**: 265, 266, 269, 274
Calanus euxinus **32**: 40, *41*
Calanus finmarchicus **44**: 64–66, 248;
    32: 22, 23, 60; **26**: 280, *283*
Calanus glacialis **44**: 7, 66–68; **32**: 19, 21,
    21, 78
Calanus helgolandicus **44**: 68, 69; **44**: 69;
    32: 19, *20*, 21–23, *24*, *25*, 48, 60
Calanus marshallae **44**: 69, 70
Calanus pacificus **44**: 70, 71; **32**: 43, 269
Calanus plumchrus **32**: 265, 269, 274
Calanus propinquus **32**: 26, 27, 28; **43**: 210,
    215, 217–219
Calanus similimus **32**: 26
Calanus simillimus **44**: 71
Calanus spp. **44**: 71
Calanus tenuicornis **44**: 71
Calanus tonsus **32**: 26
Calathella mangrovei **40**: 119
California sardine *see Sardinops caeruleus*
californiensis, Morphology **27**: 31–33
  Behaviour **27**: 350
  Food/feeding **27**: 326, 330
  Life histories **27**: 307
  Moulting/growth **27**: 220, 221, **230**,
    235, **249**
  Parasites **27**: 382, 383
  Physiology **27**: **168**
  Predation **27**: 359
  Reproduction **27**: 257, **279**
  Taxonomy **27**: 97, **110**
  Zoogeography/Evolution **27**: **145**
Caligus **28**: 81; **40**: 35, 36, 54
Caligus clemensi **24**: 276, 290
  elongatus **24**: 276, 289, 304
  rapax **24**: 304
Caligus curtus **40**: 9, 32
Caligus diaphanus **40**: 9, 19, 33
Caligus elongatus **40**: 9, 32
Caligus epidemicus, Parasites **27**: 387
Caligus minimus **40**: 9, 10, 19, 33
Caligus sp. **26**: *131*
Callanthias **32**: 186, 192
Callanthias parini **32**: **173**, 189–191, **195**,
    224
Callanthias platei **32**: 224
Callanthiidae **32**: **173**

Calliactis parasitica **21**: 106; **22**: 93; **24**: 45
  polypus **22**: 68
Calliaster corynetes **24**: 52
Callinectes sapidus **44**: 213, 215, 225, 229,
  249; **29**: **133**, 134
Callinectes sapidus, Physiology **27**: 197
  Predation **27**: 360
Calliobdella nodulifera **40**: 9, 19, 32
Calliobdella vivida **24**: 273, 291
Calliodentaliidae **42**: *142, 143*
Calliodentalium **42**: *143*
  C. callipeplum **42**: 191
Callionymus lyra **28**: **118**
Callionymus sousai **30**: 327
Calliopes **25**: 8
Calliopius laeviusculus **44**: 9, 122–124;
  39: 196, 197
Calliostoma **34**: 33
  canaliculatum **34**: 33
  zizyphinum **34**: 35
Callipallene **24**: 2, 8, 26, 31
  amaxana **24**: 32
  brevirostris **24**: 28, 32, 46, 57
  phantoma **24**: 72
  spectrum **24**: 57
Calliphora **40**: 178
Callista brevisiphonata **21**: 170, 179; **37**: **10**
Callogorgia **30**: 318
Callorhinus ursinus **32**: 270; **39**: 280
Callyntrochlamys phronimae **25**: 141
Callyspongia **21**: 97
  fallax **21**: 97
Calocalanus **33**: **52**, *61*, 77, 120, 480, 483
  aculeatus **33**: 120
  adriaticus **33**: 120
  africanus **33**: 120
  alboranus **33**: 120
  antarcticus **33**: 120
  atlanticus **33**: 120
  beklemishevi **33**: 120
  contractus **33**: 120
  curtus **33**: 120
  dellacrocei **33**: 120
  elegans **33**: 120, **358**
  elongatus **33**: 120
  equilicauda **33**: 120
  fiolentus **33**: 121
  fusiformis **33**: 120
  gracilis **33**: 120
  gresei **33**: 120
  indicus **33**: 120
  kristalli **33**: 121

latus **33**: 120
lomonosovi **33**: 120
longifurca **33**: 120
longisetosus **33**: 120
longispinus **33**: 120
minor **33**: 120
minutus **33**: 120
monospinus **33**: 120
namibiensis **33**: 120
nanus **33**: 120
neptunus **33**: 120
omaniensis **33**: 120
ovalis **33**: 120
paracontractus **33**: 120
paralongatus **33**: 120
pavo **33**: **7**, **26**, 120, **358**, 506, **507**
pavoninus **33**: 120, **223**
plumatus **33**: 120
pseudocontractus **33**: 120
pubes **33**: 120
pyriformis **33**: 120
regini **33**: 120
sayademalja **33**: 120
spinosus **33**: 120
styliremis **33**: **7**, **26**, 120
tenuiculus **33**: 120
vinogradovi **33**: 120
vitjazi **33**: 120
vivesi **33**: 120
Caloglossa **40**: 125, 126
Caloglossa angustalata **40**: 125
Caloglossa leprieurii **40**: 125, 126
Caloglossa ogasawaraensis **40**: 125
Caloglossa stipatata **40**: 125
Calothrix **23**: 317; **40**: 117
  crustacea **23**: 97, 107
Calothrix viguieri **40**: 116
Calstevenus **42**: *144*
Calycophora **24**: 178
Calyptogena sp. **32**: 125, 130; **23**: 319, 328,
    332, 339,342, 343; **30**: 325, 326;
    35: 23, 50
  elongata **35**: 24
  kilmeri **34**: **373**
  magnifica **23**: 303, 315, 320, 321, 333,
      334, 335, 337–339, 344, 345;
      34: **373**, 385, 397, 418; **35**: 27, **40**,
      91
  pacifica **23**: 340; **34**: **373**
  pacifica-like **35**: **40**
  phaseoliformis-like **35**: **40**
  ponderosa **23**: 339; **35**: **40**

solidissima **34**: **373**
Calyptogena pacifica **32**: 131
Calyx nicaeensis **21**: 98
Campages **28**: **206**
Campaneria **33**: **51**, 74, 101
  latipes **33**: 101, 103, **498**
Campanile symbolicum **32**: 407
Campanularia flexuosa **22**: 50
Campanularia petiolatus **24**: 43
Campecopea **39**: 176
Campecopea hirsuta **39**: 208
Camponotus **40**: 145
Camptostemon philippinensis **40**: 94
Camptostemon schultzii **40**: 94
Campylaimus **30**: **50**
canaliculatus, Taxonomy **27**: 95
    Food/feeding **27**: **317**, 322, 330
    Moulting/growth **27**: 249
    Reproduction **27**: **258**, **270**
    Zoogeography/Evolution **27**: **144**
Cancellothyris **28**: **206**
Cancer balssii **32**: **163**, 217
Cancer chinensis (*see Penaeus chinensis*)
  kerathurus (*see Penaeus kerathurus*)
  setiferus (*see Penaeus setiferus*)
Cancer magister (Dungeness crab) **39**: 17;
      26: 260; **34**: 224
Cancer margaritarius **32**: 217
Cancer polydon **35**: **184**
Candacia **25**: 141
  ecology **33**: 480, 483, 490, 528
  morphology **33**: *16*, 36
  nutrition **33**: 160, 165, 166
  taxonomy **33**: **51**, *61*, 95, 111
Candacia aethiopica **33**: **7**, **26**, **32**, **491**
  armata **33**: **26**, 111, **302**, **376**, **512**
  bipinnata **33**: 111, **491**, 503, **512**
  bradyi **33**: 111, 160, **512**
  caribeannensis **33**: 111
  catula **33**: 111, 503, **512**
  cheirura **33**: 111, **512**
  columbiae **33**: 111
  curta **33**: 111, **491**, **512**
  curticauda **33**: **223**
  discaudata **33**: 111, **512**
  elongata **33**: 111
  ethiopica **33**: 111, **512**
  falcifera **33**: 111
  giesbrechti **33**: 111
  grandis **33**: 111
  guggenheimi **33**: 111, **512**
  guinensis **33**: 111

  ketchumi **33**: 111
  longimana **33**: 111, **512**
  magna **33**: 111
  maxima **33**: 111, **415**
  nigrocincta **33**: 111
  norvegica **33**: 111, **512**
  pachydactyla **33**: 111, **223**, **415**, **491**, 508,
      512
  paenelongimana **33**: 111
  parafalcifera **33**: 111
  pofi **33**: 111
  rotunda **33**: 111
  samassae **33**: 111, **512**
  tenuimana **33**: 111, **512**
  tuberculata **33**: 111, **512**
  varicans **33**: 111, **512**
Candacia armata **29**: **131**, 179
Candacia spp. **44**: 72
Candaciidae
  behaviour **33**: 434
  distribution **33**: **512**, 515
  ecology **33**: **491**, 526
  nutrition **33**: **152**, 160
  taxonomy **33**: **51**, 111
Candelabrum
  phrygium **34**: **365**
  serpentarii **34**: **365**
Candelabrum phrygium **32**: **126**, 130
Candelabrum serpentarii **32**: **126**
Canthocalanus **33**: **52**, 68, 118, 497
  pauper **33**: 118, 524
Cantrainea jamsteci **34**: **372**
Capelin *see Mallotus villosus*
Capillaria **40**: 8, 18
Capillaria gracilis **40**: 8, 18, 28, 31
Capillaria kabatai **40**: 18
Capitella **21**: 37; **33**: 293; **34**: **366**; **43**: 125,
      131, 132, 137, 138, 139, 144
  nr *capitata* **34**: **366**
Capitella capitata **26**: 190
Capitellidae **32**: **126**; **39**: 36
Caprella **39**: **206**
Caprella advena **39**: **206**
Caprella albifrons **39**: 125
Caprella bidentata **39**: **206**
Caprella bispinosa **39**: **206**
Caprella christibrachium **39**: **206**
Caprella danilevskii **39**: **206**
Caprella decipiens **39**: **206**
Caprella equilibra **39**: 125, 129
Caprella gorgonia **39**: 150, **206**
Caprella kroyeri **39**: **206**

Caprella laeviuscula **39**: 146, 206
Caprella monoceros **39**: 181, 206
Caprella mutica **39**: 206
Caprella okadai **39**: 206
Caprella penantis **39**: 206
Caprellidae **39**: 36
Caprellidea **39**: 157, 186, 187, 189, 206
Caprodon **32**: 186
Caprodon longimanus **30**: 334, 338; **32**: **172**, 218, 224
Caproidae **32**: **172**
caramote (*see Penaeus kerathurus*)
Carangidae **32**: 62, **173**
Carangoides fulvoguttatus **44**: 231
Caranx bucculentus, Predation **27**: 361, **362**, 368
  ignobilis, Predation **27**: **362**, 366
  melampygus, Behaviour **27**: **352**
    Predation **27**: 366
  papuensis, Predation **27**: **362**, 366
  sem, Predation **27**: **362**
  sexfasciatus, Predation **27**: **362**, 366
Caranx ignobilis **30**: 334; **20**: 107
Caranx melampygus **44**: 222
Carapid fish (*Encheliophis vermicularis*) **41**: 188, *189*
Carapidae **32**: **171**
Carapus (fish) **41**: *189*
Carapus bermudensis **21**: 164
Carassius auratus **21**: 99; **24**: 347; **26**: 128, 129
Carcharhinus cautus, Predation **27**: 366
  fitzroyensis, Predation **27**: 367
  leucas, Predation **27**: 366
  maenas, Physiology **27**: 196, 202
  plumbeus, Predation **27**: 366
Carcharhinus spp., Predation **27**: 367
Carcharinidae **32**: 308
Carcinonemertes epialti **43**: 70
Carcinoscorpinus rotundicaudata **21**: 140
Carcinus
  maenas **34**: 289
  mediterraneus **34**: 81
Carcinus maenas **22**: 93; **29**: **133**, 134–136, 138, 142; **40**: 140; **43**: 70
Carcinus, Moulting/Growth **27**: 220
Cardinella, Introduction **27**: 1
Cardiophilus baeri **39**: 197
Cardisoma guanhumi **21**: 115
Cardium (Cerastroderma) edule **22**: 103, 104, 108, 110, 124, 125, 127, 128, 132, 134, 150, 156

  glaucum **22**: 103, 108, 124, 125, 127
Cardium **38**: 18
Cardium edule **35**: **168**
Careproctus hyaleius **34**: **381**
Caretta caretta **24**: 208
Caridea **32**: **162**
Caridean shrimps **43**: 98, 135, 142, 143, 148
Caridean sp. **26**: *131*
Caridion gordoni **38**: 179
Carinaria cristata **24**: 209
carinatus (*see Penaeus monodon*)
Carp *see Cyprinus carpio*
Carpodon longimanus **32**: **192**, **195**
Carybdea alata **21**: 105
  marsupialis **21**: 105
  rastoni **21**: 105
Caryotoma bernardi **25**: 127, *128*
Casco bigelowi **39**: 197
Caspievadne **31**: 86
Cassidula nucleus **40**: 147
Cassiopea xamachana **21**: 105
Cassiopeia **25**: 125; **40**: 130
Cassis tuberosa **34**: 24
Castanidium longispinum **25**: 119
Cataetyx laticeps **32**: **129**, 130; **34**: 380
Catanella **40**: 125
Catenella caespitosa **40**: 125
Catenella impudica **40**: 126
Catenulida **43**: 22
Catfish, Arabian Gulf *see Arius thalassinus*
Catfish, channel *see Ictalurus punctatus***40**: 287, 321
Catfish, European *see Silurus glanis*
Catfish, Indian *see Heteropneustes fossilis*
Catharacta maccormicki **20**: 297
  skua **20**: 236
Catoblema vericarium **32**: 72
Catostylus mosaicus **21**: 105
Caudofoveata **32**: 369
Caulerpa **21**: 71, 92; **40**: 126; **42**: 78
  racemosa **21**: 92
Caulerpa racemosa **43**: 282
Cavolina longirostris **32**: **452**
Cavolina uncinata **32**: **452**, **453**
Cavolinia **25**: 152; **42**: 73
Cebus apella apella **40**: 157
Cellana **23**: 97; **34**: 33
  grata **34**: 21, 28, 40
  toreuma **34**: **6**, 22
Centraugaptilus **33**: **51**, 71, 74, 103, 486
  cucullatus **33**: 103, **435**
  horridus **33**: 103, *422*, *423*, **435**

lucidus **33**: 103
porcellus **33**: 103
pyramidalis **33**: 68, 103
rattrayi **33**: 103, **435**
Centrechinus artillarium **21**: 169
Centroberys affinis **30**: 334
Centroceras clavulatum **23**: 108
Centrodraco **32**: 167, 186
Centrodraco gegonipa **32**: **173**, 188
Centrodraco nakaboi **32**: **173**, 188, 190,
    194, 218
Centrodraco nanus **32**: **173**
Centrodraco striatus **32**: **173**, 189, 190,
    193, **194**
Centrolophidae **32**: **174**
Centrolophus niger **35**: 34
Centronotus gunnellus **26**: 121
Centropages **25**: 147, 151; **28**: 100; **30**: **164**;
    33: 218, 298
  behaviour **33**: 402, **411, 427**, 428
  distribution **33**: **507**
  ecology **33**: 468, 480, 481, 490, 497, 528
  furcatus **29**: **131**, 175, *177*
  hamatus **29**: *23*
  life history **33**: 347, 389
  morphology **33**: 31
  nutrition **33**: **152, 194**
  reproduction **33**: 46, 257, 261, **270**
  size **33**: *332*
  taxonomy **33**: **52**, *61*, 68, 73, 93, 112
  typicus **29**: **131**, 175, *177*
Centropages abdominalis **33**: *22*, **26**, 112
  acutus **33**: 112
  alcocki **33**: 112
  aucklandicus **33**: 112, **512**
  australiensis **33**: 112, **512**
  brachiatus **33**: 112, **376, 484, 485, 512**
  bradyi **33**: 112, **512**
  brevifurcus **33**: 112
  calaninus **33**: 112
  caribbeanensis **33**: 112
  chierchiae **33**: **26**, 112, **303, 446, 484, 512**
  dorsispinatus **33**: 112
  elegans **33**: 112
  elongatus **33**: 112
  furcatus **33**: **7, 26**, *32*, 112
    biochemistry **33**: 251
    chromosomes **33**: **47**
    distribution **33**: **512**
    growth/development **33**: **303, 310**
    nutrition **33**: 159, **184, 200**, 203, **208**
    reproduction **33**: **268, 269**

gracilis **33**: 112
halinus **33**: 112
hamatus **33**: **7, 26**, 112, 527
  behaviour **33**: 403, 409, **410, 415**, 419,
    444
  distribution **33**: 512
  ecology **33**: **358**, 369, **376, 396**
  growth/development **33**: **303**, 316, *317*,
    320, 342
  life history **33**: 388
  nutrition **33**: 143, 161, 164, 177, **184,
    186, 190**, 197, **206**
  physiology **33**: 216, 217
  reproduction **33**: **270**, 282, **286**
  size/weight **33**: **230**, 330, 333
karachiensis **33**: 112
kröyeri **33**: **26**, 112, **376, 396**
longicornis **33**: 112
mcmurrichi **33**: 112
natalensis **33**: 112
orsinii **33**: 112, **426**, 429, **430**
ponticus **33**: 112, **270, 288**
sinensis **33**: 112
  behaviour **33**: **431**
  biochemistry **33**: **239, 241**
  distribution **33**: 506, **512**
  ecology **33**: **358**
  growth/development **33**: **303, 310, 312**
  nutrition **33**: **200**
  reproduction **33**: **269**, 279, **286**
  size/weight **33**: **230**, 330
tenuiremis **33**: 112, **270**
trispinosus **33**: 112
typicus **33**: **7, 26**, 112, 236
  behaviour **33**: 406, **411, 415, 426**, 432,
    433
  biochemistry **33**: 251
  distribution **33**: **507**, 508, **512**
  ecology **33**: **358**, 364, 366, **377, 396**,
    468, 475
  growth/development **33**: **303, 310**, 315,
    316, **320, 342**, 344
  longevity **33**: **345**
  morphology **33**: **32**, *40*
  nutrition **33**: 145, 150, 161, 169, **183,
    186, 188, 190**, 195, **206, 208**
  reproduction **33**: 43, 259, 260, **270**,
    281, 282, 283, **287, 288**, 291,
    293
  size/weight **33**: 223, **230, 331**, 333, 335
velificatus **33**: 112
  behaviour **33**: 403, **408, 411**

distribution **33**: **512**
growth/development **33**: **304**, **342**
life history **33**: 383
nutrition **33**. 143, 144, 164, **190**, **192**
reproduction **33**: 291
violaceus **33**: 112, **223**, **396**, **415**, **512**
yamadai **33**: **26**, 112, **270**, 506
Centropages abdominalis **44**: 72–75
Centropages brachiatus **44**: 75
Centropages hamatus **44**: 75–77, 303;
    30: 245; **31**: 195, 199
Centropages typicus **44**: 77, 78, 297
Centropages velificatus **44**: 79
Centropagidae **33**: 3, 6, 434
  ecology **33**: 473, **491**, **498**
  reproduction **33**: 44, 46, 257, 261, 268,
    269
  taxonomy **33**: **52**, 111–113
Centropagoidea **33**: 50, **51**, **52**, *54*, *55*,
  110–118, 389
Centropomus unidecimalis **40**: 153
Centropristis philadelphica,
    Predation **27**: **364**
Centropyge **31**: 28
Centroscymnus **32**: 186, 198
Centroscymnus coelolepis **34**: **380**
Centroscymnus owstoni **32**: **168**
Centrostephanus **32**: 175, 180
Centrostephanus nitidus **32**: 182
Centrostephanus rodgersi **21**: 145
Centrostephanus rodgersii **32**: 223
Centrostephanus sylviae **32**: **165**, 223
Cepaea nemoralis **34**: 22, 37, **41**
Cephalaspidea **42**: 68, 71, 75, 105, 118
  see also *Bulla*; Diaphanidae;
    Philinidae; Runcinacea;
    Scaphandridae
  digestive system **42**: 96, 98
  mantle cavity **42**: 92, 94, 95
  morphology **42**: *64*, 75–77, *77*
  shell **42**: 82, 83, 86, 87, 90, 91
  taxonomic history **42**: 73, *89*, 108,
    110–117
  passim
Cephalodiscus **42**: 36
Cephaloidophora
  clausii **25**: 143
  firoloidae **25**: 143
Cephalolobus lavali **25**: 141
Cephalopage trematoides **24**: 209
Cephalophanes **33**: **53**, 132, 486
  frigidus **33**: 132

refulgens **33**: *4*, 31, **32**, 132, 444
  tectus **33**: 132
Cephalopoda **31**: 39; **32**: 357; **42**: 145
Cephalopyge **25**. 157
  trematoides **25**: 157
Cepphus columba (pigeon guillemot)**39**: 57,
  69–70, 77, 80
Cepphus columba **34**: 270
Cepphus grylle **20**: 282
Ceramaster sp. **32**: **165**
Ceramaster misakiensis **34**: **379**
Ceramium fastigiatum **23**: 108
Cerastoderma **29**: 252
Cerastoderma edule **31**: 4; **34**: 13, 24, 211;
  37: **10**; **43**: 68, 69
Cerastoderma glaucum **31**: 4
Cerathophysa ceratopyga **32**: 346
Ceratium sp. **32**: 555; **23**: 236; **25**: 122;
  31: 91, 92; **40**: 124
Ceratium arcticum **32**: 537, *544*, 555–557
Ceratium belone **32**: 535, *537*, 554, 555
Ceratium carriense **32**: 535, *536*, 554
Ceratium coarctatus **32**: 554
Ceratium deflexum **32**: *533*, 535, 554
Ceratium falcatum **32**: 535, *537*, 554, 555
Ceratium fusus **37**: 125
Ceratium incisum **32**: 535, *539*, 554
Ceratium lunula **32**: 535, *540*, 544
Ceratium tripos **21**: 73
Ceratobothrium xanthocephalum **40**: 6, 13,
  21
Ceratocephale pacifica **34**: **367**, **373**
Ceratocorys **32**: 560
Ceratocymba **24**: 176
  dentata **24**: 132
  indica **24**: 121
  leuckarti **24**: 132
  sagittata **24**: 132, 158
Ceratoidei **32**: 62, 64
Ceratomyxa arcuata **40**: 5, 13, 20, 32
Ceratomyxa auerbachi **24**: 285, 287
Ceratomyxa drepanopsettae **26**: 38
Ceratoscopelus warmingi **32**: 61
Ceratoseralis trilobitoides **39**: 208
Ceratoserolis **43**: 149
Cerberilla **28**: *393*, 412
Cercaria doricha **24**: 284, 291, 294, 295,
  297, 298, 305
  pythionike **24**: 276, 284, 291, 294, 295,
    297, 298
  villosum **24**: 284
Cercaria owreae **25**: 150

Cercopagis **31**: 86
Cerebratulus **21**: 135–137
    lacteus **21**: 134, 135
Cereus pedunculatus **22**: 72; **24**: 45
Cerianthus **34**: **365**
Cerianthus lloydii **22**: 50, 74
Ceriops **40**: 99, 103, 107, 113, 135
Ceriops decandra **40**: 95, 108, 121, 158, 167, 176
Ceriops tagal **40**: 95, 109, 110, 112–114, 136, 164, 165
Ceriops tagal australis **40**: 112
Ceritellidae **42**: 87
Cerithiacea **32**: **127**
Cerithium nesioticum **28**: *392*
Cerorhinca monocerata **39**: 280; **20**: 246
Certoidei **32**: 63
Cervonema **30**: **50**
Cervus duvauceli **40**: 157
Ceryle rudis, Predation **27**: 372
Cestida **32**: 9
Cestoda **43**: 17, *32*, 33, 54, 59, 60, 62, 66
Cestus veneris **25**: 137, 138
Cetacea **40**: 26; **43**: 227, 228–231, 242
Cetengraulis **20**: 14, 38
Cetengraulis edentulus **20**: 180
    mysticetus **20**: 6, 7, 36, 39, 107, 128
Cetomimidae **32**: 63
Cetonurus **32**: 186
Cetonurus crassiceps **32**: **170**
Ceuthoecetes
    acanthothrix **34**: **375**
    aliger **34**: **375**
    cristatus **34**: **375**
    introversus **34**: **375**
Ceuthoecetes aliger **23**: 329
Chaceon **32**: 223
Chaceon affinis **32**: **129**; **34**: **378**
Chaceon chilensis **32**: **163**, 176, 216, 223
Chaenichthys rhinoceratus **30**: 334
Chaenocephalus **24**: 334
    aceratus **24**: 334, 346, 349, 371
Chaenocephalus aceratus **35**: **34**, 43, 103
Chaenogobius isaza **26**: **120**, 129
Chaetoceros (alga) **41**: 199
    C. calcitrans **41**: 36, 182
    C. ceratosporum **41**: 182
    C. muelleri **41**: 182
Chaetoceros **29**: 79, *106*; **31**: 92; **40**: 124; **20**: 27, 36
    calcitrans **20**: 325
Chaetoceros affinis **32**: 540, *548*

Chaetoceros bulbosus **32**: *534*, 542, 555
Chaetoceros calcitrans **37**: 39, 65
Chaetoceros coarctatus **32**: 535, *538*
Chaetoceros didymus **32**: 540, *548*
Chaetoceros neglectus **32**: *531*, 541, 555
Chaetocerus **43**: 233
Chaetodon trifasciatus **44**: 232
Chaetogammarus hyrcanus **39**: 197
Chaetogammarus ischnus **39**: 197
Chaetogammarus marinus **39**: 197
Chaetogammarus palcidas **39**: 197
Chaetogammarus pauxillus **39**: 197
Chaetogammarus stoerensis **39**: 197
Chaetogammarus warpachowskyi **39**: 197
Chaetognatha **32**: **56**, 57; **40**: 27; **43**: *21*
Chaetomorpha linum **23**: 110, 134
Chaetopteridae **32**: 98, **126**
Chaetozone
    n. sp.1 **34**: **368**
    n. sp.2 **34**: **368**
Chalinura **35**: 52
Chamaesipho columna **22**: 209
Chamaesiphoneae **42**: 212
Champsocephalus gunnari **30**: 334; **35**: **34**, 45, 52, 103
Chanichthys rhinoceratus **24**: 352; **35**: 52
Chanos chanos (milkfish) **38**: 33, 54
Chaoborus americanis **25**: 15
Charonia lampas **34**: 7
Charr *see Salvelinus alpinus*
Chascanopsetta **32**: 186
Chascanopsetta megagnatha **32**: **174**, 188, **194**
Chasmatopontius thescalus **34**: **375**
Chatonella calani **25**: 138
Chattonella **31**: 356
Chauliodontidae **32**: 62
Chaunacidae **32**: **171**
Chaunax **34**: **380**
Chaunax latipunctatus **32**: **171**, 189
Cheatomorpha minima **21**: 92
Cheilodactylidae **32**: **173**
Cheirimedon femoratus **39**: 180, 197
Cheirimedon fougneri **39**: 197
Chelidonichthys kumu **35**: **39**, 103
Chelinus trilobatus **34**: 230
Chelon vaigiensis **44**: 231
Chelonia mydas, Predation **27**: 371
Chelonobia **22**: 233
    patula **22**: 233
    testudinaria **22**: 203, 233

Chelophyes **24**: 100, 108, 139, 176, 179–183, 199, 207, 208; **25**: 131
  appendiculata **24**: 135, 142, 153, 157, 166, 169–172, 174, 179, 192, 225–227; **25**: *155*, 156
  contorta **24**: 153, 154
Chelura terebrans **39**: 197
Cheramyzon abyssale **34**: 375
Cherax destructor, Physiology **27**: 180
Cherbonniera utriculus **43**: 106
Cheumatopsyche **35**: **163**, **169**
Chevroderma hadalis **32**: 369
Chilina **42**: 104, 105
Chimaeridae **32**: **129**, **168**
chinensis **27**: 1
  Behaviour **27**: 336, 347
  Life histories **27**: 296, 308, 312, **313**, 314
  Moulting/growth **27**: **228**, **230**, 235, 246, **249**
  Parasites **27**: 380
  Predation **27**: 375
  Reproduction **27**: 264, **270**, **279**
  Taxonomy **27**: 96, **110**
  Zoogeography/Evolution **27**: 134, **144**, 157
Chionelasmatida **32**: **409**
Chionelasmus darwini **32**: **409**
Chionoecetes **30**: 325
  bairdi **29**: 144
  opilio **29**: **133**, 139, 140
  tanneri **30**: 334
Chionoecetes angulatus **32**: **446**, **448**, **452**
Chionoecetes bairdi **31**: 19, 51; **32**: **449**
Chionoecetes opilio **31**: 19, 51; **32**: **448**
Chiracanthium **40**: 177
Chiridiella
  ecology **33**: 486
  taxonomy **33**: **53**, 66, 77, 88, 90, 93, 125, 126
Chiridiella abyssalis **33**: 125
  atlantica **33**: 125
  bichela **33**: 125
  bispinosa **33**: 125
  brachydactyla **33**: 125
  brooksi **33**: 125
  chainae **33**: 125
  gibba **33**: 125
  kuniae **33**: 125
  macrodactyla **33**: 125
  megadactyla **33**: 125
  ovata **33**: 125

pacifica **33**: 125
reducta **33**: 125
sarsi **33**: 125
smoki **33**: 125
subaequalis **33**: 125
trihamata **33**: 125
Chiridius **25**: 138
  ecology **33**: 486, **499**
  nutrition **33**: **152**, 166
  taxonomy **33**: **53**, 88, 89, 92, 126
Chiridius armatus **33**: **32**, 32, 126
  carnosus **33**: 126
  gracilis **33**: 126, 267
  longispinus **33**: 126
  mexicanus **33**: 126
  obtusifrons **33**: 126
  pacificus **33**: 126
  polaris **33**: 126
  poppei **33**: 126
  subantarcticus **33**: 126
  subgracilis **33**: 126
    ecology **33**: **358**, **499**
    nutrition **33**: **206**
    reproduction **33**: 267
Chiridota aff. *laevis* **32**: **129**
Chiridota **35**: 22
Chiridotea **30**: 100; **39**: 158
Chiridotidae **32**: **129**
Chirocentrus **20**: 34
Chirodropus gorilla **21**: 105
Chiromanthes bidens **40**: 142
Chirondota n. sp. **34**: **379**
Chironex **21**: 117, 119
  fleckeri **21**: 104, 105, 116–118, 120, 130, 131
Chironex fleckeri **24**: 221
Chironomus **35**: 166, **169**, 176, *180*, **181**
Chiropsalmus buitendijki **21**: 105
  quadrigatus **21**: 104, 105, 120, 130
  quadrumanus **21**: 105
Chiroteuthidae **32**: 62
Chirundina **33**: **53**, 92, 93, 126, 486
  antarctica **33**: 126
  indica **33**: 126
  streetsi **33**: 126
Chirundinella **33**: 53, 87, 90, 126
  magna **33**: 126
Chistikovia **42**: *144*
Chitaristes **32**: 560
Chiton pelliserpentis **34**: 19, 37, *38*, 40, **41**
Chlamydomonas (chlorophyte) **41**: 62; **31**: 149; **20**: 28

Chlamys **35**: 171
  patagonica **35**: *175*
  varia **35**: *175*
Chlamys asperrima **37**: **10**, 37
  bioassay methodology **37**: 43, **44**, 48
  bioassay procedures **37**: 50, 531, 59
  toxicity testing **37**: **91**, **100**, **110**, **111**
Chlamys farreri nipponensis **20**: 326
  nobilis **20**: 325, 326
Chlamys opercularis **22**: 103, 134
Chlamys rubida (scallop) **39**: 9
Chlidonophora chuni **28**: 200, 211, 335, *353*
Chlidonophora incerta **28**: 211
Chloeia **21**: 137, 138; **32**: **161**
  euglochis **21**: 138
  flava **21**: 138
Chlorella **37**: 39; **20**: 183
Chlorella pyrenoidosa **21**: 71
Chlorodiella longimana **40**: 141
Chloromyxum **24**: 278, 291
Chlorophthalmidae **32**: 167, **169**
Chlorophthalmus **32**: 186
Chlorophthalmus ichthyandri **32**: **169**, **194**
Chlorophthalmus zvezdae **32**: **169**
Chlorophycota **32**: **160**
Chlorophyta **43**: 281, *283*, 293
Choanoflagellates **43**: 213
Cholrodon anchorago **44**: 231
Chondrilla nucula **21**: 98
Chondrophellia coronata **34**: **365**
Chondrus crispus **23**: 13–16, 39, 45, 58, 86, 87–89, 120, 135
Chone ecaudata **26**: **179**
Chordaria flagelliformis **23**: 3, 11, 63
Chordata **43**: *21*
  see also Tunicata
Chorismus antarcticus **38**: 214; **43**: 143, 148
Chorocaris sp. **32**: 107, 110, 111, 114, 120, **129**, 130, 131
Chorocaris **34**: **377**
  chacei **34**: **377**
  vandoverae **34**: **377**
Chorocaris aff. *fortunata* **35**: **40**
Chorocaris chacei **32**: 99, *101*, 102, 105, 107, 108, 110, 111, 114, 116, **117**, 118, **129**, *134*
Chorocaris vandoverae **32**: 110, 116
Choromytilus meridionalis **22**: 103, 165,168
Chrionema **32**: 186
Chrionema chryseros **32**: **173**, **218**
Chrionema pallidum **32**: **173**, 188, 190, **194**
Chromadorita sp. **32**: **126**

Chromaspirina okemwai **40**: 135
Chromatium **40**: 118
Chromidina sp. **32**: 308
Chrysaora helvola **21**: 106
  hysoscella **21**: 106
  melanaster **21**: 106
  quinquecirrha **21**: 106, 120, 121
Chrysaora hysoscella **24**: 208
Chrysaora melanaster **25**: **35**; **26**: 145
Chrysaora quinquecirrha **44**: 22, 121, 122
Chrysochromulina **29**: 79; **31**: 309, 317
  polylepis **29**: 79
Chrysochromulina leadbeateri **31**: 317
Chrysochromulina polylepis **31**: *317, 321, 326, 327*, **327**, *334*, 347, 348, 356, 357
Chrysochromulina polylepis **37**: 125
Chrysophyceae **41**: 49
Chrysophytes **43**: 199, 201
Chthalamus dalli (barnacle) **39**: 18, 24, 26, 29, 72, 78
Chthamalophilus delagei **22**: 219
Chthamalus **22**: 207, 216, 223, 228, 232, 234; **23**: 45, 59
  depressus **22**: 209, 224
  hembeli **22**: 204, 232
  intertextus **22**: 204
  stellatus **22**: 204, 213, 230, 232
Chthamalus anisopoma **34**: 49
Chuniphyes multidentata **24**: 132, 145, 151
Chydorus sphaericus **31**: 87
Chytriodinium **25**: 125, 127, 133, 135
Cibicides weullerstorfi **30**: 53
Ciliata **43**: 7, 198, 212, 213, 215, 216
  see also Ctenophora; Turbellaria
Ciliata mustela **26**: 138; **40**: 15, 311, 314
ciliates (*Balanion*) **41**: *34*, 38
Ciliophora **40**: 4; **43**: *20*, 41
Cilunculus alcicornis **24**: 8, 11, 16
  cactoides **24**: 11
Cinidae *see* Lower Heterobranchia
Ciona intestinalis **31**: 317
Cirolana sp. **32**: **162**
Cirolana **39**: 157
Cirolana borealis **40**: 11
Cirolana harfordi **39**: 208
Cirolana imposita **39**: 208
Cirolana parva **39**: 208
Cirolanidae **32**: **409**; **39**: 157
Cirrenalia pygmea **40**: 120
Cirripathes **32**: **160**, 178

Cirripedia **32**: **161**, **210**, 329, 333, **335**, 344, 347, 349, 366, **409**, **438**, 467, 468, **469**, 470, **483**, **486**, **492**, **504**; **43**: 45
Cirripedia Thoracica **32**: 347
Cirrophorus branchiatus **26**: **179**
Cirroteuthis magna **34**: **380**
Cistothorus palustris (marsh wren) **39**: 45
Citharichthys arctifrons **26**: 146
Citharichthys stigmaeus **39**: 279
Cladocera **31**: 80–152
Cladophora **43**: 282
Cladophora columbiana **23**: 110, 134
Cladophora, Food/Feeding **27**: 321, 327
Cladopyxis **32**: 560
Cladopyxis brachiolata **32**: 545
Cladorhiza **32**: 495
Cladorhiza longipinna **32**: 503
Cladorhiza rectangularis **32**: 503
Cladorhizidae **32**: 467
Cladosiphon okanuranus **21**: 92
Cladosporium marinum **40**: 121
Cladotanytarsus **35**: 176, *180*, **181**
Clam
 coot *see Mulinia lateralis*
 hard *see Mercenaria mercenaria*
 surf *see Spisula*
Clam *see Humilaria*; *Macoma*; *Mya*; *Protothaca*; *Saxidomus*; *Serripes*; *Yoldia*
Clam, surf (*Spisula* spp.) **43**: 104
Clangula hyemalis **39**: 15
Clarias macrocephalus **26**: **92**
Clathromorphum circumscriptum **23**: 22
Clausia **25**: 138
Clausocalanidae **33**: 53, 129, 130
Clausocalanoidea **33**: 50, **53**, *55*, 124–138, 502
Clausocalanus **25**: 122, 133, 137, 138, 142, 147
 distribution **33**: **513**
 ecology **33**: 377, 480, 483, **485**, 490, 528
 reproduction **33**: 265
 taxonomy **33**: **53**, 85, 129
Clausocalanus arcuicornis **33**: **7**, 129, **223**
 brevipes **33**: 129
 dubius **33**: 129
 farrani **33**: 129, **512**
 furcatus **33**: *24*, **26**, 129, **223**
  behaviour **33**: 443
  ecology **33**: **358**, 480

 ingens **33**: 129
 jobei **33**: 129, **513**
 laticeps **33**: 129, **513**
 latipes **33**: 129
 lividus **33**: 129, **513**
 mastigophorus **33**: 129, **194**
 minor **33**: 129, **513**
 parapergens **33**: 129
 paululus **33**: 129, **223**, 262, **358**, **513**
 pergens **33**: 129, 443, 480
  behaviour **33**: **415**
  distribution **33**: 506, **507**, **512**
  ecology **33**: **358**, **484**
  nutrition **33**: **198**
Clausocalanus spp. **44**: 93, 94, 109
Clausophyes massiliana **24**: 121
 moserae **24**: 151
 ovata **24**: 145, 151
Clava multicornis **24**: 43
Clavella adunca **40**: 9, 13–14, 33, 36, 37, 39, 53, 54
Clavularia **21**: 127
Clibanarius laevimanus **40**: 140
Clibanarius longitarsus **29**: **132**, *147*
Climacodium **32**: 560
Climacodium frauenfeldianum **32**: 535, *541*, 554, 558
Clingfish *see Gobiesox maeandricus*
Clingfish, Connemara *see Lepadogaster candolei*
Clinocardium nuttalli **37**: **10**
Clinocardium nuttallii (cockle) **39**: 15
Clio **25**: 152
 cuspidata **25**: 139
Clio pyramidata **32**: **453**
Clione **32**: 302
Clione limacina **44**: 22, 31, 34, 43, 132; **32**: 266
Clupea **20**: 14, 15, 21, 53, 87
 (*Strangomera*) *bentincki* **20**: 180
 harengus **25**: *3*, 4, *6*, 7, 9, 12, 13, *13*, 14, 15, 17–20, *20*, 21, 22, 25, 29, 30, **31**, **34**, **35**, 37, 41, 44, 45, 47, 50, **52**, 58, 59, 64
 harengus **30**: 220; **34**: 264; **20**: 5, 6, 18, 39, 86, 131, 279
 harengus harengus **20**: 29
 pallasi **30**: 234; **34**: 265; **25**: 11, *13*, 20, 22, 29, 32, 43, 47; **20**: 5, 6, 169, 240
 sprattus **25**: 12, **34**

Clupea harengus (Atlantic herring) **38**: 3, 4, 6, 53
Clupea harengus **28**: 8, 9, 19, **26**, 28, 29, **30**, **31**, 33, **34**, **36**, 42, **44**, *45*, 68–85, 97, **99**, **107**, 116, 121, 122, 128–130; **24**: 263–319; **44**: 215, 216, 248; **23**: 234; **26**: 33, 75, 78, **92**, **120**, **121**, 122, 251; **31**: 178; **40**: 15
   h. harengus **24**: 274, 285
   h. pallasi **24**: 274, 285
Clupea harengus maris-alba **28**: 82, *88*
Clupea harengus membras **28**: 37, 82
Clupea harengus pallasi **26**: 73, **92**
Clupea pallasi **28**: 28, 67, 82, 88, 89, **93**; **39**: 52–54, 56, 57, 68–70, 72, 77, 80
Clymenura columbiana **26**: **179**
Clypeaster australasiae **32**: 223
Clypeaster isolatus **32**: **165**, 223
Clypeoniscus **39**: 184
Clypeosectidae **32**: **127**
Clypeosectus
   curvus **34**: **371**
   delectus **34**: **371**
Cnidaria **32**: 329, 330, 333, 344, 360, 467, 468; **43**: *21*, 110
Coccidinium **25**: 129
Coccodentalium **42**: *143*
   C. carduum **42**: 212
coccolithophores **41**: 32, 71
   see also *Emiliania huxleyi*
Coccoloba uvifera **40**: 177
Cocculina cf. *craigsmithi* **34**: **370**
Cocculina craigsmithi **35**: 24
Cochlodinium **21**: 72
   catenatum **21**: 72
   heterolobatum **21**: 72
Cochlodinium heterolobatum **25**: 120; **37**: 69, 125
Cochlosyninium **25**: 127
Cockle see *Clinocardium nuttalii*
Cockles (*Austrovenus, Cerastoderma*) **43**: 34, 35, 68, 69
Cocyzus minor **40**: 157
Cod see *Gadus macrocephalus*
Cod see *Gadus morhua*
Codium **43**: 282
   dimorphum **23**: 116, 133
   fragile **23**: 28, 33–35
Codonella **25**: 129
   campanula **25**: *130*

Coelacanth see *Latimeria chalumnae*
Coelenterata **32**: 57, 306, **335**; **40**: 27, 130; **43**: 287
Coelopleurus sp. **32**: **165**, 176, 180, 182
Coelorhinchus cf. *labiatus* **34**: **380**
Coeloseris mayeri **21**: 127
Coenobita
   rugosa **29**: **132**
   rugosus **29**: 146, 147, *147*
Coenobita cavipes **40**: 140
Coenobita rugosus **40**: 140
Coenobitoidea **42**: 216
Coenothecalia **43**: *291*
Colia **20**: 24, 34
Colidotea rostrata **39**: 208
Collettea minima **32**: **161**
Collisella
   (*Macclintockia*) *scabra* **34**: 28, 46
   (*Tectura*) *digitalis* **34**: 28
   digitalis **34**: **6**
   scabra **34**: **6**
Collosphaera **25**: 125
Collozoum **25**: 125, 127
   longiforme **25**: 153, 156
Colobonema sp. **32**: *72*
Cololabis saira **39**: 267
Colossendeis **24**: 2, 4, 9, 14, 16, 22, 26, 31, 33, 47, 53, 54, 64, 68, 69, 71
   angusta **24**: 31, 68
   arcuata **24**: 54
   australis **24**: 20, 31
   bicincta **24**: 5, 7, 8
   colossea **24**: 15, 22, 31, 33, 55, 75
   gigantea **24**: 33
   glacialis **24**: 16
   m. megalonyx **24**: 46, 54
   macerrima **24**: 75
   megalonyx **24**: 70
   melancholicus **24**: 15
   proboscidea **24**: 33
   robusta **24**: 32, 46
Colpomenia sinuosa **23**: 113
Colpophylla natans **22**: 11, 30
Colpophyllia natans **43**: 299–302, *301*
Columotelson **39**: 118
Colus **32**: **159**
Comantenna **33**: 53, 88, 90, 126
   brevicornis **33**: 126, **499**
   crassa **33**: 126, **499**
   curtisetosa **33**: 126
   recurvata **33**: 126, **499**
Combretaceae **40**: 94

Compressidens **42**: *144*
Compressidentalium **42**: *143*
Conchoderma **22**: 200
Conchoecia **32**: 266
Concholepas concholepas **34**: 22, 37, 41, **41**
Conchophrys **25**: 137
   davidoffi **25**: 137
Condylactis gigantea **21**: 106, 123; **22**: 93;
      31: 252
Condylonucula maya and *C. cynthiae* **42**: 3
Conger conger **40**: 19–20, 25, 299
Congridae **32**: 167, **169**, 211
Conidae **32**: **127**
Conocara fiolenti **32**: **169**
Conocardioida **42**: 145
Conocarpus erectus **40**: 94, 161, 177, 178
Constantinea
   simplex **23**: 80
   subulifera **23**: 80
Contracaecum **25**: 151, 158; **38**: 181;
      40: 7–8, 10, 41
   aduncum **25**: 151
Contracaecum aduncum **26**: 38; **24**: 277
Contracaecum osculatum **40**: 7, 18, 26, 31,
      41, 43, 44
Contracaecum septentrionale **40**: 8
Conus **21**: 170, 171, 174–176; **42**: 93
   ammiralis **21**: 175
   arenatus **21**: 175
   aulicus **21**: 175, 188
   catus **21**: 175, 188
   eburneus **21**: 175
   emaciatus **21**: 175
   episcopus **21**: 175
   figulinus **21**: 175
   flavidus **21**: 175
   geographus **21**: 174, 175, 188
   gloria-maris **21**: 188
   imperialis **21**: 175, 188
   leopardus **21**: 175
   litteratus **21**: 188
   lividus **21**: 175, 188
   magus **21**: 174, 175
   marmoreus **21**: 175, 188
   miles **21**: 175
   millepunctatus **21**: 175
   obscurus **21**: 175, 188
   omaria **21**: 175, 188
   planorbis **21**: 175
   pulicarius **21**: 175, 188
   quercinus **21**: 175, 188
   rattus **21**: 175

sponsalis **21**: 175, 188
stercusmuscarius **21**: 175
striatus **21**: 174, 175, 188
tessulatus **21**: 175
textilis **21**: 175, 188
tigrinus **21**: 175
tulipa **21**: 175, 188
virgo **21**: 175
Conus abbreviatus **28**: 412
Conus agassizi **28**: 412
Conus coronatus **28**: *393*
Conus mahogani **28**: *393*
Conus mediterraneus **28**: 404, 411
Conus profundorum **32**: *163*
Conus pulicarius **28**: *393*, **403**
Conus tessulatus **28**: 420
Cookeolus boops **32**: 188
Cookeolus japonicus **32**: **173**
Coolia monotis **31**: 315
Coot clam *see Mulinia lateralis*
copepod (crustacean) **41**: 83, *189*
Copepoda **32**: **56, 452, 453, 454**, 496; **40**: 9,
      25–28, 130
   see also Ectoparasites;
         Lepeophtheirus; Lernaeocera
   fecundity **43**: 93, *94*, 99
      environmental factors **43**: 142, 149
      parasitism **43**: 7, 11, 17, *18, 19*, 31, 41,
         46, *55, 60*
      Southern Ocean pack ice **43**: 208, 209,
         210–212, 215, 216–219, 221
Copepods **37**: 131
Copidognathus
   alvinus **34**: **374**
   nautilei **34**: **374**
   papillatus **34**: 374, 402
Copidognathus alvinus **32**: **127**
Copidognathus papillatus **23**: 331
Copilia **25**: 142
Coptothyris grayii **28**: 189, *213*, 275, 295,
      313, 347, 348, *353*
Coracinus capensis **25**: 195
Coral
   see also Pigmentation *in subject index*
   fecundity **43**: 149, 150
      life-history strategies **43**: 130, 133, 134
      quantification methodologies **43**: 105,
         110, 111
   parasitism **43**: *94, 95*
Corallina **23**: 133
   mediterranea **23**: 45
   officinalis **23**: 134, 135

vancouveriensis **23**: 133
Corallina mediterraenea **24**: 55
Corallinaceae **32**: 176
Corallium rubrum **43**: 281, *291*
Corbicula **35**: **163**
Corbicula fluminae **34**: 24
Cordagalma **24**: 107, 117, 178
  cordiformis **24**: 105, 117, 128, 191, 200,
    220, 221, 225
Cordylophora **24**: 175
Coregonus **25**: *13*; **30**: 277
  artedii **25**: 58
  clupeaformis **25**: 9, 58
  kiyi **25**: 58
  muksun **30**: 232
  pollan **30**: 237
Coregonus albula **26**: 72, 79, 83, **121**, 146
Coregonus artedii **26**: 147
Coregonus lavaretus (whitefish) **38**: 54
  stock enhancement studies **38**: **31**
Coregonus lavaretus **28**: 108
Coris julis **34**: 255
Cormorant *see Phalacrocorax*
Cornigerius **31**: 86
Cornirostridae *see* Lower Heterobranchia
Cornucalanus **33**: **53**, 81, 82, 132, 486
  antarcticus **33**: 132
  chelifer **33**: 132
  indicus **33**: 132
  notabilis **33**: 132
  robustus **33**: 132
  sewelli **33**: 132
  simplex **33**: 132
Corolla spectabilis **34**: 31
Coronaster sp. **32**: **165**
Coronula diadema **22**: 203
Corophiidae **32**: 463; **39**: 36
Corophium **34**: 44; **35**: **163**, **169**, 174, **177**,
  178
Corophium **39**: 151
  volutator **35**: **163**, **168**
Corophium acherusicum **39**: 197
Corophium arenarium **39**: 197
Corophium bonnellii **39**: 197
Corophium chelicorne **39**: 197
Corophium curvispinum **39**: 197
Corophium insidiosum **39**: 197
Corophium lacustre **39**: 152, 197
Corophium nobile **39**: 197
Corophium robostum **39**: 197
Corophium triaenonyx **39**: 191
Corophium volutator **39**: 174, 197

Corrallium **30**: 338
Corvicia barnesi **21**: 105
Corvus caurinus (northwestern crow)
  39: 45–48, 71
Corycaeus **25**: 5, *6*, 7, 147
  anglicus **25**: 7, 48
Corycaeus amazonicus **44**: 79
Corycaeus subulatus **44**: 79
Corydalus cornutus **35**: **168**, **185**
Corymorpha pendula **34**: 221
Corynactis **22**: 92
Corynactis australis **24**: 45; **21**: 106
Coryne **24**: 43
  muscoides **24**: 43
Corynophrya **25**: 138, 139
Corynosoma **24**: 284; **40**: 8, 10, 18, 24
  semerme **24**: 273, 277, 289, 293
  strumosum **24**: 277, 289, 293
  villosum **24**: 274
Corynosoma semerme **40**: 8, 18, 23, 26, 32,
  41
Corynosoma strumosum **40**: 8, 18, 23, 26,
  33, 41
Corynosoma wegeneri **40**: 8, 18, 26, 32
Coryphaenoides sp. **32**: 97; **32**: 198
Coryphaenoides **35**: 52, 69, **70**, 84
  acrolepis **35**: 35
  armatus **35**: 26, **35**, 84
  colon **35**: 35
  filifer **35**: 35
  leptolepis **35**: 35
  mexicanus **35**: 35
  pectoralis **35**: 35
  profundicolus **35**: 35
  rupestris **35**: 5, 16, **35**, 61, 65, 84,
    103–107, *104*
  yaquinae **35**: **36**, 84
Coryphaenoides acrolepis **32**: **297**
Coryphaenoides paradoxus **32**: **170**, **218**
Coryphaenoides pectoralis **32**: **297**
Coryphoblennius galerita **26**: 137, *139*
Coscinodiscus **31**: 92
Coscinodiscus **38**: 161; **40**: 124
Coscinodiscus oculus-iridis **32**: *533*, 536,
  549, 555, 556
Cosmocalanus **33**: **52**, 67, 119, 483
  caroli **33**: 119
  darwini **33**: 7, 119, 262, **377**
Cosmocephalus **40**: 8, 26, 29
Cosmorodium albus **40**: 155
Cossura longocirrata **26**: **179**
Cossura modica **26**: **179**

Costentalina **42**: *144*
Costentalina tuscaroraesubcentralis **32**: 503
Cothurnia maritima **29**: 105, *106*
Cottus **25**: 58
Cottus extensus **26**: 136
Coulmania frigida **24**: 54
Crab
  Californian (*Hemigraspus oregonensis*) **43**: 70
  European green (*Carcinus maenas*) **43**: 70
  Paradorippe granulata **43**: 133
crab, pea *see Pinnotheres halingi*
Cranchiidae **32**: 62, 64
Crangon **25**: 32, 33; **34**: 287
  crangon **25**: 64
Crangon affinis, Predation **27**: 367
Crangon allmanni **38**: 179
Crangon crangon **29**: **132**, 173, 174; **34**: 211; **38**: 214, 221; **44**: 222
Crangonidae **32**: 455
Crania, see *Neocrania anomala*
Crassadoma gigantea **35**: **185**
Crassantenna **33**: **53**, 88, 126
  comosa **33**: 126, **499**
  mimorostrata **33**: 126, **499**
Crassarietellus **33**: **51**, 71, 101, **498**
Crassostrea **21**: 4, 52, 179; **22**: 148; **43**: 132
  C. *gigas* **43**: 69
  C. *virginica* **43**: *22*, 65, 66, 69
  commercialis **22**: 103, 169, 171
  denselamellosa **21**: 52
  echinata **21**: 52
  gigas **21**: 4, 5, 52, 170, 179, 189; **22**: 103, 130, 165, 168, 169, 171; **34**: 17
  margaritacea **22**: 103, 165, 168
  rhizophorae **34**: **183**
  rivularis **21**: 4
  virginica **22**: 103, 129, 166–169, 171: 15, 28, **183**
Crassostrea angulata **37**: **10**, 11, 18
  bioassay procedures **37**: 50–51
  toxicity testing **37**: **113–114**
Crassostrea cucculata **37**: **10**, 16
  toxicity testing **37**: **90, 92, 93**
Crassostrea generally **37**: 5, 87, 135
  bioassays **37**: 45, 53, 116
  reproduction, biology of **37**: 22, *25*
Crassostrea gigas **31**: 23, 51, 52; **37**: 2, **8–10**, 10, 131
  bioassay methodology **37**: 45, 46, 48
  seawater quality **37**: 41, 42, 43, **44**

bioassay procedures **37**: 50, 65, 66
  embryo-larval **37**: 51, 52, *54, 56, 57–59,* 60
  larval physiology tests **37**: 68, *69*, 71, 73
  described **37**: 10–13
  larval rearing in laboratory **37**: 26
  nutritional requirements **37**: 38, 39
  physical requirements **37**: 36, **36, 37**
  spawning and fertilization **37**: **29, 30**, 29, **32**, 32
  reproduction
    biology of **37**: 18, 20, 21, *24*
    spawning and fertilization **37**: **29, 30**, 29, **32**, 32, 46, 48
  toxicity testing **37**: 77, 87, 118, 128
    biocides **37**: **98–106**, 97, **107**
    detergents and oil **37**: **110, 112–115**
    metals **37**: **90, 92–95, 120**
Crassostrea iredalei **37**: **10, 91**
Crassostrea madrasensis **40**: 148
Crassostrea margaritacea **37**: **10**, 16
  toxicity testing **37**: **90, 92, 93**
Crassostrea rhizophorae **37**: **10**, 16, 37; **40**: 149, 173
Crassostrea sikamea **31**: 23, 51, 52
Crassostrea virginica **37**: 2, 6, **8–10**, 10
  bioassay methodology **37**: 42, **44**, 46
  bioassay procedures **37**: 63
    embryo-larval **37**: 52, 54–56
    larval physiology tests **37**: 66, 67, 69, 70, 71, 73
  described **37**: 12, 13
  larval rearing in laboratory **37**: 25, 26,
  physical requirements **37**: **36, 37**, 37
  spawning and fertilization **37**: 27, **29, 30, 32**, 32–36
  reproduction
    biology of **37**: 16, 17, 19, 20, 21, 23
    spawning and fertilization **37**: 27, **29, 30**, 29, **32**, 32–36, 46, 48
  toxicity testing **37**: *122*,125, 135
    biocides **37**: **98–106, 107**
    detergents and oil **37**: **110, 112–114**, 116
    metals **37**: **90, 94, 95**, *95*
Crenimugil crenilabis **44**: 231
Crepidoodinium **25**: 127, 134
Crepidula fornicata **34**: 31
Crescent gunnels *see Pholis laeta*
Creseis **21**: 170; **42**: 73
Cribrosoconcha **32**: 211

Cribrosoconcha alephtinae **32**: **164**
Cribrosoconcha elegans **32**: **164**
Cricotopus **35**: **163**, *180*, **181**
Crinoidea **31**: 39; **32**: *97*, **335**, 344, 357, **409**,
    468, **483**
Criseis **25**: 152
Crocodylus porosus **40**: 154
Crossota sp. **32**: *72*
Crossota brunnea **32**: 55, *72*
Crotalus **21**: 64
Crow, northwestern *see Corvus caurinus*
Crustacea **40**: 25–28, 136–144; **41**: 83, 188,
    189
  see also Amphipoda; Copepoda;
      Decapoda; Isopoda
  fecundity **43**: 93
    environmental factors **43**: 142, 143,
        148, 149
    life-history strategies **43**: 130, 131,
        135, 140
    parasitism **43**: 64, 70
    Southern Ocean pack ice **43**: 209, 210,
        215
Crustaceans *see under subject index*
Cryptocaryon irritans **43**: *22*
Cryptocope **39**: 160
Cryptocopoides artica **32**: **161**
Cryptocotyle lingua **24**: 277, 287, 294,
    295, 304; **40**: 5, 16, 25, 31, 35, 36,
    43, 45, 46, 51, 55–58
Cryptodonta **42**: 2, 4
Cryptolithoides **31**: 48
Cryptolithus **31**: 48
Cryptomonas **20**: 28
Cryptomonas batica, cultivation of
    31: 148
Cryptopleura violacea **23**: 28
Cryptopora **28**: 211
Cryptosporidium parvum **43**: 68, 69
Cryptovalsa halosarceicola **40**: 119
Crysopetalum **32**: **161**
Ctena **32**: **159**
Ctenocalanus **33**: **53**, 85, 129, 480, 528;
    43: 216
  campaneri **33**: 129
  citer **33**: 6, 129, **304**, **358**, **493**, 494
  huysi **33**: 101
  tageae **33**: 129
  vanus **33**: 7, **26**, 129, **223**, **513**
    ecology **33**: **358**, 377, 396, 480, **484**,
        493, 494
    growth **33**: **304**

Ctenodiscus crispatus **32**: **462**
Ctenodontidae **42**: 43
Ctenopelta porifera **34**: **370**
Ctenophora **31**: 39; **32**: 306; **40**: 27; **43**: *21*
Cubiceps baxteri **35**: **36**
Cubomedusae **21**: 132
Cuclopoapsuedes **39**: 159
Cucubalus **25**: 124
Cucullanus **40**: 28
Cucullanus cirratus **40**: 8, 13–14, 31
Cucullanus heterochrous **40**: 8, 13, 21
Cucumaria abyssorum **32**: **461**, **469**
Cucumaria echinata **21**: 146
  fraudatrix **21**: 160
  japonica **21**: 146
Culeolus sp. **32**: **469**
Culeolus murrayi **32**: **469**
Culeolus robustus **32**: **469**
Culeolus tenuis **32**: **469**
Culex quinquefasciatus **40**: 106, 116
Culex tritaeniorhynchus **40**: 106
Cumacea **32**: **335**; **39**: 107, 161, 186, 187,
    189, 193, 217–219
Cumopsis goodsiri **39**: 217
Curcuteria australis **43**: 34, 35
Curtipathes **32**: **160**
Cuspidaria sp. **32**: **164**
Cuspidaria macrorhynchus **32**: **164**
Cuspidariidae **32**: 468
Cut-throat trout *see Oncorhynchus clarki*
Cyanagraea praedator **34**: **378**
Cyanathea hydrothermala **34**: **365**
Cyanea **21**: 132; **20**: 53
  capillata **21**: 106
  ferruginea **21**: 106
  lamarcki **21**: 106
  nozaki **21**: 106
  purpurea **21**: 106
Cyanea sp. **26**: 145
cyanobacteria **41**: 31–33, 45, 50
  Synechococcus **41**: *36*, 47, 58, 64, 65
  Trichodesmium **41**: 40, 49, 58
Cyanograea **34**: 385
Cyanograea praedator **23**: 330
Cyanophora paradoxa **25**: 120
Cyanophyceae **40**: 124
Cyanophycophyta **32**: **160**
Cyathermia naticoides **34**: **369**
Cyathidium foresti **32**: **409**, 410, *411*
Cyathura carinata **39**: 114, 144, 208
Cyathura polita **39**: 114
Cyclocaris tahitensis **32**: **162**

Cyclope neritea **28**: *392*, 397, 412, **414**, 416, 417, 423, 424, 426
Cyclophyllidea **40**: 10
Cyclopoida **32**: **452**; **33**: 1, **2**, *3*; **40**: 130
Cycloporus papillosus **21**: 133
Cyclopozoa **43**: 210
Cyclops
  americanus **29**: **131**, 181
  castor **29**: 174
  longicornis **33**: 49
  marina **33**: 49
Cyclops bicuspidatus **25**: 5
Cyclopterus lumpus **26**: 76; **40**: 19, 22, 56
Cyclosalpa affinis **44**: 22, 132
Cyclosalpa bakeri **44**: 22, 132, 133
Cyclosalpa virgula **25**: 143
Cyclotella **31**: 92
Cyclotella nana **37**: 65; **20**: 325
Cyclothone spp. **32**: 63
Cyclothone **30**: 317; **35**: 50
  alba **35**: 87
Cyclothone acclinidens **32**: 64
Cyclothone atraria **32**: 57, 64
Cyclothone braueri **32**: 64
Cyclotrichium **23**: 221
Cydippe **25**: 150
Cyemidae **32**: 63
Cyerce **42**: *116*, 133, 135
Cylichna **42**: 75, *116*, *132*, 135
  C. cylindracea **42**: 94
Cylichnidae **42**: 76, 87, 90
Cylindrobulla **42**: 71, 95, 98, 103, 133,135
  C. beauii **42**: 94
  morphology **42**: 78, *79*
  shell **42**: 83, *84*, *85*
  taxonomic history **42**: 115, *116*
Cylindrobullidae **42**: 80, 87
Cymatogaster aggregata **24**: 276
Cymodetta gambosa **39**: 146, 154
Cymodocea **41**: *150*
Cymodocea serrulata **24**: 60
Cymodocella **39**: 176
Cymodocella acuta **39**: 208
Cymodocella tubicauda **39**: 208
Cymudasa compta **39**: 198
Cynoglossus browni **29**: 220
Cynometra iripa **40**: 94
Cynometra ramiflora **40**: 94
Cynoscion arenarius, Predation **27**: **363**
  nebulosus, Predation **27**: 361, **363**, 368, 369, 375
  nothus, Predation **27**: **363**, **364**

Cynoscion nebulosus (spotted seatrout) 27, 28, 54
Cynoscion regalis **44**: 258
Cyphastrea **21**: 127; **22**: 10, 17
  microphthalma **22**: 5, 11
Cyphocaris sp. **32**: 266, 303
Cyphocaris richardi **32**: 71, *73*
Cypraeopsis **32**: 412
Cypraeopsis superstes **32**: **409**, 412
Cyprideis litoralis, Reproduction **27**: 275
Cyprilepas **22**: 200
Cyprinodon **40**: 153
Cyprinodon baconi **21**: 164
Cyprinus carpio **26**: *72*, **74**, 150; **30**: 233; **34**: 84; **40**: 279, 297, 319
Cyrtoconella **42**: *144*
Cyrtocrinus **32**: 410
Cyrtomaja danieli **32**: **163**
Cyrtomaja platypes **32**: **163**
Cystaphora **21**: 93
Cystidicola farionis **24**: 273
Cystidicoloides uniseriata **40**: 8, 12–13, 28, 31
Cystisoma **25**: 141; **39**: 157, 179
Cystodinedria **25**: 126
Cystodinium **25**: 126
Cystoseira
Cystoseira **24**: 55
  crinata **24**: 55, 57
  fimbriata **24**: 55, 57
  mediterranea **23**: 45
  osmundacea **23**: 20, 116
  stricta **24**: 55
Cytarocylis **25**: 129
  ehrenbergi **25**: *130*
Cyttomimus **32**: 186, 198
Cyttomimus stelgis **32**: *172*, **195**, **218**

Dab, long rough (*Hippoglossoides platessoides*) **43**: 67
Dactylogyrus **43**: 29, 46
Dactylometra **25**: 153
Dactylopsaron **32**: 198, 211
Dactylopsaron dimorphicum **32**: **173**, **194**
Dactylospora haliotrepha **40**: 120
Dahlella caldariensis **23**: 329; **34**: **376**
Dajus **25**: 154
Dallina **28**: *351–353*
Dallithyris murrayi **28**: 211; **32**: **165**
Dalyellida **43**: 22
Damkaeria **33**: **52**, 87, 89, 123
  falcifera **33**: 123, **499**

Damsel fish **43**: 16
Danoctopus sp.cf. *hoylei* **32**: **164**
Danoctopus **32**: 206, 211
Danoctopus hoylei **32**: 211, 220
Daphnella (Eubela) ichthyandri **32**: 163
Daphnia **21**: 139; **29**: 34; **30**: 183; **31**: 87,
    97–99, 102, 103, 109, 125, 145,
    151; **20**: 141
  magna **25**: 37, 119, 121; **30**: 200
  pulex **25**: 15
Daphnia laevis **31**: 132
Daphnia magna **33**: 248, 344; **26**: 135;
    31: 111, 122, 148, 151; **34**: 253
Daphnia pulex **31**: 99, 115, 126, 127, 151
Daphnia schödleri **31**: 103
Daphniopsis **31**: 87
Daptonema **30**: 48, **50**, 50
Dardanus **29**: 146
  asper **29**: **132**, *147*, 148, 149
  pugilator **29**: *147*
  punctulatus **29**: **132**, 146
Dascyllus aruanus **44**: 218
Dasyatis akajei **21**: 62
Dasyatis sabina, Predation **27**: 367
Decamastus gracilis **26**: **179**
Decapoda **31**: 39; **32**: **56**, 57, *97*, 333, **335**,
    344, **409**, **438**, **446**, **447**, **454**;
    40: 28, 139–144; **42**: 216
Decapoda and fecundity
  environmental factors **43**: 148, 149
  life-history strategies **43**: 131, 133, 135
  quantification methodologies **43**: 92, 93,
    94, 96–99
Decapterus **32**: 186
  longimanus **30**: 331
  muroadsi **30**: 334
  russellii **30**: 334
  tabl **30**: 334
Decapterus muroadsi **32**: 146, **173**, 189,
    192, **196**, 227
Decolopoda **24**: 11, 26, 46
  australis **24**: 70
Deer, Sitka black-tailed *see Odocoileus
    hemionus sitkensis*
Deima **43**: 106
Deimatidae **32**: 466
Delius **33**: **52**, 77, 121
  nudus **33**: 121
  sewelli **33**: 121
Delphinapterus leucas (white whale) **38**: 175
Deltocyathus parvulus **32**: 506
Demania **21**: 140

Deminucula atacellana **42**: 26, 35
Demospongiae **32**: **160**, **437**, 495
Dendrochirota **32**: **461**
Dendrochirotida **32**: **438**, 468, **469**, 470
Dendrocoelum lacteum **21**: 133
Dendrocygna arborea **40**: 155
Dendrocygna autumnalis **20**: 231
Dendroica discolor paludicola **40**: 156
Dendroica petechia **40**: 156
Dendroica petechia gundlachi **40**: 156
Dendronephthya hemprichi **31**: 242
Dendropoma maximum **34**: 31
Dentaliida **42**: 139, *143*, *144*, 146, 155, 156,
    162, *203*
  see also Antalis; *Dentalium*;
    Fissidentalium
  biogeography **42**: *206*, 207
  diversity through geologic time **42**: 217,
    218
  feeding and digestion **42**: 165, *170*,
    173, 175, 176, 178
Dentaliidae **42**: *177*
  see also Dentaliida; Gadilinidae;
    Rhabdidae
Dentalium **42**: *143*, 159, 194, 196, 197
  D. antillarum **42**: 212
  D. conspicuum **42**: 169, 196
  D. dentalis **42**: 211, 212
  D. inversum **42**: 205
  D. laqueatum **42**: 152, *154*, 215
  D. leptoskeles **42**: 205
  *D. mutabile inaequicostatum***42**: 212
  D. octangulatum **42**: 216
  D. oerstedii **42**: 205
  D. pretosium **42**: *Plate 4*
  D. splendidum **42**: 205
  D. vulgare **42**: 200
Depressigyra globulus **34**: **370**
Derbesia vaucheriaeformis **23**: 108
Derjuginia **33**: 127
Dermasterias imbricata (seastar) **39**: 32, 38,
    41, 42
Derocheilocaris typicus **29**: 131, 186
Derogenes **25**: 148, 150; **32**: 308
  varicus **25**: 148, *149*, 158
Derogenes varicus **24**: 271–273, 277, 284,
    288, 293, 295, 299
Derogenes varicus **26**: 38; **32**: 306, 308
Derogenes varicus **40**: 6, 16, 22, 23, 27, 33,
    35–6, 39
Deropristis inflata **40**: 46
Derris scandens **40**: 115

Desbruyeresia
  cancellata **34**: 372
  marianaensis **34**: 372
  melanioides **34**: 372
  spinosa **34**: 372
Desmarestia **23**: 46
  ligulata **23**: 116
Desmidocercella numidica **24**: 273
Desmonema gaudichaudi **25**: 155
Desmoscolex **30**: 50, **50**
Desulfococcus **40**: 114
Desulfosarcina **40**: 114
Desulfotomacula **40**: 114
Desulfovibrio **40**: 114
Deutella californica **39**: 206
Deutella penantis **39**: 206
Deuterostomia **43**: *21*
  see also Chordata; Echinodermata;
    Hemichordata
Deutroiton **32**: **162**
Dexamine spinosa **39**: 198
Dexaminidae **32**: 468
Diacria trispinosa **32**: **453**
Diacria trispinosa major **32**: **452**
Diacyclops thomasi **25**: 5
Diadema antillarum **21**: 145; **23**: 101, 103,
    105; **34**: 231
  paucispinum **21**: 145
  setosum **21**: 145
Diadumene cincta **21**: 106
Diaixidae **33**: **53**, 130, 470, **499**
Diaixis
  ecology **33**: **499**
  taxonomy **33**: **53**, 78, 81, 88, 89, 130
Diaixis asymmetrica **33**: 130, **499**
  centrura **33**: 130
  durani **33**: 130
  gambiensis **33**: 130
  helenae **33**: 130
  hibernica **33**: 130, **499**
  pygmaea **33**: 130
  tridentata **33**: 130
  trunovi **33**: 130
Diaphana **42**: *89*
Diaphanidae **42**: 112
Diaphanoidea **42**: 112
Diaphanosoma **31**: 86
Diaphanosoma celebensis **31**: 86, 149
Diaphus **32**: 61, 167
Diaphus adenomus **32**: **170**, 187, **195**
Diaphus confusus **32**: **170**, 188
Diaphus pacificus **35**: **36**

Diaphus parini **32**: **170**
Diaphus suborbitalis **30**: 316, 317
Diaptomidae **33**: 3, 6, **52**, 55, 113, 473
Diaptomus (Cyclops) castor **29**: **131**
Diaptomus **33**: 257
  clavipes **33**: 367
  minutus **33**: **433**
Diaptomus kenai **28**: 119
Diarthrodes cystoecus **29**: **131**, 179
Diasoma **42**: 145, 204
Diastoma melanoides **32**: 407
Diastylis laevis **39**: 217
Diastylis lucifera **39**: 217
Diastylis polita **39**: 217
Diastylis quadrispinosa **39**: 217
Diastylis rathkei **35**: **163**, **168**; **39**: 152, 161,
    217
Diastylis sculpta **39**: 217, 218
Diastylis tumida **39**: 218
Diastyloides biplicata **39**: 218
Diatoms **41**: 32, 33, 47, 52, 53, 55, 57, 58,
    64, 77; **43**: 190, 198, 203, 205,
    206
  see also Thalassiosira
  blooms **43**: 197, 232
  secondary production **43**: 207, 208, 210,
    212, 213
  sediment traps **43**: 233–235
Diaules **42**: 118
Dibranchus sp. **32**: **171**, 186
Dicentrarchus labrax (sea bass) **38**: 33, 54
Dicentrarchus labrax **34**: 310
Dicentrarus labrax **26**: 85
Dichelopandalus **38**: 105, 106
Dichelopandalus bonnieri **38**: 105
Dichelopandalus leptoceros **34**: 221
Dichocoenia stokesii **22**: 12, 30, 38
Diclidophora **40**: 11
Diclidophora morrhae **40**: 5, 11
Dicmlene nigra **32**: **171**, 220
Dicologoglossa
  cuneata **29**: 229
  hexophthalmus **29**: 229
Dicrolene **32**: 206
Dictyota **23**: 103; **43**: *Plate 4*
Dicyathifer manni **40**: 149
Didymozoid *see* Trematodes
Digenea **24**: 278; **40**: 5–6, 25, 27; **43**: 16,
    17, 22, 23, 33, 52, *53*, 54, 55
Digramma alternans **40**: 7, 11
Dihemistephanus fragilis **40**: 11
Dikerogammarus aralensis **39**: 198

Dikerogammarus caspius **39**: 198
Dikerogammarus haemobaphes **39**: 198
Dikonophora **39**: 160
Dimophyes arctica **24**: 133, 142, 147, 167,
    168, 171, 208
Dimorphostylis asiatica **39**: 152
Dinobothrium **40**: 7, 10, 17, 25
Dinoflagellates **41**: 32, 47, 77; **43**: 198, 199,
    201, 213, 284, 285
Dinophilidae **29**: 195
Dinophilus **37**: 132
Dinophilus gyrociliatus **43**: 137, 144, 145
Dinophyceae **40**: 124; **41**: 49; **43**: 284
    see also Dinoflagellates
Dinophysis **31**: 309, 314
Dinophysis acuminata **33**: 157
Dinophysis fortii **21**: 72; **20**: 369
Diogenes pugilator **29**: **132**, *147*, 159
Diogenichthys laternatus **26**: 120; **35**: **36**
Diogodias littoralis **39**: 198
Diomeda exulans **39**: 281
Diomedia chrysostoma **20**: 295
    exulans **20**: 295
    melanophris **20**: 295
Diothona oculata **40**: 130
Diphyes **24**: 139, 202, 207, 234, 239
    antarcticus **24**: 141
    bonjami **24**: 167
    chamissonis **24**: 154, 166
    dispar **24**: 142, 166, 207, 214, 227, 234
    sieboldi **24**: 174
Diphyllobothrium **40**: 7, 10, 17, 26
Diplacanthopoma **23**: 328, 331
Diplectrona **35**: **163**
    modesta **35**: **169**
Diplectrum bivattatum, Predation **27**: **364**
Diploblasts **43**: 22
Diplocotyle olrikii **40**: 7, 10, 18, 28, 30, 36,
    37
Diplodus sargus, Predation **27**: **364**
Diplomonadida **43**: *20*
Diplopeltula **30**: 48, **50**
Diplopteraster multipes **35**: **31**
Diplora clivosa **22**: 45
    labyrinthiformis **22**: 10, 11, 21, 23, 30,
        33
    strigosa 10, 12, 21, 22, 30, 33, 37, 43–45,
        48, 49
Diploria labyrinthiformis **43**: *301*, 302
Diplostomum spathaceum **24**: 273, 295,
    302; **40**: 5, 16, 25, 34, 36, 43
Dirivultidae **32**: 124, 130

Dirivultus dentareus **23**: 329
Dischides **42**: *144*
Discinisca lamellosa **28**: 202, *213*, **228**, 236,
    244, 246, 249, 259, 274, 279, **280**,
    340
Disco **33**: **51**, 76, 106, 486
    atlanticus **33**: 106
    caribbeanensis **33**: 106
    crealus **33**: 106
    curtirostris **33**: 106
    elephantus **33**: 106
    erythraeus **33**: 106
    fiordicus **33**: 106
    hartmanni **33**: 106
    inflatus **33**: 106
    intermedius **33**: 106
    longus **33**: 66, 106
    marinus **33**: 106
    minutus **33**: 106
    oceanicus **33**: 106
    oviformis **33**: 106
    peltatus **33**: 106
    populosus **33**: 106
    robustipes **33**: 106
    tropicus **33**: 106
    vulgaris **33**: 106
Discoarachne (Tanystylum) brevipes **24**: 71
Discoidae **33**: **51**, 106
Disconectes sp. **32**: **162**
Dispora **25**: 121
Disseta **33**: **51**, 70, 73, 106, 107, 486
    coelebs **33**: 106
    grandis **33**: 106
    magna **33**: 106
    palumboi **33**: 106, **435**
    scopularis **33**: 106
Dissodinium **25**: 129
    pseudocalani **25**: 133
Dissostichus
    eleginoides **35**: 52
    mawsoni **35**: **36**
Dissostichus eleginoides **30**: 334
Dissostichus mawsoni **24**: 327, 328, 329,
    333, 334, 336, 347, 348, 353,
    355–358, 361, 363, 369–371, 374
Distichopora **43**: 292
    D. coccinea, D. nitida and *D. violacea*
        43: *291*
Distolasterias nipon **20**: 365
Ditylum **20**: 27, 28, 40
    brightwelli **20**: 40, 69
Dodecolopoda mawsoni **24**: 20, 26, 69

Dogfish *see Scyliorhinus*
Dogielinotus loquax **39**: 198
Dogielodinium **25**: 126, 129
Dolabella auriculasia **21**: 181
Dolabrifera **42**: 88
Dolabrifera dolabriefa **21**: 181
Dolabriferidae **42**: 78, 88
Dolichandrone spathacea **40**: 94
Dolioletta gegenbauri **23**: 264; **44**: 22, 31,
    135, 136
Dolium **28**: *392*
Dolly Varden char *see Salvelinus malma*
Donax **35**: 167, 171
    denticulatus **25**: 199, 208, 211
    incarnatus **25**: 209
    serra **25**: 180–238, *182*, *195*, *199*, *202*,
            204, *206*, **209**, **211**, **213**, *215*, **217**,
            220, **220**, *222*, **226**, **231**, *232*, *233*
    sordidus **25**: 180, 189, 208
    vittatus **35**: 167, **173**, *174*
Donax deltoides, Food/Feeding **27**: 327
Donax denticulatus **22**: 103, 110, 131
    juliane (D. trunculus) **22**: 103, 156
    serra **22**: 103, 116, 131, 150
    trunculus **22**: 103, 135
Donax serra **28**: 390, 398, 415, 421, 422
Doriopsilla albopunctata **34**: 33
Dorosoma **30**: 256; **20**: 160
    cepedianum **20**: 6, 39
    petenense **20**: 6, 39, 100
Dorosoma cepedianum **26**: **120**, **121**
Dosidicus gigas **39**: 271, 277, 280, 283, 285,
    288, 289
Dosinia japonica **21**: 170
Draconettidae **32**: 167, **173**, 211
Dreissena polymorpha **34**: 17
Drepanophorus crassus **21**: 134
Drepanopsis **25**: 138
Drepanopsis orbus **33**: 125
Drepanopus
    ecology **33**: 492
    nutrition **33**: 160
    taxonomy **33**: **53**, 81, 82, 85, 89, 129
Drepanopus bispinosus **33**: 129, **358**, **387**,
    390, **493**
    bungei **33**: 129, 473
    forcipatus **33**: 6, 129
    pectinatus **33**: 129, 174, **230**, 251, **331**, **358**
Drescheriella glacialis **43**: 210–212
Drill *see Nucella lamellosa*
Dromalia **24**: 178, 198
    alexandri **24**: 200

Dromedopycnon acantus **24**: 8
Dromia dehaani **32**: **163**
Dromia personata **29**: **133**, 136, 139, 141
Drosophila **39**: 165
Drosophila melanogaster **21**: 164
Drupella **34**: 230, 232
Duboscquella **25**: 126, 127, 129–131, *130*
    caryophaga **25**: 131
    tintinnicola **25**: 129, *130*
Duboscquodinium **25**: 127, 131
    collini **25**: *130*
    kofoidi **25**: *130*
Dugesia tigrina **21**: 164
Dunaliella **38**: 159; **20**: 28
Dunaliella euchlora **37**: 66
Dunaliella salina **41**: 199
Dungeness crab *see Cancer magister*
duorarum, Morphology **27**: 44
    Behaviour **27**: 333, 334, 335, 336,
        337, 338, 339, 340, 343, 346,
        347, 348, 349, 350, 352, 354,
        355
    Food/feeding **27**: **317**, 319, 320, 321,
        322, 324, 327, 330, 331, 332
    Life histories **27**: 285, 294, 296, 298,
        302, **304**, 306, 307, 308, 310, 311,
        313
    Moulting/growth **27**: 219, 220, 221,
        223, 225, 226, 227, **228**, **231**, 233,
        234, 237, 242, 243, *244*, 245, 246,
        249
    Parasites **27**: 380, 383, 384, 386, 387
    Physiology **27**: 171, 187, **204**
    Predation **27**: 357, 366, 375, 377
    Reproduction **27**: 253, 257, **258**, **270**,
        279
    Taxonomy **27**: 98, **110**, **111**, 123, 124
    Zoogeography/Evolution **27**: 139, **146**
Durvillea antarctica **23**: 30; **24**: 60
Dusa, Zoogeography/Evolution **27**: 150
    monocera,
            Zoogeography/Evolution **27**: 151
Dyaster insignis **35**: **31**
Dynamena pumila **24**: 45
Dyopedos monacanthus **39**: 198
Dysidea etheria **21**: 97, 98
Dysidea fragilis **34**: 33
Dytaster grandis **43**: 147
Dytaster spinosus **32**: **453**

Eastern oyster *see Crassostrea virginica*
Ebalia sculpta **32**: **163**

Ebalidae *see* Lower Heterobranchia
Ecarines **32**: **409**
Ecbathyrion prolixicauda **34**: **376**
Ecdytolopha **40**: 178
Echinactinomyxon **25**: 146
Echinaster sepositus **21**: 144
Echinidae **32**: **129**
Echinimunna **32**: **162**
Echinobothrium **25**: 151
Echinocardium cordatum **34**: 213; **20**: 355
Echinocyamus incertus **32**: **159, 165**
Echinodea **31**: 39
Echinodermata **32**: 330, 355, 360, **409, 446**,
        467, 497
Echinodermata
    see also Asteroidea; Echinoidea;
            Holothuroidea; Ophiuroidea
    fecundity
        environmental factors **43**: 146, 147
        life-history strategies **43**: 126, 136
        quantification methodologies **43**: 100,
            101, 103, 106, 107
    parasitism **43**: *21*
Echinogammarus pirloti **39**: 198
Echinoidea **32**: 158, **165, 210**, 333, **335**, 344,
        347, 364, 404, 429, **438, 462, 483**,
        486, 492
Echinoidea and fecundity **43**: *94*, 100, 103
    environmental factors **43**: 146, 147, 149
    life-history strategies **43**: 115, 126–128
Echinometra
    lacunter **23**: 101
    viridis **23**: 105
Echinopelta fistulosa **34**: **370**
Echinopora **31**: 247
Echinopora gemmacea **31**: 247, 264
Echinorhinidae **32**: **168**
Echinorhinus **32**: 198, 206
Echinorhinus cookei **32**: **168**
Echinorhynchus **32**: 308; **40**: 10
Echinorhynchus corrugatus **25**: 152
Echinorhynchus cotti **32**: 306
Echinorhynchus gadi **24**: 271, 277, 289;
        26: 38; **40**: 9, 18, 22–3, 28, 32,
        36–39, 41–44, 49, 51, 52
    salmonis **24**: 273
Echinorhynchus salmonis **40**: 9, 19, 28, 34
Echinosigra **32**: 366, 466
Echinothrix calamaris **21**: 145, 150, 169
    diadema **21**: 145, 150, 169
Echinus sp. **32**: **165**
    acutus **35**: 110

affinis **35**: 16, **31**
    esculentus **35**: 110, **184**
Echinus acutus **21**: 145
Echinus affinis **43**: 147
Echinus alexandri **32**: **129**; **34**: **379**
Echinus esculentus **37**: 7
Echiura **32**: 429, 506; **43**: *21*
Echiuroidea **32**: 333, **335, 438, 460**, *506*
Ecklonia **30**: 113, 318
Ecklonia maxima **25**: 185
Ecklonia radiata **23**: 10, 11, 30
Ecnomiosa **28**: *353*
Ecteinascidia turbinata **40**: 132, 133
Ectenagena sp. **32**: 120, 130
Ectenagena extenta **35**: 50; **34**: **373**
Ectenurus lepidus **25**: 150
Ectocarpus siliculosus **23**: 17
Ectoprocta **43**: *21*
Edotia **39**: 178
Edotia oclopetiolata **39**: 158
Edotia oculata **39**: 158, 178
Edwardsia callimorpha **22**: 77, 83, 84
Eel *see Anguilla anguilla*
Egeria radiata **35**: **184**
Egregia **23**: 56
    laevigata **23**: 5, 6, 11, 12, 30, 116
Egretta alba, Predation **27**: 372
Eigenmannia virescens **40**: 308
Eimeria **40**: 4
Eimeria sardinae **24**: 275, 279, 287,
        293–295, 298, 303, 305
Eisenia arborea **23**: 109, 110
Elasipoda **32**: 333, 335, *372*, **460**, 498
Elasipodida **32**: **438**, 466
Elasmobranchia **43**: 17
Elasmobranchs **32**: 167
Elasmopus levis **39**: 198; **43**: 142
Elassorhis laterospina **24**: 11, 13
Electrona antarctica **43**: 220, 226
Eledone **21**: 173, 184
    aldrovandi **21**: 184, 191
    cirrosa **21**: 173
    cirrosa **25**: 104
    moschata **21**: 184, 191; **25**: 103
Eleginus **24**: 276; **40**: 14
    gracilis **24**: 274
    navaga **24**: 274
Ellipinion **32**: 366
Ellisella **30**: 318
    flagellum **30**: 330
Ellobiocystis **25**: 136
Ellobiopsus **25**: 136, *136*

Elminius **22**: 227, 233; **40**: 138
  kingii **22**: 233
  modestus **22**: 209–211, 213, 215, 219,
    220, 224, 225, 234, 235
  plicatus **22**: 230, 233
  simplex **22**: 230, 233
elongatus **33**: 7, 130
  behaviour **33**: **412**, **417**, 444
  biochemistry **33**: 237
  chromosomes **33**: 47
  distribution **33**: **507**, **515**
  ecology **33**: **361**, 366, **378**, **397**, 398
  growth/development **33**: **307**, **311**, **313**,
    320, 321, **343**
  life history **33**: **387**, 390
  nutrition **33**: 145, 150, 161, 163, 166, 171,
    187, **191**, **198**, **201**, **206**, **284**
  reproduction **33**: 260, **288**, 289, 293
  size/weight **33**: **231**, 233
Elops **20**: 34
Elops machnata, Predation **27**: **363**, 367
Elpidia **32**: 366, 369, 371, 372
Elpidia birsteini **32**: **460**
Elpidia decapoda **32**: 371
Elpidia glacialis **32**: 358
Elpidia gracilis **32**: 371
Elpidia hanseni **32**: **461**
Elpidia kurilensis **32**: **461**
Elpidia lata **32**: 369
Elpidia nimae **32**: 369
Elpidiidae **32**: 330, 350, 362, 366, 371,
    460, 466
Elymus **39**: 43
Elysia **42**: 71, *116*, 133, 135
Emerita **29**: 146
  asiatica **29**: **133**, 150, 152, *151*, 153, **156**,
    157, **158**, 159, 160, 197, 198
Emerita analoga **21**: 140
Emerita portoricensis **25**: 209
Emiliana huxleyi **23**: 271
Emiliania huxleyi (coccolithophore) **41**: *31*,
    34, 36, *36*, 47, 65, 66, 69, *70*, 78
Emiliania huxleyi **33**: 159, 204, 528; **31**: *343*,
    344
Emmelichthyidae **32**: 167, **173**, 211
Emmelichthys **32**: 186, 189, 193
Emmelichthys cyanescens **32**: 146, **173**, 189,
    190, 191, **192**, **196**, 227
Emmelichthys cyaneus **32**: 224
Emmelichthys elongatus **32**: 146, **173**, 188,
    192, 193, **196**, 227
Emmelichthys stuhsakeri **32**: 217

Enantiosus **33**: **51**, 70, 73, 99
  cavernicola **33**: 99, **489**
Encephaloides armstrongi **35**: **32**,
    81,82–83, 110
Encheliophis (fish) **41**: *189*
  E. (Jornanicus) gracilis **41**: *189*
  E. vermicularis **41**: 188, *189*
Encope **32**: *214*
Endeis **24**: 11, 14, 18, 21, 25, 27
  australis **24**: 60, 70, 71
  mollis **24**: 75
  nodosa **24**: 66
  picta **24**: 66
  spinosa **24**: 16, 19, 21, 30, 33, 35, 38, 46,
    57, 64, 75
  straughani **24**: 66
  viridis **24**: 16, 63, 70
Endocladia **23**: 133
Endodinium
  chattoni **25**: 125
  nutricola **25**: 125
Engraulis **34**: 257; **20**: 15, 17, 20, 21, 38,
    177, 178, 179
  anchoita **20**: 6, 18, 29, 37, 103,180
  australis **20**: 173
  capensis **25**: 15, 21, *22*, 25, 36, 37, 48, 49;
    20: 6, 14, 37, 84, 171, 240
  encrasicholus **20**: 6, 86, 134, 136,
    140, 171
  japonicus **20**: 6, 29, 84
  mordax **25**: 7, 9, 11, 12, *13*, 14–16, *17*,
    18, 19, 21, *22*, 25, 30, **31**, **35**,
    36–38, 40, 42, 43, 48, 53, 57, 61,
    66; **34**: 276; **20**: 6, 11, 13, 18, 25,
    26, 39, 84, 86, 87, 100, 101, 102,
    103, 108, 131, 171, 246
  ringens **25**: **35**, 48; **34**: 205; **20**: 3, 6, 29,
    84, 171, 259
Engraulis anchoita **28**: 48–51, **50**, *89*
Engraulis capensis **28**: 54, **55**, *89*
Engraulis encrasicholus **24**: 276; **23**: 226
Engraulis japonica **28**: 87, 89
Engraulis japonicus **26**: 137
Engraulis mordax **26**: 137, 144; **28**: 10, 19,
    20, 25, **26**, 29, 59–61, **60**, **62**, 63,
    65, 84, 86, *87*, *89*, 93, 104, 123,
    125, 128; **29**: 280, 286, 287;
    30: 223; **31**: 178, 186, 196; **39**: 277
Engraulis ringens (Peruvian
    anchoveta) **38**: 3, 45, 54; **28**: *89*,
    90; **31**: 196
Engraulis ringens Jenyn **26**: 262

Engyproposon regani **32**: **174**
Enhalus (seagrass) **41**: *150*, 175, 199
 E. acoroides **41**: 149, 174, 175
Enhalus acoroides **40**: 127
Enhydra lutris (sea otter) **39**: 15–17,
       41–43, 60–62, 73, 76, 77, 80, 280
Enhydra lutris **34**: 270
Enhydrina schistosa, Predation **27**: 371
Enigmapercis **32**: 198
Enigmapercis acutirostris **32**: **173**
Enigmonia aenigmatica **40**: 139, 148
Enneagonum **24**: 133
 hyalinum **24**: 139
 searsae **24**: 121
Enoplometopus occidentalis **29**: **132**, 161,
       163, 168
Enoploteuthidae **32**: 62, 63
Enoploteuthis chuni **32**: 266, 303
Ensis siliqua **34**: 211
Entalimorpha **42**: 140, *144, 203*
Entalina **42**: *144*, 191
 E. quinquangularis **42**: 193, 194
Entalinidae **42**: *142, 144*; **42**: *144*
Entalinopsis **42**: *144*
Entemnotrochus **32**: 410
Entemnotrochus adansonianus **32**: **409**
Entemnotrochus rumphii **32**: **409**
Enteromorpha **23**: 87, 88; **24**: 47
 intestinalis **23**: 88, 134
 linza **23**: 36, 37
 prolifera **23**: 108
Enteromorpha intestinalis **22**: 219
Entobdella hippoglossi **26**: 38, *40*, 41, 54, 55
Entomoneis **43**: 203
Entophysalis **32**: **160**
Entoprocta **43**: *21*
Enypniastes diaphana **32**: **77**
Enypniastes eximia **32**: 97
Eochionelasmus
 n. sp. **34**: **375**
 ohtai **34**: **375**
 paquensis **34**: **375**
Eochionelasmus ohtai **32**: **409**
Eodentalium **42**: *143*
Eogammarus confervicolus **39**: 147, 198
Eogonatus **32**: 245
Eogonatus tinro **32**: 248
Eopsetta jordani **39**: 279
Eosipho
 desbruyeresi **34**: **373**
 n. sp. **34**: **373**
Epacteriscidae **33**: 17, **51**, 99–100, **489**

Epacteriscioidea **33**: **51**, 54, 55, 99–100
Epacteriscus **33**: **51**, 68, 99
 rapax **33**: 99, **489**
Ephelota **25**: 138, 139
 gemmipara **25**: 139
Ephelota sp., Parasites **27**: 385
Ephemera (*Navicula*)
       planamembranaceae **32**: *532*,
       539, 555
Epibionts **43**: 35
 see also Barnacles; Polychaeta
Epibulia chamissonis **24**: 112
 ritteriana **24**: 112, 113
Epigonidae **32**: **173**, 211
Epigonus **32**: 167, 186
Epigonus atherinoides **32**: **173**, 188, **218**
Epigonus elegans **32**: 146, **173**, 189, 190,
       191, **196**, 227
Epigonus notacanthus **32**: **173**, 189, **196**
Epigonus telescopus **34**: **381**
Epilabidocera **33**: **52**, 66, 95, 113
 amphitrites **33**: **26**, 113
 ecology **33**: **358**
 morphology **33**: **32**, 38, 39, 46
 reproduction **33**: 43, **270**
 longipedata **33**: **270**, **513**
Epilabidocera longipedata **44**: 93
Epimenia verrucosa **42**: 184
Epimeria **32**: 463, 466
Epinephelus **30**: 334
 aeneus **30**: 331
 E. coioides **43**: 65
 E. merra **43**: 37
 guttatus **34**: 254
 quernus **30**: 334
 striatus **34**: 254
Epinephelus striatus **44**: 226, 230
Epinephelus tauvina **44**: 231
Epipenaeon elegans, Parasites **27**: 388, 389
 ingens, Parasites **27**: 388, 389
Epirhabdoides **42**: *143*
 E. ivanovi **42**: 191
Epischura **33**: **52**, 116
 lacustris **33**: **411**
Episiphon **42**: *143*
 E. subtorquatum **42**: 147
Episiphoninae **42**: *143*, 147
Epistominella exigua **30**: 53
Epistylis sp., Parasites **27**: 384
Epixanthus denatus **40**: 141
Eptatretus stouti **40**: 273
Erebonaster protentipes **34**: **375**

Eremicaster sp. **32**: **462**
Eremicaster tenebrarius **32**: **462**
Eremicaster vicinus **32**: **462**, 464
Erenna **24**: 117
  bedoti **24**: 117
  richardi **24**: 117, 199
Eretmocrinus magnificus **31**: 47
Eretmocrinus praegravis **31**: 47
Ergasilus sieboldi **24**: 273
Erignathus barbatus (bearded seal) **38**: 175
Erignathus barbatus **40**: 25
Erilepsis zonifer **30**: 334
Eriopisa chilkensis, Food/Feeding **27**: 327
Eriphia sebana **21**: 140
Erobonectes **33**: **51**, 67, 100
  macrochaetus **33**: 100, **489**
  nesioticus **33**: 100, **489**
Ersea elongata **24**: 121
Erythrocles schlegeli **30**: 334
Erythrocles scintillans **32**: **173**, **218**
Erythrops **25**: 151, 153
Erythrops serrata **39**: 214
Erythrotrichia carnea **23**: 17
Escarpia **23**: 325; **34**: 405, 407, 408, 410;
    35: 50
  laminata **34**: *409*
  spicata **34**: **366**, 393, 405, *409*
Escherichia coli **21**: 37; **31**: 269
esculentus, Morphology **27**: *16*, 19, 3
    Behaviour **27**: 333, 334, 335, 336, 337,
      340, 341, 343, 344, *344*, 345, 348,
      349, 350
    Food/Feeding **27**: **317**, 320, 322, 323,
      324, 325, 326, 327, 328, 329, 330
    Life histories **27**: 286, 290, 296, 312,
      313, **313**, 314
    Moulting & growth **27**: 214, 219, 221,
      222, 225–6, **228**, **231**, 236, 240,
      249
    Parasites **27**: 380, 389
    Physiology **27**: 163, 177, 178, 179,
      181, 182, 183, **183**, 191, **204**, 206,
      208
    Predation **27**: 375, 377
    Reproduction **27**: 257, **258**, 262, 264,
      270, **279**
    Taxonomy **27**: 96, **111**, 124, 125, 126
    Zoogeography/Evolution **27**: **144**
Etelis
  carbunculus **30**: 334
  coruscans **30**: 334
Etheostoma **26**: 80

Ethmalosa fibriata **20**: 179
Ethmidium **20**: 177
Ethmodiscus **32**: 560
Etmopterus **32**: 167, 186
  mirabilis **34**: **380**
  princeps **34**: **380**
  pusillus **34**: **380**
Etmopterus litvinovi **32**: **168**
Etmopterus lucifer **32**: **168**
Etmopterus pycnolepis **32**: **168**
Etrumens **20**: 22
  teres **20**: 6, 39, 247
Euaetideus **33**: 486
  acutus **33**: **485**
Euandania aff. *ingens* **34**: **377**
Euandania ingens **32**: **128**
Euapta lappa **21**: 147
Euaugaptilidae **32**: 54
Euaugaptilus **25**: 138; **33**: 236
  ecology **33**: 486, 528
  nutrition **33**: **152**
  taxonomy **33**: **51**, 56, 61, 71, 73, 74, 76,
    103, 104
Euaugaptilus affinis **33**: 103
  aliquantus **33**: 103
  angustus **33**: 103
  antarcticus **33**: 103
  atlanticus **33**: 103
  austrinus **33**: 103
  brevirostratus **33**: 103
  brodskyi **33**: 103
  bullifer **33**: 103, **435**
  clavatus **33**: 103
  curtus **33**: 103
  digitatus **33**: 103
  diminutus **33**: 103
  distinctus **33**: 103, 104
  elongatus **33**: 103
  facilis **33**: 103
  fagettiae **33**: 103
  farrani **33**: 103, **435**
  fecundus **33**: 103
  filigerus **33**: *103*, **435**
  fundatus **33**: 103
  gibbus **33**: 103
  gracilis **33**: 103
  graciloides **33**: 103
  grandicornis **33**: 103
  hadrocephalus **33**: 104
  hecticus **33**: 68, 104
  hulsemannae **33**: 104
  humilis **33**: 104

hyperboreus **33**: 104
indicus **33**: 104
laticeps **33**: 104, 435
latifrons **33**: 104
longicirrhus **33**: 104
longimanus **33**: 104
longiseta **33**: 104
luxus **33**: 104
magnus **33**: 104, 267, **435**, 436, 438, 444
malacus **33**: 104
marginatus **33**: 104
matsuei **33**: 104
maxillaris **33**: 104
mixtus **33**: 104
modestus **33**: 104
niveus **33**: 104
nodifrons **33**: 104, **435**
nudus **33**: 104
oblongus **33**: 104
pachychaeta **33**: 104
pacificus **33**: 104
palumbii **33**: 104
parabullifer **33**: 104
paroblongus **33**: 104
pencillatus **33**: 104
perasetosus **33**: 104
periodosus **33**: **435**
placitus **33**: 38, 40, 104
propinquus **33**: 104
pseudaffinis **33**: 104
quaesitus **33**: 104
rectus **33**: 104, **435**
rigidus **33**: 104
roei **33**: 104
sarsi **33**: 104
similis **33**: 104
squamatus **33**: 104, **435**
sublongiseta **33**: 104
tenuicaudis **33**: 104
tenuispinus **33**: 104
truncatus **33**: 104, **435**
unisetosus **33**: 104
validus **33**: 104
vescus **33**: 104
vicinus **33**: 104, **435**
Eubostrichus africanus **40**: 135
Eubothrium **24**: 289; **43**: 65
Eucalanidae **32**: 11; **33**: **52**, 122, 340, 434
Eucalanoidea **33**: **52**, 54, 55, 122
Eucalanus **25**: 150; **33**: 30, 215
  behaviour **33**: 407
  biochemistry **33**: 245

  ecology **33**: 480, 483, **484**, 486, 487, 528
  nutrition **33**: **152**
  taxonomy **33**: **52**, 62, 62–63, 76, 122
Eucalanus attenuatus **33**: **7**, **26**, 122, **223**
  bungii **33**: **7**, **26**, 47, 63, 122
    distribution **33**: **513**
    ecology **33**: **358**, 468, 479, 481
    growth/development **33**: **341**, **342**
    life history **33**: **387**
    nutrition **33**: 167, 172, 203
    reproduction **33**: **287**
  californicus **33**: 63, 122, **485**, **513**
  crassus **33**: **7**, **26**, 122, **223**
    behaviour **33**: **446**
    chromosomes **33**: 47
    distribution **33**: **513**
    ecology **33**: **485**
    nutrition **33**: 148
  dentatus **33**: 122, **485**
  elongatus **33**: **7**, **26**, **32**, 32, 122
    behaviour **33**: 455
    chromosomes **33**: 47
    distribution **33**: **513**
    nutrition **33**: 183
    reproduction **33**: 42, 43
  hyalinus **33**: 122, 181
    behaviour **33**: **408**
    biochemistry **33**: 226, 243, 248, 250
    distribution **33**: **513**
    growth/development **33**: **304**
    longevity **33**: **345**
    nutrition **33**: 181, **183**
  inermis **33**: 122, 226, 243, 248, **485**, **513**
  langae **33**: 122, **513**
  longiceps **33**: 122
  monachus **33**: **47**, 122
    behaviour **33**: **446**
    distribution **33**: **513**
    ecology **33**: **387**, **485**
    physiology **33**: 217
  mucronatus **33**: **47**, 122, **485**, **513**
  muticus **33**: 122
  parki **33**: 122, **513**
  peruanus **33**: 122
  pileatus **33**: **26**, 122
    behaviour **33**: **408**, **411**, **446**
    distribution **33**: **513**
    growth/development **33**: **304**
    longevity **33**: **345**
    nutrition **33**: 162, 165, **190**, **198**, 199,
      200
  pseudoattenuatus **33**: **223**, **377**

quadrisetosus **33**: 122
sewelli **33**: 122, **513**
subcrassus **33**: **47**, 122
subtenuis **33**: 62, 122, **194**, **223**, **485**, **513**
  behaviour **33**: **416**
  chromosomes **33**: **47**
  distribution **33**: **513**
  nutrition **33**: **194**
Eucalanus bungii **32**: 11, 14–15, *15*, **16**, 17, 55, 266
Eucalanus californicus **44**: 79, 80
Eucalanus inermis **32**: 9, 34, *35*, 43–46, *44*, *45*
Eucalanus spp. **32**: 17, 29, 43, 44, 303
Eucalathis murrayi **28**: 211
Eucestoda **40**: 6–7, 25; **43**: *22*
Euchaeta spp. **32**: 37
Euchaeta **25**: *6*, 8, 14, 138; **29**: 176; **33**: 236
  behaviour **33**: 405
  ecology **33**: **460**, 480, 483, 486, 528
  elongata **25**: 23, 27, *28*, **34**, 43
  morphology **33**: 22
  norvegica **25**: 18, **31**, **34**; ;**29**: **131**, 174, *177*, 178–180, 181
  nutrition **33**: 140, 148, 149
  reproduction **33**: 266, 267, 274, **275**, 277, *279*, *281*
  taxonomy **33**: 53, *61*, 61, 63, 92, 130
Euchaeta acuta **33**: **7**, 130, **275**, **513**
  concinna **33**: 130, **230**, 524
    biochemistry **33**: **239**, **241**
    distribution **33**: **513**
    reproduction **33**: 275
  elongata **33**: **411**
  indica **33**: 130, **275**, **513**
  longicornis **33**: 130, **275**, **513**
  magniloba **33**: 130
  marina **33**: **7**, 49, 130, 297
    behaviour **33**: **411**, **416**, **418**, **446**
    distribution **33**: **513**
    ecology **33**: **377**, **485**
    growth/development **33**: **304**, **342**
    morphology **33**: *24*, 24, **32**, 47, 48
    reproduction **33**: 275
    size/weight **33**: **223**
  marinella **33**: 130
  media **33**: 130, **275**, 445, **513**
  norvegica **33**: 411
  paraacuta **33**: 130, **513**
  paraconcinna **33**: 130, **223**
    behaviour **33**: 445, **446**
    ecology **33**: **484**

growth **33**: **304**
reproduction **33**: **275**
plana **33**: 130, **230**, **239**, **241**, **513**
pubera **33**: 130, **275**, **513**
rimana **33**: 130
  behaviour **33**: 402, **411**, **416**, **418**, 432, **433**
  nutrition **33**: 149, 163
spinosa **33**: 130, **275**, **513**
tenuis **33**: 130, **513**
wrighti **33**: 130
Euchaeta antarctica **43**: 217
Euchaeta marina **44**: 81
Euchaeta spp. **44**: 81
Euchaetidae **32**: 54; **33**: 222, 434
  ecology **33**: **500**
  growth/development **33**: 321
  morphology **33**: 21, 27
  nutrition **33**: 140, **152**
  reproduction **33**: 265, 266, 277, **278**, **280**
  taxonomy **33**: **53**, 130, 131
Eucheuma **23**: 33–35
  acanthocladum **23**: 36
  gelidium **23**: 36
  isoforme **23**: 36
  nudum **23**: 36
Eucheuma cottonii (seaweed) **41**: 200
Euchirella **25**: 138; **33**: **6**, 236, 486, **499**
  behaviour **33**: 407, **416**
  ecology **33**: 486
  nutrition **33**: 140, 149, **152**
  reproduction **33**: 263, 267
  taxonomy **33**: **53**, 92, 93, 126
Euchirella amoena **33**: 126
  bella **33**: 126
  bitumida **33**: 126
  curticauda **33**: 126, **223**
    behaviour **33**: **411**, **416**, **418**, 444
    biochemistry **33**: 250
  formosa **33**: 126
  galeata **33**: 126
  grandicornis **33**: 126
  latirostris **33**: 126
  lisettae **33**: 126
  maxima **33**: 126
  messinensis **33**: **7**, 29, 126, 149, 263
  orientalis **33**: 126
  paulinae **33**: 126
  pseudopulchra **33**: 126
  pseudotruncata **33**: 126
  pulchra **33**: 126, **223**

rostrata **33**: **7**, 126, **223**
   behaviour **33**: 407, **411**, **418**
   distribution **33**: **513**
   nutrition **33**: 206
rostromagna **33**: 126, 153, 246, 247
similis **33**: 126
speciosa **33**: 126
splendens **33**: 126, **224**
tanseii **33**: 126
truncata **33**: 126
unispina **33**: 126
venusta **33**: 126, **416**
Euchone incolor **26**: **179**
Eucitheria stromboides **28**: *393*, **403**
Euclio **25**: 139; **32**: 302
   pyramidata **25**: 150, 151
Euclio sulcata **32**: 269
Euclymene **26**: **179**
Euclymene zonalis **26**: **179**
Euclymeninae **26**: **179**
Eucopia **25**: 142, 154; **39**: 160
Eucopia australis **32**: 57, *73*
Eucopia grimaldi **32**: *73*; **32**: 55, 57;
   39: 214
Eudentalium **42**: *143*
Eudocimus ruber **40**: 155
Eudorella emarginata **39**: 218
Eudorella pusilla **39**: 218
Eudorella truncatuta **39**: 218
Eudoxia dohrni **24**: 119, 120
   tenuis **24**: 119
   vasconiensis **24**: 119, 120
Eudoxoides **24**: 139
   mitra **24**: 154, 170, 171
   spiralis **24**: 133, 138, 139, 142, 147, 156,
      157, 165, 166, 169, 170
Eudyptes chrysolophus **20**: 275; **39**: 281
   chrosolophus schlegeli **20**: 295
Eudyptes pachyrhynchus **39**: 279
Eudyptula minor (little penguin) **39**: 282
Euglena gracilis **21**: 164
Euglenozoa **43**: *20*, 213, 232
Eugoria ampla **43**: *291*, 292
Eukrohnia fowleri **32**: 55
Eukrohnia hamata **32**: 26, 55
Eulalia (Protomystides) papillosa **34**: **367**
Eulas pusiolous **38**: 179
Eulepetopsis vitrea **34**: **370**, 397
Eulimnogammarus obtusatus **39**: 198
Eumatopias jabatus **20**: 246
Eumetiopias jubatus **34**: 274

Eumetopias jubatus (Steller sea lion)
   39: 62
Eunice **32**: **161**
   masudai **34**: **366**
   northioidea **34**: **366**
   pulvinopalpata **34**: **366**
Eunicella verrucosa **34**: 223
Eunoe nodosa **32**: **446**, **448**
Eunoyx chelatus **39**: 198
Euonyx mytilus **34**: **377**
Eupagurus bernhardus **26**: 27
Eupagurus bernhardus **29**: **133**, *147*, 148,
   149
Euphausia **22**: 91; **25**: 6, 22, 150, 151;
   29: 183, 184; **30**: **164**; **32**: 303;
   43: 210; **20**: 53
   crystallorophias **30**: *166*, **170**, 178
   E. crystallorophias and *E. frigida*
      43: 219
   E. pacifica **43**: 208
   E. superba **43**: 208, 219, 221–225, 239
   krohni **25**: 152, 154
   lucens **30**: **170**, 176
   pacifica **25**: *28*, **34**, 36, 43; **30**: **170**, 176,
      313; **20**: 52; **33**: 372
   sibogae **25**: 154
   superba **30**: 156, 159, 160, *167*, 167,
      170, 172, 173, 176, *177*,
      178–180, 187, 198–201, 203;
      20: 271; **33**: 528
   vallentini **30**: **170**
Euphausia distinguenda **32**: 46
Euphausia eximia **32**: 46
Euphausia japonica **32**: 266
Euphausia mucronata **32**: 34
Euphausia pacifica **44**: 124; **32**: 266, 269,
      274; **44**: 124, 125; **32**: 26, 28;
      34: 264; **32**: 34
Euphausia vallentini **32**: 26
Euphausiacea **32**: **56**, **452**, **453**, **454**
Euphausiidae **40**: 26
Euphilomedes climax **34**: **376**
Euphorbiaceae **40**: 94
Euplectella sp. **32**: **160**
Eupronoe **24**: 205
Euprymna **44**: 171
Euprymna scolopes **44**: 169, 172
Euraphia **40**: 138
Eurete (Pararete) farreopsis **32**: **160**
European oyster *see Ostrea edulis*
Eurycoipidae **39**: 159
Eurycope **32**: **162**, 463

Eurycope brevirostris **39**: 208
Eurycope cornuta **39**: 209
Eurycope magna **32**: **446**
Eurycope murrei **32**: 71, *73*
Eurycyde raphiaster **24**: 14
Eurydice **39**: 157, 178
Eurydice affinis **39**: 157, 209
Eurydice longicornis **39**: 209
Eurydice natalensis **39**: 209
Eurydice pulchra **39**: 209
Eurymera monticulosa **39**: 198
Eurypharingidae **32**: 63
Eurypleuron **32**: 186
Eurypleuron cinereum **32**: **171**, 218
Eurytemora **29**: 252; **33**: **52**, 93, 117, 298
    behaviour **33**: 405
    ecology **33**: 365, 475, 497, 528
    growth/development **33**: 299
    nutrition **33**: 154
    physiology **33**: 218
    reproduction **33**: 257, 265, 267, 279
    size/weight **33**: 333
Eurytemora affinis **33**: **7**, **26**, 60, 117, 346
    americana **33**: **26**, 117, **271**, **306**, **396**, 475
    anadyrensis **33**: 117
    arctica **33**: 117
    asymmetrica **33**: 117
    bilobata **33**: 117
    canadensis **33**: 117
    composita **33**: 117
    foveola **33**: 117
    gracilis **33**: 117
    graculicauda **33**: 117
    grimmi **33**: 117
    herdmani **33**: **7**, **26**, 63, 117
        behaviour **33**: **433**
        ecology **33**: 346, 374, **377**, 477, **493**, 495
        growth/development **33**: **306**, **310**, **312**, **320**, **342**
        reproduction **33**: 279, 282, **284**, 293
        size/weight **33**: **230**, 328, 333
    hirundo **33**: **26**, 60, 117
    hirundoides **33**: **26**, 60, 117, **288**
    inermis **33**: 117
    kieferi **33**: 117
    kurenkovi **33**: 117
    lacustris **33**: 117
    pacifica **33**: **26**, 117, **271**, 429, **431**
    raboti **33**: 117
    richingsi **33**: 117
    behaviour **33**: 407, **408**, **411**, **418**, 420, 432, **433**, 452
    biochemistry **33**: 239
    chromosomes **33**: **47**
    ecology **33**: **359**, 366, 367, 367, 369, 371, **377**, **396**, 398, 473, 475, 476, 525, 527
    growth/development **33**: 296, **304**, 305, **310**, **320**
    life history **33**: **345**, 388
    nutrition **33**: 155, 165, 185, **186**, **190**, **206**
    physiology **33**: 215, 216
    reproduction **33**: 260, 262, **271**, 271, 272, 280, 281, **284**, **288**, 293
    size/weight **33**: 227, **230**, 335
    thompsoni **33**: 117
    transversalis **33**: 117
    velox **33**: **26**, 117
        ecology **33**: **359**
        growth/development **33**: **306**
        physiology **33**: 216
        reproduction **33**: 260, **271**, **284**, **288**
        size/weight **33**: **231**, 332
    wolteckeri **33**: 117
    yukonensis **33**: 117
Eurytemora affinis **44**: 7, 103, 104; **31**: 92
Eurytemora herdmani **44**: 7; **38**: 161
Eurytemora spp. **44**: 303
Eurytemora velox **25**: 121
Eurythenes gryllus **30**: 330, 331; **32**: **77**, 116; **35**: 28, **32**, **42**, 43, 50, 78, *78*, 81, 87, 96; **39**: 144, 157, 198
Eurythenes obesus **39**: 198
Eurythoe **21**: 137
    aphrodite **21**: 138
    complanata **21**: 138
Euselenops **42**: 92, 112, 113, *114*, *116*, 132, 133, 135
Eusiridae **32**: **128**
Eusirus antarcticus **43**: 220
Eusirus perdentatus **39**: 198
Eusmalia fastigiata **22**: 11, 34
Eusyllis assimilis **26**: **179**
Euterpina **40**: 129
Euthyneura **42**: 70, 103, 119
    see also Opisthobranchia
Euthynnus
    affinis **30**: 334
    lineatus **30**: 228, 259, 279
Euthynnus affinis **43**: 84
Euthyonidium **21**: 147

Eutypa **40**: 122
Eutypa bathurstensis **40**: 119
Evadne **31**: 84, 86, 93, 98, 105, 116, 124,
    125, 130
Evadne anonyx **31**: 90, 91, 94, 102, 104,
    106, 117–120, 125, 128
Evadne nordmanni **23**: 225, 226; **31**: 87,
    88–89, 89–93, 101, **101**, 102, 103,
    106, 107, **109**, 111, 112, 115, 116,
    119, 120, 121, 128, 130, **133**, 136,
    137, *136*, 138–142, **144**, 145–148,
    152
Evadne spinifera **31**: 87, *136*, 137, 142, 148
Evadne tergestina **31**: 86
Evasterias **39**: 42
Evasterias troschelii (seastar) **39**: 41
Evechinus chloroticus **23**: 112; **35**: **184**;
    43: 147
Exallopus jumarsi **34**: **366**
Excirolana **39**: 157, 178
Excirolana braziliensis **39**: 209
Excirolana chiltoni **39**: 155, 209
Excirolana japonica **39**: 157, 209
Excirolana kumari **39**: 157
Excoecaria **40**: 101–103, 108, 111, 145,
    178
Excoecaria agallocha **40**: 90, 94, 98, 105,
    112, 121, 158, 167, 178, 181
Excoecaria dallachyana **40**: 94
Excoecaria indica **40**: 94
Exocoetidae **32**: 62
Exoediceroides maculosus **39**: 199
Exoediceros fossor **39**: 199
Exogone verugera **26**: **179**
Exosphaeroma truncatitelson **39**: 209
Exrima
    dolichopus **34**: **375**
    singula **34**: **375**
Exumella **33**: **51**, 68, 73, 101, **498**
    mediterranea **33**: 101, **489**
    polyarthra **33**: 101
    tuberculata **33**: 101

Facciolella castlei **32**: **168**, 191, **194**, 208,
    211
Falcidens **23**: 324
Falcidens n. sp. **34**: **369**
Falciformispora lignatilis **40**: 119
Fallax dalliniformis **28**: 192, 250, 254
Famelica nitida **32**: **164**
Farfantepenaeus duorarum **44**: 258
Farfantepenaeus, subgenus **27**: 95

Farfantepenaeus, Taxonomy **27**: 125
Farrania **33**: **53**, 81, 82, 129
    frigida **33**: 129
    lyra **33**: 129
    orbus **33**: 80, 125, 129
    pacifica **33**: 129
Fasciospongia cavernosa **21**: 100
Fauveliopsidae **32**: 463, 466
Favella **40**: 129
Favella ehrenbergii **31**: 313
Favia **22**: 11
    fragum **22**: 24, 27, 37
Faviidae **43**: 288
Favites **22**: 11
Fecampiida **43**: 20, 22
Fellodistomum furcigerum **40**: 21
Fenneropenaeus, subgenus **27**: 95
Fibrocapsa japonica **31**: 361
Filodinium hovassei **25**: 133, 134
Firoloida **25**: 142
    desmaresti **25**: 143
Fish **41**: 188, *189*; **43**: 121, 281
    see also Scomber; Scomberomorus;
        Teleost
    as hosts *see* Parasites *in subject index*
    Southern Ocean pack ice **43**: 226, 227,
        229
Fissidentalium **42**: *143*
    F. actinophorum **42**: Plate *3*, 164, 213
    F. exuberans **42**: 213
    F. floridense **42**: 212, *212*
    F. magnificum **42**: 216
    F. majorinum **42**: 163
    F. megathyris **42**: 164, 213
    F. meridionale **42**: 205
    F. scamnatum **42**: 212
    F. vernedei **42**: 152, 213, 215, 216
Fissuricola caritus **34**: **376**
Flabellifera **39**: 157, 158
Flagellata **43**: 7, 198, 208, 232
flagellates **41**: 32, 47, 49, 77
Flathead sole *see Hippoglossoides elassodon*
Flavobacterium **40**: 116
Floribella aldrichi **42**: 88
Flota sp. **32**: 71
Flounder (*Platichthys flesus*) **43**: 68
Flounder *see Platichthys flesus*
Flustra foliacea **24**: 45
fluviatilis (*see Penaeus setiferus*)
Foraminifera **32**: 357; **43**: 7, 209, 214
Fordilla **42**: 41
Fordillidae **42**: 41

Forskalia **24**: 103, 184, 190, 195, 196, 197,
        200, 207, 225, 226, 234, 239;
        25: **31**, 131
    cuneata **24**: 117
    edwardsi **24**: 118, 178, 218, 236
    leuckarti **24**: 118
    tholoides **24**: 118, 205
Fossaria modicella **34**: 20
Fosshagenia **33**: **51**, 77, 109
    ferrarii **33**: 50, 109, **489**
Fosshageniidae **33**: **51**, 109, **489**
Fosshagenioidea **33**: 50, **51**, *54*, 109
Foxtonia **33**: **52**, 66, 123
    barbatula **33**: 123, 502
Fragilaria **31**: 311; **20**: 36
Fragilaria striatula **21**: 72, 90, 142
Fragilariopsis (Nitzschia) kerguelensis
        33: 528
Fragilariopsis
    F. curta **43**: 198, 233
    F. cylindrus **43**: 198, 233, 237
Fragilariopsis curta **32**: *552*
Fragilariopsis cylindrus **32**: 535, *551*
Fragilariopsis kerguelensis (diatom) **41**: 55
Fratercula arctica **34**: 268; **20**: 282
Freileia helli **28**: 265, 275, 347, 348
Frenulata **32**: 408
Frenulina sanguinolenta **28**: 216, 249, 259,
        264, 265, 267–270, 275
Freyella **32**: 362
Frieleia halli **35**: **32**
Frillagalma vitiazi **24**: 116, 117
Fritillaira **25**: *128*, 134
Fritillaria borealis **32**: 22
Fritilliaria **44**: 34
Fritilliaria borealis sargassi **44**: 131
Fritilliaria haplostomai **44**: 131, 132
Fucaria **34**: 416, 417
    n. sp. **34**: **371**
    striata **34**: **371**
Fucus **23**: 4, 5, 19, 20, 21, 26, 45, 65,88,
        120; **24**: 47
    ceranoides **23**: 137
    distichus **23**: 26, 135
    evanescens **23**: 26, 63
    serratus **23**: 26, 109, 136, 139, 140, 144
    spiralis **23**: 26, 108, 123, 124, 133,
        136–141, 144
    vesiculosus **23**: 11, 47, 109, 116, 124,
        137, 135, 140
Fucus distichus **34**: 49

Fucus gardneri **39**: 17–19, 23, 25, 26, 44,
        72, 78
Fucus serratus **31**: 16
Fucus spiralis × *Fucus vesiculosus* **31**: 25
Fucus spp. **39**: 18–24, 28, 29, 43, 52, 72
Fucus vesiculosus **31**: 16
Fugu **25**: **52**
Fulmaris glacialis **20**: 226; **34**: 285; **34**: 296;
        39: 280
Funchalia spp., Morphology **27**: 12
    Taxonomy **27**: 91
    danae, Taxonomy **27**: 68
    taaningi, Taxonomy **27**: 68
    villosa, Taxonomy **27**: *67, 68*
    woodwardi, Taxonomy **27**: 68, **108**
Funchalia, Taxonomy **27**: 61, 66–68, 88,
        108, **113**, 117
    (*see also Pelagopenaeus*)
        Life histories **27**: 287
        Physiology **27**: 193, 203, 255
        Zoogeography/Evolution **27**: 141, 150,
        154
Fundulus heteroclitus **26**: 128; **40**: 286
Fundulus heteroclitus, Predation **27**: 374
Fungi **40**: 118–122; **43**: *20*, 207
Fungia **31**: 246
Fungia fungitis **22**: 43
    scutaria **22**: 23, 26, 28
Fungia Plate 4: **43**
Fungia scutaria **31**: 253
Funiculina quadrangularis **34**: 223
Fusarium, Parasites **27**: 383
    solani, Parasites **27**: 383
Fustiaria **42**: *143, 177*, 191
Fustiariidae **42**: *142, 143*

Gadella obscurus **32**: **170**, 188
Gadidae **32**: **129**, **170**, 308
Gadidius variabilis **32**: 274
Gadila **42**: *144*
    G. aberrans **42**: *Plate* 2, 162, 164
        feeding and digestion **42**: 163, *166*,
        169, 176, 179
    G. metivieri **42**: 156
Gadilia aberrans **42**: 215
Gadilida **42**: 139, *144*, 146, 156, 162, *203*,
        217
    see also Gadilidae; Pulsellidae
    biogeography **42**: 204, 205, *206*, 207
    feeding and digestion **42**: 165, 173, 175,
        176, 178
Gadilidae **42**: *142, 144*, 146

see also in particular Cadulus
Gadilimorpha **42**: 140, *144, 203*
Gadilina **42**: *143, 144*
  G. insolita **42**: 191, 215
Gadilinae **42**: 146
Gadilinidae **42**: *142, 143*; **42**: *143*
  see also Gadilina
Gadilopsis **42**: *144*
Gadinalia nivea **34**: 31
Gadomus **32**: 186
Gadomus melanopterus **32**: **170**
Gadus
  macrocephalus **34**: 265
  merlangus **29**: 280, 325
  morhua **29**: *27*, 286, 350
  morrhua **34**: 205
Gadus esmarkii **26**: *27*
Gadus luscus **38**: 175
Gadus macrocephalus (Pacific cod) **38**: 178, 210
Gadus macrocephalus (Pacific cod) **39**: 17, 49, 50, 53, 73
  predation on *P. eous* **38**: 218
Gadus macrocephalus **26**: **121**; **32**: 267, 304; **40**: 11, 14
Gadus merlangus **26**: 145; **38**: 175
Gadus minutus **26**: *27*
Gadus morhua (Atlantic cod) **38**: 4, *11, 20, 22,* 53, 164, 175, 178, 198, 210
  egg production **38**: 6
  in Scandinavia **38**: 18–23
  predation on *Pandalus borealis* **38**: 218
  relationship with *Pandalus borealis* **38**: 175
  survival and size at release **38**: **38**
  survival from hatchery releases **38**: **41**
Gadus morhua **44**: 219, 223, 224, 250, 252; **24**: 271, 274; **21**: 122; **25**: *3*, 10, 17–19, *20*, 21, 22, **31**, **34**, 41, 50, 57, 64; **26**: *73*, **75**, *77, 86*, **92, 120, 121**, *126, 131, 132*, 141, 252, 270; **28**: 24, 27, 37, 46, 49–51, *89*, **94**, 101, 102, **103–105**, **112**, 121, 129, 130; **30**: 220; **31**: 178; **35**: 5, 104; **40**: 1–59; **20**: 291
Gadus morhua see cod, Atlantic *under subject index*
Gadus ogac **40**: 14
Gadus poutassou **26**: 33
Gaetanus **32**: 303
  behaviour **33**: **416**

ecology **33**: 486
morphology **33**: 28
nutrition **33**: 140, **152**
taxonomy **33**: **53**, 87, 90, 127
Gaetanus antarcticus **33**: 127
  armiger **33**: 127
  brachyurus **33**: 127
  brevicaudatus **33**: 127
  brevicornis **33**: 127
  campbellae **33**: 127
  curvicornis **33**: 127
  divergens **33**: 127
  intermedius **33**: 127
  kruppi **33**: 127, 250, 444
  latifrons **33**: *4*, 127, 328, 422
  microcanthus **33**: 127
  miles **33**: 127, 422, *422*
  minispinus **33**: 127
  minor **33**: 127
  paracurvicornis **33**: 127
  pileatus **33**: 127, 250, *422, 422*
  recticornis **33**: 127
  tenuispinus **33**: 127
  wolfendeni **33**: 127
Gaetanus intermedius **32**: 274
Gaidiopsis **33**: **53**, 88, 127
  crassirostris **33**: 127
Gaidius **32**: 303
  ecology **33**: **499**
  nutrition **33**: **152**
  taxonomy **33**: **53**, 88, 90, 92, 93, 127
Gaidius affinis **33**: 127
  brevirostris **33**: 127
  brevispinus **33**: 127
  columbiae **33**: 127
  inermis **33**: 127
  intermedius **33**: 127
  minutus **33**: 127
  pungens **33**: 127, **499**
  robustus **33**: 127
  tenuispinus **33**: **224, 231, 234, 416**
  variabilis **33**: 127
Gaidropsus sp. **32**: **129**
Gaidropsarus mediterraneus **40**: 313
Gaidropsarus n. sp. **34**: **380**
Gaidropsarus parini **32**: **170**, 206, 208, 211
Galactosomum phalaerocoracis **24**: 277, 287, 294
Galapagomystides aristata **34**: **367**
Galathealinum brachiosum **34**: *409*
Galatheanthemidae **32**: 349
Galatheidae **32**: 97, **129**, 455

Galathowenia oculata **26**: **179**
Galaxea fascicularis **31**: 260
Galeichthys felis (*see Arius felis*)
  caerulescens, Predation **27**: 371
Galeolaria **25**: 138
Galetta pacifica **24**: 119
Galiteuthis glacialis **39**: 284
Gambierdiscus toxicus **21**: 73; **31**: 315
Gambusia affinis **21**: 126
Gammaracanthus **39**: 188
Gammaracanthus caspius **39**: 199
Gammaracanthus loricatus **39**: 199
Gammaridae **32**: 463; **40**: 28; **39**: 157, 186,
  187, 189, 193, 195–205
Gammaropsis nitida **39**: 190, 199
Gammarus **25**: 8; **31**: 12; **35**: **163**, **169**, 174,
  177, *178*; **38**: 18; **39**: 145, 185, 188
Gammarus aequicauda **39**: 199
Gammarus angulosus **39**: 199
Gammarus chevreuxi **39**: 199
Gammarus crinicornis **39**: 199
Gammarus duebeni **39**: 112, 146, 148, 183,
  199
Gammarus duebeni duebeni **39**: 113
Gammarus finmarchicus **39**: 199
Gammarus inaequicauda **39**: 199
Gammarus insensibilis **39**: 199
Gammarus lacustris **39**: 120
Gammarus lawtencianus **39**: 199
Gammarus locusta **39**: 199
Gammarus mucronatus **39**: 199–200;
  43: 142
Gammarus obtusatus **39**: 200
Gammarus oceanicus **39**: 200
Gammarus olivii **39**: 200
Gammarus palustris **39**: 200
Gammarus pulex **30**: 125; **39**: 131, 165, 170
Gammarus roeselii **39**: 165
Gammarus salinus **39**: 143, 200
Gammarus setosus **39**: 145, 200
Gammarus squamosa **39**: 200
Gammarus subtypicus **39**: 200
Gammarus tigrinus **39**: 200
Gammarus wilkitzii **39**: 200
Gammarus zaddachi **39**: 200
Ganchosia **33**: **52**, 66, 117
  littoralis **33**: 117
Ganymedes **25**: 141, 142
  vibiliae **25**: *142*
Gar *see Lepidosteus platyrhyncus*
Gasterosteus aculeatus L **26**: 262; **40**: 15
Gastopteridae **42**: 109, 110, 129

Gastopteron **42**: 92, 105
Gastroclonium coulteri **23**: 136
Gastropoda **31**: 39; **32**: **163**, **210**, **335**, 357,
  409, 445; **40**: 25, 27; **41**: *189*;
  42: 145; **43**: 101, 103, 145, 146
  see also Opisthobranchia
Gastrosaccinae **39**: 161
Gastrosaccus **25**: 154; **39**: 161
Gastrosaccus lobatus **39**: 214
Gastrosaccus psammodytes **39**: 214
Gastrosaccus vulgaris **39**: 214
Gastrotricha **43**: *21*
Gattyana cirrosa **26**: **179**
Gaussia **33**: **51**, 67, 108, 486
  asymmetrica **33**: 108, 261
  princeps **33**: 108, 211, **435**, 436, 437, 438
  scotti **33**: 108
  sewelli **33**: 108
Gavia arctica **39**: 279; **20**: 241
Gegania **42**: 102, 112, 115, *116*, 135
Gelidum coulteri **23**: 109, 114
Geloina erosa **40**: 148
Geloina expansa **40**: 148
Gelyelloida **33**: 1, **2**, *3*
Gemmulat (*Ptychosyrinx*) *bisinuata* **32**: **164**
Gemmulat (*Ptychosyrinx*) *naskensis* **32**: **164**
Gempylidae **32**: 167, **174**
Genarches muelleri **40**: 31
Gennadas **25**: 144
Gennadas borealis **32**: 55, 265
Gennadus, Physiology **27**: 190
Genyonemus lineatus **25**: 7, 48; **39**: 279
Genypterus blacodes **35**: **35**, 86, 103, 108
Geodia **21**: 98
  cydonium **21**: 98
  gibberosa **21**: 98
  mesotriaena **21**: 98
Geoduck clam *see Panope generosa*
Gephyrothuriidae **32**: **462**, 466
Gerardia **30**: 318
Gerres abbreviatus **44**: 231
Geryon fenneri **29**: **133**, 134
Geryonidae **32**: **129**
Geukensia demissa **24**: 436
Ghynnascaris sp. **32**: 306
Giffordia ralfsiae **23**: 108
Gigartina **23**: 56; **30**: 113
  canaliculata **23**: 114–116, 120
  intermedia **23**: *63*
  leptorhynchos **23**: 114
  papillata **23**: 22, 109, 114
  stellata **23**: 32

Gigartina stellata **24**: 47
Gilbertidia sigalutes **26**: 153
Gillichthys mirabilis **40**: 287, 288
Ginglymostoma cirratum, Predation **27**: 367
Gippslandia **33**: **52**, 67, 112
    estuarina **33**: 112
Girella tricuspidata **24**: 371
Gitanopsis alvina **32**: **128**; **34**: **376**
Glacidorbidae *see* Lower Heterobranchia
Gladioferens **33**: **52**, 67, 112, 475, **498**
    imparipes **33**: 30, 112, **331**
        ecology **33**: 469, 501
        growth **33**: **306**
        reproduction **33**: 280
    inermis **33**: 112
    pectinatus **33**: **26**, 112, 280, **306**, 365, 447
    spinosus **33**: 112
    symmetricus **33**: 112
Gladioferens imparipes **44**: 298
Glaucias **40**: 178
Glaucilla marginata **24**: 209
Glaucophyta **43**: *20*
Glaucous gull *see Larus glaucescens*
Glaucus atlanticus **24**: 209
Gleba cordata **34**: 31
Glenodinium foliaceum **25**: 120
Glenodinium polaceum **21**: 72
    rubrum **21**: 72
Globarcturus angelikae **32**: 372
Globicephala macrorhynchus **32**: 267;
        39: 279
Globicephala melaena **39**: 280, 281
Globigerinoides **25**: 125
    sacculifer **25**: 129
Gloeotrichia echinulata **21**: 71
Gloiopeltis furcatis **39**: 18
Glossanodon **30**: 334; **32**: 187
Glossanodon danieli **32**: **169**, 190, 191,
        195, 217
Glossanodon lineatus **32**: 217
Glossanodon nazca **32**: **169**, 188, 206, 217
Glossanodon struhsakeri **32**: 217
Glottidia albida **28**: **183**, 200, 249
Glottidia pyramidata **28**: 182, 188, 193,
        203, 217, 221, 246, 250, 265, 271,
        274, 289, 290, 302, 318, 325, 326,
        327, 329, 342
Glycera **21**: 138
    convoluta **21**: 138
Glycera capitata **26**: **179**
Glycera profundi **34**: **366**
Glyceridae **32**: **126**

Glycinde armigera **26**: **179**
Glycymeris **32**: **159**
Glyphanostomum pallescens **26**: **179**
Glypheidae **32**: **409**
Glyphocrangon wagini **32**: **162**
Glyptocephalus **31**: 27
Glyptonotus **30**: 91; **39**: 188
Glyptonotus acutus **39**: 209
Glyptonotus antarcticus **24**: 32, 329;
        39: 158, 209
Glyptotendipes **35**: *180*, **181**
Gmelina brachyura **39**: 201
Gmelinopsis tuberculata **39**: 201
Gnathia **32**: **162**
Gnathia elongata **40**: 9
Gnathonemus petersii **40**: 308
Gnathophausia **25**: 119; **39**: 160
    zoea **25**: 137
Gnathophausia gigas **32**: 55, *73*, 191
Gnathophausia gracilis **39**: 214
Gnathophausia ingens **39**: 214
Gnathophausia longispina **30**: 316; **39**: 214
Gnathophausia zoea **39**: 214
Gnathophis **32**: 167, 191
Gnathophis andriashevi **32**: **169**
Gnathophis codoniphorus **30**: 327
Gnathophis parini **32**: **169**, 188
Gnathophis smithi **32**: **169**, **194**
Gnathostomulida **43**: *21*
Gnorimosphaeroma insulare **39**: 209
Gnorimosphaeroma naktongense **39**: 114
Gnorimosphaeroma noblei **39**: 209
Gobiesox maeandricus **40**: 260, 262
Gobiidae **40**: 27
Gobius microps **26**: **121**
Gobius scorpioides **26**: **121**
Golden mullet *see Liza aurata*
Goldeneye *see Bucephala*
Golfingia (Nephasoma) improvisa **32**: 503
Golfingia margaritacea **32**: **460**
Golfingiidae **32**: 466
Gonactinia **22**: 73
Gonastomatidae **32**: 62
Gonatidae **32**: 248, *250*, 289, 295, 298, *298*,
        299, 302
Gonatopsis (Boreoteuthis) **32**: 295
Gonatopsis (Boreoteuthis) borealis **32**: 247,
        299
Gonatopsis (Gonatopsis) makko **32**: 249
Gonatopsis (Gonatopsis) octopedatus
        32: 299
Gonatopsis **32**: 244, 245, 288, 295, 308

Gonatopsis borealis **32**: *250*, 270–275, 283, 290, **290**, 291–296, **297**, 299, 300, 301, 302, 303, 304, 308, 309
Gonatopsis borealis makko **32**: 275
Gonatopsis japonicus **32**: 248, 249, 275, 278, 290, **290**, 291–295, 299, 303, **307**
Gonatopsis makko **32**: 275, 276, 278
Gonatopsis octopedatus **32**: 248, 278, 279, 290, **290**, 291, 292, 294, 295, 303, **307**
Gonatopsis okutanii **32**: 249, 286, 288
Gonatopsis s.str **32**: 295, 299, 304
Gonatopsis sp. **32**: 277
Gonatopsis ursabrunae **32**: 249
Gonatus (Eogonatus) tinro **32**: 299
Gonatus (Gonatus) fabricii **32**: 247, 262, 290, 297
Gonatus **32**: 244, 245, *246*, 258, 283, 288, 292, 293, 295, 299, 302, 304, 308
Gonatus antarcticus **32**: 247, 290, 297, 299; **39**: 284
Gonatus berryi **32**: 249, 284, 285, 290, **290**, 291, 294, 295, 299, **307**
Gonatus californiensis **32**: 249, 284, 290, **290**, 291, 293–297, **307**
Gonatus kamtschaticus **32**: 249, 282, 283, 290, **290**, 291, 293–296, 299, 301, 303, **307**
Gonatus madokai **32**: 249, *250*, 262, 276, 286–289, 290, **290**, 291, 294–296, 306, **307**, 308
Gonatus middendorffi **32**: 249, 282
Gonatus onyx **44**: 177; **32**: 249, *250*, 280–282, 290, **290**, 291–296, **307**
Gonatus oregonensis **32**: 293
Gonatus pyros **32**: 248, 249, 285, 286, 290, **290**, 291, 294–296, **307**
Gonatus steensrupi **32**: 290, 297
Gonatus tinro **32**: 279, 280, 283, 290, **290**, 291, 294–296, 301, **307**
Gonatus ursabrunae **32**: 249, 291
Goniada annulata **26**: **179**
Goniasteridae **32**: **165**, 445, 455
Goniastrea **22**: 33
  G. aspera **43**: 105, 133, 134
  G. australensis **43**: 134
  reitiformis **22**: 5
Goniastrea aspera **43**: 285
Gonionemus vertens **21**: 105
Goniopora **21**: 106, 127, 128; **22**: 10; **31**: 230
Goniopora pandoraensis **31**: 250, 251

Goniopora stokesi **31**: 251
Gonipora djiboutiensis **31**: 250, 251, *251*
Gonocerca macroformis **40**: 6, 16, 27
Gonocerca phycidis **40**: 6, 16, 33
Gonospora **25**: 139
  beloneides **25**: *140*
Gonostoma **32**: 61
Gonostomatidae **32**: 62, 63
Gonyaulax **21**: 60, 73, 75, 78, 79, 140; **23**: 233; **28**: 100; **20**: 27
  acatenella **21**: 73
  catanella **21**: 73, 76; **20**: 369
  cohorticula **21**: 73
  dimorpha **21**: 73
  excavata **21**: 73
  fratercula **21**: 73
  monilata **21**: 73, 85, 86
  polyedra **21**: 73; **20**: 45
  polygramma **21**: 73
  tamarensis **21**: 78, 79
Gonyaulax catenella **25**: 129
Gonyaulax grindleyi **33**: 165
Gonyaulax tamarensis see Alexandrium tamarense
Gorgoleptis
  emarginatus **34**: **372**
  patulus **34**: **372**
  spiralis **34**: **372**
Gorgonacea **43**: *291*
Gorgonaria **32**: *97*, **160**, *333*
Gorgonia **35**: **184**
Gorgonia ventalina **22**: 49
Gossleriella **32**: 560
Gotocotylidae **43**: 9
Gould's arrow squid *see Nototodarus gouldi*
Gourmya gourmyi **32**: 407
Goussia clupearum **24**: 277, 279, 287, 294, 295, 299, 305
Goussia gadi **40**: 4, 12–13, 30, 49
Goussia spraguei **40**: 4, 12–13, 30
Gracilaria (seaweed) **41**: 200
Gracilaria changii **40**: 126
Gracilaria corticata **40**: 126
Gracilaria salicornia **40**: 126
Gracilaria tikvahiae **23**: 142
Gracillaria, Life histories **27**: 287
Grammatonotus laysanus **32**: **173**, 188, **218**
Grammatorcynus **43**: *18*, 56, 59, *60*, *61*
Grammoplites scaber, Predation **27**: **363**, 368, *369*
Graneledone **34**: **380**
Graneledone spp. **44**: 150

Granuloreticulosia **32**: 465
Graphicomassa ligula **28**: *392*
Graptacme **42**: *143*
  G. *calamus* **42**: *168*, 171
  G. *eboreum* **42**: 169
Grassleia hydrothermalis **34**: **369**
Green algae (Chlorophyta) **43**: 281, 293
Green sea urchin *see Strongylocentrotus droebachiensis*
Gregarina
  annulata **25**: 140
  clausii **25**: 143
Griffithsia flosulosa **24**: 47, 57, 67
Grillotia erinaceus **24**: 277, 288; **26**: 38, 40, 41; **40**: 7, 17, 25, 33
Grimalditeuthidae **32**: 63
Grouper, orange-spotted (*Epinephelus coioides*) **43**: 65
Grubea **43**: 11, *29*
  G. *australis* **43**: 9 and G. *cochlear* **43**: 24
Gryllus **31**: 36
Gryphus vitreus **28**: 188, 224, 229, 244, 245, 248, 249, 265, 266, 269, 313, *315*, 335, 345, *351–353*
Guillecrinus **32**: 410
Guillecrinus neocaledonicus **32**: **409**
Guillecrinus reunionensis **32**: **409**
Guillemot *see Uria aalge*
Guinarda striata **32**: 555
Guinardia delicatula **31**: *346, 347*, 347
Gulls *see Larus* spp.
Guppy *see Lebistes*
Guschanskiana **25**: 148
Gwynia capsula **28**: 211, 265, 272
Gymnellis **26**: 33
Gymnobela **32**: 166
Gymnobela altispira **32**: **164**
Gymnobela brachypleura **32**: **164**
Gymnobela brunnistriata **32**: **164**
Gymnobela chisticovi **32**: **164**
Gymnobela crassilirata **32**: **164**
Gymnobela eugeni **32**: **164**
Gymnobela gracilis **32**: **164**
Gymnobela granulisculpturata **32**: **164**
Gymnobela laticaudata **32**: **164**
Gymnobela nivea **32**: **164**
Gymnobela rotundata **32**: **164**
Gymnobela turrispina **32**: **164**
Gymnocrinus richeri **32**: **409**, 410, *411*
Gymnodinium **21**: 37, 72, 82, 83; **25**: 125, 129; **28**: 84, 100; **31**: 309; **38**: 180;

43: 284, 285; **20**: 31, 40, 43, 45, 182
breve **21**: 73, 77, 82–86, 91
flavum **21**: 72
galatheanum **21**: 72
mikomoto **21**: 72
sanguineum **21**: 73
splendens **25**: 120
splendens **20**: 27, 40, 43, 182
ubberrimum **25**: 118, 119
veneficum **21**: 73, 83
Gymnodinium breve see Ptychodiscus breve
Gymnodinium catenatum **31**: 307, 355
Gymnodinium sanguineum **31**: 92, 93, 101, 102, 148
Gymnodinium splendens **26**: 261; **37**: 125
Gymnodraco acuticeps **24**: 328, 329, 332, 346, 349, 353, 355, 356
Gymnogongrus furcellatus **23**: 111
Gymnoplea **33**: 2
Gymnoressella **32**: 346
Gymnosomata **42**: 111
Gymnotiformes **40**: 306
Gynaecotyla adunca **43**: 44
Gyptis **26**: **179**
Gyrocotylidea **43**: 22
Gyrodactylidae **43**: 51, *52*, 84, 85
Gyrodactyloides andriaschewi **24**: 274, 291
  baueri **24**: 285, 291
  petruschewskii **24**: 274, 291
Gyrodactylus **40**: 5, 12, 29, 35, 39, 48, 55
Gyrodactylus arcuatus **40**: 5, 10, 15, 34
Gyrodactylus callariatis **40**: 5, 12–13, 31, 35, 36, 38, 55, 56
Gyrodactylus cryptarum **40**: 5, 12–13, 29, 36
Gyrodactylus cyclopteri **24**: 274, 291
  flesi **24**: 274, 291
  gerdi **24**: 274, 291
  groenlandicus **24**: 274, 291
  harengi **24**: 286, 287
  pterygialis **24**: 274, 291
  pungitii **24**: 273
  robustus **24**: 274, 291
  unicopula **24**: 274, 291
Gyrodactylus emembrenatus **40**: 5, 12–13, 30, 36
Gyrodactylus errabundus **40**: 15
Gyrodactylus marinus **40**: 5, 13–14, 31, 35, 36, 56
Gyrodactylus perlucidus **40**: 15

Gyrodactylus pharyngicus **40**: 5, 13–14, 31, 36
Gyrodactylus pterygialis **40**: 5, 15, 31, 36
Gyrodinium **23**: 236, 252, 263, 267, 268; **31**: 148
   aureolum **23**: 214, 232–236, 238, 239, 241, 246, 256, 257, 259, 261, 267, 278
   breve **23**: 234
   instriatum **25**: 120
   resplendens **25**: 118
Gyrodinium aureolum **21**: 73; **31**: 317, 318, 348, 351, 355, 357, 364
Gyrodinium aureolum **37**: 125
Gyrodinium dorsum **33**: 164
Gyrothyris **28**: *353*

H. scabra versicolor **41**: 134, 176, 201
   anatomy **41**: 143, 146, *147, 148*
      fisheries **41**: 188, 190, 191, 192, 193, 194, 195
      reproductive cycle **41**: 158, 159, 160, 165, 167
      spawning **41**: 168, 169
      H. tigris **41**: 134
Habrocidaris argentea **32**: **165**
Haddock *see Melanogrammus aeglefinus*
Hadropenaeus lucasii **32**: **162**
Haematodinium **25**: 127
Haematopus **43**: 34
Haematopus bachmani (black oystercatcher) **39**: 15–17, 64–66, 78, 80
Haematopus ostralegus **34**: 271
Haemogregarina aeglefini **40**: 4, 12–13, 30
Haemohormidium terraenovae **40**: 4, 15, 30
Haemulon flavolineatum **44**: 223
Haemulon plumieri, Predation **27**: 368
Hagfish *see Eptatretus stouti*
Hagfish *see Myxine*
Hake *see Merluccius hubbsi*
Halacarellus
   alvinus **34**: **374**
   auzendei **34**: **374**
Halacarellus alvinus **32**: **127**
Halacaridae **32**: **127**
Halacarus prolongatus **32**: **127**
Halalaimus **30**: 48, **50**, 50
Halargyreus **32**: 186
Halargyreus johnsoni **32**: **170**
Halcampa **22**: 73
Halecium beani **21**: 105
Haliaeetus albicilla **20**: 282

Haliaeetus vocifer, Predation **27**: 372
Haliaetus leucocephalus (bald eagle) **39**: 17, 45, 46, 71
Haliaetus leucogaster **40**: 155
Haliaetus leucorhyphus **40**: 155
Halice hesmonectes **34**: **377**
Halice quarta **32**: 74
Halice subquarta **32**: 74
Halichoeres poecilopterus **26**: 129
Halichondria okadai **21**: 102
Halichondria panicea **24**: 63
Haliclona **21**: 98, 99
   doria **21**: 98
   erina **21**: 98
   okadai **21**: 98
   paricea **21**: 98
   rubens **21**: 98, 101
   viridis **21**: 97, 98
Haliclona magniconulosa **40**: 133
Halidrys dioica **23**: 109, 110
Halimeda macroloba **40**: 126
Halimeda opuntia **23**: 100; **40**: 126
Halimeda Plate 4: **43**, 282, 305
Haliotis **21**: 170, 176; **28**: 397; **35**: **163**
   discus hannai **21**: 176
   discus hannai **34**: 48
   iris **35**: **185**
   japonica **21**: 176
   midae **35**: **184**
   sieboldi **21**: 176
   tuberculata **34**: 23, 37, *38, 39*, 40, **41**
   tuberculata **35**: **184**
Haliotis tuberculata **43**: 101
Haliotrema **43**: *29*
Haliplanella (Diadumene) luciae **22**: 68
Halisiphonia arctica **32**: **126**, 130
Halistema **25**: 124
Halistemma **24**: 105, 117, 125, 126, 188, 190, 200, 207
   (Stephanomia) amphytridis **24**: 117
   rubrum **24**: 103, 117, 178, 200
Hallezia **25**: 138
Halobaena caerula **20**: 295
Halobates **30**: 193
Halochlorococcum operculatum **40**: 125
Halodeima grisea **21**: 159
Halodule (turtlegrass) **41**: *151*
Halodule wrightii **40**: 127
Halophila (turtlegrass) **41**: *151*; **24**: 60
Halophila baccarii **40**: 127
Halophila ovalis **40**: 127
Halophytophthora **40**: 120

Halophytophthora kandeliae **40**: 119
Halophytophthora spinosa **40**: 119, 120
Halophytophthora vesicula **40**: 119, 120
Halopteris buski **24**: 16
  scoparia **24**: 55
Haloptilus **25**: 138
  behaviour **33**: 422
  ecology **33**: 483
  nutrition **33**: **152**
  taxonomy **33**: **51**, 56, 71, 73, 74, 104, 106
Haloptilus aculeatus **33**: 104
  acutifrons **33**: 104, **513**
  angusticeps **33**: 104
  austini **33**: 104
  bulliceps **33**: 104
  caribbeanensis **33**: 104
  chierchiae **33**: 104, 105
  fertilis **33**: 104
  fons **33**: 104
  furcatus **33**: 104
  longicirrus **33**: 104, **435**
  longicornis **33**: **7**, 104, **359**, **416**, **513**
  major **33**: 104
  mucronatus **33**: 104
  ocellatus **33**: 104
  orientalis **33**: 105
  ornatus **33**: 105
  oxycephalus **33**: 105
  pacificus **33**: 105
  paralongicirrus **33**: 105
  plumosus **33**: 105
  pseudooxycephalus **33**: 105
  setuliger **33**: 105
  spiniceps **33**: 105
  tenuis **33**: 105
  validus **33**: 105
Halopus rangi **32**: 409
Halosaccion glandiforme **23**: 135
Halosarpheia marina **40**: 122
Halosarpheia minuta **40**: 119
Halosarpheia retorquens **40**: 122
Halosauridae **32**: **168**
Halosbaena acanthura **39**: 155, 175
Halosphaera **25**: 127; **20**: 28
Haminea **28**: *393*, 412, 419
Haminoea **42**: 93, 97, 98, 106, *116*, 132, 135
  H. elegans **42**: 86
  H. taylorae **42**: 86
Haminoeidae **42**: 75
Hannonia **24**: 8, 26, 55, 71
  spinipes **24**: 13, 71
  typica **24**: 71

Hansenulus **25**: 153
Hapalochaena, see also, *Octopus*
  maculosa **21**: 173, 184–187, 191
Hapalonotus reticulatus (crustacean)
  41: *189*
Haplocope angusta **39**: 219
Haplomesus quadrispinosus **39**: 159, 209
Haplopharyngida **43**: *22*
Haplosporidium nelsoni **43**: 69
Haplozoon **25**: 127, 131
Hapsidascus hadrus **40**: 119
Haptenchelys texis **32**: *129*, 130
Harbour seal *see Phoca vitulina*
Harbour seals *see Phoca borealis*
Harengula **20**: 97, 103, 108, 109
  clupeola **20**: 96
  humeralis **20**: 96, 102
  jaguana **20**: 180
  pensacolae **20**: 6, 39
  thrissina **20**: 6, 107, 128
Harengula abbreviata, Predation **27**: *362*
Hargeria rapax **39**: 219
Harlequin duck *see Histrionicus histrionicus*
Harmothoe **26**: 179
Harmothoe derjugini **32**: 444, **446**, **448**, **452**
Harmothoe lunalata **26**: 179
Harmothoe macnabi **34**: **367**
Harmothoe rarispina **32**: **446**, **448**
Harp seal *see Phoca groenlandica*
Harpacticoida **32**: **128**; **33**: **2**, 2, *3*
Harpacticozoa **43**: 208–210, 216
Hastigerina pelagica **25**: 129
Hastula solida **28**: *393*
Hatschekia hippoglossi **26**: 38
Haustoriodes japonicus **39**: 201
Haustorius arenarius **39**: 201
Haustorius canadensis **39**: 201
Haustorius saginatus **39**: 201
Hedgpethia **24**: 26
Hedophyllum sessile **23**: 75, 76, 132, 135
Helcion pellucidum **34**: 35
Heleocius cordiformis **40**: 140
Helicolenus **32**: 186, 218; **35**: 108
  dactylopterus **35**: **37**, 43, 52, 103
  hilgendorfi **35**: **38**
Helicolenus dactylopterus **30**: 334; **32**: 219
Helicolenus lengerichi **32**: **172**, 185, 189,
    190, 191, 193, **194**, 218, 219
Helicolenus mouchezi **32**: 219
Helicolenus percoides **32**: 218, 219
Helicoradomenia
  juani **34**: 369

n. sp.1 **34**: **369**
n. sp.2 **34**: **369**
n. sp.3 **34**: **369**
n. sp.4 **34**: **369**
n. sp.5 **34**: **369**
n. sp.6 **34**: **369**
Helicrenion reticulatum **34**: **372**
Heliopora **43**: 292
  H. coerulea **43**: *291*
Heliopora coerulea **32**: 407
Helix **25**: 210
Helix aspersa **22**: 103, 145; **34**: *10*, 32, 33
Helleria brevicornis **39**: 110
Helmet crab *see Telmessus cheiragonus*
Helminths **43**: 14, 36, 64
Hemectyon **21**: 98
  ferox **21**: 97
Hemiarthrus abdominalis **38**: 181, 182
Hemiasteridae **32**: 466
Hemiaulus spp. **32**: 555, 560
Hemiaulus hauckii **32**: *534*, 535, *551*,
     554, 558
Hemiaulus sp. **31**: 92
Hemiboeckella **33**: **52**, 112
Hemichordata **43**: *21*
Hemicrinidae **32**: **409**
Hemicrinus **32**: 409
Hemidiaptomus
  ingens **33**: 29
  ingens provinciae **33**: 41
Hemidiscus **32**: 560
Hemieutreptia antigua **21**: 37, 72
Hemigraspus oregonensis **43**: 70
Hemigymnus melapterus **43**: 37, 45
Hemilamprops calfornica **39**: 175
Hemilamprops rosea **39**: 218
Hemimysis lamornae **39**: 214
Hemioniscus balani **22**: 219
Hemirhabdus **33**: **51**, 70, 74, 107, 486
  falciformis **33**: 107
  grimaldii **33**: 107, **435**
  latus **33**: 107, 434, **435**
  truncatus **33**: 107
Hemiselmis
  aff *anomala* **29**: 78
  virescens **29**: 78
Hemisepius **25**: 87
Hemithiris psittacea **28**: 211, *213*, 223, 227,
     231, 232, 244, 248, 265, 271, *275*,
     283, 295, 306, 312, 316, 319, 345
Hemiurus **25**: 147, 148; **28**: 82
  communis **25**: 149

  levinseni **25**: 148
  levinsi **25**: 158
Hemiurus communis **40**: 6, 16, 27, 33, 35,
     36, 39, 43, 46
Hemiurus levinseni **24**: 277, 285, 288
  luehei **24**: 276, 285, 291, 292, 296; **40**: 6,
     16, 22, 23, 27, 31, 35, 36, 42, 43,
     45
Hemiurus luehei **40**: 6, 13, 20, 46
Henricia spp. **32**: **165**, 175, 176
Henricia **43**: 93, 100
Hepatoxylon trichuiri **40**: 7, 25, 33
Heritiera **40**: 99
Heritiera fomes **40**: 96, 102, 158
Heritiera globosa **40**: 96
Heritiera littoralis **40**: 96
Herklotsichthyes spp. **44**: 231
Herklotsichthys castelnaui **28**: 108
Hermodice **21**: 137; **22**: 92
  carunculata **21**: 138
Herniflora, Food/feeding **27**: 319
Herring *see Clupea harengus*
Herring *see Clupea pallasi*
Hesiocaeca hessleri **34**: **366**
Hesiodeira glabra **34**: **366**
Hesiolyra bergi **34**: **367**
Hesionidae **29**: 195
Hesiospina vestimentifera **34**: **367**
Heteralepas mystacophora **32**: **159**, **161**, 178
Heteramalla **33**: **53**, 82, 134
  dubia **33**: 134
  sarsi **33**: 134
Heterarthrandria **33**: 17
Heterobranchia **42**: 106
  see also Lower Heterobranchia;
     Opisthobranchia; Pulmonata
  characters and evolutionary
     pathways **42**: 95, 99, 100, 102, 104
  taxonomic history **42**: 112, 115, 116
Heterocapsa triguetra **21**: 73
Heterocarpus
  ensifer **30**: 334
  laevigatus **30**: 334
Heterocarpus fenneri **32**: **162**
Heterocarpus laevigatus **32**: **162**
Heterocarpus sibogae **32**: **162**
Heterocentrotus mammillatus **21**: 145
Heterocoma hyperparasitica **25**: 137
Heterocope **33**: **52**, 66, 117
  appendiculata **33**: 117
  borealis **33**: 117
  saliens **33**: 117

septentrionalis **33**: 117, **426**
Heterocrypta epibranchialis **32**: **163**
Heterocyathus japonicus **42**: 213, 214, 216
Heterodonta **42**: *37*; **43**: *20*
Heterokrohnia furnestinae **32**: 77
Heterolobosea **43**: *20*
Heteromysis **39**: 160
Heteromysis armoricana **39**: 214
Heteromysis beetoni **39**: 214
Heteromysis filitelsona **39**: 214
Heteromysis formosa **39**: 214
Heteromysis tuberculospina **39**: 214
Heteronymphon **24**: 26
  exiguum **24**: 60
Heteropenaeus, Taxonomy **27**: 68, 69, 107,
    113
    Physiology **27**: 203
    Zoogeography/Evolution **27**: 136, 141,
      150
  longimanus, Taxonomy **27**: 69, *69*
    Life histories **27**: 285
    Zoogeography/Evolution **27**: **142**
Heterophryxus **25**: 154
  appendiculatus **25**: 154
Heteropneustes **26**: 103
Heteropneustes fossilis **40**: 258, 259, 261,
      263–265, 267, 268, 272, 280, 281,
      283–285, 298, 299
Heteropoda **32**: **452**, **454**
Heteropsammia **42**: 216
Heteroptilus **33**: **51**, 70, 74, 76, 105
  acutilobus **33**: 105, **435**
  attenuatus **33**: 105
Heteropyramis alcala **24**: 121
  maculata **24**: 176
Heterorhabdidae
  behaviour **33**: **435**, 436
  ecology **33**: 502
  morphology **33**: 20, 36
  nutrition **33**: **152**
  taxonomy **33**: **51**, 106, 107
Heterorhabdidae **32**: 54, 55
Heterorhabdus **33**: 6
  behaviour **33**: **411**
  ecology **33**: 486, 528, 529
  morphology **33**: 19
  nutrition **33**: 160, 164
  taxonomy **33**: **51**, 70, 74, 107
Heterorhabdus abyssalis, 107, **513**
  atlanticus **33**: 107
  austrinus **33**: 107, **231**
  brevicornis **33**: 107

caribbeanensis **33**: 107
clausi **33**: 107
compactoides **33**: 107
compactus **33**: 107
egregius **33**: 107
farrani **33**: 107, **224**, **231**, **234**
fistulosus **33**: 107
lobatus **33**: 107
longispinus **33**: 107
medianus **33**: 107
nigrotinctus **33**: 107
norvegicus **33**: 107, **435**, **513**, 522
pacificus **33**: 107
papilliger **33**: **7**, 107, **435**, **513**
proximus **33**: 107
pustulifera **33**: 107
robustoides **33**: 107
robustus **33**: 107, **435**
spinifer **33**: 107
spinifrons **33**: **7**, 107, **435**
spinosus **33**: 107
sub-spinifrons **33**: 107
tanneri **33**: 107, 479
tenuis **33**: 107
tropicus **33**: 107
vipera **33**: 107
Heterosaccus lunatus **43**: 70
Heteroschismoides **42**: *144*
Heteroschismoidinae **42**: *144*
Heterosigma akashiwo **31**: 317, 364
Heterostropha *see* Lower Heterobranchia
Heterostylites **33**: **51**, 70, 73, 107
  longicornis **33**: 107, **435**
  major **33**: 107
Heterosygma **21**: 15
Heterotanais **39**: 151, 160
Heterotanais oerstedi **39**: 114, 148, 153,
    154, 219
Heteroteuthis sp.n.aff.
    dispar **32**: **164**25: 101; **32**: 205,
    220
Heteroteuthis dispar **32**: 204, 220
Heteroxemia fuscescens **22**: 46
Hexabranchus ("Spanish Dancer") **42**: 106,
    116, 133, 135
Hexacorallia **32**: **160**
Hexactinellida **32**: **160**
Hexactinellidae **32**: 335, 345, 495
Hexagrammos decagrammus (kelp
    greenling) **39**: 48, 49
Hexagrammos octogrammus (masked
    greenling) **39**: 49

Hexaminius **40**: 138, 139
Hexaminius foliorum **40**: 139
Hexaminius popeiana **40**: 139
Hexanchidae **32**: **168**
Hexanchus **32**: 198, 206
Hexanchus griseus **32**: **168**
Hiatella orientalis **24**: 52; **20**: 366
High cockscomb *see Anoplarchrus purpurescens*
Hildenbrandia **23**: 111
  rubra **23**: 113
Hilsa **20**: 160
Himanthalia elongata **23**: 111
Himantothallus grandifolius **24**: 60
Himantura uarnak, Predation **27**: 367
Himasthla elongata **40**: 6, 13, 20
Himasthla tensa **40**: 20
Hime **32**: 186
Hime damasi **32**: 217
Hime macrops **32**: **169**, 217
Hippa pacifica **29**: **133**, 150
Hippoglossoides elassodon (flathead sole) **39**: 60
Hippoglossoides elassodon **28**: 106; **38**: 178
Hippoglossoides platessoides **34**: 289; **40**: 20; **43**: 67
Hippoglossus **28**: 25
Hippoglossus americanus **26**: 3
Hippoglossus hippoglossus **44**: 297; **26**: 2, 3, 5, 76, 78, 86, **92**; **34**: 240; **40**: 19
Hippoglossus hippoglossus stenolepis **26**: 2
Hippoglossus platessoides **26**: 257
Hippoglossus stenolepis **44**: 260; **26**: 2; **39**: 279
Hippoglossus vulgaris **26**: 3
Hippolyte varians **38**: 179
Hippomedon propinquus **39**: 201
Hippomedon whero **39**: 201
Hippopodius **24**: 107, 119, 173, 176, 179, 181, 183, 188, 191, 192, 199, 202, 203
  hippopus **24**: 139, 158, 159, 165, 208, 212, 226, 227, 232
Hirame *see Paralichthys olivaceus*
Hirondella brevicaudata **32**: **128**
Hirondella gigas **39**: 201
Hirondellea gigas **32**: 74
Hirondellea glutonis **34**: **377**
Hirtopelta hirta **34**: **370**
Hirudinea **40**: 9, 24; **43**: 55
Hirudo medicinalis **34**: *409*
Hispanirhynchildae **32**: **409**

Histiodraco velifer **24**: 328, 332
Histioteuthidae **32**: 62
Histioteuthis eltaninae **39**: 284
Histoneis **32**: 545, 560
Histrionicus histrionicus (harlequin duck) **39**: 15, 17, 64, 65, 67, 68, 78, 80
Histrionicus histrionicus **34**: 270
Hochbergia sp. **32**: 306, 308
Hochbergia moroteuthensis **32**: 289
Hodgsonia **24**: 26
Holichthys citrinus **32**: **172**
Holmesimysis costata **39**: 214
Holobomolochus confusus **40**: 9, 19, 33
Holophryxus **25**: 154
Holopodidae **32**: **409**
Holopus alidis **32**: **409**
Holopus rangi **32**: **409**
Holothuria (sea cucumber)
  H. (Halodeima) scabra **41**: 134
  H. (Holothuria) scabra **41**: 134
  H. (Metriatyla) **41**: 133
  H. (Metriatyla) scabra **41**: 134
  H. cadelli **41**: 134
  H. fuscogilva **41**: 190
  H. gallensis **41**: 134
  H. nobilis **41**: 190
  H. scabra see sea cucumber *in subject index*
Holothuria atra **21**: 147, 148, 150, 169; **35**: **184**
  axiologa **21**: 147, 148, 169
  forskåli **21**: 147, 159
  imitans **21**: 147
  impatiens **21**: 147
  kefersteini **21**: 147
  leucospilota **21**: 147
  lubrica **21**: 147, 158, 166, 167
  monacaria **21**: 147
  nobilis **21**: 146
  poli **21**: 147, 159
  scabra **21**: 147
  tubulosa **21**: 147, 159
  unitans **21**: 147
  vagabunda **21**: 147, 157, 158, 166, 167
Holothuria lubrica moebii **24**: 52
Holothuria scabra **43**: 102
Holothuriidae **41**: 133
Holothurioidea **32**: 97, 333, 335, **335**, 344, 357, 407, 429, **438**, 445, 468, **469**, **483**, **486**, **492**
Holothurioidea Elasipoda **32**: 347

Holothuroidea **31**: 39; **41**: 133; **43**: 95, 100,
    105, 106, 126
Homalophiura madseni **32**: **453**
Homarus **29**: 189; **37**: 132
    americanus **29**: **132**, 140, 161, 164, 165,
        168, 190, **192**, 194
    gammarus **29**: **132**, 190, **192**
Homarus americanus **44**: 125, 229, 250;
    34: 224
Homarus, Physiology **27**: 176
    americanus, Physiology **27**: 162
Homola barbata **29**: **133**, 136
Homolampas sp. **32**: **165**
Homologenus orientalis **32**: **163**
Homotrema **32**: **159**
Hoplichthydae **32**: **172**
Hoplichthys citrinus **32**: 188, 190, **194**, **218**
Hoplostethus (orange roughy) **38**: 3, 54
Hoplostethus **30**: 337
    atlanticus **30**: *333*, 334, 336
Hoplostethus atlanticus **34**: 210; **35**: **39**, **45**,
        47, 48, 50, 51, 85, 103
Hoplostethus atlanticus **43**: 62, *63*, *64*
Hormantia **22**: 92
Hormathia pectinata **42**: 213
Hormosira banksii **40**: 126
Hornellia marina **21**: 72
Hozukius emblemarius **30**: 334
Humilaria kennerleyi (clam) **39**: 39
Humiriaceae **40**: 95
Huxleyia **42**: 4
    H. cavernicola **42**: 41
Hyale barbicornis **39**: 201
Hyale nilssoni **39**: 201
Hyale pugettensis **39**: 201
Hyalella **35**: 174, **177**, *178*
    azteca **35**: 166, **168**
Hyalella azteca **39**: 150
Hyalella dentata **39**: 146
Hyalinoecia **32**: 445
Hyalogyra vitrinelloides **34**: 372
Hyalogyrina **34**: 415
    grasslei **34**: 372
    n. sp. **34**: 372
Hyalogyrinidae *see* Lower Heterobranchia
Hyalonema sp. **32**: *97*
Hyalonematidae **32**: *97*
Hyalopontius boxshalli **34**: **376**
Hyalospongiae **32**: *97*, **210**, 346
Hyanassa obsoleta **43**: 44
Hyas **26**: 27
Hyas araneus **44**: 125

Hyas coarctatus **44**: 125, 126
Hyas coarctatus alutaceus **32**: **446**, **448**
Hydatina **42**: *74*, 75, 132, 135
    character and evolutionary pathways
        42: *84*, *85*, *89*, 97, 98, 99, 105
    H. amplustre **42**: 90
    H. physis **42**: *72*, 73
    H. verrilli **42**: 90
    morphology **42**: *74*, 75
    taxonomic history **42**: 112, *116*
Hydatinidae **42**: 75, 109, 118, 129
Hydra **21**: 109, 158; **22**: 66, 68; **24**: 213, 226
Hydra vulgaris **31**: 250
Hydrasterias sp. **32**: **165**, 178
Hydrobates pelagicus **20**: 282
Hydrobia **32**: 458
    ulvae **34**: 24, 31
    ventrosa **34**: 37, **41**
Hydroidea **32**: *333*
Hydroides **21**: 15, 32, *35*, 37, 41
    norvegica **21**: 15
Hydrolagus sp. **32**: **168**
    mirabilis **34**: **380**
    pallidus **34**: **380**
Hydrolagus colliei **35**: **34**
Hydrolagus mirabilis **32**: **129**, 130
Hydroprogne tschegrava, Predation **27**: 372
Hydropsyche **35**: **163**, **169**
Hydrozoa **40**: 134; **43**: 287, *291*, 292
Hydrozoa sp. **32**: *97*
Hydrurga leptonyx **43**: 228, 242, 243
Hyella **42**: 212
Hyla geographica **30**: 173, 203
Hymeniacidon **21**: 100
    amphilecta **21**: 98, 99
    perlevis **21**: 98
Hymeniacidon perleve **24**: 63
Hymenocephalus **32**: 167, 186, 187
Hymenocephalus aterrimus **32**: **170**
Hymenocephalus gracilis **32**: **170**
Hymenocephalus neglectissimus **32**: **170**,
        188
Hymenocephalus semipellucidus **32**: **170**,
        189, **195**
Hymenocephalus striatulus **32**: **171**, 189,
        195, **218**
Hymenodora sp. **32**: **452**
Hymenodora acanthitelsonis **32**: 77
Hymenodora frontalis **32**: 55
Hymenodora glacialis **32**: 57, **77**
Hymenopenaeus halli **32**: **162**
Hyperbionychidae **33**: **51**, 108, **498**

Hyperbionyx **33**: **51**, 71, 74, 108
  pluto **33**: 108, **498**
Hyperia **25**: 154, 156; **32**: 303
  galba **25**: 155
  schizogeneios **25**: 156
Hyperia galba **32**: 269, 274; **39**: 207
Hyperietta **25**: 156
Hyperiidae **32**: **452**, **453**
Hyperiidea **39**: 157, 186, 187, 189, 207
Hyperoche **39**: 182
Hyperoche medusarum **25**: 8, 22, **34**;
      39: 207; **20**: 52
Hyperoglyphe **30**: 334
  japonica **30**: 334
Hyperoglyphe antarctica **35**: 5, **34**, **42**, 43,
      86, 103, 108
Hyperoodon planifrons **39**: 281
Hyphalaster inermis **43**: 147
Hyphalion **34**: **375**
  captans **34**: **375**
Hyphomonus **23**: 317
Hypnea **23**: 108
Hypocoma **25**: 138
Hypodytes rubripinnis **28**: 89, 95
Hypoechinorhynchus thermaceri **34**: **365**
Hypoglossus stenolepis **38**: 178
Hypoxylon oceanicum **40**: 119
Hypselodoris webbi **34**: 33
Hysterothylacium **32**: 308
Hysterothylacium aduncum **24**: 273, 277,
      285, 288, 294, 299, 304
Hysterothylacium aduncum **40**: 8, 18, 23,
      24, 26, 28, 36, 37, 39, 41, 43, 45,
      49, 51, 52
Hysterothylacium aduncum aduncum **40**: 18
Hysterothylacium aduncum gadi **40**: 8, 18,
      31
Hysterothylacium auctum **40**: 8, 21, 41
Hysterothylacium habena **40**: 8, 11
Hysterothylacium reliquens **32**: 306
Hysterothylacium rigidum **40**: 8, 10, 18,
      26, 33
Hysterothylacium sp., Parasites **27**: 387

Iais pubescens **39**: 209
Ianthella **21**: 98
Ianthina prolongata **24**: 209
Ichthyobdella borealis **25**: 147
Ichthyodinium **25**: 127
  chabelardi **25**: 134
Ichthyodinium chabelardi **28**: 81, 82
Ichthyophaga **43**: 20, 22

Ichthyophonus **25**: 144
  hoferi **25**: 122
Ichthyophthirius multifiliis **43**: 41
Ichthyosporidium **25**: 122, 144
Icichthys lockingtoni **26**: 145
Ictalurus melas **30**: 223
Ictalurus nebulosus **40**: 284, 308
Ictalurus punctatus **40**: 287, 321
Idas (Idasola) washingtonia **34**: **374**
Idasola **23**: 338; **34**: 384
Idasola washingtonia **32**: 131
Idedella achaeta **32**: **460**
Idiolychnus **32**: 187
Idiolychnus urolampus **32**: **170**, 188, **196**
Idiosepius **44**: 152, 157, 167, 182; **25**: 87
Idiosepius pygmaeus paradoxus **44**: 150
Idotea **39**: 29, 118, 185, 210
Idotea baltica **39**: 110, 158, 209, 210
Idotea baltica basteri **39**: 111, 139, 140,210
Idotea baltica tricuspidata **39**: 111
Idotea chelipes **39**: 210
Idotea granulosa **39**: 210
Idotea marginata **39**: 158
Idotea neglecta **39**: 146, 158, 210
Idotea ochotensis **39**: 210
Idotea pelagica **39**: 210
Idotea resecata **39**: 210
Idotea viridis **39**: 144, 158
Idoteidae **39**: 158
Idothea **23**: 230
Ifremeria **32**: 120
Ifremeria nautilei **34**: **372**
Iheyomytilidicola tridentatus **34**: **367**
Ihlea **25**: 156
Ihlea asymmetrica **44**: 22, 133
Ilionia prisca **34**: 35
Ilisha elongata, Predation **27**: **362**
Illex argentinus **44**: 151, 156, 160, 175;
      32: 263
Illex argentinus **39**: 272, 273, 280, 283, 284,
      286–289
Illex coindetii **39**: 274
Illex illecebrosus **32**: 263; **39**: 273, 274
Ilyanassa **42**: 204
Ilyanassa obsoleta **34**: *10*, 18, 22, 26, 37, *38*,
      39, 40, **41**
Ilyaranchna sp. **32**: **162**
Ilyophis blachei **32**: **168**; **34**: **380**
Imbracoverruca n. sp. **34**: **375**
Imbricaria **28**: 419
Imbricaria conularis **28**: *393*
Imbricaria olivaeformis **28**: *393*, **403**

Imocaris, Zoogeography/Evolution **27**: 149
Inachus phalangium **29**: **133**, 145
Incertae **32**: **409**
indicus, Morphology **27**: 44
  japonicus **27**: 1
    Behaviour **27**: 333, 334, 336, 339, 340,
      341, 346, 348, 352
    Food/feeding **27**: **317**, 326, 327, 328,
      331
    Life histories **27**: 293, 296, 302, **304**,
      313
    Morphology **27**: 41, 43
    Moulting/growth **27**: 224, 225, 226,
      235, 237, 240
    Parasites **27**: 380, 381, 383, 389
    Physiology **27**: 160, **161**, 162, 164,
      165, **168**, 171, 175, 176, 183, 184,
      193, 194, 195, 197, **204**, 206, 208,
      209
    Predation **27**: 375, 376
    Reproduction **27**: 252, 255, **258**, 260,
      261, 262, 263, 264, **270**, 275, 277,
      279
    Taxonomy **27**: 96, 107, **111**, 124
    Zoogeography/Evolution **27**: **144**, 147
  kerathurus **27**: 2
    Food/feeding **27**: **317**
    Morphology **27**: 44
    Moulting/growth **27**: **232**, 239, 245,
      246, **249**
    Physiology **27**: 160, **161**
    Reproduction **27**: 252, 253, 255, 257,
      259, 263, 268, **270**, 272, *275*, 276,
      277, **279**
    Taxonomy **27**: 98, 107, **111**
    Zoogeography/Evolution **27**: 139, **147**
  latisulcatus, Taxonomy **27**: 96, **111**, 124
    Behaviour **27**: 336, 340, 343, 344, *344*,
      345, 348
    Food/feeding **27**: **317**
    Life history **27**: 286, 296
    Moulting/growth **27**: **228**, **231**, 235,
      236, 246, **249**
    Parasites **27**: 389
    Physiology **27**: 163, 206
    Predation **27**: 375
    Reproduction **27**: **259**, 260, **279**
    Zoogeography/Evolution **27**: **144**, 147,
      152, 154, 157
    Behaviour **27**: 336, 341, 343, *344*, 347,
      354, 355

    Food/Feeding **27**: **317**, 319, 320, 322,
      326, 331
    Life histories **27**: 292, 293, 294, 302,
      304, 307, 309, 310, 314
    Moulting/growth **27**: 221, 227, **228**,
      231, 237, **249**
    Parasites **27**: 389
    Physiology **27**: **161**, **168**, 181, **204**, 205
    Predation **27**: 367, 375
    Reproduction **27**: 252, 253, 257, **258**,
      263, 268, **270**, **279**
    Taxonomy **27**: 96, **111**, 124
    Zoogeography/Evolution **27**: **144**
Indomysis annandalei **39**: 214
Inopsetta ischrya (= *Parophrys vetulus* ×
      Platichthys stellatus) **31**: 8, 20
Insecta **40**: 144–147
Iorania spp. **32**: 107, 114, 120, 130
Iorania concordia **32**: 99, 102, 107, 110, 111,
      114, 116, **117**, 118, 124, **129**, *131*
Iorania concordia **34**: **377**
Iotrochota birotulata **21**: 98
Iphigenella andrussowi **39**: 201
Iphinoe serrata **39**: 218
Iphinoe tenella **39**: 218
Iphinoe trispinosa **39**: 218
Iphionella
    n. sp. **34**: **367**
    risensis **34**: **367**
Ircinia campana **21**: 98
    felix **21**: 97, 98
    strobilina **21**: 98
Ireinia **21**: 99
Iridaea **23**: 80
    boryana **23**: 111
    cordata **23**: 131
    cornucopiae **23**: 135
Iridoteuthis sp.n.aff. *maoria* **32**: **164**
Iridoteuthis **32**: 205, 220
Iridoteuthis iris **30**: 316
Iridoteuthis maoria **32**: 220
Isaacscalanus paucisetus **23**: 328, 329
Isaacsicalanus **33**: **52**, 87, 123
    paucisetus **33**: 123, **499**, 502
Isaacsicalanus paucisetus **34**: **375**
Isaeidae **39**: 36
Ischnocalanus **33**: **52**, 77, 121, 483
    equalicauda **33**: 121
    gracilis **33**: 121
    plumulosus **33**: 121, **359**
    tenuis **33**: 121
Ischnomesidae **32**: **460**, 463

Ischyromene **39**: 176
Isias **33**: **52**, 70, 112, 405
  clavipes **33**: **7**, 112, **491**, **513**
  tropica **33**: 112
  uncipes **33**: 112, 447
Isochrysis galbana (alga) **41**: 182, 199;
    31: 93, 149, 150; **37**: 6, 39
Isognomon alatus **22**: 103, 107, 108, 124,
    127, 150
Isognomon californicum **37**: **10**, 16, **30**, 47,
    48
  bioassay procedures **37**: 51, 52, 53
  toxicity testing **37**: **91**, **93**, **94**, **98**, **99**
Isokerandria **33**: 17
Isopoda **31**: 39; **32**: **161**, 329, 333, *334*, 335,
    335, 344, 355, 404, **409**, **438**, **446**,
    460, 463; **39**: 107, 157–160, 186,
    187, 189, 193, 208–213; **40**: 9, 27;
    43: 93, 149
  flabelliferids **39**: 157
  parasitism **43**: 17, 34, 55
    gnathiid **43**: 33, 35, 36, 45
  valviferids **39**: 158
Istiophoridae **39**: 281
Isuridae **32**: 308
Isurus oxyrhincus **39**: 281
Ivellopsis **33**: **52**, 66, 114
  elephas **33**: 114

Jaera **31**: 12, 26, 37, 38; **39**: 29, 118, 154,
    155, 159
Jaera albifrons **31**: 24, 35, 38; **39**: 110, 148,
    154, 210
Jaera albifrons × *ischiosetosa* **31**: 24
Jaera hopeana **39**: 155
Jaera ischiosetosa **31**: 24; **39**: 210, 211
Jaera italica **39**: 155
Jaera marina **39**: 154, 163, 211
Jaera nordmani **39**: 155, 211
Jaera praehirsuta **31**: 38
Jaera praehirsuta × *ischiosetosa* **31**: 23
Jaeropsis **32**: **162**
Jakobia birsteini **32**: **460**
Jania capillacea **23**: 107
Janickina **25**: 122
  chaetognathi **25**: 122
  pigmentifera **25**: 122, *123*
Janira gracilis **39**: 211
Janira maculosa **39**: 211
Janthina **34**: 35
Japanese fly flounder *see Paralichthys
    olivaceus*

Japanese oyster *see Crassostrea gigas*
Japanese sardine *see Sardinops
    melanostictus*
Jaschnovia **33**: **53**, 78, 127
  johnsoni **33**: 127
  tolli **33**: 127
Jasmineira pacifica **26**: **179**
Jassa **39**: 151
Jassa falcata **39**: 201
Jassa marmorata **39**: 116, 150, 201
Jasus **35**: **45**
  edwardsii **30**: 202
  tristani **30**: 318, 329, 331
Jasus lalandii **29**: **132**, 161, **163**, 165, 166
Jaxea nocturna **34**: 219
Jellyfish (*Spirocodon* and *Aequorea*)
    43: 102, 287
Jeppsia calani **25**: 138
Johanssonia arctica **40**: 9, 19, 30, 55
Julella avicenniae **40**: 119, 122
Julia **42**: 95
Juliidae **42**: 79, 80

Kadosactis commensalis **35**: **33**
Kalliapseudes **39**: 159
Kallichroma tethys **40**: 122
Kalyptrorhynchia **43**: *22*
Kandelia **40**: 107, 111
Kandelia candel **40**: 95, 98, 99, 101, 102,
    108, 110, 142, 178, 184
Kareius bicoloratus **30**: 224
Katsuwonus pelamis **28**: 89, 99; **30**: 335
Kefersteinia cirrata **26**: **179**
Kelp greenling *see Hexagrammos
    decagrammos*
Keppenodinium **25**: 126, 127, 129
Keratoisis **30**: 318, 320
Khuskia oryzae **40**: 119
Killer whale *see Orcinus*
Killifish *see Fundulus heteroclitus*
Kinorhyncha **43**: *21*
Knifefish *see Xenomystus nigri*
Kofoidinium **25**: 126; **32**: 545, 560
Kogia breviceps **39**: 281
Kolga **32**: 366
Kolga hyalina **32**: 358
Komokiidae **32**: 465
Konosirus **20**: 22
  punctatus **20**: 6, 11, 13, 39
Kophobelemnon **32**: 362
Kophobelemnon biflorum **32**: 373
Kraussina rubra **28**: *351–353*

Krill (*Euphausia*) **43**: 174, 226–229
  exploitation **43**: 229
  faecal pellets **43**: 233, 234
  responses to sea ice changes **43**: 240–244
  secondary production **43**: 208, 218–220, 221–225
  swarms *Plate 1*: **43**
  UV radiation **43**: 239
Krohnitta **25**: 151
  pacifica **25**: 141
Kudoa **40**: 5
Kudoa clupeidae **24**: 275, 290–291, 293, 305
Kudoa thyrsites **40**: 5, 10, 15, 32, 55
Kuhnia **43**: *9, 29*
  K. scombercolias **43**: 24
  K. scombri **43**: *10, 15*, 23, *24*
  K. sprostonae **43**: 23, *24*
Kunihulsea **33**: **52**, 87, 123
  arabica **33**: 123
Kurchatovus sp. **32**: 506
Kuronezumia **32**: 186
Kuronezumia pallida **32**: **171**, 189, 217
Kuroshiodaphne phaeacme **32**: **164**
Kyphosus sectatrix **44**: 218, 270

Labidocera **23**: 226; **29**: 181, 182
  aestiva **29**: **131**, 176, 179
  behaviour **33**: 428
  chromosomes **33**: 47
  distribution **33**: 506, **513**
  ecology **33**: 480, 483, **491**
  jollae **25**: **34**
  jollae **29**: **131**, 181
  morphology **33**: 23
  reproduction **33**: 46, 257, 266, **270**
  retroversa **23**: 264
  taxonomy **33**: **52**, 55, 61, 66, 93, 95, 114
  trispinosa **25**: 48
Labidocera acuta **33**: **32**, **47**, 114, **430**, **513**
  acutifrons **33**: **26**, 114, **224**
    behaviour **33**: 409, **416**
    chromosomes **33**: 47
    distribution **33**: **513**
  aestriva **33**: **26**, 64, 114
    distribution **33**: **513**
    ecology **33**: 490
    life history **33**: 389
    nutrition **33**: **190**, 203
    physiology **33**: 217

  reproduction **33**: 42, 43, 45, 259, 259, **268**, 268, **270**, 272, 291, 293
  agilis **33**: 506
  antiguae **33**: 114
  barbadiensis **33**: 114
  barbudae **33**: 114
  bataviae **33**: **47**, 114
  bengalensis **33**: **26**, **47**, 114
  bipinnata **33**: 114, **270**, 506
  brunescens **33**: **26**, 114, **288**
  carpentariensis **33**: 114
  caudata **33**: 114
  cervi **33**: 114
  dakini **33**: 114
  detruncata **33**: 114, **507**, **514**
  diandra **33**: 114, 262, **514**
  discaudata **33**: **47**
  euchaeta **33**: **26**, **47**, 114
  farrani **33**: 114
  fluviatilis **33**: **26**, 114
  insolita **33**: 114
  jaafari **33**: 114
  japonica **33**: 114
  johnsoni **33**: 114
  jollae **33**: **26**, 114
    behaviour **33**: 405, **411**, 432
    distribution **33**: **514**
    nutrition **33**: **188**
  kolpos **33**: 114
  kröyeri **33**: **47**, 114
  laevidentata **33**: **47**, 114
  lubbockii **33**: 114, **514**
  madurae **33**: **47**, 114, **507**
  minuta **33**: **26**, **47**, 114
  mirabilis **33**: 114
  moretoni **33**: 114
  nerii **33**: 114, 490
  orsinii **33**: 114, 506
  panamae **33**: 114
  papuensis **33**: 114
  pavo **33**: **26**, **47**, 114, **427**, 429, **507**
  pectinata **33**: **47**, 114
  pseudacuta **33**: **47**, 114
  rotunda **33**: **26**, 114
  scotti **33**: 114, **270**
  tenuicauda **33**: 114
  trispinosa **33**: **26**, 114
    behaviour **33**: **412**, **418**, 421
    distribution **33**: **514**
    growth/development **33**: **306**
    nutrition **33**: 159, 163
    reproduction **33**: **270**

size/weight **33**: **224**
wilsoni **33**: 114
wollastoni **33**: **7**, 114
  behaviour **33**: 406, **412**, 420
  distribution **33**: **514**
  ecology **33**: 490
  growth/development **33**: **306**
  life history **33**: 388
  morphology **33**: 30, **32**
  nutrition **33**: 150
  reproduction **33**: **270**
Labidocera jollae **20**: 52
Labidocera pavo **30**: **164**
Labridae **43**: 25, 35
Labroides miniatus **43**: 35
Labrus bergylta **40**: 19
Lacazella mediterranea **28**: 236, 244, 248,
  250, 265, 271, 340
Lacistorhynchus **24**: 294, 297, 298
  tenuis **24**: 277, 292
Lacistorhynchus tenuis **40**: 7, 17, 25, 33
Lacuna **34**: 25; **40**: 27
  variegata **34**: 25
  vincta **34**: 25
Lacunidae **39**: 36
Lacunoides exquisitus **34**: **369**
Lacydonia n. sp. **34**: **367**
Laemonema rhodochir **32**: **170**, **218**
Laemonema yuvto **32**: **170**, 188
Laetmogonidae **32**: 466
Laetmonice wyvillei **32**: **446**, **448**, **452**
Laevidentaliidae **42**: *143*, 205
Laevidentalium **42**: *143*
  L. lubricatum **42**: 205
Laeviphitus japonicus **34**: **372**
Lafystius **40**: 19
Lafystius morhuanus **40**: 9, 19, 32
Lafystius sturionis **40**: 9, 19, 32
Lagenidium sp., Parasites **27**: 383
Lagenorhynchus obliquidens **39**: 280
Lagodon rhomboides **44**: 226; **26**: **121**
Lagodon rhomboides, Predation **27**: 374,
  375, 376, 377
Laguncularia **40**: 111, 114
Laguncularia racemosa **40**: 94, 104, 146,
  159, 161, 162, 174, 178
Lahmeyeria **33**: 52, 93, 117
  turrisphari **33**: 117
Laitmatobius crinitus **34**: **375**
Lamellibrachia **23**: 324, 339; **34**: 405, *406*,
  407, 408, 410
Lamellibrachia **35**: 50

barhami **23**: 324, 329, 339; **34**: **366**, 394,
  405, *409*
  columna **34**: **366**, 405, *409*
  luymesi **23**: 324
Lamellibrachia sp. **32**: 125, 130
Lamellibranch **42**: 30
Lamellibranchs **42**: 36
Lamellileda **42**: *6*
Lametila **42**: 25
Lametilidae **42**: *5*
Laminaria **23**: 29, 46, 65, 76, 79, 80, 82–85,
  95; **24**: 47; **30**: 113; **38**: 116
  angustata **23**: *5*
  digitata **23**: 30, 31, 90, 111, 136, 137,
    139, 140
  groenlandica **23**: 81, 82
  hyperborea **23**: 23, 24, 40, 111, 143
  longicruris **23**: 24, 30, 31, 37, 38, 90;
    24: 432
  longipedalis **23**: 64
  longissima **23**: 63
  religiosa **23**: 64
  saccharina **23**: 4, 5, 10, 23, 24, 39, 143
  setchelli **23**: 80
  solidungula **23**: 143
Laminaria bongardiana **39**: 40
Laminaria japonica **20**: 320
Laminaria pallida **25**: 185
Laminaria saccharina **39**: 40
Laminatubus alvini **34**: **368**
Laminiscus dogieli **24**: 285, 291
Lamnidae **32**: 308
Lampactena sp. **32**: *72*
Lampadena urophaos **32**: **170**
Lampanyctodes hectoris **20**: 249
Lampanyctus **32**: 61
Lampechinus nitidus nascaensis **32**: **165**
Lampenus lampetriformis **26**: **121**
Lampetra fluviatilis **40**: 276, 278, 279
Lamprey, sea *see Petromyzon marinus*
Lampropidae **39**: 116
Lamprops **39**: 115
Lamprops fasciata **39**: 148, 218
Lamprops quadrispinosa **39**: 218
Landrumius **33**: **53**, 85, 134
  antarcticus **33**: 134
  gigas **33**: 134
  insignis **33**: 134
  sarsi **33**: 134
  thorsoni **33**: 134
Lanice **32**: **161**
Lankesteria **25**: 141

leuckarti **25**: 141
Laptoseris fragilis **31**: 262
Laqueus californianus **28**: 203, 221, 237, 300, 309, 312, 316, 319, 326, 330, 352
Laqueus rubellus **28**: 351
Larus **34**: 285; **39**: 17
 audouinii **34**: 286
Larus argentatus **43**: 44; **20**: 281
 canus **20**: 281
 fuscus **20**: 281
 glaucescens **20**: 241
 marinus **20**: 281
 ridibundus **20**: 281
Larus californicus **39**: 280
Larus canus (mew gull) **39**: 47, 280
Larus glaucescens (glaucous gull) **39**: 47, 64, 67, 280
Larus heermanni **39**: 280
Larvacea **32**: 306
Lateroprotomeritus **25**: 142
Lates calcarifer **38**: 53
 survival from hatchery releases **38**: **41**
Lates calcarifer, Predation **27**: 361, **363**, 368, 369, *369*, 377
Latimeria chalumnae **40**: 305
Latreillia phalangium **32**: **163**
Laubieriellus grasslei **34**: **368**
Laudema **20**: 28
Laurencia nidifica **21**: 180
 nipponica **21**: 180
 okamurai **21**: 180
 pacifica **21**: 180
Laurencia spectabilis **23**: 80
Lawrenicia papillosa **24**: 60
Leathesia difformis **23**: 3, 5, 7, 12–14, 113, 132
Lebbeus **34**: **378**
 carinatus **34**: **377**
 washingtonianus **34**: **378**
Lebbeus carinatus **23**: 330
Lebistes reticulus **40**: 290
Lebrunia **22**: 90
Lebrunia danae **21**: 106
Lecithaster **24**: 287
 confusus **24**: 265, 277, 285, 288
 gibbosus **24**: 265, 277, 285, 288, 293, 296, 304
Lecithaster confusus **40**: 46
Lecithaster gibbosus **40**: 6, 16, 27, 31, 35, 36
Lecithochirium **25**: 150; **40**: 5, 20

Lecithochirium rufoviride **40**: 6, 13, 20
Lecithocladium **25**: 148
Lecithocladium excisum **40**: 6, 10, 20
Lecithoepitheliata **43**: 22
Lecudina **25**: 140, 141
 danielae **25**: 140
Ledella
 L. pustulosa **42**: 5, 24, 26
 L. ultima **42**: 39
Ledella crassa **32**: 503
Ledellidae **32**: 371, 373
Ledellinae **42**: 5, 8
Leech **43**: 56
Leiosella **21**: 95
Leiostomus xanthurus **26**: **121**, 140; **28**: 64; **30**: 222
Leitoscoloplos pachybranchiatus **34**: **366**
Lembos **39**: 151
Lembos websteri **39**: 190, 201
Lensia **24**: 107, 135, 139, 180, 207
 achilles **24**: 135
  bigelowi **24**: 119
 assymmetrica **24**: 119
 baryi **24**: 119, 200
 campanella **24**: 120, 121, 154
  campanella **24**: 121
  cossack **24**: 120
  elongata **24**: 119
  petrovskyi **24**: 119
 canuposi **24**: 119
 conoidea **24**: 107, 120, 145, 147, 160, 163, 167, 169, 172
 conoides pacifica **24**: 119
 cordata **24**: 119
 cossack **24**: 120, 121
 eltanin **24**: 119
 eugenioi **24**: 119
 exeter **24**: 135
 fowleri **24**: 120
 gnanamuthui **24**: 119
 grimaldi **24**: 135
 havock **24**: 120, 121, 135, 145
 hostile **24**: 135, 141
 hotspur **24**: 121
 landrumae **24**: 119
 lebedevi **24**: 119
 lelouveteau **24**: 120, 135
 meteori **24**: 135, 154, 169
 minuta **24**: 119
 multicristata **24**: 120, 135
 multilobata **24**: 119
 nagabhushanami **24**: 119

panikkari **24**: 119
patritii **24**: 119
peresi **24**: 119, 121
reticulata **24**: 120
roonwali **24**: 119, 121
subtilis **24**: 154, 169
subtiloides **24**: 154
tiwarti **24**: 119
zenkevitchi **24**: 119
Leocrates **32**: **161**
Lepadogaster **40**: 260
Lepadogaster candolei **40**: 258, 260, 283
Lepadomorpha **32**: 349, 468
Lepas **22**: 200, 204, 205, 207
Lepeophtheirus hippoglossi **26**: 38, 41
Lepeophtheirus pectoralis **29**: **131**, 181;
    40: 9, 13, 21
Lepeophtheirus pollachius **24**: 291
Lepeophtheirus salmonis **43**: 44
Lepetodrilacea **32**: *101*
Lepetodrilus **34**: **372**, 415, 417
    corrugatus **34**: **371**
    cristatus **34**: **371**
    elevatus **34**: **371**, 402, 416, 418
    fucensis **34**: **371**
    guaymasensis **34**: **371**
    japonicus **34**: **372**
    ovalis **34**: **372**
    pustulosus **34**: **372**, 397, 418
    schrolli **34**: **372**
    tevnianus **34**: **372**
Lepetodrilus sp. **32**: **127**
Lepidactylus dytiscus **39**: 201
Lepidapedon elongatum **40**: 6, 13–14, 27,
    31, 35, 36
Lepidapedon gadi **40**: 6
Lepidapedon microcotyleum **40**: 6
Lepidapedon rachion **40**: 6, 16, 27, 33
Lepidepecreum cingulatum **39**: 202
Lepidepecreum longicorne **40**: 11
Lepidion ? schmidti **34**: **380**
Lepidion **32**: 186, 198
Lepidion guentheri **32**: **170**
Lepidisis olapa **30**: 338
Lepidochelys olivacea **40**: 154
Lepidocybium flavobrunneum **32**: **174**
Lepidomysidae **39**: 160
Lepidonotopodium
    fimbriatum **34**: **367**
    minutum **34**: **367**
    piscesae **34**: **367**
    riftense **34**: **367**

williamsae **34**: **367**
Lepidonotothenia
    larseni **35**: **37**
    squamifrons-group **35**: **37**
Lepidopus caudatus **30**: 330
Lepidosteus platyrhyncus **40**: 269
Lepocraedium album **24**: 208; **25**: 150
Lepomis
    gibbosus **25**: 14
    macrochirus **25**: 9, 15, 21, **35**, 37
Lepomis gibbosus **26**: 129
Lepomis macrochirus **26**: 127
Lepomis sp. **26**: **121**
Leptasterias **43**: 93
    L. epichlora **43**: 105, 147
    L. hexactis **43**: 147
    L. ochotensis **43**: 100
Leptasterias epichlora **31**: 20, 37
Leptasterias hexactis **31**: 20, 37
Leptocheirus pilosus **39**: 201
Leptocheirus pinguis **39**: 182
Leptochelia dubia **39**: 219
Leptochelia savigngi **39**: 219
Leptochiton sp. **32**: **163**
Leptochiton tenuidontus **34**: **369**
Leptocylindrus **20**: 27
Leptodiaptomus minutus **33**: 272
Leptodiscus **25**: 129
Leptodora **31**: 97
Leptognathia brevimanus **39**: 219
Leptognathia breviremis **32**: **161**;
    39: 219
Leptognathia robusta **32**: 373
Leptognathia ventralis **34**: **377**
Leptogyra **34**: **371**
    inflata **34**: **371**
Leptohalysis **30**: 52
Leptolaimus **30**: 48, **50**
Leptomonas, Parasites **27**: 385
Leptomysis apiops **39**: 214
Leptomysis gracilis **39**: 214
Leptomysis lingvura **29**: **131**, 185; **39**: 214
Leptonychotes weddelli **24**: 329
Leptonychotes weddellii **43**: 228, 229, 242,
    243
Leptoplana tremellaris **21**: 133
Leptoseris **32**: **159**
Leptostrea purpurea **22**: 11, 14, 33
Leptostylis longimana **39**: 218
Leptotheca **24**: 278, 290, 294
Lernaeenicus sprattae **24**: 278, 291
Lernaeocera **24**: 274, 279, 293, 304

L. branchialis **43**: 11
L. lusci **43**: 11
Lernaeocera branchialis **40**: 9, 19, 24, 28, 32, 36, 37, 39, 43, 44, 48, 52, 53, 55, 56
Lessoniopsis littoralis **23**: 75, 76
retroversa **23**: 264
Lestrigonus schizogeneios **25**: 156
Lesueurigobius friesii **26**: 127
Lethmaster rhipidophorus **32**: 369
Lethrinidae **43**: *29*
Lethrinus **34**: 219
Lethrinus miniatus **43**: *29*
Lethrinus spp. **44**: 231
Leucicorus **32**: 359
Leuckartiara gardineri **21**: 105
Leucon jonesi **39**: 161, 218
Leucon nasica **39**: 218
Leucon profundus **39**: 218
Leuconidae **39**: 116
Leucosyrinx turridus **32**: **164**
Leucothoe spinicarpa **39**: 182
Leucothoe uschakovi **32**: 357
Leucothrix **24**: 54
Leucothrix mucor, Parasites **27**: 382
Leuresthes sardina **31**: 194
Leuresthes tenuis **28**: 20; **30**: 220
Leuroglossus **32**: 265, 306
Leuroglossus schmidti **32**: 265, 266, 274, **297**, 303, 305, 306
Leuroglossus stilibius **26**: **120**, 121
Leuroleberis zeylandica **39**: 202
Levensteiniella **34**: **367**
intermedia **34**: **367**
kincaidi **34**: **367**
raisae **34**: **367**
Libinia emarginata **29**: **133**, 134–136, 152, 172
Libinia emarginata, Physiology **27**: 196
Lichia amia, Predation **27**: 366
Lichothuria mandibularis (crustacean) **41**: *189*
Licnophora luidiae **25**: 139
Ligia **30**: 114, 134; **39**: 136
oceanica **30**: 121
Ligia exotica **39**: 120, 122, 127, 168
Lignincola laevis **40**: 120, 122
Ligumia subrostrata **22**: 103, 130
Lilyopsis **24**: 107, 186, 191
rosea **24**: 119
Lima scabra **22**: 103, 134, 136
Limacina **23**: 264

Limacina helicina **33**: 160; **32**: 269, **453**
Limacina sp. **26**: *131*; **32**: 302, **452**
Limacina trochiformis **28**: 107
Limanda **29**: 324; **29**: 237
Limanda aspera **38**: 178
Limanda ferruginea **26**: 116, 257
Limanda limanda **26**: 270; **34**: 287; **40**: 23
Limanda yokohama **26**: 73, **75, 92**, *95*, 102
Limax
maximus **34**: 8, 31
pseudoflavus **34**: 26
Limnocalanus
behaviour **33**: 405
nutrition **33**: **152**
reproduction **33**: 257, 266
taxonomy **33**: **52**, 67, 112
Limnocalanus grimaldi **33**: **26**, 112, 473
johanseni **33**: 112
macrurus **33**: 112
behaviour **33**: 406, **412**
biochemistry **33**: **239**
ecology **33**: **359, 396**, 473
growth **33**: **306**
life history **33**: 384
reproduction **33**: 262
Limnochironomus **35**: *180*, **181**
Limnoria **39**: 119, 122, 157
Limnoria andrewsi **39**: 211
Limnoria lignorum **39**: 169; **39**: 157, 169, 211
Limnoria tripunctata **39**: 151, 211
Limnoriidae **39**: 157
Limnothrissa **20**: 160
Limnothrissa miodon **26**: 130
Limopsidae **32**: 469, 470
Limpet **43**: 101; **24**: 23
polyphemus **24**: 35
Limulus polyphemus **29**: **188**
Lindaspio
dibranchiata **34**: **368**
southwardorum **34**: **368**
Lineus corrugatus **24**: 329
Lineus lacteus **21**: 134
marinus **21**: 134
ruber **21**: 134
viridis **21**: 134
Lineus sp. **32**: **126**
Lingula anatina see subject index
Lingula reevi **28**: 178, 318, 324–326, 330
Lingula unguis **28**: **232**
Linuche unguiculata **44**: 122
Linuparus trigonus **29**: **132**, 161, **162**, 165

Liocarcinus **34**: 287
Lionurus **35**: 52
Liopsetta glacialis **24**: 274
Liothyrella **28**: *352, 353*
Liothyrella blochmanni **28**: 192
Liothyrella neozelandica **28**: 249, 338, *351*
Liothyrella uva **28**: 188, **206**, 212, 276, 316, 318, 320, 321, *324*, 348, *357*
Liothyrella uva antarctica **28**: 265, 271
Liothyrella uva notorcadensis **28**: 347
Liparidae **32**: 304
Liparis **40**: 28
Liparis coheni **40**: 21
Liparis sp. **26**: **121**
Lipodonta **42**: 4
Lipolagus **32**: 266, 306
Lipolagus ochotensis **32**: 266, 274, **297**, 303, 305, 306
Lirapex
 granularis **34**: **370**
 humata **34**: **370**
Liriope tetraphylla **21**: 105; **23**: 235
Lissocarcinus orbicularis (crustacean) **41**: *189*
Lissodendoryx aff *kyma* **21**: 98
Lissodentoryx isodictyalis **40**: 133, 134
Lithodes
 aequispina **30**: 334
 couesi **30**: 331
Lithognathus
 lithognathus **25**: 195
 mormyrus **25**: 195
Lithophyllum **23**: 56; **32**: **160**, 176
 incrustans **23**: 7, 25, 27, 45
 tortuosum **23**: 45
Lithophyton arboreum **31**: 260
Lithoscalpellum discoveryi **24**: 54
Lithothamnion **32**: **160**, 176; **34**: 217
Litopenaeus **29**: **132**, 169, *170*, 171, **172**, **173**, 190, 191, **192**, 193, 194
Litopenaeus, subgenus, Taxonomy **27**: 95, 125
Litoscalpellum piliferum **32**: **161**
Littleneck clam *see Protothaca staminea*
Littoraria pallescens **40**: 139, 147, 148
Littorina **23**: 86; **40**: 25, 27; **42**: *142*
 aspera **34**: 9
 irrorata **34**: 26, 45
 littorea **23**: 87, 88, 90, 113; **34**: *10*, 13, *14*, 19, *20, 29*, 45,50
 sitkana **34**: 24

Littorina littorea **40**: 56, 58
Littorina rudis **31**: 8
Littorina saxatilis **31**: 8; **40**: 27; **43**: 44
Littorina scutulata **39**: 18, 28
Littorina sitkana **39**: 18, 24, 28
Livona **21**: 170
Liza aurata (golden mullet) **38**: 33, 54
Liza spp. 33
Lobantale **42**: *143*
Lobantalinae **42**: *143*
Lobata **32**: 9
Lobatostoma manteri **43**: 43
Lobianchella beloneides **25**: 139
Lobiger **42**: *116*, 133, 135
Lobigeridae **42**: 80
Lobodon carcinophagus **43**: 227, 242, 243; **20**: 271
Lobonema mayeri **21**: 106
 smithi **21**: 106
Lobophora Plate 4, **43**: 282
Lobosa **43**: *20*
Lohmaniella spiralis **29**: 24, *26*
Loliginidae **32**: 259, 289, 295
Loligo (squid) **39**: 267, 268, 274, 276, 279, 282, 289
Loligo **30**: 157; **20**: 184
 opalescens **30**: 199
 opalescens **20**: 54
 pealei **25**: *92*, 93
 vulgaris **25**: 93, 101, 103, *104*, 105
 vulgaris reynaudii **30**: 188
Loligo chinensis **39**: 267, 269, 279, 289
Loligo duvauceli **39**: 269, 274
Loligo duvaucili, Predation **27**: 359
Loligo forbesi **44**: 179; **39**: 274
Loligo gahi **39**: 272, 285
Loligo opalescens **39**: 270, 271, 279, 289
Loligo pealei **44**: 151, 173; **39**: 274; **22**: 103, 150
Loligo plei **39**: 274
Loligo reynaudi **39**: 274
Loligo singhalensis **39**: 274
Loligo vulgaris **34**: 33; **39**: 274
Loliguncula brevis **39**: 274
Lolligneula brevis **22**: 103, 150
Loma branchialis **40**: 4, 10, 13, 14, 30, 43, 44
Loma morhua **40**: 4, 10, 50, 55
Longicyatholaimus **30**: **50**
longistylus, Taxonomy **27**: 95, 124
 Behaviour **27**: 340
 Life history **27**: 286, 303

Parasites **27**: 389
Physiology **27**: 203
Reproduction **27**: **259**
Zoogeography/Evolution **27**: **144**
Lophiidae **32**: **171**
Lophiodes **32**: 186
Lophiodes mutilus **32**: **171**, **218**
Lophiostoma asiana **40**: 119
Lophiostoma mangrovei **40**: 121
Lophius piscatorius **29**: 295; **40**: 25, 26
Lophogaster **39**: 160
Lophogaster typicus **39**: 214, 215
Lophogastrida **39**: 160
Lophogorgia chilensis **43**: *291*
*Lopholatilus chamaeleonticeps***34**: 255
Lophophorates **31**: 39
Lophor gorgia rigida **21**: 113
Lophothrix **33**: **53**, 82, 85, 134, 135, 136, 486
 frontalis **33**: *40*, 40, 134, 444
 humilifrons **33**: 134
 latipes **33**: 134
 quadrispinosa **33**: 135
 similis **33**: 135
 simplex **33**: 135
 varicans **33**: 135
Lophotrix **25**: 138
Lophozozymus pictor **21**: 140
Loricata **32**: **163**
Lota lota **26**: 129; **40**: 14
Lottia **34**: 48
 gigantea **34**: **6**, 28, 46
Lower Heterobranchia **42**: 68, 70, 100, 115, 116, 118, 119
 see also Acteon; Gegania
Loxothylacus panopei **43**: 70
Lucernaria n. sp. **34**: **365**
Lucicutia
 behaviour **33**: **412**, **416**, 423
 ecology **33**: 486, 528
 taxonomy **33**: **51**, *61*, 70, 71, 73, 76, 108
Lucicutia anisofurcata **33**: 108
 anomala **33**: 108
 aurita **33**: 108, **435**
 bella **33**: 108
 bicornuta **33**: 108
 biuncata **33**: 108
 challengeri **33**: 108
 cinerea **33**: 108
 clausi **33**: 108, **435**
 curta **33**: 108
 curvifurcata **33**: 108
 flavicornis **33**: **7**, 108, **223**, **359**, **435**, **485**

formosa **33**: 108
gaussae **33**: 108
gemina **33**: 108, **435**
gigantissima **33**: 108
grandis **33**: 108, **435**
intermedia **33**: 108
longicornis **33**: 108
longifurca **33**: 108
longiserrata **33**: 108
longispina **33**: 108
lucida **33**: 108
macrocera **33**: 108
magna **33**: 108, **435**
major **33**: 108
maxima **33**: 108
oblonga **33**: 108
orientalis **33**: 108
pacifica **33**: 108
pallida **33**: 108
paraclausi **33**: 108
parva **33**: 108
pellucida **33**: 108
pera **33**: 108
polaris **33**: 108
profunda **33**: 108
pseudopolaris **33**: 108
rara **33**: 108
sarsi **33**: 108, **435**
sewelli **33**: 108
uschakovi **33**: 108
wolfendeni **33**: 108, **435**
Lucicutiidae **33**: **51**, 108, **152**, **435**, 436, 502
Lucina pectinata **40**: 148
Lucinidae **42**: 23, 24
Lucinoma atlantis **23**: 339
Lucioperca lucioperca **26**: 83
Luckia **32**: 130
Luckia striki **32**: **128**; **34**: **377**
Luidia sarsi **25**: 139
Luidia sp. **32**: **165**, 177
Lulworthia **40**: 122
Lumbriconereis brevicirea **21**: 138
 heteropoda **21**: 138
Lumbrineridae **39**: 36
Lumnitzera × *rosea* **40**: 94
Lumnitzera **40**: 99, 103, 178
Lumnitzera littorea **40**: 94, 107
Lumnitzera racemosa **40**: 94, 107, 108, 176
Lunatia **28**: *392*
Lunatia pallida **40**: 27
Lunda cirrhata **20**: 245; **34**: 270

Lungfish *see Neoceratodus; Protopterus*
Lutamator **33**: **53**, 88, 127
  elegans **33**: 127, **499**
  hurleyi **33**: 127, **499**
Lutjanidae **32**: **173**
Lutjanus **34**: 219
  kasmira **30**: 334
  lutjanus **30**: 334
  synagris **30**: 239
Lutjanus argentimaculatus **44**: 231, 297
Lutjanus griseus **40**: 269
Lutra canadensis (river otter) **39**: 17, 61, 62, 72
Lutra perspicillata **40**: 157
Lyaea pulex **39**: 207
Lycaea **25**: 155, 156
Lycaeopsis **25**: 156
  themistoides **25**: *155*, 156
Lycaeopsis themistoides **24**: 205, 206
Lycengraulis **20**: 34
  olidus **20**: 180
  simulator **20**: 180
Lychnagalma (Agalmopsis)
    utricularia **24**: 113, 117, 193, 197, 198
Lychnorhiza **25**: 153
Lycodapus mandibularis **32**: 282
Lycodes **26**: 33
Lycodes seminudus **38**: 175
Lycodes soldatovi **32**: 297
Lycodes vahli **38**: 175
Lycogramma brunnea **32**: 297
Lycogrammoides schmidti **32**: 306
Lycothrissa **20**: 34
Lymnaea **35**: **163**
  peregra **34**: 37, **41**
  stagnalis **34**: 19, 37
  truncatula **34**: 4, 6, 19
Lymnaea peregra **40**: 25
Lyngbya majuscula **21**: 67, 71, 88; **31**: 316
Lyreidus tridentatus **29**: **133**, 141
Lyrodus pedicellatus **40**: 149
Lyrodus singaporeana **40**: 149
Lysianassidae **32**: 75, **128**, 474
Lytechinus variegatus **21**: 146, 155; **43**: 147
Lythraceae **40**: 95
Lytocarpus pennarius **21**: 105
  philippinus **21**: 105
  phoeniceus **21**: 105

M Acartia tonsa **37**: 131, 132
Maackia amurensis **40**: 259

Macaca mulatta **40**: 157
Macandrevia americana **23**: 338
Macandrevia cranium **28**: 178, 211, 229, 230, 244, 246, 248, 249, *353*
Macandrewella **33**: **53**, 78, 81, 84, 135
  agassizi **33**: 135
  asymmetrica **33**: 135
  chelipes **33**: 135
  cochinensis **33**: 135
  joanae **33**: 135
  mera **33**: 135
  scotti **33**: 135
  sewelli **33**: 135
Macellicephala galapagensis **34**: **367**
Macellicephaloides alvini **34**: **367**
Mackerel
  Atlantic *see Scomber scombrus*
  blue (slimy) *see Scomber australasicus*
  Spanish *see Scomberomorus regalis*
Macoma (clam) **39**: 39; **31**: 26; **35**: **163**, **169**
  balthica **35**: **168**, **173**, *174*
  M. balthica **42**: 16, 17
  M. tenta **42**: 33
Macoma balthica **22**: 103, 12; **31**: 35; **34**: 28; **37**: **10**; **43**: 101, 145
Macomona liliana **42**: 35
Macroalgae **43**: 281–284, 306, *382*
Macrobrachium **29**: 171, 190
  acanthurus **29**: **132**, **192**
  acanthus **29**: **192**
  asperulum **29**: **132**, 192
  carcinus **29**: **132**, **192**
  formosense **29**: **132**, **192**
  nipponsense **29**: **132**, **192**
  rosenbergii **29**: **132**, **173**, **187**, 188, 192, **194**
  shokitai **29**: **132**, **192**
Macrobrachium lamarrei **40**: 137
Macrobrachium, Introduction **27**: 1
  Physiology **27**: 162
Macrocoeloma trispinosum **29**: **133**, 134
Macrocyctis pyrifera **31**: 25
Macrocystis **23**: 21; **30**: 113, 114
  pyrifera **23**: 5, 8, 12, 28, 116; **30**: 113
Macrocystis pyrifera **24**: 60
Macronectes giganteus **39**: 281
Macronectes halli **39**: 281
Macronema carolina **35**: **169**
Macropetasma, Taxonomy **27**: 70, **108**, **113**, 119, 121, 122, **122**
  africana (*see africanus*)
  africanum (*see africanus*)

africanus, Taxonomy **27**: 70, *70*, **108**
  Behaviour **27**: 340
  Food/Feeding **27**: **317**, 322, 323, 326
  Life histories **27**: 296
  Physiology **27**: 181
  Reproduction **27**: **258**, 267, **269**, **278**
  Zoogeography/Evolution **27**: 135, 147,
    142, **148**
  Food/feeding **27**: **317**
  Life histories **27**: 285
  Physiology **27**: 203
  Reproduction **27**: **269**, **278**
  Zoogeography/Evolution **27**: 139, 148,
    149, 150, 154, 156, 157
Macrophthalmus **40**: 141
Macrophyes **25**: 124
Macrorhamposidae **32**: **172**
Macrorhamposus **32**: 186, 187, 192
Macrorhamposus scolopax **32**: **172**, 188,
  189, **192**, **195**
Macroronus **32**: 206
Macrosetella **40**: 129; **44**: 81
Macrostomida **43**: 22
Macrouridae **32**: 75, *97*, **129**, 167, **170**, 211,
  267
Macrouroides **32**: 186
Macrouroides inflaticeps **32**: **171**
Macrourus halotrachys **35**: **36**
Macrura **20**: 160
Macrura **32**: *97*, **210**
Macruridae **32**: 57
Macruronus magellanicus **32**: **170**, 209, 220
Macruronus novaezelandiae **35**: **36**, **45**, 50,
  84, 85, 103, 107–109, *108*
Mactra chilensis **37**: **10**
Mactra chinensis **43**: 104
Madracis asperula **22**: 36, 37
  decactis **22**: 24, 27, 36, 38
  mirabilis **22**: 10, 16, 23, 27
Madreporaria **32**: *333*, 468, 470, **483**, 500,
  504
Magadina cumingi see Anakinetica cumingi
*Magasella sanguinea see Terebratella
  sanguinea*
Magellania **28**: **206**, 231, 238, 336, 342
Magellania venosa **28**: 178
Maja squinado **44**: 218, 229, 252
Majidae **32**: 455
major **33**: 130
Malacocephalus **32**: 186, 191
Malacocephalus laevis **32**: **171**, 188, **195**
Maldanella **32**: 362, 372

Maldanella antarctica **32**: 362, 372
Maldanidae **32**: **126**, 362, 463; **39**: 36
Malletia **35**: **32**; **42**: 24, **26**, **36**, **39**
  M. gigantea **42**: 17
Malletia **32**: **437**
Malletiidae **32**: **492**; **42**: *5*
Mallocephaloides grandelythris **32**: 369
Mallomonas **25**: 119
Mallotus sp. **26**: **120**
Mallotus villosus (capelin) **39**: 52, 70, 72,
  80; **24**: 274; **20**: 174; **25**: *3*, 9, 10,
  11, 29, 32, 36, 44, 50, **52**, 53, 57;
  26: 33, **121**, 130,257; **28**: 15, 28,
  43, *88*, 92; **30**: 268; **31**: 203;
  34: 257
Malpighiaceae **40**: 97
Mammalia **40**: 25, 26, 157, 158; **43**: 68, 69,
  226–230, 242, 243
Manaia **33**: 52, 66, 117
  velificata **33**: 117
Mancocuma stellifera **39**: 152, 218
Mandritsa **32**: **172**
Manicina areolata **22**: 39, 41, 47, 49
Mantoniella squamata **29**: 79
Manzanellacea and Manzanellidae (formerly
  Nucinellidae) **42**: 4, *5*, *6*, 7, 9
Maresearsia sphaera **24**: 118
Margarites shinkai **34**: **371**
Marginaster sp. **32**: **165**
marginatus, Taxonomy **27**: 95
  Physiology **27**: 162, 203
  Reproduction **27**: 380
  Zoogeography/Evolution **27**: **144**
Marginella **28**: *393*, 419
Marianactis bythios **34**: **365**
Marinogammarus marinus **39**: 202
Marinogammarus obtusatus **39**: 202
Marinogammarus stoerensis **39**: 202
Marinosphaera mangrovei **40**: 120, 122
Marmara **40**: 178
Marrus antarcticus **24**: 116, 141, 200
  orthocanna **24**: 117, 141
  pacifica **24**: 117, 141
Marsh wren *see Cistothorus palustris*
Marsupenaeus, subgenus, Taxonomy **27**: 95
Martesia nari **40**: 149
Martesia striata **40**: 149
Marthasterias glacialis **21**: 144, 153, 168
Martialia hyadesi **39**: 272, 274, 283–286,
  288, 289
Masked greenling *see Hexagrammos
  octogrammus*

Massarina armatispora **40**: 119
Massarina ramunculicola **40**: 119
Massarina velatospora **40**: 119
Mastacembelus erythrotaenia **40**: 283
Mastigias **30**: **164**
Mastigophora **40**: 4
Mastigoteuthidae **32**: 63
Mastigoteuthis sp. **32**: 288
Mataeocephalus **32**: 186
Mataeocephalus acipenserinus **32**: **171**, **218**
Mathildidae *see* Lower Heterobranchia
Maurolicus **32**: 186, 187, 193
  muelleri **30**: 316, 329, 330
Maurolicus japonicus **32**: 266
Maurolicus rudjakovi **32**: **169**, 193, **195**
Mazocraeoides georgei **24**: 273, 291
Mazocraes harengi **24**: 264, 276, 291
Meandrina meandrites **31**: 271; **43**: 285
Mecynocera **33**: **52**, 77, 119, 120, 483
  clausi **33**: **7**, 119, **224**, **359**, **514**
Mecynoceridae **33**: 30, **52**, 119
Mediterranean mussel *see Mytilus*
    galloprovincialis
Megacalanidae **33**: 30, **52**, 119, 277, 435
Megacalanoidea **33**: **52**, *54*, 55, 118–121
Megacalanus **33**: **52**, 71, 119, 486
  longicornis **33**: 119
  princeps **33**: 119, **435**, 436, 444
Megadesmolaimus sp. **32**: **126**
Megadyptes antipodes **39**: 279
Megaentalina **42**: *144*
Megalasma (Glyptelasma) caudata **32**: **161**
Megalasma (Megalasma) elegans **32**: **159**,
    161
Megalops **20**: 34
Meganyctiphanes **25**: 141; **30**: 198
  norvegica **25**: 137, 138, 139, 154; **30**: **164**,
    170, 172, 176, 178
Megaptera novaeangliae **34**: 275; **43**: 229,
    230
Megathiris **28**: 249, 265, *353*
Megayoldia **42**: 7
Megerlia truncata **28**: **206**, 249, 313, *315*,
    351–353
Melamphaeidae **32**: 57, 62
Melamphaes spp. **32**: 63
Melampus ceylonicus **40**: 147
Melanitta (scoter) **39**: 15
Melanitta perspicillata (surf scoter) **39**: 15,
    17, 64, 80
Melanodryamia aurantiaca **23**: 321
Melanodrymia

  aurantiaca **34**: **370**
  brightae **34**: **370**
  n. sp. **34**: **370**
Melanogrammus aeglefinus (haddock) **38**: 4,
    54
Melanogrammus aeglefinus **26**: **120**, *126*,
    131, *132*, *148*; **28**: 24, 25,
    46, **47**, 49–51, *88*, **104**, **105**,
    109, **112**; **29**: 325; **30**: 245; **31**: 189;
    34: 262; **35**: 5; **37**: 208, 220; **40**: 12;
    20: 51, 291; **44**: 250, 311; **25**: *3*, 4,
    10, 30, 41, 43, 57
Melanoplus sanguinipes **29**: 184
Meliaceae **40**: 95
Melicertus, subgenus, Taxonomy **27**: 95
Melita appendiculata **39**: 202
Melita nitida **39**: 202
Melita palmata **39**: 202
Melita pellucida **39**: 202
Melita zeylandica **39**: 143, 202
Mellita quinquiesperforata **35**: **185**
Melo **28**: *392*
Melobesia **32**: **160**, 176
Melophysa melo **24**: 126
Menidia beryllina **26**: 130
Menidia menidia **33**: 293; **44**: 229
Menippe adina **31**: 32
Menippe mercenaria **29**: **133**, 143; **31**: 32;
    35: 6
Menippe mercenaria × *Menippe*
    adina **31**: 27
Menticirrhus americanus, Predation
    27: **364**
Mercenaria (Venus) mercenaria **22**: 103,
    109, 131, 151, 152, 154, 156–158,
    167, 169, 170
Mercenaria (Venus) mercenaria **37**: 7
Mercenaria **28**: *353*
Mercenaria campechiensis **31**: 17, 21, 27,
    35, 49–51; **37**: 14
Mercenaria generally **37**: **115**
Mercenaria hybrid zones **31**: 22
Mercenaria mercenaria **31**: 17, 21, 27, 35,
    49–51; **34**: 211; **35**: **168**, **185**;
    37: 6, **8**, **9**, 10, 131
  bioassay methodology **37**: 42, **44**, 46,
    47, 48
  bioassay procedures **37**: 70
    embryo-larval **37**: 52, 53, 54–56
    larval physiology tests **37**: 63, 64, 65
  biology **37**: 16, 17, 18, 19, 20
  described **37**: 14, 15

larval rearing in laboratory **37**: 26–27
  physical requirements **37**: 36, **36**,
    37,'37
  spawning and fertilization **37**: 27, **29**,
    30, 29, **32**, 32
  reproduction **37**: 46, 47, 48
  toxicity testing **37**: 135
    biocides **37**: **98–100, 102–114, 106,
    107**
    detergents and oil **37**: **109, 111, 112**
    metals **37**: **91, 93, 94**
Mercenaria mercenaria × *Mercenaria
    campechiensis* **31**: 38
Meretrix lusoria **37**: **10**, 46
Merganser *see Mergus*
Merguia oligodon **40**: 138
merguiensis, Morphology **27**: 11, 13–15,
    37, 50
  Behaviour **27**: 336, 337, 338, 339, 340,
    341, *344*, 347, 352, 354, 355
  Food/feeding **27**: **317**, 318, 319, 320,
    321, 322, 323, 324, 325, 326, 329,
    330, 331
  Life histories **27**: 284, 285, 291, 292,
    293, 294, 296, 299, 301, 303, **305**,
    306, 307, 309, 310, 311, 312, 313,
    313, 314
  Moulting/growth **27**: 214, 221,
    222–223, 225, 226, **231**, 234, 235,
    237, 239, 243, *245*, **250**
  Parasites **27**: 380, 386, 388, 389, 390
  Physiology **27**: 163, 164, **168**, 172,
    183, 193, 194, **204**, 205, 206, *206*,
    208
  Predation **27**: 359, 373, 375, 376
  Reproduction **27**: 252, 253, 254, **259**,
    260, 261, 262, 263, **270, 279**
  Taxonomy **27**: 97, **111**, 124
  Zoogeography/Evolution **27**: **144**, 152
Mergus merganser (merganser) **39**: 69
Mergus serrator (red-breasted
    merganser) **39**: 69
Merismopedia **29**: 80
Merlangius merlangus (whiting) **38**: 4, 54
Merlangius merlangus **26**: 86, 270; **28**: *89*,
    118; **34**: 262; **40**: 11; **20**: 291;
    32: 170
Merluccius
  australis **35**: 108
  capensis **35**: **36, 45**, 103
  gayi **35**: **36**, 103
  hubbsi **35**: **36**, 103

  merluccius **25**: **34**, 43, 59
  paradoxus **35**: **36, 45**, 103
  productus **25**: 23, 27, 28, *28*, **52**; **35**: **36**,
    103
Merluccius bilinearis **26**: **120**; **38**: 175
Merluccius capensis **20**: 171
  gayi **20**: 171
  merluccius **20**: 171
  productus **20**: 171
Merluccius gayi **26**: **120**
Merluccius hubbsi (hake) **39**: 277, 280,
    283
Merluccius merluccius **26**: 33
Merluccius novaezelandiae **26**: 130
Merluccius polylepis **39**: 277
Merluccius productus (Pacific hake) **38**: 178,
    210
  predation on *P. jordani* **38**: 218
Merluccius productus **26**: 260; **28**: 59, *87*
Merodinium **25**: 127
Merulina **22**: 10
Mesaiokeras **33**: **53**, 87, 88, 89, 131
  heptneri **33**: 131
  kaufmanni **33**: 131, **500**
  nanseni **33**: 131, **500**
  semiplenus **33**: 131
  tantillus **33**: 131
Mesaiokeratidae **33**: **53**, 131, **500**
Mesidotea entomon **39**: 211
Mesidotea sp. **32**: 356
Mesocalanus **33**: **52**, 119, 140
  lighti **33**: 119
  tenuicornis **33**: **7**, 119, **359, 396, 514**
Mesocomantenna **33**: **53**, 88, 127
  spinosa **33**: 127
Mesodesma donacium **35**: **185**
Mesodinium **23**: 221, 225, 226
Mesodinium rubrum **33**: 157; **29**: 50
Mesodopsis slabberi **25**: 151
Mesophyllum lichenoides **24**: 57
Mesopodopsis orientalis **39**: 117, 120, 121,
    140, 188, 215
Mesopodopsis slabberi **39**: 215
Mesorchis denticulatus **24**: 273, 302
Mesorhabdus **33**: **51**, 70, 73, 107, 486
  angustus **33**: 107
  brevicaudatus **33**: 107
  gracilis **33**: 107
Mesosignum **32**: **162**, 359
Mesovelia polhemusi **40**: 145
Mesozoa **43**: 8
Metacalanus **33**: **51**, 71, 74, 102, **489, 498**

acutioperculum **33**: 102
aurivilli **33**: 102
curvirostris **33**: 102
inaequicornis **33**: 102
Metacercariae **43**: 67
Metaleptamphopus pectinatus **39**: 202
Metamysidopsis elongata **44**: 126; **39**: 151, 215
Metamysidopsis swifti **39**: 215
Metapenaeopsis acclivis, Taxonomy **27**: 74, 78
   Zoogeography/Evolution **27**: **142**, 147
andamanensis, Taxonomy **27**: 77, **108**
   Zoogeography/Evolution **27**: **142**
angusta, Taxonomy **27**: 77
   Zoogeography/Evolution **27**: **142**
assimilis, Taxonomy **27**: 75
   Zoogeography/Evolution **27**: **142**
barbata, Taxonomy **27**: 74, **108**
   Food/feeding **27**: **317**
   Moulting/growth **27**: **229**
   Zoogeography/Evolution **27**: **142**, 147
beebei, Taxonomy **27**: 78
   Zoogeography/Evolution **27**: **145**
borradailei, Taxonomy **27**: 74
   Zoogeography/Evolution **27**: **142**
caliper, Taxonomy **27**: 78
commensalis, Taxonomy **27**: 74
   Zoogeography/Evolution **27**: 136, **142**
coniger, Taxonomy **27**: 77
   Zoogeography/Evolution **27**: **142**
crassissima, Taxonomy **27**: 73
   Zoogeography/Evolution **27**: **142**
dalei, Taxonomy **27**: 76
   Predation **27**: 367
   Zoogeography/Evolution **27**: **142**
distincta, Taxonomy **27**: 75
   Zoogeography/Evolution **27**: **142**
dura, Taxonomy **27**: 73
   Moulting/growth **27**: 241, 245, **248**
   Zoogeography/Evolution **27**: **142**
enhraea, Taxonomy **27**: 77
   Zoogeography/Evolution **27**: **142**
evermanni, Taxonomy **27**: 74
   Zoogeography/Evolution **27**: **142**
faouzii, Taxonomy **27**: 76
   Zoogeography/Evolution **27**: **142**
fusca, Taxonomy **27**: 74
   Zoogeography/Evolution **27**: **142**, 147, **148**
gerardoi, Taxonomy **27**: 78
   Zoogeography/Evolution **27**: **145**

goodei, Taxonomy **27**: 78
   Food/Feeding **27**: 317
   Zoogeography/Evolution **27**: **146**
hilarula, Taxonomy **27**: 75
   Zoogeography/Evolution **27**: **142**
hobbsi, Taxonomy **27**: 78
   Zoogeography/Evolution **27**: **146**
incompta, Taxonomy **27**: 75
   Zoogeography/Evolution **27**: **142**
insona, Taxonomy **27**: 75
   Zoogeography/Evolution **27**: **142**, 147
kishinouyei, Taxonomy **27**: 78
   Zoogeography/Evolution **27**: **145**
kyushuensis, Taxonomy **27**: 76
   Zoogeography/Evolution **27**: **142**
lamellata, Taxonomy **27**: 74
   Zoogeography/Evolution **27**: **142**
lata, Taxonomy **27**: 76, 78
   Zoogeography/Evolution **27**: **142**
lindae, Taxonomy **27**: 74
   Zoogeography/Evolution **27**: **142**, 147, **148**
liui, Taxonomy **27**: 77
   Zoogeography/Evolution **27**: **142**
mannarensis, Taxonomy **27**: 75
   Zoogeography/Evolution **27**: **142**
martinelta, Taxonomy **27**: 78
   Zoogeography/Evolution **27**: **146**
miersi, Taxonomy **27**: 77
   Zoogeography/Evolution **27**: **146**
mineri, Taxonomy **27**: 78
   Zoogeography/Evolution **27**: **145**
mogiensis, Taxonomy **27**: 75
   Zoogeography/Evolution **27**: **108**, **142**
novaeguineae, Taxonomy **27**: *71*, 74, **108**
   Parasites **27**: 389
   Zoogeography/Evolution **27**: **143**
palmensis, Taxonomy **27**: 73, **108**, **117**, **123**
   Reproduction **27**: 264
   Zoogeography/Evolution **27**: **143**
philippii, Taxonomy **27**: 77
   Zoogeography/Evolution **27**: **143**, 147
provocatoria, Taxonomy **27**: 77
   Zoogeography/Evolution **27**: **143**
quinquedentata, Taxonomy **27**: 75
   Zoogeography/Evolution **27**: **143**, 147
rosea, Taxonomy **27**: 73
   Zoogeography/Evolution **27**: **143**
scotti, Taxonomy **27**: 77
   Zoogeography/Evolution **27**: 135, **143**, 147

sibogae, Taxonomy **27**: 76
    Zoogeography/Evolution **27**: **143**
sinica, Taxonomy **27**: 78
sinuosa, Taxonomy **27**: 73
    Zoogeography/Evolution **27**: **143**
smithi, Taxonomy **27**: 78
    Zoogeography/Evolution **27**: **146**
stridulans, Taxonomy **27**: 73
    Food/feeding **27**: **317**
    Moulting/growth **27**: **229**
    Zoogeography/Evolution **27**: **143**
tarawensis, Taxonomy **27**: 76
    Zoogeography/Evolution **27**: **143**
tenella, Taxonomy **27**: 78
toloensis, Taxonomy **27**: 73
    Zoogeography/Evolution **27**: **143**
velutina, Taxonomy **27**: 74
    Zoogeography/Evolution **27**: **143**
wellsi, Taxonomy **27**: 74
    Zoogeography/Evolution **27**: **143**
Metapenaeopsis spp. **27**: 4, 72, 136, 141,
    286, 287, 367
Metapenaeopsis stokmani **32**: **162**
Metapenaeopsis, Morphology **27**: 42
    Food/feeding **27**: **317**, 321
    Life histories **27**: 285
    Moulting/growth **27**: 229, *247*
    Physiology **27**: 159, 203
    Reproduction **27**: 255
    Taxonomy **27**: 71–78, 107, **108**, **113**,
      119, *120*, 121, 123
    Zoogeography/Evolution **27**: 154
Metapenaeus **40**: 138
Metapenaeus affinis **44**: 220
Metapenaeus affinis, Taxonomy **27**: 81, **108**
  anchistus, Taxonomy **27**: 81
    Zoogeography/Evolution **27**: **143**
  bennettae, Morphology **27**: 19, 40
    Behaviour **27**: 334, 339, 340, 346, 351,
      352, 353, 354
    Food/feeding **27**: 325, 326, 327, 328
    Life histories **27**: 284, 297, 307,
    Moulting/growth **27**: 214, 218, 219,
      220, 222, 226, **228**, 237
    Physiology **27**: 163, 176, 197, **204**,
      205, *205*, 208, 209, **210**
    Reproduction **27**: **269**, 273, **278**
    Taxonomy **27**: 82, **108**, 113
    Zoogeography/Evolution **27**: **143**, 147,
      148, 152, 154
  brevicornis, Taxonomy **27**: 83, **109**
    Food/feeding **27**: 317

    Life histories **27**: 284, 294
    Moulting/growth **27**: **229**, 239, 245,
      248
    Predation **27**: 372
    Reproduction **27**: 263, 269, **278**
    Zoogeography/Evolution **27**: **143**
  burkenroadi (*see Metapenaeus moyebi*)
  conjunctus, Taxonomy **27**: 82
    Life histories **27**: 284
    Zoogeography/Evolution **27**: **143**
  dalli (*see Metapenaeus moyebi*)
  demani, Taxonomy **27**: 83
    Zoogeography/Evolution **27**: **143**
  dobsoni, Taxonomy **27**: 83, **109**, 113
    Behaviour **27**: 347
    Food/feeding **27**: **317**, 321, 325, 327,
      331
    Life histories **27**: 294, 307, 310, **313**
    Moulting/growth **27**: **228**, **229**, 239,
      246, **248**
    Physiology **27**: **204**
    Reproduction **27**: **258**, 268, **269**, 273,
      278
    Zoogeography/Evolution **27**: **143**
  eboracensis, Taxonomy **27**: 83
    Zoogeography/Evolution **27**: **143**
  elegans, Taxonomy **27**: 82
    Life histories **27**: 284
    Zoogeography/Evolution **27**: **143**
  endeavouri **27**: 4
    Behaviour **27**: 340, 341, 343, *344*, 348
    Food/feeding **27**: **317**
    Life histories **27**: 286, 314
    Parasites **27**: 389
    Physiology **27**: 163
    Reproduction **27**: 264
    Taxonomy **27**: 81, **109**
    Zoogeography/Evolution **27**: **143**
  ensis **27**: 4
    Behaviour **27**: 340, 341, 343, *344*, 348
    Food/feeding **27**: **317**, 322
    Moulting/growth **27**: **229**, 235
    Parasites **27**: 389
    Physiology **27**: **204**
    Reproduction **27**: 278
    Taxonomy **27**: 8*0*, 82, **109**, *123*
  insolitus, Taxonomy **27**: 82
    Zoogeography/Evolution **27**: **143**
  intermedius **27**: 4
    Moulting/growth **27**: **229**
    Taxonomy **27**: 81
    Zoogeography/Evolution **27**: **143**

joyneri, Taxonomy **27**: 83, **109**
  Moulting/growth **27**: 235
  Reproduction **27**: **269**, **278**
  Zoogeography/Evolution **27**: **143**
krishnatrii, Taxonomy **27**: 82
  Zoogeography/Evolution **27**: **143**
kutchensis, Taxonomy **27**: 81
  Moulting/growth **27**: 245, **248**
  Zoogeography/Evolution **27**: **143**
lysianassa, Taxonomy **27**: 80
  Moulting/growth **27**: **229**
  Zoogeography/Evolution **27**: **143**
macleayi, Taxonomy **27**: 83, **109**
  Behaviour **27**: 334, 337, 339, 340, 343,
    345, 347, 348
  Food/feeding **27**: **317**, 318, 326, 328,
    329
  Life histories **27**: 297, 306, 307, 309,
    313
  Physiology **27**: **168**
  Reproduction **27**: **269**, 273, **278**
  Zoogeography/Evolution **27**: 135, **143**,
    148, **148**
mastersii (*see Metapenaeus bennettae*)
monoceros, Taxonomy **27**: 81, **109**, 114
  Behaviour **27**: 341, 343, 346
  Food/feeding **27**: **317**, 320, 322, 326,
    331
  Life histories **27**: 294, 307, 310, **313**,
    314
  Moulting/growth **27**: 218, **229**, 235,
    239
  Physiology **27**: 160, **161**, **168**, 187,
    200, **204**, 209
  Predation **27**: 372,
  Reproduction **27**: **269**, **278**
  Zoogeography/Evolution **27**: **143**, **146**
moyebi, Taxonomy **27**: 82, **109**
  Life histories **27**: 291
  Moulting/growth **27**: **229**, 235, 236,
    237
  Physiology **27**: 176, 184
  Reproduction **27**: 269, 278
  Zoogeography/Evolution **27**: **143**, 147,
    154
mutatus, Moulting/growth **27**: **229**
  Food/feeding **27**: **317**
papuensis, Taxonomy **27**: 82
  Zoogeography/Evolution **27**: **143**
philippinensis, Taxonomy **27**: 82
  Zoogeography/Evolution **27**: **143**
stebbingi, Taxonomy **27**: 82, **109**

  Moulting/growth **27**: **229**
  Reproduction **27**: **269**
  Zoogeography/Evolution **27**: **143**, **146**
suluensis, Taxonomy **27**: 81
  Zoogeography/Evolution **27**: **143**
tenuipes, Taxonomy **27**: 83
  Zoogeography/Evolution **27**: **143**
  Food/feeding **27**: **317**, 329, 331
  Life histories **27**: **313**, 314
  Moulting/growth **27**: **228**, **229**, 235,
    245, 246, **248**
  Reproduction **27**: **269**, **278**
  Zoogeography/Evolution **27**: **143**
Metapenaeus dobsoni **40**: 137
Metapenaeus macleayi **34**: *120*, 146
Metapenaeus monoceras **40**: 137; **29**: **132**,
  170
Metapenaeus spp., Morphology **27**: 7
  Life histories **27**: 284
  Moulting/growth **27**: *236*, 237
Metapenaeus, Morphology **27**: 14, 31, *36*,
  40, 41, 42
  Food/feeding **27**: **317**
  Life histories **27**: 286, 289, **313**
  Moulting/growth **27**: **228**, **229**, 235,
    236, 246, *247*
  Physiology **27**: 168, 183, 196, 203, **204**
  Predation **27**: 358, 360
  Reproduction **27**: 255, 267, 268, **269**,
    277, **278**, **281**
  Taxonomy **27**: 79–83, 99, 107, **108**,
    109, 113, **113**, 118, 19, 120, *120*,
    121, 123, 124, 125
  Zoogeography/Evolution **27**: **141**, 150,
    154, 155
Metaphrya **25**: 137
  sagittae **25**: 138, 158
Metapseudes **39**: 159
Metaverruca pallida **32**: **161**
Metaverruca tarasovi **32**: **161**, 216
Metazoa **43**: *20*, 193, 207, 208, 209, 214
  see also Cnidaria; Coelenterata
Metazooplankton **43**: 183, 220, 221
Meterythrops microphthalma **32**: 266
Methanococcus jannaschii **23**: 318
Methanopyrus kandleri **35**: 18
Metharyzium anisopliae **37**: 107
Methylococcus capsulatus **23**: 317
Metopa
  alderii **25**: 156
  borealis **25**: 156
Metopograpsus messor **40**: 140

Metridia **25**: 138
  ecology **33**: 480, 486, 528
  longa **25**: 138
  lucens **25**: 138
  taxonomy **33**: **51**, 64, 67, 109
Metridia alata **33**: 109
  andraeana **33**: 109
  asymmetrica **33**: 109, 454, 502
  bicornuta **33**: 109
  boecki **33**: 109
  brevicauda **33**: 109
  calypsoi **33**: 109
  curticauda **33**: 109
  discreta **33**: 109
  effusa **33**: 109
  gerlachei **33**: 6, 109
    behaviour **33**: **416**, **433**, **435**
    biochemistry **33**: 237, 240, 242, 243,
      247
    ecology **33**: **359**, 492, **493**, 494
    life history **33**: 384
    nutrition **33**: 159, 164, 166, **191**, **206**,
      208
    physiology **33**: 210
    size/weight **33**: **224**, **231**, **234**
  gurjanovae **33**: 109
  ignota **33**: 109
  longa **33**: **7**, 109
    behaviour **33**: 403, **412**, **416**, 419, **435**,
      435, 438, 439, **446**, 447
    biochemistry **33**: 237, 240, 243
    distribution **33**: **514**
    ecology **33**: **359**, 364, **493**, 495
    growth/development **33**: **310**
    life history **33**: 384, 385
    nutrition **33**: **187**, **191**
    physiology **33**: 216
    reproduction **33**: 254, **255**
    size/weight **33**: **231**, **234**
  lucens **33**: **7**, **26**, 37, 109, **231**
    behaviour **33**: **412**, **416**, 419, **435**, **446**,
      447
    distribution **33**: **514**
    ecology **33**: 352, **359**, **377**, **396**, 468,
      484
    nutrition **33**: **188**
  macrura **33**: 109, **435**
  okhotensis **33**: 109, 407
  ornata **33**: 109
  pacifica **33**: 48, 109
    behaviour **33**: 403, 406, 407, **412**, 424,
      435

  ecology **33**: 352, **359**, 503
  growth/development **33**: **306**, **342**
  life history **33**: **387**
  nutrition **33**: 160, 188, **191**
  reproduction **33**: **268**
  princeps **33**: 109, **435**
  similis **33**: 109
  trispinosa **33**: 109
  venusta **33**: 109
Metridia gerlachei **44**: 81; **32**: 26, 29;
      43: 210, 217, 218, 219
Metridia longa **32**: 21, 22
Metridia lucens **44**: 81, 82; **32**: 22, 269
Metridia okhotensis **32**: 274, 306
Metridia pacifica **32**: 17, 265, 266, 274
Metridia spp. **32**: 29, 303
Metridiidae **32**: 193; **33**: 222
  behaviour **33**: **435**, 436
  ecology **33**: 347, 502
  nutrition **33**: **152**
  taxonomy **33**: **51**, 108, 109
Metridium **22**: 73
  senile **22**: 66, 72, 84, 92
Metridium senile **21**: 123; **31**: 248
Metschnikowia **25**: 121
Mew gull *see Larus canus*
Mica micula **24**: 116
Micrallecto **25**: 153
Micraster coranguinum **31**: 47
Micraster glyphus **31**: 47
Micraster senonensis **31**: 47
Micraster stolleyi **31**: 47
Microalgae **43**: 213, 221, 222
Microcalanus
  ecology **33**: 480
  nutrition **33**: 160, **188**
  size **33**: *332*
  taxonomy **33**: 53, 88, 89, 129
Microcalanus pigmaeus **32**: 22
Microcalanus pusillus **33**: **26**, 129, 254
  pygmaeus **33**: 129, 254, **306**, **359**, **493**,
      494
Microchirus **28**: **118**, **120**
  azevia **29**: 220
  boscanion **29**: 220
  ocellatus **29**: 220
  variegatus **29**: 220, 229
Microciona parthena **21**: 98
  prolifera **21**: 97
  spinosa **21**: 98
Microcoleus lyngbyaceus **21**: 88, 89
Microcystis **29**: 80

Microcystis aeruginosa **21**: 71, 87
  toxica **21**: 71
Microdeutopus **39**: 151
Microdeutopus danmoniensis **39**: 202
Microdeutopus gryllotalpa **39**: 151, 202
Microdisseta **33**: **51**, 70, 73, 107
  minuta **33**: 107
Microgloma **42**: *6*, 35; **43**: 115
  M. pusilla **42**: 24, 25, 28
Microhyla **40**: 154
Microlaimus **30**: 48, **50**, 50
Micromelo **42**: 90
Micromesistius australia **39**: 277
Micromesistius poutassou (blue whiting)
  38: 3, 53
Micromesistius poutassou **24**: 305;
    25: **52**, 53; **30**: 335; **35**: **35**, 52;
    20: 173
Micromomalthus debilis **31**: 90
Micromonas pusilla **25**: 119; **29**: 79
Micronekton **43**: 219
Microphallus piriformis **43**: 44
Microphthalmus depressus **40**: 141
Micropogonias undulatus **26**: **121**, 140
Micropogonias undulatus,
        Predation **27**: **364**, 374, 376
microps **23**: *330*
Micropterus salmoides **25**: *13*, 19
Microspio mecznikowianus **29**: 195
Microspio spp. **44**: 132
Microsporida **24**: 278; **40**: 4; **43**: 17, *20*
Microsporidium **38**: 180
Microstomatidae **32**: 303
Microstomus kitt **34**: 310
Microthambema **32**: **162**
Microthrissa **20**: 160
Mictacea **39**: 107
Mictis **40**: 178
Midshipman *see Porichthys notatus*
Mikadotrochus **32**: 410
Mikadotrochus notialis **32**: **409**
Mikadotrochus salmianus **32**: **409**
Mikadotrochus schmalzi **32**: **409**
Milax sowerbii **34**: 23, 24, 33
Milkfish *see Chanos chanos*
Millepora **21**: 130; **22**: 12, 24, 27
  alcicornis **21**: 105, 113; **22**: 30
  complanata **22**: 30, 33
  complanta **21**: 105
  dichotoma **21**: 105, 113
  dichotoma **22**: 5, 11
  platyphylla **21**: 105

  tenera **21**: 105, 113
Millericrinidae **32**: **409**
Mimocalanus **33**: **52**, 87, 89, 123
  brodskii **33**: 123
  crassus **33**: 123
  cultrifer **33**: 123
  damkaeri **33**: 123
  heronae **33**: 123
  inflatus **33**: 123
  major **33**: 123
  nudus **33**: 123
  ovalis **33**: 123
  sulcifrons **33**: 123
mimus **33**: 130, **485**
Minutocellus polymorphus **31**: 361
minutus **33**: **7**, **26**, 130
  behaviour **33**: **412**, 429, 451, 454
  biochemistry **33**: **239**
  chromosomes **33**: **47**
  ecology **33**: **361**, **378**, **397**, 475, **493**
  growth/development **33**: 298, **307**, **311**,
      312, 338
  life history **33**: **387**
  nutrition **33**: 145, 169, 172, **183**, **186**, **187**,
      188
  reproduction **33**: 260, 282, **284**, **288**
  size/weight **33**: **231**, 328, 329, **331**
Miostephos **33**: **53**, 85, 89, 137
  cubrobex **33**: 137
  leamingtonensis **33**: 137, **489**
Miracia efferata **25**: 153
Mirocalanus **43**: 216
Mirocarididae **32**: 110, 111, 120, **129**, 130
Mirocaris
  (Chorocaris) fortunata **34**: **377**
  keldyshi **34**: **377**
Mirocaris fortunata **32**: *101*, 110, 111, 116,
      129, *134*, 135
Mirocaris keldyshi **32**: 110, 111, 114, 116,
      117, **129**, *134*
Mirocaris spp. **32**: 110, 111, 114, 118, 120,
      130
Mirounga angustirostris **39**: 280; **20**: 246
Mirounga leonina **39**: 281
Misophrioida **33**: 1, **2**, *3*
Mitromorpha (Mitrolumna) maculata
      32: **164**
Mitromorpha aspera **32**: 217
Mitromorpha maculata **32**: 217
Mixodiaptomus
  kupelwieseri **33**: 41
  laciniatus **33**: **433**

Mixtocalanus **33**: **53**, 80, 84, 135
   alter **33**: 135
   robustus **33**: 135
   vervooti **33**: 135
Miyadella (*see Atypopenaeus stenodactylus*)
Mizuhopecten yessoensis **37**: **10**, 45
Mnemiopsis **25**: 17, 30; **31**: 146
   leidyi **25**: 7, **31**, **34**, 36, 43, 44
Mnemiopsis leidyi **31**: 146
Mnemiopsis maccradyi **31**: 178
Mnemiopsis mccradyi **44**: 22, 129
Modiolus **21**: 178; **22**: 109
   demissus **22**: 103, 108, 124–126,
      150, 165
   modiolus **22**: 103, 112, 125, 127, 165
Moelleriopsis n. sp. **34**: **371**
Moina **30**: 199; **31**: 87, 95, 97
Moina brachiata **31**: 102
Moina macrocopa **31**: 115
Moina salina **31**: 87
Mola mola **40**: 11
Mollisquama **32**: 211
Mollisquama parini **32**: **168**, 208, 211
Mollusca **32**: 330, **409**; **40**: 25, 147–149;
      42: 145, 188, 219
   see also Amphineura; Bivalvia;
      Gastropoda
   see also Scaphopoda
   fecundity
      environmental factors **43**: 145, 146
      life-history strategies **43**: 125, 132,
         133
      quantification methodologies **43**: *94,*
         99, 100, 101, 103–105
      parasitism **43**: 6, *21*
Molpadiidae **32**: **461**, 466
Molpadiodemas sp. **32**: **462**
Molpadonida **32**: **438**, **461**, 466
Monacilla **33**: **52**, 78, 80, 123
   gracilis **33**: 123
   tenera **33**: 123
   typica **33**: 123
Monas **21**: 5
Monastraea Plate 4: **43**
   M. annularis **43**: 133
Monastrea annularis **22**: 11, 14, 18, 21–30,
      32–34, 36–38, 41–45, 48
   cavernosa **22**: 11, 17, 26, 34, 39, 41,
      43, 44
Monhystera **30**: 50, 53
Monhysteroida **43**: 214
Monilicaecum **25**: 148

Monocentridae **32**: **171**
Monocentris **32**: 186
Monocentris reedi **32**: **171**, 188, **192**,
      195, 224
Monochirus hispidus **29**: 221
Monochrysis lutheri **44**: 302; **37**: 39; **20**: 325
   bioassay procedures **37**: 65, 66, 72, 73
Monoculodes edwardsi **39**: 202
Monoculodes gibbosus **39**: 202
Monoculodes packardi **39**: 203
Monoculoides **25**: 48
Monoculus finmarchicus **33**: 49
monodon **27**: 1, 2, 4
      Behaviour **27**: 336, 337, 341, 343, *344,*
         346, 353, *353*, 354, 355, 356
      Food/feeding **27**: **317**, 318, 319, 320,
         320, 322, 325, 326, 327, 328, 329,
         331, 332
      Life histories **27**: 293, 294, 305, 308
      Morphology **27**: 20, 40
      Moulting/growth **27**: 225, **228**, **231**,
         235, 237, 239
      Parasites **27**: 380, 381, 382, 383, 384,
         388, 389
      Physiology **27**: **161**, 163, 164, 168,
         171, 175, 181, 187, 190, 195, **204**,
         205, 206, 207
      Predation **27**: 375
      Reproduction **27**: 257, **259**, 260, 261,
         261, 263, 264, **271**, 276, **279**
      Taxonomy **27**: 96, **111**, **112**, 124
      Zoogeography/Evolution **27**: **144**
Monodonta turbinata **34**: 22
Monogenea **40**: 5, 24; **43**: 65, 69, 84, 85
   see also Dactylogyrus;
      Monoopisthocotylea;
      Polyopisthocotylea
   distribution **43**: *51, 53, 55, 60,* 62
   diversity **43**: 8, 9, *11*, 14, *15*, 16, 17,
      18, 19
   ecology **43**: 23, *24*, 27, *28*, *29*, 31, 33, 34
Monoopisthocotylea **43**: *22*, 46
Monophyes japonica **24**: 121
Monoplacophora **31**: 39; **32**: **335**, 408
Monospora **25**: 121
Monostroma oxyspermum **40**: 126
Monstrilloida **33**: **2**, 2, *3*
Montastraea cavernosa **43**: 285, 297, *298*
Montastrea **31**: 241
Montastrea annularis **31**: 242, 256
Montastrea faveolata **31**: 250, 256
Montipora **31**: 246

Montipora patula **22**: 11
  verrucosa **22**: 11, 23, 26, 28
Montipora verrucosa **31**: 246, 266
Mora moro **34**: **380**
Moridae **32**: 167, **170**, 211
Mormonilloida **33**: 1, **2**, *3*
Mormyridae **40**: 304
Mormyriformes **40**: 306
Morone
  americana **25**: 5
  saxatilis **25**: 5, 21, **35**, 37, 57, 58
Morone chrysops **26**: **120**
Morone saxatilis (striped bass) **38**: 28, 29,
    54; **44**: 258; **28**: 78; **30**: 222, 240;
    31: 186
Moroteuthis knipovitchi **39**: 284
Morum **28**: *392*, **403**, 419
Moseria similis **24**: 116
moultoni **33**: 130, **284**, 292, **311**, **313**, **361**
Mucronalia (gastropod) **41**: *189*
Mudskipper *see Periophthalmodon*
    schlosseri
Mudsucker *see Gillichthys mirabilis*
Muggiaea **23**: 235; **24**: 100, 101, 102, 108,
    120, 121, 133, 139, 168, 178, 207
  atlantica **24**: 121, 151–153, 168, 169,
    171, 205, 208, 217, 225, 227, 230,
    236, 239, 241, 244
  bargmannae **24**: 141, 151
  cantabrica **24**: 119
  delsmani **24**: 151, 154
  kochi **24**: 121, 151, 152–153, 169, 170,
    212
Mugil cephalus (striped mullet) **38**: 29, 30,
    33, 54
  survival and size at release **38**: **38**
  survival from hatchery releases **38**: **41**
Mugil cephalus **44**: 221, 222; **26**: 103
Mugil soiuy (mullet) **38**: 54
  studies in Hawaii **38**: 29, 30
Mugiloididae **32**: **173**
Mulinia lateralis **37**: 7, **8**, **9**, 10, 131
  bioassay applications **37**: 77
  bioassay procedures **37**: 51, 52, 56, 60,
    63, 64, 68
  described **37**: 15, 16
  future research **37**: 137
  larval rearing in laboratory **37**: 26
    physical requirements **37**: **36**, **37**, 37
    spawning and fertilization **37**: 27, **29**,
      30, 29, **32**, 32
  reproduction **37**: 17, 18

  spawning and fertilization **37**: 27, **29**,
    30, 29, **32**, 32
  toxicity testing **37**: **91**, **94**, **101**, 107, 135
    detergents and oil **37**: **110**, **113–115**,
      116
Mullet *see Mugil so-iuy*
Mulloides flavolineatus **44**: 218, 223
Mulloidichthys flavolineatus **44**: 231
Multicotyle purvisi **43**: 43
Multitentacula (Multitentacula)
    admirabilis **32**: **164**
Mummichog *see Fundulus heteroclitus*
Munida **26**: 27
Munida gregaria **30**: 201
Munida magniantennulato **34**: **378**
Munidopsis **23**: 303; **34**: **378**, *404*; **35**: 24,
    65, 69, **70**
  alvisca **34**: **378**
  cf. *crassa* **34**: **378**
  crassa **35**: 32
  diomedeae **35**: 32
  hamata **35**: 32
  lauensis **34**: **378**
  lentigo **23**: 330; **34**: **378**
  marianica **34**: **378**
  parfaiti **35**: 32
  scobina **35**: 32, 81, *82*
  spinihirsuta **35**: 32
  starmer **34**: **378**
  subsquamosa **23**: 330, 338; **34**: **378**;
    35: 32
Munidopsis antoni **32**: 367
Munidopsis beringana **32**: **446**
Munidopsis crassa **32**: 105, **129**, 130
Munidopsis sp. **32**: 99, *100*, 102, *104*, 108,
    109, **129**, **437**
Munnogonium **32**: **162**
Munnopsis typica **39**: 211
Munnopsurus atlanticus **39**: 211
Mupus imperialis **30**: 331
Muraenichthys chilensis **32**: 224
Muraenichthys profundorum **32**: **168**, 208,
    224
Muraenolepis marmoratus **35**: 52
Murex **21**: 170, 176, 189
  brandaris **21**: 176, 189
Muricea californica **43**: *291*
Muricidae **42**: 106
Murre *see Uria aalge*
Mursia aspera **32**: **163**
Mursia gaudichaudii **32**: **163**, 216
Musculis discors **40**: 28

Musculus **39**: 32, 37, 49, 52, 53, 73
Mussa angulosa **22**: 11, 34, 36
Mussel
  blue *see Mytilus edulis*
  Mediterranean *see Mytilus*
    galloprovincialis
Mussel *see Musculus*; *Mytilus*
Mussels *see Mytilus*
Mustelus mustelus **40**: 15
Mya **21**: 178
Mya arenaria (clam) **39**: 39
Mya arenaria **22**: 103, 105, 111, 113, 115,
    121, 122, 124, 125, 127, 130, 134,
    150, 156, 157, 171; **31**: 311;
    34: 217; **35**: **185**; **37**: **10**, *55*, 71,
    101, 102
Mycale **21**: 98
  lingua **21**: 98
Mycetophyllia aliciae **43**: 285
Mycetozoa **43**: *20*
Mycteroperca
  microlepis **34**: 254
  venenosa **34**: 254
Myctophidae **32**: 46, 62, 63, 167, **170**, 193,
    266, 274
Myctophum **32**: 63
Mylio berda **26**: 103
Myliobatiformes **40**: 309
Myliobatus aquila **25**: 195
Myoxocephalus scorpius **24**: 274
Myriotrochidae **32**: 362
Myrsinaceae **40**: 95
Myrtaceae **40**: 95
Myrtales **40**: 93, 96
Mysida **39**: 160, 213–217
Mysidacea **32**: **56**, 362, **452**, **454**; **39**: 107,
    160, 161, 186; **40**: 28
Mysidae **39**: 160
Mysidella **39**: 124
Mysidetes posthon **39**: 215
Mysidion **25**: 153
  abyssorum **25**: 153
  commune **25**: 153
Mysidium columbiae **30**: 179, 180, 199, 200;
    39: 170, 215
Mysidium integrum **39**: 215
Mysidobdella **25**: 147
  borealis **25**: 147
Mysidopsis **39**: 161, 188, 190
Mysidopsis bahia **37**: 131
Mysidopsis didelphys **39**: 215
Mysidopsis gibbosa **39**: 215

Mysinae **32**: 362; **39**: 160
Mysis **25**: 147, 151
  mixta **25**: 147
  oculata **25**: 147
Mysis gaspensis **30**: 181
Mysis litoralis **39**: 215
Mysis mixta **39**: 215
Mysis oculata **39**: 117
Mysis relicta **29**: **131**, 186; **39**: 215
Mysis stenolepis **39**: 215
Mysticotalitrus **39**: 184, 190
Mytilidae **32**: **127**, 130, 468; **39**: 36; **42**: 24,
    42
Mytilidiphila
  enseiensis **34**: **367**
  okinawaensis **34**: **367**
Mytilus **21**: 178; **23**: 133; **31**: 22; **43**: 132;
    20: 184
  californianus **21**: 76, 79; **23**: 17
  corscum **20**: 366
  edulis **23**: 343; **20**: 366
  edulis galloprovincialis **21**: 32
  grayanus **20**: 366
  M. edulis **43**: 68, 101, 133, 145
  M. galloprovincialis **43**: 66, 68, 69
Mytilus californianus **24**: 46, 52; **37**: **9**, 15,
    17, 52
  edulis **22**: 50, 54, 103–113, 120–126,
      128–135, 138–141, 144, 146, 148,
      150–152, 154, 156, 161, 163,
      165–170, 172–181; **24**: 46;
      22: 103, 110, 124–126, 131, 146,
      150, 166, 168
  edulis planulatus **22**: 103, 105, 171
  galloprovincialis **22**: 103, 124–126, 165
  larval rearing in laboratory **37**: 32,48
  viridis **22**: 103, 105
Mytilus edulis (blue mussel) **39**: 13, 18, 29;
    25: 214, 215, 221, 229; **31**: 7, 10,
    17, 19, 21, 35, 36, 37, 311, 317;
    34: 13–15, *16*, 17, *17*, 18, 24, 27,
    35; **35**: 97, **163**, **168**; **37**: 2, **8–10**,
    10, 125, 131; **38**: 226; **40**: 25;
    42: 161, 175
  bioassay methodology **37**: 42, **44**, 45
  bioassay procedures **37**: 52, 56, 59, 64,
      65, 66, 69
  described **37**: 13, 14
  larval rearing in laboratory
    physical requirements **37**: 36, *36*, **35**,
        36, 37

spawning and fertilization **37**: 27, **29**,
  30, **32**, 32–36, 46, 47, 48
reproduction
  biology of **37**: 18, 19, 20, 21
  spawning and fertilization **37**: 27, **29**,
    30, **31**, 32–36, 46, 47, 48
  toxicity testing **37**: **113**, **127**, 135
  bioasay applications **37**: 77
  biocides **37**: **98–100**, **102**, **104**, **105**,
    107
  detergents and oil **37**: **110**, **112–114**
  metals **37**: **91**, **93**, **94**, 95, **120**
Mytilus edulis × *Mytilus*
    galloprovincialis **31**: 26, **38**
Mytilus edulis, Physiology **27**: 168
Mytilus galloprovincialis **31**: 7, 10, 17, 19,
    21, 35, 36; **37**: 2, **8**, **9**, 131
  bioassay methodology **37**: 42, 43, **44**
  described **37**: 13, 14
  larval rearing in laboratory **37**: 46
    nutritional requirements **37**: 38
    physical requirements **37**: 36, **36**, **37**,
      37
    spawning and fertilization **37**: 27, **29**,
      30, **32**, 32–36
  reproduction **37**: 18
    spawning and fertilization **37**: 27, **29**,
      30, **32**, 32–36
Mytilus generally **37**: 5, 29, 46, 135
  bioassay procedures **37**: 51, 72
  larval rearing in laboratory **37**: 39
  reproduction
    biology of **37**: 22
    toxicity testing **37**: **100**, 136
Mytilus trossulus (west coast mussel) **39**: 13,
    15, 18; **31**: 35, 37; **37**: **9**, 14, 16, 45,
    56
Myxicola infundibulum **34**: 219
Myxidium **40**: 15, 43
Myxidium bergense **40**: 5, 15, 30
Myxidium gadi **40**: 5, 15, 30, 31, 50
Myxidium oviforme **40**: 5, 15, 32, 36, 37, 43
Myxine **40**: 256
Myxine glutinosa **38**: 175
Myxobolus aeglefinus **40**: 5, 15, 30
Myxobolus bergense **40**: 30
Myxocephalus octodecemspinosus **40**: 20
Myxocephalus scorpius **40**: 18, 23, 28
Myxoderma sacculatum ectenes **35**: **31**
Myxodinium **25**: 127
Myxosphaera **25**: 127
Myxosporea **40**: 5, 24

Myxozoa **43**: 17

Nacella concinna **24**: 47; **34**: 5, **6**, 21
Naesa bicentata **39**: 110, 180, 211
Naiades cantrainii **25**: 139, *140*
Najas graminea, Food/feeding **27**: 321
Nannallecto **25**: 153
Nannocalanus **25**: 138
Nannocalanus minor **44**: 71
Nannochioropsis oculata **41**: 182
Nannochloris **37**: 38
Nannochloropsis **29**: 79, 91
Nanocopia **33**: **51**, 93, 99
  minuta **33**: 5, 99, **489**
Nanomia **24**: 101, 102, 103, 164, 165, 173,
    174, 179, 181, 183–186, 188, 190,
    191, 194–197, 200–203, 225
  (Stephanomia) bijuga **24**: 100, 103, 104,
    117, 147, 164, 165, 176, 194, 195,
    208, 209, 214, 216, 232
  cara **24**: 117, 141, 147, 165, 167, 168,
    173, 178, 191, 194, 201, 202, 226
Nanomia bijuga **25**: 157
Naskia axiplicata **32**: **164**
Nassarius **25**: 200
Nassarius distortus **28**: *392*, 412, 423
Nassarius festivus **43**: 145, 146
Nassarius granifer **28**: *392*, **403**, 412, 415,
    419, 423, 426
Nassarius kraussianus **28**: 397, 423, *425*
Nassarius pygmaeus **40**: 27
Nassarius reticulatus **40**: 27
Nasutitermes nigriceps **40**: 145
Natantia **32**: *97*
Natica **25**: 148, *149*; **28**: *399*, 409; **40**: 27;
    42: 92
  maculosa **25**: 225
Natica alderi **40**: 27
Natica grayi **28**: *392*
Natica gualteriana **28**: *392*, **403**, 415;
    34: 31
Natica hebraea **28**: *392*, 407, *417*
Natica millepunctata **28**: 407, **414**
Nausitora fusticula **40**: 149
Nautiloidae **32**: 407
Nautilus **44**: 146, 148, 151, 153, 155, 159,
    164; **25**: 87, 90, 99, *100*, 101, 108
Nautilus macromphalus **44**: 150
Nautilus sp. **32**: 407
Navanax inermis **34**: **41**
Navicula **31**: 352; **40**: 124
Navicula granii **32**: *534*, 537

Nearchaster aciculosus **32**: *104*; **35**: **31**
Neasticilla **39**: 159
Nebalia **25**: 135
Nebaliidae **32**: 408
Nebaliopsis typica **32**: *73*
'*Nectalia loligo*' **24**: 105, 117, 125, 178
Nectocarmen antonioi **24**: 119
Nectocrangon dentata **32**: **446, 448**
Nectocrangon lar **32**: **446, 448**
Nectonema **38**: 181
Nectopyramis **24**: 135
Negaprion brevirostris **44**: 218
Negaprion brevirostris, Predation **27**: 366
Neilonella hadalis **42**: *6*
Neilonellidae **42**: *5, 6*
Nekton **43**: 216
Nellia tenella **32**: 407
Nemadactylus macropterus **30**: 334; **35**: 30, 34, 45, 48, 85, 103
Nematocarcinus **23**: 30
Nematocarcinus pseudocurso **32**: **162**
Nematocarcinus undulatipes **32**: **162**
Nematoda **32**: 496; **40**: 7–8, 26, 28, 136, 137; **43**: 66, 68
   distribution **43**: 54, 55, 59, 60, 62
   diversity **43**: 4, 7, 17, 21
   ecology **43**: *32*, 36
   Southern Ocean pack ice **43**: 208, 214, 216
Nematomorpha **43**: *21*
Nematonurus **35**: 52
Nematonurus armatus **32**: 108, **129**, 130; **34**: **380**
Nematopsis penaeus, Parasites **27**: 383
Nematoscelis **25**: 119, 154
  difficilis **25**: 154
  megalops **25**: 142, 154
Nembrotha **34**: 33
Nemertea **43**: *21*, 70
Nemertodermatida **43**: *22*
Nemipterus personii **30**: 334
Nemoderma tingitanum **23**: 45
Neoancistrocrania norfolki **32**: **409**, 410, *411*
Neoaugaptilus distinctus **33**: 104
Neoboeckella **33**: **52**
Neobrachylepas relica **34**: **374**
Neocalanus
   behaviour **33**: 405
   ecology **33**: 374, 480, 486, 487, 503
   morphology **33**: 30
   nutrition **33**: 172

   reproduction **33**: **255**, 283
   taxonomy **33**: **52**, 68, 119
Neocalanus cristatus **33**: **7**, 119, 336
  flemingeri **33**: 35, 119
   biochemistry **33**: 237, 240, 242
   ecology **33**: **360**, 468, 528
   growth/development **33**: 316, **342**
   life history **33**: **387**
  gracilis **33**: **7**, **26**, **32**, 119, **224**
   behaviour **33**: **416**
   distribution **33**: **514**
   ecology **33**: **360**
   nutrition **33**: **194**
  minor **33**: 516
  plumchrus **33**: **7**, 35, 119
   behaviour **33**: **426**, 450, 454
   biochemistry **33**: 235, 237, 240, 243, **244**
   distribution **33**: **514**
   ecology **33**: **360**, 370, 372, **377**, 468, 479, 481, 502, 523, 528
   growth/development **33**: **306**, 317, **342**
   life history **33**: **387**
   nutrition **33**: 153, 156, 181, **186, 191**, *193*, 196, 197, **200**, 203, **206**
   reproduction **33**: 267, 281, **287, 288**
  robusitor **33**: **47**, 119, **224, 416**
   behaviour **33**: **416**, 424, **426**
   distribution **33**: 508
   ecology **33**: **359**, 370, 468, 481, 502, 524, 528
   growth/development **33**: **342**
   life history **33**: **387**
   nutrition **33**: 140, 181, **191, 206**
  tonsus **33**: **26**, 119
   behaviour **33**: **417**, 424, **426**
   biochemistry **33**: 237, 240, 242, 247, 251
   nutrition **33**: 181, **191, 192**
   reproduction **33**: **288**, 292
Neocalanus cristatus **32**: 9, 11–15, *12*, *13*, *14*, **16**, 17, 23, 55, 78
Neocalanus marshallae **32**: 11
Neocalanus plumchrus **26**: 136; **32**: 11, 14, 15, *15*, **16**, 17, 23, 55
Neocalanus plumchrus f. plumchrus **32**: 17
Neocalanus spp. **32**: 17
Neocalanus tonsus **44**: 71
Neoceratodus forsteri **40**: 305
Neochela **32**: 463
Neocopepoda **33**: 2
Neocrania **28**: **180, 195**, 197, **230**, 234

Neocrania anomala **28**: 178, 200, 226, 235, 236, 244, 248, 249, 256, **258**, 259, 265, 274, 277–279, **280**, 281, 283, 293, 312, 313
Neocrania californica **28**: 312
Neocyttus **32**: 186; **35**: 37
 helgae **35**: 37
 rhomboidalis **35**: 37
Neocyttus helgae **34**: **381**
Neocyttus rhomboidalis **32**: **172**
Neodenticula **32**: 560
Neodenticula seminae **32**: *532*, 538, 539, *549*, 555, 556
Neodermata **43**: 20, *21*
Neoechinorhynchus rutili **24**: 273; **40**: 9, 19, 28, 34
Neofibularia nolitangere **21**: 96, 97, 99
Neoglyphea inopinata **32**: **409**, 410
Neogoniolithon **24**: 57
Neohaustorius biarticulatus **39**: 203
Neohaustorius schmitzi **39**: 203
Neolaenira racemosa **34**: 368
Neolebouria tinkerbellae **38**: 180
Neolepas
 n. sp.1 **34**: **374**
 n. sp.2 **34**: **374**
 n. sp.3 **34**: **374**
 rapanuii **34**: **374**
 zevinae **34**: **374**
Neolepas zevinae **23**: 329; **32**: 408
Neolepetopsis
 densata **34**: **370**
 gordensis **34**: **370**
 occulata **34**: **370**
 verruca **34**: **370**
Neolithodes diomedeae **23**: 338
Neoloboquadrina pachyderma **43**: 209
Neomenia **23**: 324
Neomiodon angulata **31**: *306*
Neomphalus fretterae **23**: 303, 321, 322, 334, 346; **34**: **369**
Neomysis **25**: 147, 151; **39**: 131
 americana **25**: 147
Neomysis americana **26**: *131*, 143
Neomysis awatschensis **39**: 123, 128
Neomysis integer **33**: 527; **29**: **131**, 185; **39**: 215, 216
Neomysis japonica **39**: 117
Neon flying squid *see Ommastrephes bartrami*
Neonotoporus **25**: 150
Neonymphon **24**: 26

Neopallene **24**: 9
Neophasis burti **40**: 6, 13, 20
Neophasis oculatus **40**: 6, 20
Neophocaena, Predation **27**: 373
 sphocanoides, Predation **27**: 373
Neopontella **33**: **52**, 95, 113
 typica **33**: 113
Neorhabdus **33**: 107
Neosarmatium meinerti **40**: 142
Neoscalpellum eltanini **32**: **469**
Neoscolecithrix
 behaviour **33**: *422*
 ecology **33**: **501**
 taxonomy **33**: **53**, 78, 87, 90, 138
Neoscolecithrix antarctica **33**: 138
 catenoi **33**: 138
 farrani **33**: 138
 koehleri **33**: 138
 magna **33**: 138
 watersae **33**: 138
Neoscopelidae **32**: **169**
Neoscopelus **32**: 186
Neoscopelus macrolepidotus **32**: **170**, **195**
Neosmilaster georgianus **43**: 147
Neotanais **39**: 160
Neothyone gibbosa **21**: 146
Neothyris lenticularis **28**: **204**, 249, *265*, 276, 285, 305, 307, 309, *311*, 316, 317, 319, 335, 338, 342, 345, 351, *357*
Neotrigonia **32**: 407
Neotrigonia margaritacea **22**: 103, 105
Neoturris pileata **42**: 36
Neoverruca brachylepadoformis **32**: 408; **34**: **375**
Nephpterix syntaractis **40**: 178
Nephrops norvegicus **26**: 34
Nephrops, Physiology **27**: 208
Nephropsidae **32**: **129**
Nephroselmis minuta **29**: 79
Nephtys **35**: 166, **169**
 hombergi **35**: **163**, 166, **168**
Nephtys hombergi **43**: 106, 139
Neptunea **21**: 170; **40**: 27
 antigua **21**: 177
 arthritica **21**: 177, 189
Neptunea antigua **40**: 27
Neptunea insulalis **34**: **373**
Nereidae **39**: 36
Nereimyra alvinae **34**: **367**
Nereis (Neanthes) succinea **44**: 132
Nereis **34**: **367**

piscesae **34**: **367**
  sandersi **34**: **367**
Nereis diversicolor **25**: 209
Nereis sandersi **23**: 328
Nereis virens **35**: **185; 43**: **136**
Nereocystis (kelp) **39**: 52
Nereocystis **39**: 52
Nereocystis leutkeana **23**: 24, 80; **39**: 40;
    38: 227
Neresheimeria catenata **25**: 134
Neritopsina **42**: 181
Neritopsis radula **32**: 407
Nerocula **39**: 135
Nesiarchus **32**: 206
Nesiarchus nasutus **32**: **174**
Nettastoma **32**: 186, 191
Nettastoma falcinaris **32**: **168**, 185, **194**
Nettastomatidae **32**: **168**, 211
newmani **33**: 130
  ecology **33**: **361**, 366, 368, **378**, **493**
  growth/development **33**: **311**, **313**
  nutrition **33**: 182, **183**, 197, **198**, **201**,
    203
  reproduction **33**: **284**, 292
Nezumia **32**: 186, 206
Nezumia convergens **32**: **171**, 220
Nezumia propinqua **32**: **171**, 189, **218**
Nibea soldado, Predation **27**: **364**
Nicolea zostericola **43**: 137
Nicomache
  arwidssoni **34**: **366**, 402
  venticola **34**: **366**
Nilva torifera **34**: **375**
Niphargoides derzhavini **39**: 203
Niphargoides grimmi **39**: 203
Niphates erecta **21**: 98
Nitzschia
  N. frigida **43**: 200
  N. sturionis **43**: 69
Nitzschia bicapitata **32**: *550*
Nitzschia closterium **40**: 124
Nitzschia frigida **32**: 537, *547*, 557
Nitzschia lineola **32**: *552*
Nitzschia pseudoseriata **31**: 311
Noctiluca **23**: 175, 225–227, 232, 241, 256;
    25: 124, 129; **30**: **164**; **32**: 560
  miliaris **25**: 124
  scintillans **25**: 4
Noctiluca miliaris **21**: 73; **32**: 40
  scintillans **21**: 37, 73
Noctiluca scintillans **20**: 53

Nodopelta
  heminoda **34**: **370**
  subnoda **34**: **370**
Nodularia spumigena **33**: 155; **31**: 315, 316
Nomeus gronovii **24**: 203, 205
Nosema (*see Ameson*)
Nostoc spumigena **21**: 71
Notacanthus **32**: 186
Notacanthus chemnitzi **32**: **168**, 185, 190,
    193, **194**
Notarchidae **42**: 78, 88
Notarchus **42**: 104
Notaspidea **42**: 68, 70, 71
  see also Umbraculum
  characters and evolutionary
    pathways **42**: 83, 88, 91, 93,
      94–96, 100
  morphology **42**: 80, 81, *80*
  taxonomic history **42**: 73, *89*, 107, 108,
    110, 112–117 *passim*
Notemigonus crysoleucas **31**: 186
Nothria **32**: **161**
notialis, Taxonomy **27**: 98
  Behaviour **27**: 336, 338
  Food/feeding **27**: **318**, 319
  Life histories **27**: 289, 292, 293, **305**,
    307, 309
  Moulting/growth **27**: 237, **240**, 243,
    250
  Reproduction **27**: 260
  Zoogeography/Evolution **27**: **146**, **147**
Notobranchaea **25**: 153
Notocrangon antarcticus **43**: 143, 148;
    38: 214
Notodiaphanidae **42**: 112
Notonecta **30**: 183
Notophryxus **25**: 154
  globularis **25**: 154
  lateralis **25**: 154
Notopogon **32**: 218; **32**: **172**, 223
Notopogon fernandinus **32**: 218
Notosaria **28**: **183**, **206**, 231, 234
Notosaria nigricans **28**: 187, 221, 222, 224,
    249, 250, 261, 265, 271, 272, 276
Notoscopelus elongatus **26**: 116
Notothenia **24**: 328, 340, 348
  angustata **24**: 333, 338, 339, 353
  coriiceps **24**: 334, 346
  gibberifrons **24**: 346; **35**: **37**, 43, 103
  neglecta **24**: 368, 371; **35**: **37**, 43, 103
  rossii **24**: 371; **35**: **37**, 43, 52, 103
Notothenia rossii **30**: 334

Notothenia squamifrons **30**: 334
Nototodarus **32**: 206, 220; **39**: 270, 279, 288
Nototodarus gouldi (Gould's arrow
        squid) **39**: 269, 270, 279, 282,
        283, 289
Nototodarus hawaiiensis **32**: **164**, 205,
        206, 220
Nototodarus sloanii (Wellington flying
        squid) **39**: 269, 270, 279, 282,
        283, 289
Nototropis guttatus **39**: 203
Notropis atherinoides **25**: 58
Notropus hudsonius **20**: 55
Nucella **34**: 49
   emarginata **34**: 4, **6**, 24, 28
   lamellosa **34**: 48
   lapillus **34**: 25
Nucella lamellosa **39**: 24, 27, 28, 30, 73
Nucella lapillus **35**: 52
Nucella, see also, *Thais* **21**: 177
Nucinella **42**: 4, 43
   N. serrei **42**: 6
Nucinellidae *see* Manzanellacea
Nucinellina (Solemyoida) **42**: 5
Nucula **35**: **163**, **169**; **42**: *142*
   anatomy and morphology **42**: *6*, 7, *10*, 11,
      12, 17
   development **42**: 28, 29
   ecology **42**: 31, 32, 33, 34, 35, 36
   evolution **42**: 42, 43, 44
   feeding **42**: 18, *18*, 20, 21
   life mode **42**: *18*
   N. annulata **42**: 20, 29, 31, 32
   N. bicornis **42**: 26
   N. darella **42**: 26
   N. delphinodonta **42**: 16, 28
   N. exigua **42**: 6
   N. hartvigiana **42**: 26
   N. nitidosa **42**: 17
   N. notobenthalis **42**: 40
   N. nucleus **42**: *10*, 17, 26, 29, 36,43
   N. proxima **42**: 20, 21, 28, 31, 33, 34, 42
   N. pusilla **42**: 26
   N. subovata **42**: 25
   N. sulcata **42**: 25, 36
   N. turgida **42**: 20
      development **42**: 29, 30
      ecology **42**: 31, 32, 34, 35
      reproduction **42**: 24–26
   physiology **42**: 30, 31
   reproduction **42**: 24, 25, 26
   zoogeography **42**: 30, 36, 39

Nuculacea **32**: 463
Nuculana **23**: 340; **39**: 39; **42**: 17, 30, 36,
   41
   anatomy and morphology **42**: *7*, *10*, 11,
      12, 13, 14, 16
   development **42**: 26, 27, 29
   life mode **42**: *18*
   N. commutata **42**: 17
   N. grasslei **42**: 27, 30, 41
   N. minuta **42**: *10*, 11, *12*, 13, 18, 29,36
   N. pontonia **42**: 26
Nuculana grasslei **34**: **374**
Nuculana pontonia **35**: 32
Nuculanidae **32**: 466, **492**
Nuculoida **42**: 2, 4, *5*, 11, 18, 26
   Nuculacea, Nuculidae and
      Nuculinae **42**: 4, *5*, 8–13, *10*, *12*,
      40
   Nuculanacea, Nuculanidae and
      Nuculaninae **42**: 4, *5*, *6*, 8–12, *10*,
      12, 15, 16, 37–39, 40
      see also Nuculana
   Nuculidae **42**: *5*, *6*, 37–39
      see also Nucula; Nuculoma
   Nuculominae **42**: *5*
Nuculoidea pernambucensis **42**: 9
Nuculoma
   N. similis **42**: 25
   N. tenuis **42**: 21, 26, 28, 33, 36
Nudibranchia **42**: 68, 70, 71, 105; **43**: 140,
   146
   see also Armina; Bathydoris;
      Hexabranchus; Nudipleura;
      Tritonia
   characters and evolutionary
      pathways **42**: 87, 88, 94, 95, 98,
      99, 102, 103, 104
   taxonomic history **42**: 73, *89*, 107, 108,
      110, 111, 113, *114*, 115, 117
Nudipleura **42**: 68, 113, 117
Nybelinia **25**: 151
Nybelinia sp. **32**: 308
Nybelinia surmenicola **24**: 274, 290; **32**: 306,
   308
Nycticorax nycticorax **40**: 156; **27**: 372
Nyctiphanes capensis **30**: 176
Nyctiphanes simplex **25**: 48
Nymphon **24**: 4, 9, 11, 13, 20, 22, 26, 31, 38,
   39, 44, 71
   adareanum **24**: 60
   australe **24**: 18, 28, 60, 69, 70, 71
   brevicaudatum **24**: 60, 71

brevirostre **24**: 16, 21, 27, 35, 40, 46, 63
caementarum **24**: 14
discorsicoxae **24**: 14
gracile **24**: 21, 22, 23, 28, 30–37, 45, 46,
    53, 57, 63, 66
gracilipes **24**: 70
grossipes **24**: 63
hampsoni **24**: 4, 10, 12
hirtipes **24**: 28, 42, 55
hirtum **24**: 63
inerme **24**: 8
laterospinum **24**: 75
leptocheles **24**: 34, 35, 41
longitarse caecum **24**: 64, 75
macronyx **24**: 18
maculatum **24**: 16
orcadense **24**: 28
parasiticum **24**: 52
pixellae **24**: 20
procerum **24**: 75
robustum **24**: 22, 43
rubrum see *N. brevirostre*
sluiteri **24**: 42
stroemi **24**: 54
tripectinatum **24**: 64, 75
tubiferum **24**: 14
Nymphonella **24**: 9, 13, 16
  lambertensis **24**: 64
  lecalvezi **24**: 63
  tapetis **24**: 9, 14, 17, 32, 38, 44, 52, 63,
    64, 72
Nymphopsis acinacispinatus bathursti **24**: 10
Nypa **40**: 92; **40**: 167
Nypaceae **40**: 93
Nytiphanes australis **30**: **170**, 170, 176, 203

Oactopoteuthidae **32**: 62
Oasisia **23**: 325, 335; **34**: 408
  alvinae **34**: **366**, 405, *409*
Obturata **32**: 408
occidentalis, Taxonomy **27**: 97, 123
  Zoogeography **27**: **145**
Oceanobdella sexoculata **40**: 9
Ocenebra **21**: 32
  japonica **21**: 31
Ochiomonas malhamensis **21**: 164
Ochromonas (flagellate) **41**: 49
Ochromonas danica **21**: 72, 90
  malhamensis **21**: 72, 90
Octocorallia **32**: **160**
Octolasmis **22**: 200
Octomitus intestinalis **40**: 4, 11

Octopodidae **32**: 289
Octopus
  bimaculatus **25**: *106*
  burryi **25**: *97*
  cyanea **25**: *106*
  tetricus **25**: *106*
  vulgaris **25**: 98, *98*, 104, *106*
Octopus apollyon **21**: 183, 186, 191
  australis **21**: 173, 191
  bimaculatus **21**: 183, 191
  dofleini **21**: 186, 191
  edulis **21**: 5
  fitchi **21**: 191
  flindersi **21**: 191
  lunulatus **21**: 191
  macropus **21**: 184, 191
  maculosa, see also, *Octopus*
    maculosa **21**: 187
  rubescens **21**: 173, 191
  vulgaris **21**: 173, 182–184, 191
Octopus cf. *aegina/kagoshimensis* **44**: 180
Octopus macropus **44**: 178
Octopus vulgaris **44**: 149, 169; **39**: 278
Octosporea effeminaus **39**: 113, 114
Oculina diffusa **22**: 24, 27
Ocypoda platytarsis **29**: **133**, 134
Ocypode
  ceratophthalmus **30**: 97
  kuhlii **30**: 97
Ocythoe **25**: 89, 90
Odocoileus hemionus sitkensis (Sitka black-
    tailed deer) **39**: 17, 43, 44
Odontaster validus **24**: 329
Odontoglaja guamensis **42**: 97
Odostomia (gastropod) **41**: *189*
Odostomia eulimoides **24**: 285
  trifida **24**: 285
Odostomia eulimoides **40**: 27
Oediceros saginatus **39**: 203
Oenopota ogasawarana **34**: **373**
Ogcocephalidae **32**: **171**
Oiketicus kirbyi **40**: 178
Oikopleura **44**: 34; **23**: 225; **25**: 134; **26**: 251,
    270; **31**: 92
  albicans **25**: 143
Oikopleura dioica **44**: 31, 33, 43, 130, 131;
    28: 100
Oikopleura labradorienis **32**: 22
Oikopleura longicauda **44**: 131
Oikopleura rufescens **44**: 34
Oikopleura spp. **32**: 23
Oikopleura vanhoeffeni **32**: 22

Oiphiozonoida sp. **32**: **453**
Oithona **28**: 113; **33**: 160, **184**, 429, 473;
    38: 18; **43**: 216; **20**: 30
    davisae **33**: 257
    O. similis **43**: 217, 218
    oculata **33**: **430**
Oithona aruensis **44**: 82
Oithona atlantica **32**: 22, 23
Oithona borealis **32**: 19–22
Oithona davisae **44**: 82–84
Oithona nana **44**: 7, 84
Oithona plumifera **44**: 84
Oithona similis **44**: 7, 84, 85; **30**: 250; **32**: 11, 19, 22, 23
Oithona simplex **44**: 84
Oithona spp. **44**: 85, 86
Oithonal/Acartia **30**: **164**
Okadaia elegans **42**: 184
Oldsquaw *see Clangula hyemalis*
Olgaconcha **32**: 120
Olgasolaris tollmanni **34**: **370**
Oligocaris **38**: 16
Oligocottus maculosus **26**: 138, 142, 148
Olindias sambaquiensis **21**: 105
    singularis **21**: 105
Olindioides formosa **21**: 105
Olisthodiscus **20**: 28
Oliva **28**: *393*, 395, 401, 402, 412, 419;
    42: 92
Oliva annulata **28**: *393*, 415
Oliva gibbosa **28**: 411, 412
Oliva miniacea **28**: *393*, **403**, 415
Olivella columellaris **28**: *393*, 411; **34**: 31
Olivella semistriata **28**: *393*
Olivella volutella **28**: *393*
Olividae **39**: 36
Omalaxidae *see* Lower Heterobranchia
Omalogyridae *see* Lower Heterobranchia
Ommastrephes bartrami (neon flying
    squid) **39**: 266, 267, 273, 274,
    276, 279, 282, 283, 288, 289
Ommastrephes bartramii **32**: 247, 270, 271
Ommastrephes sloani pacificus **21**: 191
Ommastrephidae **32**: 259, 295
Omniglypta **42**: *143*, 147
Omniglyptidae **42**: *143*, 147
Oncaea **33**: 160; **43**: 216
Oncaea notopus **32**: 22
Oncaea praeclara **34**: **375**
Onchidoris muricata **43**: 140
Onchocalanus **25**: 138; **33**: **53**, 81, 82, 132, 486

affinis **33**: 132
cristatus **33**: 132
hirtipes **33**: 132
latus **33**: 132
magnus **33**: 132
paratrigoniceps **33**: 132
scotti **33**: 132
subcristatus **33**: 132
trigoniceps **33**: 132
wolfendeni **33**: 132
Onchorhynchus **37**: 132
Oncomegas **25**: 150
Oncorhynchus clarki (cut-throat trout)
    39: 17, 59
Oncorhynchus gorbuscha (pink salmon)
    39: 17, 52, 55–59, 74, 78
Oncorhynchus gorbuscha **32**: 267, 269, 283
Oncorhynchus keta (chum salmon) **39**: 58
Oncorhynchus keta **32**: 267, 269, 305
Oncorhynchus kisutch (coho salmon)
    39: 279
Oncorhynchus kisutsch **32**: 267, 269, 283, 305
Oncorhynchus mykiss **30**: 229; **40**: 287
Oncorhynchus nerka **26**: 144, 152, 252;
    32: 267, 269, 283, 286; **43**: 60
Oncorhynchus sp. **32**: **297**
Oncorhynchus spp. **44**: 210
Oncorhynchus tschawytscha **32**: 267, 269, 283, 304, 305
Oncorhynchus tshawytscha (Chinook
    salmon) **39**: 279
Oncorhynchus tshawytscha **44**: 252
Oneirophanta **43**: 106
Oniscus **30**: 90
Onissimus **22**: 92
    normani **22**: 92
Onoba aculeus **40**: 27
Onsimus lotoralis **39**: 203
Onuphidae **32**: 362, 445
Onuphis lepta **32**: 362
Onychoteuthis borealijaponica **32**: 271
Oodinioides **25**: 127
Oodinium **25**: 126, 127, 131, 134, 135
    amylaceum **25**: 134
    dogieli **25**: 131
    fritillariae **25**: *128*, 134
    poucheti **25**: 129
Opaepele spp. **32**: 114, 131
Opechona **25**: 148
    bacillaris **25**: 149

Opechona bacillaris **24**: 274, 288; **40**: 6, 16, 27, 33
Opecoeloides fimbriatus, Parasites **27**: 387
Opercularella lacerata **24**: 54
Opheliidae **32**: 467; **39**: 36; **32**: 168; **32**: 171
Ophiocoma
  doderleini **24**: 47
  erinaceus **24**: 47
  pica **24**: 47
Ophiocomina nigra **21**: 145
Ophiocten gracilis **35**: 32
Ophioctenella **32**: 130
Ophioctenella acies **32**: 99, *101*, 102, 103, *104*, **129**, 135; **34**: **379**
Ophiodon elongatus **38**: 178
Ophioglypha bullata **35**: 31
Ophioleuce brevispina **32**: **166**
Ophiomusium
  lymani **35**: **31**, 79, 80
  planum **35**: 32
Ophiomusium multispinum **32**: **453**
Ophiomusium planum **32**: **453**
Ophiomusium sp. **32**: **446**, **452**, **453**
Ophiomyces sp. **32**: **166**
Ophiophagus hannah **40**: 154
Ophiophyces sp. **32**: **166**
Ophiopyrgus wyvillethomsoni **32**: 503
Ophiotholia **32**: *214*
Ophiotholia multispina **32**: **166**
Ophiura
  sarsi **35**: 32
  signata **35**: 32
Ophiura bathybia **32**: **446**, **452**
Ophiura flagellata **32**: **452**
Ophiura irrorata **32**: **453**
Ophiura leptoctenia **32**: **452**
Ophiura ljungmani **43**: 107, 147
Ophiura loveni **32**: **446**, **452**
Ophiura sarsi vadicola **32**: **452**
Ophiura sarsii **20**: 355
Ophiura sp. **32**: **166**, **166**, **453**
Ophiuridae **32**: **447**, **453**
Ophiuroidea **31**: 39; **32**: 97, 99, **129**, **166**, 329, 333, **335**, 357, 366, 429, **438**, **446**, **447**, **453**, **462**, 463, **504**; **39**: 36; **43**: *95*, 100, 106, 107, 126, 147
Ophryotrocha **34**: **366**
  akessoni **34**: **366**
  globopalpata **34**: **366**
  O. labronica **43**: 130, 132, 145
  O. puerilis puerilis **43**: 131
  platykephale **34**: **366**

Ophryotrocha akessoni **23**: 328, 340
Opiodon elongatus **39**: 279
Opisa eschrichti **40**: 9, 19, 32
Opisthobranchia **42**: 68, 82, 102, 115, 119; **43**: 125, 146
  see also Shelled Opisthobranchs in *subject index*
  major taxa *see* Anaspidea; Architectibranchia; Cephalaspidea; Notaspidea; Sacoglossa
Opisthonema **20**: 34, 177
  oglinum **20**: 6, 180
Opisthoteuthis sp. **32**: 301
Opisthotrochopodus
  alvinus **34**: **368**
  japonicus **34**: **368**
  marianus **34**: **368**
  segonzaci **34**: **368**
  trifurcus **34**: **368**
  tunnicliffeae **34**: **368**
Oplophoridae **32**: 62
Oplophorus spinosus **32**: **162**
Opsanus tau **24**: 346; **40**: 11
Opuntiella californica **23**: 80
Oradarea longimana **34**: **376**
Orange roughy (*Hoplostethus atlanticus*) **43**: 62
Orange roughy *see Hoplostethus atlanticus*
Oratosquilla oratoria, Predation **27**: 360
Orbiniella
  aciculata **34**: **366**
  hobsonae **34**: **366**
Orbitestellidae *see* Lower Heterobranchia
Orchesetta **29**: 148
Orchestia **39**: 129, 147, 184, 190
Orchestia bottae **39**: 203
Orchestia cavimana **39**: 110, 125,165–167
Orchestia gammarella **39**: 109, 110, 112, 126, 127, 132, 137–142, 144, 145, 147, 203
Orchestia mediterranea **39**: 125
Orchestia montagui **39**: 109
Orchestia platensis **39**: 146, 185, 203
Orchestia scutigerula **39**: 203
Orchomene
  abyssorum **34**: **377**
  distinctus **34**: **377**
Orchomene cavimanus **39**: 203
Orchomene gerulicorbis **39**: 203
Orchomene montana **32**: **162**
Orchomene nanus **39**: 203

Orchomene plebs **24**: 329, 375; **39**: 203
Orchomene rossi **39**: 203
Orchomenella minuta **39**: 203
Orchomenella pinguis **39**: 203
Orchomenella proxima **39**: 204
Orcinus orca (killer whale) **39**: 63; **34**: 286; **43**: 228, 229
Orconectes limosus **35**: **185**
Orconectes virilis **21**: 123
Orconectes, Physiology **27**: 176
   limosus, Physiology **27**: 202
Oregonia bifurca **32**: **446, 448**
Oreochromis
   aureus **34**: 85
   mossambicus **34**: 83, 84, 253
Oreochromis mossambicus **40**: 286, 291
Oreochromis niloticus **40**: 290
Oreosoma atlanticum **35**: 37
Oreosomatidae **32**: **172**
orientalis (*see Penaeus chinensis*)
Ornithocercus **32**: 560; **43**: 35
Oropallene **24**: 9, 26
Orseis grasslei **34**: **367**
Ortholinea orientalis **24**: 276, 289
Orthopagurus minimus Plate 4, **42**: 216
Orthopristis chryosoptera, Predation
   27: **363**
Oryzias latipes **26**: 103
Osbornia **40**: 102
Osbornia octodonta **40**: 95, 165
Oscillatoria **21**: 90
   nigroviridis **21**: 71, 89
Oscillatoria submembranacea **22**: 48
Osmerus mordax **25**: 58; **26**: **120**, 140; **28**: 121
Osopsaron **32**: 186, 198
Osopsaron karlik **32**: **173**
Osteichthyes **41**: 189
Ostracoda **32**: **56**, 496
Ostrea edulis **22**: 103, 111, 132, 171; **37**: **8**, 131
   bioassay methodology **37**: 41, 42, **44**
   bioassay procedures **37**: 68, 71
   described **37**: 13
   larval rearing in laboratory **37**: 26
      nutritional requirements **37**: 38, 39
      physical requirements **37**: 37
      spawning and fertilization **37**: 27, **29**, 32
   metals **37**: **92, 94**
   reproduction
      biology of **37**: 16, 17, 18, 21, *22, 23*

      spawning and fertilization **37**: 27, **28**, 32
   toxicity testing **37**: 97
      detergents and oil **37**: **109, 111, 113**, 116
Ostrea plicatula **37**: **8, 9**, 16
Ostreopsis siamensis **31**: 315
Otala lactea **34**: 32
Otodistomum **40**: 5, 16, 25, 33
Otodistomum veliporum **40**: 5
Otolithes ruber, Predation **27**: **364**
Otopleura nodicincta **28**: *393*
Otter *see Enhydra lutris*; *Lutra canadensis*
Ovalipes **25**: 196
   punctatus **25**: 195
Ovalipes ocellatus **29**: **133**, 134
Owenia **34**: 44
Owenia fusiformis **35**: **168**
Oxycephalus **25**: 141, 156
Oxycephalus sp. **32**: 269
Oxynaspis michi **32**: **161**
Oxynoe
   O. antillarum **42**: 86
   O. azuropunctata **42**: 86
Oxynoidea **42**: 80, 88, 102, 108, 109, 112, 115, 129
Oxytonostoma typica **40**: 9
Oxytoxum **32**: 545, 560
Oxytoxum milneri **32**: 545
Oxyuroidea **24**: 278
Oxyurostylis smithi **39**: 218
Oyster
   see also Saccostrea
   American/Eastern *see Crassostrea virginica*
   eastern (*Crassostrea virginica*) **43**: 22, 65, 66, 69
   European *see Ostrea edulis*
   Hawaïan *see Isognomon californicum*
   Japanese/Pacific *see Crassostrea gigas*
   Pacific (*Crassostrea gigas*) **43**: 69
Oystercatcher (*Haematopus*) **43**: 34
Oystercatcher *see Haematopus bachmani*
Oziotephusa senex senex **40**: 105

Pachycara
   gynmium **34**: **381**
   thermophilum **34**: **381**
Pachycara thermophilum **32**: **129**, 130
Pachydermia
   laevis **34**: **370**
   sculpta **34**: **370**

Pachygrapsus crassipes **44**: 248
Pachygrapsus, Moulting/growth **27**: 224
Pachyptila desolata **20**: *295*
Pachyptilus
  ecology **33**: **498**
  nutrition **33**: **152**
  taxonomy **33**: **51**, 70, 76, 105, 106
Pachyptilus abbreviatus **33**: 105
  eurygnathus **33**: 105, **435**
  lobatus **33**: 105
  pacificus **33**: 105
Pachypygus gibber **29**: **131**, 179, 182, 183
Pacific cod *see Gadus macrocephalus*
Pacific hake *see Merluccius productus*
Pacific herring *see Clupea pallasi*
Pacific oyster *see Crassostrea gigas*
Pacific saury *see Cololabis saira*
Padiculariidae **32**: **409**
Padiculariina **32**: 412
Padina **43**: 308
Padina pavonica **24**: 55, 57
Pagellus bogaraveo **26**: 73, **75**
Pagellus centrodontus **25**: 15, 17, 21, 33, **35**, 37; **26**: 33
Pagetopsis macropterus **24**: 328, 332, 350, 352
Pagirapseudes largoensis **39**: 219
Pagothenia
  bernachii **35**: 52
  hansoni **35**: 52
Pagothenia borchgrevinki **24**: 323–329, 334, 335, 337, 340, 347, 348, 353, 355, 356, 358, 359, 361, 363, 364, 366–369, 374, 375
Pagrus major (red sea bream) **38**: 26, 28, 39, 54
  in Japan **38**: 24, 25
  survival and size at release **38**: **38**
  survival from hatchery releases **38**: **41**
Pagrus major **26**: *101*, **121**; **28**: 66, *89*, 95, **96**
Pagrus pagrus **39**: 280
Paguridae **32**: 455
Paguristes seminudus **42**: 215
Pagurus
  bernhardus **29**: **133**, 159
  novae-zealandiae **29**: **133**, *147*, 149
  prideauxi **29**: **133**, 146, 159
Pagurus bernardus **34**: 288
Pagurus imafukui **42**: 215
Pagurus larvae **26**: *131*
Pagurus pubescens **32**: **448**

Pagurus sp. **26**: *131*
Pagurus, Morphology **27**: 37
Paivella **33**: **53**, 92, 127
  inaciae **33**: 127
  naporai **33**: 127
Palaeisopus **24**: 77, 78
  problematicus **24**: 76, 77
Palaemon, Morphology **27**: 37
  elegans, Moulting/growth **27**: 226
  serratus, Physiology **27**: 166, 179
Palaemonetes **29**: **132**, 190
  pugio **29**: **132**, **192**
  vulgaris **29**: **132**, **192**
Palaeoacanthocephala **25**: 152
Palaeomonetes, Physiology **27**: 190
  pugio, Predation **27**: 374
  Parasites **27**: 388
Palaeopantopus **24**: 77
  maucheri **24**: 76, 77
Palaeothea **24**: 77
  devonica **24**: 76, 77
Palaepalaemon newberryi,
  Zoogeography/Evolution **27**: 149
Paleodentalium **42**: 143
Paleotaxodonta **42**: 2, 4
Palinurus gilchristi **29**: **132**, 161, **162**, 165
Palinurus marginatus **34**: 224
Pallasiella quadrispinosa **39**: 204
Pallenoides spinulosa **24**: 66
Pallenopsis **24**: 8, 15, 26, 32, 46, 64, 71, 76; **25**: 152
  calcanea **24**: 52
  cidaribatus **24**: 5, 10
  mollissima **24**: 68
  patagonica **24**: 31
  pilosa **24**: 13
  scoparia **24**: 52; **25**: 152
  tydemani **24**: 68
  vanhoeffeni **24**: 60
Palythoa **21**: 127
Palythoa **22**: 30
  caribaeorum **21**: 127
  caribbea **21**: 107
  mammilosa **21**: 107
  toxica **21**: 107, 126, 127
  tuberculosa **21**: 107, 126, 127
  vestitus **21**: 107, 127
Palythoa caribaeorum **31**: 248, 257
Pancolus **39**: 160
Pandalid shrimp **43**: 148
Pandalidae **32**: **129**
Pandalina **38**: 106

Pandalina brevirostris **38**: 179
Pandalina nana **32**: **162**
Pandalopsis **38**: 62, 106, 152
Pandalopsis ampla **35**: **32**
Pandalopsis coccinata **38**: 106, **107**
Pandalopsis dispar **38**: 106, **107**, 206
  fishing **38**: 224, 225
Pandalus (shrimp) **39**: 17; **26**: *27*; **38**: 105
Pandalus bonnieri (*Dichelopandalus bonnieri*) **38**: 105
Pandalus borealis **32**: **448**; **35**: **32**; **38**: 57, 58, 68–70
  abundance **38**: 209, 210, 216
  aquaculture **38**: 226
  catch per unit effort (CPUE) **38**: 209
  competitors **38**: 179
  depth distribution **38**: 116, 117
  distribution **38**: *69*, 108, 109, **110**, 112, 113, 114
  duration of oogenesis **38**: 153, 154
  duration of ovigenous period **38**: 153
  duration of reproductive cycle **38**: 153
  egg mortality **38**: 214
  embryonic development **38**: 155, 156
  eye **38**: 62
  fecundity **38**: 212, 213
  fisheries/fishing **38**: 218, 220, 221, 222, 223, 224, 225
  food preferences **38**: 174, **176**
  genetic studies **38**: 205
  gonad development **38**: 143–145, *144*, 153–157
  growth curve **38**: 186, **188, 189**
  growth models **38**: 193
  growth rate **38**: 196
  growth rate and temperature **38**: 201, 202
  hatching behaviour **38**: 152
  juvenile growth **38**: 197, 198
  larval behaviour **38**: 157, 158
  larval growth **38**: 162, 163, 164
  larval nutrition **38**: 159, **160**, 161, 162
  larval stage **38**: **102**, 103, 105
  larval survival **38**: 165, 166
  length/weight relationship **38**: **184**
  mating in **38**: 148
  migration **38**: 169–171, 172
  mortality **38**: 217
  parasites **38**: 180, 181, 182, **183**
  phenotypic plasticity **38**: 214
  pleonites in **38**: 63
  population studies **38**: 205
  post-larval behaviour **38**: 167, 168
  predation **38**: 218
  predators **38**: 175, 178
  reproductive cycles **38**: 135, **136, 137**, 139, 141, 143
  reproductive system **38**: 122, **122**, 123
  salinity/temperature ranges **38**: 119, 120
  schooling behaviour **38**: 168, 169
  seasonal variations in growth **38**: 201
  sex allocation **38**: 128, 129, 130, 131
  sex change **38**: 133, 135
  sexual dimorphism **38**: 135
  size estimation **38**: 190
  spatial distribution **38**: 215
Pandalus chani **38**: 66, 92–94
  distribution **38**: 110, 112, 113
Pandalus curvatus **38**: 66, 80–82
  distribution **38**: *66*, 112, 113
Pandalus danae **38**: 66, 84–86, 105
  aquaculture **38**: 227
  depth distribution **38**: 115, 116
  distribution **38**: 85, 112, 114
  fishing **38**: 224
  food preferences **38**: **177**
  gonad development **38**: 143
  growth rate **38**: 196
  hatching behaviour **38**: 152
  larval behaviour **38**: 157
  larval nutrition **38**: **160**
  larval stage **38**: **102, 107**
  length/weight relationship **38**: **185**
  migration **38**: 169
  oviposition in **38**: 151
  parasites **38**: **183**
  post-larval behaviour **38**: 168
  predators **38**: 178
  reproductive cycles **38**: 135, **138**
  reproductive system **38**: 122, **122**, 123
  salinity/temperature ranges **38**: 121
  sex allocation **38**: 132
Pandalus eous **38**: 57, 76–79
  abundance **38**: 210
  aquaculture **38**: 226
  depth distribution **38**: 116, 117
  distribution **38**: 108, 109, **110**, 112, 113
  duration of ovigerous period **38**: 153
  egg mortality **38**: 214
  embryonic development **38**: 155
  fisheries/fishing **38**: 218, 220, 224, 25
  food preferences **38**: 174, **176**
  genetic studies **38**: 205
  growth curve **38**: **188**

growth rate **38**: 196
growth rate and temperature **38**: 202, 203
hatching behaviour **38**: 152
larval behaviour **38**: 157, 158
larval growth **38**: 163
larval nutrition **38**: **160**, 161, 162
larval stage **38**: 100, **102**, 103, **107**
larval survival **38**: 165
length/weight relationship **38**: **184**
migration **38**: 169, 170, 172
mortality **38**: 217
parasites **38**: 180, 181, 183
population studies **38**: 205
post-larval behaviour **38**: 168
predators **38**: 175, 178, 218
reproductive cycle **38**: **136**, 139, 141
reproductive system **38**: 122, **122**
salinity/temperature ranges **38**: 119, 120
schooling behaviour **38**: 169
sex allocation **38**: 129
Pandalus formosanus **38**: 66, 94, 95
    distribution **38**: *88*, 110, 112, 113
Pandalus goniurus **32**: 266, 303; **38**: 71, 72
    aquaculture **38**: 226
    depth distribution **38**: 116
    distribution **38**: *71*, 112, 113
    fishing **38**: 225
    food preferences **38**: 174, **177**
    hatching behaviour **38**: 152
    larval behaviour **38**: 157
    larval nutrition **38**: **160**
    larval stage **38**: 100, **102**, **107**, **110**
    length-frequency relationship **38**: 192
    nocturnal migration **38**: 172, 173
    parasites **38**: **183**
    pleonites in **38**: 63
    post-larval behaviour **38**: 167, 168
    predators **38**: 178
    reproductive cycle **38**: **138**, 139, 141
    reproductive system **38**: **122**
    salinity/temperature ranges **38**: 121
Pandalus gracilis **38**: 60, 88, 89
    distribution **38**: *88*, **110**, 112, 113, 118
    food preferences **38**: **177**
    reproductive system **38**: **122**
Pandalus gurneyi **38**: 66
    distribution **38**: 112
Pandalus hypsinotus **38**: 82–84, 105
    aquaculture **38**: 226
    depth distribution **38**: 116, 117
    distribution **38**: *83*, 112, 113
    fishing **38**: 225, 226

food preferences **38**: 174, 177
gonad development **38**: 143
growth rate **38**: 196
hatching behaviour **38**: 152
juvenile growth **38**: 197
larval behaviour **38**: 157, 158
larval growth **38**: 162, 163, 164
larval nutrition **38**: **160**, 162
larval stage **38**: 100, **102**, **107**, **110**
mating in **38**: 149–151, *150*
migration **38**: 170
nocturnal migration **38**: 172
oviposition in **38**: 151, 152
pleonites in **38**: 63
post-larval behaviour **38**: 168
reproductive cycle **38**: 135, **138**, 139, 141
reproductive system **38**: **122**
salinity/temperature ranges **38**: 121
size estimation **38**: 190
Pandalus jordani **38**: 57, 72–74
    abundance **38**: 210
    aquaculture **38**: 226
    depth distribution **38**: 116, 117
    distribution **38**: 73, 111, 112
    fecundity **38**: 213
    fisheries/fishing **38**: 218, 220, 224, 225
    food preferences **38**: **177**
    growth rate **38**: 196
    growth rate and temperature **38**: 202, 203
    larval behaviour **38**: 157
    larval nutrition **38**: **160**
    larval stage **38**: 102, *102*, 103, 107, **110**
    length/weight relationship **38**: **185**
    migration **38**: 169
    nocturnal migration **38**: 172
    parasites **38**: 180, **183**
    pleonites in **38**: 63
    population discreteness **38**: 206
    predators **38**: 178, 218
    reproductive cycle **38**: **137**, 139
    reproductive system **38**: 122, **122**
    schooling behaviour **38**: 169
    sex allocation **38**: 129, 130
    sex differences in growth **38**: 199
    size estimation **38**: 190
Pandalus latirostris **38**: 97–99, 105, 106, 107, 111
    aquaculture **38**: 226
    depth distribution **38**: 115, 116
    digestive system **38**: 173

distribution **38**: *98*, **110**, 111, 112, 113,
    114
embryonic development **38**: 156
feeding **38**: 174
fishing **38**: 220, 226
food preferences **38**: **177**
gonad development **38**: 145, 153–157
growth rate and temperature **38**: 202, 203
haemolymphic proteins in **38**: 154, 148
hatching behaviour **38**: 152
juvenile growth **38**: 197
larval behaviour **38**: 156, 157
larval growth **38**: 162, 163
larval nutrition **38**: 159, **160**
larval stage **38**: **102**, 103
length/weight relationship **38**: **185**
mating in **38**: 149
reproductive cycle **38**: **138**, 143
reproductive system **38**: **122**
salinity/temperature ranges **38**: 119, 121
sex change **38**: 135
size estimation **38**: 190
Pandalus meridionalis **38**: 64
Pandalus montagui **38**: 59, 59, 67, 68, 105,
    109
aquaculture **38**: 226
competitors **38**: 179
depth distribution **38**: 116, 117
distribution **38**: **110**, 113, 114
embryonic development **38**: 156
feeding **38**: 174, 175
fishing **38**: 220, 221
food preferences **38**: **176**
growth curve **38**: **189**
larval behaviour **38**: 157
larval nutrition **38**: **160**
larval stage **38**: **102**, 105
length/weight relationship **38**: **185**
migration **38**: 170, 173
mortality **38**: 214
parasites **38**: 181, 182, **183**
pleonites in **38**: 63
post-larval behaviour **38**: 167
predators **38**: 178
reproductive cycle **38**: **137**, 139
reproductive system **38**: **122**
salinity/temperature ranges **38**: 119, 120
schooling behaviour **38**: 169
Pandalus montagui tridens **38**: 68, 76, 139
Pandalus nipponensis **38**: 89, 90
depth distribution **38**: 116
distribution **38**: *88*, **110**, 112, 113

larval stage **38**: 102
reproductive cycle **38**: **138**, 139
reproductive system **38**: **122**
salinity/temperature ranges **38**: 121
Pandalus platyceros **38**: 60, 95–97, 105,
    106
aquaculture **38**: 226, 227
depth distribution **38**: 116
distribution **38**: *96*, **110**, 112, 113
fishing **38**: 220, 224, 225
food preferences **38**: **177**
gonad development **38**: 143
growth rate **38**: 196
hatching behaviour **38**: 152
larval behaviour **38**: 157
larval nutrition **38**: 159, **160**
larval stage **38**: **102**, **107**
length/weight relationship **38**: **185**
mating in **38**: 148
migration **38**: 169
mortality **38**: 217
parasites **38**: 179, 180, 182, **183**
pleonites in **38**: 63
post-larval behaviour **38**: 168
predators **38**: 178
reproductive cycle **38**: **137**
reproductive system **38**: **122**
salinity/temperature ranges **38**: 121
size estimation **38**: 190
Pandalus prensor **38**: 64, 86, 87
depth distribution **38**: 115, 116
distribution **38**: *87*, **110**, 112, 113, 114
fecundity **38**: 214
fishing **38**: 226
food preferences **38**: **177**
hatching behaviour **38**: 152
larval behaviour **38**: 156, 157
larval nutrition **38**: 159–161, **160**
larval stage **38**: **102**, 103
reproductive cycle **38**: **138**
reproductive system **38**: 122, **122**
salinity/temperature ranges **38**: 121
Pandalus propinquus **38**: 66, 99–101
depth distribution **38**: 116, 117
distribution **38**: *100*, **110**, 113
gonad development **38**: 143
larval nutrition **38**: **160**
larval stage **38**: **102**, 105
parasites **38**: 181, **183**
reproductive cycle **38**: **137**, 139
reproductive system **38**: 122, **122**
salinity/temperature ranges **38**: 121

Pandalus stenolepis **38**: 79, 80, 105
  depth distribution **38**: 116, 117
  distribution **38**: *79*, **110**, 112
  larval behaviour **38**: 157
  larval stage **38**: **102**, **107**
  parasites **38**: **183**
  reproductive cycle **38**: **138**, 139
Pandalus teraoi **38**: 58, 90–92
  distribution **38**: *88*, **110**, 112, 113, 118
  reproductive system **38**: 122, **122**
Pandalus tridens **38**: 74–76
  depth distribution **38**: 116
  distribution **38**: *75*, 109, **110**, 112, 113
  larval stage **38**: **102**, **107**
  parasites **38**: **183**
  reproductive cycle **38**: **137**, 139
  reproductive system **38**: **122**
  salinity/temperature ranges **38**: 119, 120
Pandina sanctae-crucis **23**: 103
Pandion haliaetus **40**: *155*
Panetela **32**: **162**
Pangasius sutchi **40**: 280
Panope generosa (geoduck clam) **38**: 227
Panopea eblanae **25**: 156
Panopeus herbstii **44**: 22, 126
Pantachogon haeckeli **32**: 72
Panthera tigris **40**: 158
Pantopipetta **24**: 26, 54, 71
  auxiliata **24**: 4
  brevicauda **24**: 75
  longituberculata **24**: 75
Pantopoda **32**: 329, **335**, 344, 347, 366, 445, **483**, **504**
Panulirus
  angulatus **29**: **132**, 161, **162**
  homarus **29**: **132**, 160, 165, **162**, **163**, 165, 167, 168, *167*, 186, 197, 198
  interruptus **29**: **132**, 161–164, 187
  penicillatus **29**: **132**, 160, **163**, 165
Panulirus argus **44**: 226; **40**: 139
Panulirus cygnus **44**: 214
Panulirus marginatus **44**: 250
Panulirus, Zoogeography/Evolution **27**: 181
  japonicus, Physiology **27**: 198
  Moulting/growth **27**: 220
Papandalopsis **38**: 111
Papillonema clavatum **40**: 135
Papillonema danieli **40**: 135
Parabasaia **43**: *20*
Parabothrium gadi-pollachii **40**: 7, 10, 13–4, 28, 33
Parabothus **32**: 186

Parabothus amaokai **32**: **174**, 188, 190, **194**
Paracalanidae **33**: 30, **52**, 120, 121, 434
Paracalanus **25**: 121, 122, 133, 137, 147, 148; **33**: 298; **40**: 130
  behaviour **33**: 405, 407
  biochemistry **33**: **239**, 249
  ecology **33**: 473, 480, 482, 486, 497, 527, 528
  growth/development **33**: 299, **307**, **311**, 313, 314, **320**
  nutrition **33**: **152**, 181, **192**, 192, **200**, **206**
  parvus **25**: 144
  reproduction **33**: 256, **268**, 280, **287**
  size/weight **33**: **331**, 333
  taxonomy **33**: **52**, 77, 121
Paracalanus aculeatus **33**: **7**, *24*, **26**, 37, 121
  brevispinatus **33**: 121
  campaneri **33**: 121
  crassirostris **33**: **446**
  denudatus **33**: 121, **485**
  gracilis **33**: 121
  indicus **33**: 121, **343**, 373, 420, 524
  intermedius **33**: 121
  mariae **33**: 121
  nanus **33**: 121, **360**
  parvus **33**: 6, **7**, **26**, 37, 47, 121, **287**
    behaviour **33**: **412**, **417**, **426**, 444
    biochemistry **33**: **239**
    ecology **33**: **360**, **378**, **397**, 475, 480, **484**, **485**, 524
    growth/development **33**: **306**, 307, **311**, **313**, **343**
    longevity **33**: **345**
    nutrition **33**: 150, 161, **183**, **186**
    reproduction **33**: **255**, 282, **288**
    size/weight **33**: **224**, **231**
  ponticus **33**: 121
  quasimodo **33**: 121, **408**, **412**
  serratipes **33**: 121
  serrulus **33**: 121
    behaviour **33**: **408**, **412**
    chromosomes **33**: **47**, 48
    ecology **33**: 475
    growth/development **33**: **306**, **320**, **343**
    life history **33**: 383
    nutrition **33**: 143, 144, **187**, 192
    size/weight **33**: **224**
  tropicus **33**: 121
Paracalanus aculeatus **44**: 87, 88
Paracalanus parvus **44**: 88–91; **28**: 100
Paracalanus spp. **44**: 91, 92, 109
Paracalliactis stephensoni **22**: 84; **42**: 216

Paracandacia **33**: **51**, 95, 111, 160, 165
  bispinosa **33**: 111, **514**
  simplex **33**: 111, **491**, **514**
  truncata **33**: 111, 503, **514**
  worthingtoni **33**: 111
Paracaudina australis **21**: 147
  chilensis **21**: 147
Paracentrotus **37**: 131
Paracentrotus lividus **21**: 145, 169; **34**: 231
Paracerceis sculpta **39**: 111, 153, 212
Parachristianella heteromegacantha,
    Parasites **27**: 386
  monomegacantha, Parasites **27**: 386
Paracineta **25**: 138
Paracomantenna **33**: **53**, 88, 127, *422*, **499**
  gracilis **33**: 127
  magalyae **33**: 127
  minor **33**: 127
Paracrangon n. sp. **34**: **378**
Paracyclopia **33**: **53**, 78, 133
  gitana **33**: 133, **489**
  naessi **33**: 133, **489**
Paradentalium **42**: *143*
Paradinium poucheti **25**: 133
Paradisco **33**: **51**, 71, 74, 106
  gracilis **33**: 106
  grandis **33**: 106
  mediterraneus **33**: 106
  nudus **33**: 106
Paradorippe granulata **43**: 133
Paraeuchaeta **25**: 138
Parafilitanais similis **32**: **161**
Paragalma birsteini **24**: 116
Paragnathia formica **39**: 153
Parahaustorius longimerus **39**: 204
Parahemiurus **25**: 148
Parahemiurus merus **24**: 276, 285, 292, 294
Paralabidocera **33**: **51**, 95, 111
  antarctica **33**: 111, **361**, **387**, 492, **493**,
      494
  grandispina **33**: 111, **493**, 494
  separabilis **33**: 111
Paralabidocera antarctica **43**: 210, 211,
    212
Paralabrax nebulifer **30**: 226
Paralebeda **40**: 178
Paralepas ichtiandri **32**: **161**
Paralepas nascai **32**: **161**, 216
Paralepas xenophore **32**: 216
Paralepetopsis **34**: 417
  n. sp. **34**: **370**
  rosemariae **34**: **370**

Paraleptosphaeroma glynii **39**: 212
Paralicella **35**: 50
  tenuipes **35**: **32**
Paralicella caperesca **32**: **77**
Paralicella tenuipes **32**: **77**
Paralichthys albigutta, Predation **27**: 366
  lethostigma, Predation **27**: 366, 375, 376
Paralichthys olivaceus (Japanese fly
      flounder; hirame) **38**: 24, 25, 54
  cost-benefit analysis **38**: 44, 64
  survival after release **38**: 36
  survival and size at release **38**: **38**
  survival from hatchery releases **38**: **41**
Paralichthys sp. **26**: **121**, 140, *141*
Paraliparis devriesi **24**: 328, 329
Paralithodes **30**: 325
Paralithodes camtschatica **32**: **449**
Paralithodes platipus **32**: **449**
Parallobiopsis coutieri **25**: 135
Paralomis **34**: **378**
  jamsteci **34**: **378**
Paralvinella **23**: 327, 335; **34**: 410, 411,
      412; **35**: 22, 30
  bactericola **34**: **368**, *413*
  dela **34**: **368**, *413*
  fijiensis **34**: **368**, *413*
  grasslei **23**: 328, 335, 338; **34**: **368**,
      410–412, *413*; **35**: 18, **41**
  hessleri **34**: **368**, *413*
  p. irlandei **34**: *413*
  p. pandorae **34**: *413*
  palmiformis **34**: **369**, *409*, 411, 412, *413*
  pandorae **34**: **369**, 410, 411, 412
  sulfincola **34**: **369**, *413*
  unidentata **34**: **369**, *413*
Paralycaea **24**: 205
Paramarteilia orchestiae **39**: 112
Paramecium **31**: 194
Paramecium caudatum **21**: 164
Paramesopodopsis rufa **30**: 157; **39**: 216
Paramicrochiron **25**: 153
Paramisophria **33**: **51**, 68, 102, 402, **498**,
      501
  ammophila **33**: 102
  cluthae **33**: 102
  fosshageni **33**: 102
  galapagensis **33**: 102, **489**
  giselae **33**: 102
  itoi **33**: 102
  japonica **33**: 102
  ovata **33**: 102
  platysoma **33**: 102, 402, 501

reducta **33**: 102, **489**
rostrata **33**: 102
spooneri **33**: 102
variabilis **33**: 102
Paramoera mohri **39**: 204
Paramoera walkeri **39**: 204; **43**: 220
Paramola **32**: 223
Paramola japonica **32**: **163**, 176, 178
Paramphithoidae **32**: 463, 466
Paramunna **32**: **162**
Paramysis arenosa **39**: 216
Paramysis bacescoi **39**: 216
Paramysis bahamensis **39**: 216
Paramysis nouveli **39**: 216
Paramyxa paradoxa **25**: *145*, 146
Paranathura insignis **32**: **161**
Paranemertes peregrina **21**: 134
Paranotothenia magellanica **35**: 52
Paranthura **39**: 122
Paranybelinia **25**: 150
Paranymphon **24**: 11
   spinosum **24**: 14, 24, 41, 54, 64, 66, 69,
      72, 75
Parapagurus dimorphus **32**: 216
Parapagurus pilosimanus **32**: **446**, **448**;
      42: 216
Parapagurus sculptochela **32**: 216
Parapallene **24**: 8, 71
   algoae **24**: 71
   australiensis **24**: 16, 23, 66
   nierstraszi **24**: 11
Parapasiphae sulcatifrons **43**: 97, 98
Parapenaeopsis acclivirostris,
      Taxonomy **27**: 87, **109**
   affinis (*see Parapenaeopsis sculptilis*)
   arafurica, Taxonomy **27**: 86
      Zoogeography/Evolution **27**: **143**
   aroaensis, Taxonomy **27**: 86
      Zoogeography/Evolution **27**: **143**
   atlantica, Taxonomy **27**: 85, 86, 87
      Life histories **27**: 284, 286
      Zoogeography/Evolution **27**: **146**
   balli, Taxonomy **27**: 86
      Zoogeography/Evolution **27**: **145**
   cornuta, Taxonomy **27**: 85, 87
      Zoogeography/Evolution **27**: **143**
   coromandelica, Moulting/growth
      27: **230**
      Food/feeding **27**: **317**
   gracillima, Taxonomy **27**: 86
      Zoogeography/Evolution **27**: **143**
   hardwickii, Taxonomy **27**: 86, **109**

      Food/feeding **27**: **317**
      Life histories **27**: 286
      Moulting/growth **27**: **230**
      Physiology **27**: 203, **204**
      Zoogeography/Evolution **27**: **144**
   hungerfordii, Taxonomy **27**: 86
      Food/feeding **27**: **317**
      Moulting/growth **27**: **230**
      Zoogeography/Evolution **27**: **144**
   incisa, Taxonomy **27**: 87
      Zoogeography/Evolution **27**: **144**
      Life histories **27**: **304**
      Reproduction **27**: **270**, **279**
      Zoogeography/Evolution **27**: **143**
   nana, Taxonomy **27**: 86
      Zoogeography/Evolution **27**: **144**
   probata (*see Parapenaeopsis uncta*)
   sculptilis, Taxonomy **27**: **84**, 86, **109**
      Food/feeding **27**: **317**
      Moulting/growth **27**: **230**
      Zoogeography/Evolution **27**: **144**
   sinica, Taxonomy **27**: 87
   stylifera, Taxonomy **27**: 85, **109**
      Life histories **27**: 286, 314
      Moulting/growth **27**: **228**, **230**, **248**
      Physiology **27**: 203, **204**
      Reproduction **27**: **270**, **279**
      Zoogeography/Evolution **27**: **144**
   tenella, Taxonomy **27**: 87
      Food/feeding **27**: **317**
      Moulting/growth **27**: **230**
      Reproduction **27**: 260
      Zoogeography/Evolution **27**: **144**
   uncta, Taxonomy **27**: 85
      Zoogeography/Evolution **27**: 14
   venusta, Taxonomy **27**: 86
      Food/feeding **27**: **317**
      Zoogeography/Evolution **27**: **144**
Parapenaeopsis spp., Taxonomy **27**: 72
      Moulting/growth **27**: **228**
      Zoogeography/Evolution **27**: 141
Parapenaeopsis, Morphology **27**: 41, 42
      Food/feeding **27**: **317**, 321
      Life histories **27**: 286, **304**
      Moulting/growth **27**: **228**, **230**, 235,
         247
      Physiology **27**: 203, **204**
      Reproduction **27**: 255, 267, **270**, **279**
      Taxonomy **27**: 83–87, 102, 107, **109**,
         113, 119, *120*, 121, **123**
      Zoogeography/Evolution **27**: **143**, 150,
         154

Parapenaeus, Taxonomy **27**: 87–90, 92, **110**, **113**, 118, 119, 121, 122, **122**
  australiensis, Taxonomy **27**: *88*, 90
    forme *nodosa*, Taxonomy **27**: 90
    Zoogeography/Evolution **27**: **144**
  fissuroides, Taxonomy **27**: 90
    erythreaus, Taxonomy **27**: 90
    indicus, Taxonomy **27**: *90*
    Zoogeography/Evolution **27**: **144**
  fissurus, Taxonomy **27**: 89, **110**
    Food/feeding **27**: **317**
    Physiology **27**: 198
    Zoogeography/Evolution **27**: **144**
  investigatoris, Taxonomy **27**: 110, 114
    Zoogeography/Evolution **27**: **144**
  lanceolatus, Taxonomy **27**: 89, 90
    Zoogeography/Evolution **27**: **144**
  longipes, Taxonomy **27**: 89, **110**
    Moulting/growth **27**: 245, *247*, **248**
    Zoogeography/Evolution **27**: **144**
  longirostris, Taxonomy **27**: 90, 107, **110**, 113, 114
    Behaviour **27**: 336, 337, 340, 343, *344*, 350
    Food/feeding **27**: **317**
    Life histories **27**: 297
    Reproduction **27**: 255, 256, **258**, 263, **270**, *273*, 275, 276, **279**
    Zoogeography/Evolution **27**: **147**
  murrayi, Taxonomy **27**: 90
    Zoogeography/Evolution **27**: **144**
  perezfarfante, Taxonomy **27**: 89
    Zoogeography/Evolution **27**: **144**
  politus, Taxonomy **27**: 90
    Zoogeography/Evolution **27**: **146**
  ruberoculatus, Taxonomy **27**: 89
    Zoogeography/Evolution **27**: **144**
  sextuberculatus, Taxonomy **27**: 90
    Zoogeography/Evolution **27**: **144**
    Food/feeding **27**: **317**
    Life histories **27**: 287, 288
    Physiology **27**: 203
    Predation **27**: 370
    Reproduction **27**: 255, 267, **270**, **278**
    Zoogeography/Evolution **27**: 141, 150, 154
Parapercis **32**: 186
Parapercis dockinsi **32**: 173, 188, 224
Paraphorys vetulus **28**: 66
Parapinnanema alii **40**: 135
Parapinnanema rhipsoides **40**: 135
Parapinnanema ritae **40**: 135

Parapodinium **25**: 125, 126
Parapontella **33**: **52**, 95, 113
  brevicornis **33**: 113, 406, **491**, **514**
Parapontellidae **33**: **52**, 113
Parapseudes **39**: 159
Parapseudocyclops giselae **33**: 102
Parapseudomma **25**: 154
Parapsyche **35**: **163**
  cardis **35**: **169**
Parasagitta **32**: 302
Parasagitta elegans **32**: 265, 266, 274
Parascaphocalanus **33**: **53**, 81, 85, 135
  zenkevitchi **33**: 135
Parasellota **39**: 159
Parasesarma plicata **40**: 142
Parasicyonis ingolfi **32**: *101*, 105, 107, **126**, 130
Parasicyonis ingolfi **34**: **365**
Parasicyonis sp. **32**: 98, *100*, 108
Parastephos **33**: **53**, 87, 89, 137
  esterlyi **33**: 137
  occatum **33**: 137
  pallidus **33**: 137
Parastichopus nitropunctatus **21**: 147
Paratelphusa hydrodromus **29**: **133**, 137, 142, 143
Parathemisto **25**: 6, 137; **39**: 181, 182
  gaudichaudi **25**: 8, **31**, **34**, 36
  japonica **25**: 8, 21, 23, 30, **34**, 40, 47, 48
Parathemisto gaudichaudi **32**: 26; **39**: 121, 154, 169, 207; **20**: 53
Parathemisto gracilipes **39**: 207
Parathemisto japonica **39**: 207
Parathemisto pacifica **39**: 207
Parathunnus obesus **30**: 335
Paraugaptiloides **33**: **51**, 74, 102
  magnus **33**: 102, **498**
Paraugaptilus **33**: **51**, 71, 74, 102, 267
  archimedi **33**: 102
  bermudensis **33**: 102
  buchani **33**: 102
  indicus **33**: 102
  magnus **33**: 102
  meridionalis **33**: 102
  mozambicus **33**: 102
  similis **33**: 102
Parayoldella **32**: 371
Parayoldiella **42**: 41
Parazoanthus **30**: 338
Parborlasia corrugatus **35**: 33
Pardalisca endeavouri **34**: **377**

Parechinus angulosus **28**: 425; **35**: **185**
Parerythrops **25**: 153
Pareucalanus **33**: 122
Pareuchaeta **33**: 236, 515
  behaviour **33**: 405
  distribution **33**: 517
  ecology **33**: 486, 492, 494, **500**, 526, 528
  morphology **33**: 22, 44
  nutrition **33**: 140, 160, 166, 174, **206**
  reproduction **33**: 258, 262, *263*, 263, 266,
     267, **275**, *277*, *279*, *281*, 294
  size **33**: *332*
  taxonomy **33**: **53**, *61*, 61, 63, 92, 130, 131
Pareuchaeta abbreviata **33**: 131, **275**
  abrikosovi **33**: 131
  abyssalis **33**: 131
  abyssaloides **33**: 131
  aequatorialis **33**: 131, **514**
  affinis **33**: 131
  alaminae **33**: 131
  altibulla **33**: 131
  anfracta **33**: 131
  antarctica **33**: 131
    behaviour **33**: **417**
    biochemistry **33**: 247
    ecology **33**: **360**, 492, 524
    nutrition **33**: 153, 162, **188**
    reproduction **33**: 263, **284**
    size/weight **33**: **224**, **231**, **234**
  austrina **33**: 131
  barbata **33**: 131, **275**, 444, **514**
  biloba **33**: 131, 409, **417**
  birostrata **33**: 131, **275**
  bisinuata **33**: 131, **275**
  bradyi **33**: 131
  brevirostris **33**: 131
  bulbirostris **33**: 131
  californica **33**: 131
  calva **33**: 131
  comosa **33**: 131, **514**
  confusa **33**: 131, **275**, **514**
  copleyae **33**: 131
  dactylifera **33**: 131
  elongata **33**: **26**, 131
    ecology **33**: 372
    growth/development **33**: **307**, **343**
    nutrition **33**: 163, 165, 166, **188**, 188
    reproduction **33**: **275**, 281
    size/weight **33**: **224**, **231**
  eminens **33**: 131
  erebi **33**: 131, 262, 263
  euryhina **33**: 131

exigua **33**: 131
flava **33**: 131
glacialis **33**: 131, **275**, 492
gracilauda **33**: 131, **514**
gracilis **33**: 131, **224**, 263, **275**, **360**
grandiremis **33**: 131
guttata **33**: 131
hanseni **33**: 131, **275**
hastata **33**: 131
hebes **33**: 131, **224**, **275**, **360**, **514**
implicata **33**: 131
incisa **33**: 131
investigatoris **33**: 131
kurilensis **33**: 131, **275**
longisetosa **33**: 131
malayensis **33**: 131, 518
megaloba **33**: 131
mexicana **33**: 131
modesta **33**: 131
norvegica **33**: **7**, 131
  behaviour **33**: **417**, 419, *441*, 441, 444,
    447
  distribution **33**: 504, **514**
  ecology **33**: 349, **350**, 351, **360**, 363,
    364, *365*, 468, 470, 478, 483, 492,
    522
  growth/development **33**: 315, 321
  life cycle **33**: 347
  morphology **33**: *16*, *22*, 30, 31, **32**, 36,
    37, 39
  nutrition **33**: 140, 146, 163, **188**, **201**
  reproduction **33**: 44, 46, *258*, *259*, 259,
    260, 263, 265, **275**, 279, 280, **284**,
    294
oculata **33**: 131
orientalis **33**: 131
papilliger **33**: 131
parabreviata **33**: 131
paraprudens **33**: 131
parvula **33**: 131
pavlovskii **33**: 131
perplexa **33**: 131
plaxiphora **33**: 131
plicata **33**: 131
polaris **33**: 131
prima **33**: 131
propinqua **33**: 131
prudens **33**: 131
pseudotonsa **33**: 131, **360**, **514**
rasa **33**: 131, **275**
regalis **33**: 131
ribicunda **33**: 131

robusta **33**: 131
rotundirostris **33**: 131
rubra **33**: 131, **275**
russelli **33**: **26**, 131, 251, **275**, 321
sarsi **33**: 131, **275**, **514**
scaphula **33**: 131
scopaeorhina **33**: 131
scotti **33**: 131, **275**, 444, **514**
sesquipedalis **33**: 131
sibogae **33**: 131
similis **33**: 131, 262, 263
simplex **33**: 131, **275**
subtilirostris **33**: 131
tonsa **33**: 37, 131, **514**
tridentata **33**: 131
triloba **33**: 131
tuberculata **33**: 131, **514**
tumidula **33**: 131
tycodesma **33**: 131
vervoorti **33**: 131
vorax **33**: 131, **514**
weberi **33**: 131
Pareuchaeta elongata **32**: 265, 266, 274
Pareuchaeta glacialis **32**: 19
Pareuchaeta japonica **32**: 266
Pareuchaeta norvegica **32**: 22
Pareuchaeta polaris **32**: 19
Pareuchaeta spp. **32**: 29, 303
Parharpinia rotundifrons **39**: 204
Parhyalella basrensis **39**: 204
Parhyalella roperi **39**: 204
Pariambus typicus **25**: 156
Parkiidae **33**: **53**, 132, **500**
Parkius **33**: **53**, 132
  karenwishnerae **33**: 132, **500**
Paronatrema **25**: 148
Parophrys vetulus **31**: 8
Parophyrus vetulus **26**: 142
Parougia wolfi **34**: **366**
Parribacus antarcticus **29**: **132**, 164
Parthenope (Platylambrus) mironovi
  32: **163**
Parthenope mironovi **32**: 217
Parthenope stellata **32**: 217
Parundinella **33**: **53**, 82, 84, 85, 87, 90, 138
  dakini **33**: 138
  emarginata **33**: 138, **501**
  manicula **33**: 138
  spinodenticula **33**: 138
Parupeneus multifasciatus **44**: 226
Parupeneus porphyreus **44**: 218, 223
Parvimysis bahamensis **39**: 161

Parvocalanus **33**: **52**, 77, 120, 121, 497, 528
Parvocalanus **40**: 130
  crassirostris **33**: **26**, 37, 121
    behaviour **33**: 445
    ecology **33**: 473
    reproduction **33**: 267, 294
  dubia **33**: 121
  elegans **33**: 121
  latus **33**: 121
  scotti **33**: 121
Parvocalanus crassirostris **44**: 88, 92, 93
Pasiphaea **25**: 136
Pasiphaea americana **32**: **162**, 216
Pasiphaea flagellata **32**: **162**
Passer domesticus **20**: 231
Patella **23**: 19, 20, 21; **25**: 200, 221; **28**: 397,
    398, *399*, 418
  caerulea **34**: 22
  depressa **34**: 13
  longicosta **23**: 22, 113; **34**: 48
  ulyssopensis (P. aspera) **34**: 13
  vulgata **34**: 4, **6**, 8, 9, *10*, *11*, 12, *12*, 13,
    19, 21, 28, *29*, *30*, 33, 34, 37, *38*,
    40, **41**, 44, *47*
Patella vulgata **42**: 200
Patinapta ooplax **21**: 147
Patinigera polaris (Nacella concinna) **24**: 46
Patinopecten caurinus **35**: *175*, **184**
Patinopecten yessoensis **21**: 14
Patinopecten yessoensis **20**: 310, 322–326
Patirella calcar **31**: 18
Patirella exigua **31**: 18
Patirella gunnii **31**: 18
Patiriella calcar **21**: 144, 154
Patuki roperi **39**: 204
paulensis, Taxonomy **27**: 98
    Food/feeding **27**: 331
    Reproduction **27**: 260, 261
    Zoogeography/Evolution **27**: 139, **146**,
      147, **148**, 157
Pavlova lutherii **41**: 199
Pavona **43**: 302
  P. decussata **43**: 290
Pavona decussata **22**: 5
Pavona obtusata **21**: 106, 127
Payosphaeria minuta **40**: 119
pea crab (*Pinnotheres halingi*) **41**: *189*
Pecten albicans **21**: 14; **20**: 325
Pecten irradians **22**: 103, 106
  maximus **22**: 103, 125, 134
Pecten maximus **30**: 230; **34**: 211

Pecten maximus **37**: **10**, 48, 64, **93**, **94**;
    43: 145
Pectenogammarus longimanus **39**: 204
Pectenogammarus planicrurus **39**: 204;
    43: 140
Pectinaria **29**: 252; **34**: 44; **35**: **169**
Pectinaster sp. **32**: **165**, 180
Pectinidae **32**: 468, 470
Pedinella tricostata Rouchijajnen **29**: 78
Pedinomonas **25**: 124
   noctulicae **25**: 124
   symbiotica **25**: 124
Pedinomonas micron **29**: 79
Pedumispora **40**: 119
Pegea bicaudata **44**: 22, 31, 43, 133
Pegea confederata **44**: 22, 133, 134
Pelagia **23**: 225
Pelagia colorata **21**: 106
   noctiluca **21**: 106
Pelagia noctiluca **24**: 208, 209; **26**: 145
Pelagia sp. **32**: 72
Pelagobia longicirrata **33**: 160
Pelagodiscus **28**: 276
Pelagodiscus atlanticus **32**: **165**, 357, 360,
    495, 503
Pelagopenaeus, Taxonomy **27**: 91, **110**, **113**
   balboae, Taxonomy **27**: 91, **110**
    Life histories **27**: 287
    Zoogeography/Evolution **27**: 150
Pelagophycus porra **23**: 5
Pelagothuriidae **32**: 97
Pelates quadrilineatus, Predation **27**: **365**
Pelecanoides georgicus **20**: 295
   urinatrix exsul **20**: 295
Pelecanus occidentalis **20**: 260
Pelichnobothrium **25**: 151
Pelliciera **40**: 92
Pelliciera rhizophoreae **40**: 95
Pellicieraceae **40**: 93, 95
Pellona ditchela **24**: 288
Pellona flavipinnis **20**: 180
Pellonula **20**: 160
Peltospira **34**: 417
   delicata **34**: **370**
   lamellifera **34**: **370**
   operculata **34**: **370**
   sp.1 **34**: **370**
   sp.2 **34**: **370**
   sp.3 **34**: **370**
Peltospira sp. **32**: **127**
Pelvetia **23**: 129, 133, 139, 140, 141

canaliculata **23**: 26, 108-110, 123, 124,
    136-138, 141
   fastigiata **23**: 5
Pemphis acidula **40**: 95, 99
Pemphis madagascariensis **40**: 95
Penaeopsis, Taxonomy **27**: 88, **110**, **113**
   Life histories **27**: 287
   Physiology **27**: 203
   Zoogeography/Evolution **27**: 141, 150,
    154
   balsii, Taxonomy **27**: 93
    Zoogeography/Evolution **27**: **144**
   challengeri, Taxonomy **27**: 93
    Zoogeography/Evolution **27**: **144**
   eduardoi, Taxonomy **27**: 92
    Zoogeography/Evolution **27**: **144**
   jerryi, Taxonomy **27**: 92, 93
    Zoogeography/Evolution **27**: **144**
   megalops (*see Penaeopsis serrata*)
   rectacuta, Taxonomy **27**: 93, **110**
    Zoogeography/Evolution **27**: **144**
   serrata, Taxonomy **27**: 92,
    Zoogeography/Evolution **27**: **146**, **147**
Penaeus **29**: **192**; **38**: 111
   aztecus **29**: 131, 171
   duorarum **29**: **131**, 171
   indicus **29**: **131**, 169, 171
   japonicus **29**: 131, 191, **192**
   kerathurus **29**: **131**, 170, 187
   monodon **29**: **131**, 170, 191, 192
   occidentalis **29**: 169, **172**, **173**
   orientalis **29**: **132**, 169
   pencillatus **29**: **132**, 191, **192**
   schmitti **29**: **132**, 169, **172**, **173**, **192**, 193
   setiferus **29**: **132**, 169, *170*, 171, **172**, **173**,
    190, 191, **192**, 193, 194
   stylirostris **29**: **132**, 169, **172**, **173**,
    192, 193
   vannamei **29**: **132**, **172**, **173**, **192**, 194
Penaeus aztecus **44**: 225
Penaeus aztecus, Morphology **27**: 19, 33, 54
    Behaviour **27**: 334, 335, *335*, 336, 337,
     338, 340, 343, 346, 347, 350, 351
    Food/feeding **27**: **317**, 318, 320, 321,
     322, 323, 327, 330
    Life histories **27**: 294, 296, **304**, 306,
     308, 309, 310, 311, **313**
    Moulting/growth **27**: **228**, **230**, 233,
     234, 237, **238**, 246, **249**
    Parasites **27**: 380, 383, 384, 385, 386,
     387

Physiology **27**: 162, 163, 168, **168**, 171, 181, 200, **204**, 209, 210
Predation **27**: 358, 370, 374, 375, 376
Reproduction **27**: **257**, 264, 265, **270**, **279**
Taxonomy **27**: 98, **110**, 124, 125
Zoogeography/Evolution **27**: 141, **146**, 147, 157
Penaeus indicus **40**: 137
Penaeus merguiensis **40**: 137, 138, 187
Penaeus monodon **40**: 137, 138
Penaeus setiferus **34**: 146
Penaeus spp., Morphology **27**: 40, 41
  Life histories **27**: **304**, **305**
  Moulting/growth **27**: *236*, 237, 241
  Physiology **27**: 168, 173, 177, 184, 190, 198, 203, 205
  Predation **27**: 367
  Reproduction **27**: 253
  Taxonomy **27**: **112**
  Zoogeography/Evolution **27**: 138, 155
Penaeus stylirostris **21**: 87
Penaeus subgenus, Taxonomy **27**: 95
Penaeus, Morphology **27**: 7, *8*, *9*, 37, *39*, 41, 42
  Food/feeding **27**: **317**, 321
  Life histories **27**: 283, 284, 285, 289, 290, 292, **313**
  Moulting/growth **27**: *216*, **228**, **230**, 235, 236, 239, 246, *247*
  Physiology **27**: *189*, **204**
  Predation **27**: 359, 360, 370
  Reproduction **27**: 253, *257*, 267, 268, **270**, 277, **279**, **281**
  Taxonomy **27**: 69, 79, 91, 93–98, 107, **110**, **113**, 118, 119, *120*, 121, 123, 124, 125
  Zoogeography/Evolution **27**: 150, 154
Pendanthura tanaiformis **39**: 212
Peneidae **32**: 62
Penguin
  Adélie (*Pygoscelis adeliae*) **43**: 227, 243
  chinstrap (*Pygoscelis antarctica*) **43**: 227, 229
  emperor (*Aptenodytes forsteri*) **43**: 228, 243
Peniagone **32**: 366
Peniagone heronardi **32**: 349
Peniagone sp. **32**: **461**
penicillatus, Morphology **27**: 44
  Food/feeding **27**: **318**
  Moulting/growth **27**: **240**

Physiology **27**: 160, **204**, 205, 206
Predation **27**: 373
Reproduction **27**: 253, 260
Taxonomy **27**: 97, **112**
Zoogeography/Evolution **27**: **144**
Penicillium **40**: 121
Penilia **31**: 84, 86, 90, 92, 93, 98, 99, 102, 107–109, 116, 121, 129, 130, 132, 150, 151, 152
Penilia avirostris **25**: 121; **31**: 85, 90, 93, *100*, 101, 106, *110*, *117*, 122, 126, *127*, 130, *131*, **133**, 136, *136*, 137, 138, 139, 140, 141, 145, 146, 147, 149
Pennaria tiarella **21**: 105
Pennatula phosphorea **34**: 223
Pennatularia **32**: *333*, 373, 468, 470, **483**, 500, **504**
Pentaceros **32**: 186
Pentaceros decacanthus **32**: 224
Pentaceros quinquespinis **32**: 146, **173**, 188, 192, **192**, 193, **195**, 211, 227
Pentaceros richardsoni **30**: 331; **35**: **42**, 43, 103
Pentacerotidae **32**: **173**
Pentacolossendeis **24**: 5, 26
Pentacta australis **21**: 146
Pentaganglionata **42**: 103
Pentalalamina corona **43**: 234
Pentanymphon **24**: 26
  antarcticum **24**: 54
Pentapalaeopycnon inconspicua **24**: 77
Pentapora foliacea **34**: 223
Pentapycnon **24**: 25
  charcoti **24**: 31
Pentidotea **39**: 158
Peracarida **39**: 105–260
Peramphithoe stypotrupetes **39**: 151
Perca
  flavescens **25**: 5, *13*, 40, 58, 59, 64, *65*
  fluviatilis **25**: 9
Perca flavescens **26**: **120**, 147; **28**: 95
Percichthydae **32**: **172**
Percophidae **32**: 167, **173**, 211
Perezella **25**: 137
  pelagica **25**: 138
Pericaryon **25**: 137
  cesticola **25**: 137
Periclimenes alcocki **32**: **162**
Periclimenes imperator (crustacean)**41**: *189*
Periconia prolifica **40**: 122
Peridinium **23**: 232; **25**: 122; **20**: 27

Peridinium polonicum **21**: 73
  trochoideum **21**: 73
Peridinium trochoidum **33**: 164
Perioculodes longimanus **39**: 204
Periophthalmodon schlosseri **40**: 154
Periophthalmus **40**: 154
Periphylla **24**: 199, 200; **25**: 152
  periphylla **24**: 52
Periphylla hyacinthina **32**: 72
Periplaneta **30**: 103
Periwinkle (*Littorina saxatilis*) **43**: 44
Periwinkles *see Littorina* spp.
Perkinsiella amoebae **25**: 122, *123*
Perkinsus marinus **43**: 65, 66
Perlophiura profundissima **32**: 463, 506
Perna canaliculatus, Food/feeding
    27: 327
Perna canaliculus **31**: 34
Perna perna **22**: 103, 110, 165, 168
Perna viridis **35**: **184**
Perotrochus **32**: 410
Perotrochus africanus **32**: **409**
Perotrochus amabillis **32**: **409**
Perotrochus atlanticus **32**: **409**
Perotrochus hirasei **32**: **409**
Perotrochus lucaya **32**: **409**
Perotrochus midas **32**: **409**
Perotrochus quoyanus **32**: **409**
Perotrochus teramachii **32**: **409**
Pertusiconcha **42**: *144*
Peruvian anchoveta *see Engraulis ringens*
Perviata **32**: 408
Pestalotiopsis agallochae **40**: 121
Pestalotiopsis versicolor **40**: 121
Petalophthalmidae **39**: 160
Petalotricha ampulla **25**: 120
Petasometra helianthiodes **21**: 152
Petrel (Procellarea) **43**: 207
Petrocelis middendorfii **23**: 6
Petromyzon maninus **40**: 256, 257, 272, 274,
    276, 280, 281, 299
Peyssonelia **32**: **160**, 176
Pfiesteria piscicida **31**: 318
Phaenna **25**: 138; **33**: **53**, 78, 132, 486
  spinifera **33**: 132, *422*
  zetlandica **33**: 132
Phaennidae
  ecology **33**: 470, **500**, 502
  morphology **33**: 28, 29
  nutrition **33**: **152**
  taxonomy **33**: **53**, 132, 133

Phaeocystis **31**: 302, *303*, 307, 309, 324,
    344, 352
Phaeocystis **33**: 528
  puchetii **33**: 155, 156
Phaeocystis antarctica **43**: 198, 237, 238
Phaeodactylum cornutum **33**: 290
Phaeodactylum tricorntum **38**: 159, 161;
    29: *38, 39*; **37**: 66
Phaeolidae **42**: *5*
Phaeophyta *Plate 4* **43**: 281, *283*
Phakellia **21**: 104
Phalacrocorax aristotelis **25**: *3*; **34**: 268;
    20: 282
  auritus **20**: 245
  bougainvillii **20**: 260
  capensis **20**: 247
  penicillatus **20**: 61, 241
Phalacrocorax pelagicus (cormorant) **39**: 69
Phalacrostemna **32**: **161**
Phalocrocorax pencillatus **39**: 279
Phanerozonia **32**: 333, 335
Phascolion lutense **32**: **460**
Phascolion strombii **42**: 215
Phelliactis **22**: 74
  robusta **22**: 74, 78, 84
Phelliactis robusta **35**: 52
Phenacoscorpius eschmeyeri **32**: **172**
Pheronema naskaniensis **32**: **160**, 184, 216
Pheronema placoideum **32**: 216
Pheronema semiglobulosum **32**: 216
Pheronematidae **32**: 97
Phialidium **25**: *28*, 32, 33, 156
Philine **28**: *393*, 412, 419, 421; **42**: 86, *89*,
    95, 97, 104
  P. aperta **42**: 72, *73*
Philinidae **42**: 76, 86, 87, 88, 106, 109,
    110, 129
Philinoglossa **42**: 105
Philinoidea **42**: 76, 82, 83, 84, 85, 98, 106,
    115, 118
Philometra **25**: 151
Philonicus (*see Pleoticus*)
Phiurases sp. **32**: **166**
Phobetria palpebrata **39**: 281
Phoca
  vitulina **34**: 270
  vitulina vitulina L. **34**: 275
Phoca borealis (harbour seal) **38**: 175
Phoca groenlandica (harp seal) **38**: 175
Phoca hispida (ringed seal) **38**: 175
Phoca vitulina **39**: 60–63, 72, 76, 77, 80
Phocaena phocaena **39**: 280

Phocarctos hookeri **39**: 279
Phocascaris **40**: 8, 26, 31, 43
Phocides pigmalion **40**: 178
Phocoena phocoena **34**: 275
Phocoenoides dali **32**: 267, 270, 282, 283, 285
Phoenix paludosa **40**: 101
Pholas **25**: 203; **26**: 27
Pholis laeta **39**: 48, 49
Pholoe **32**: **161**
Phoma **40**: 122
Phomopsis mangrovei **40**: 119
Phormosoma bursarium **21**: 145, 150, 169
Phoronida **43**: *21*
Photichthyidae **32**: **169**
Photostylus pycnopterus **32**: **169**
Phoxichilidium **24**: 26, 38
  femoratum **24**: 19, 28, 33, 39, 40, 41, 43, 44, 46, 54
  maxillare **24**: 39
Phoxiphilyra **24**: 26
Phoxocephalidae **39**: 36
Phragmites, Food/feeding **27**: 318, 319
Phronima **25**: 141, 154, 155, 156
Phronima **39**: 121, 182, 183
  atlantica **25**: *155*
  curvipes **25**: 156
  sedentaria **25**: 156
Phronima atlantica **39**: 207
Phronima sedentaria **39**: 207
Phronimella **25**: 141
Phrosina **25**: 141
Phrynorhombus norvegicus **28**: **118**
Phthirusa maritima **40**: 177
Phthisica marina **39**: 175
Phyllactis **22**: 90
  flosculifera **22**: 91
Phyllidia varicosa **21**: 100
Phyllidia varricosa **34**: 33
Phylliroe **25**: 157
  bucephalum **25**: 157
Phyllobothrium **25**: 151; **32**: 308
Phyllobothrium loliginus **40**: 7, 13, 21
Phyllobothrium sp. **32**: 306, 308
Phyllodocidae **32**: 445
Phyllopodidae **33**: **51**, 109
Phyllopus **33**: **51**, 71, 74, 109, 486
  aequalis **33**: 109
  bidentatus **33**: 109
  giesbrechti **33**: 109
  helgae **33**: 109
  impar **33**: 109

integer **33**: 109
mutatus **33**: 109
muticus **33**: 109
Phyllorhiza punctata **44**: 122
Phyllospadix **23**: 109
  scouleri **23**: 30, 133
Phylophora **23**: 46
Phymorhynchus
  moskalevi **34**: **373**
  sp.2 **34**: **373**
  sp.3 **34**: **373**
  sp.4 **34**: **373**
  starmeri **34**: **373**
  waréni **34**: **373**
Phymorhynchus moskalevi **32**: 99, *100*, 107, **127**, 130, 135
Phymorhynchus sp. **32**: 102, 103, 105, 108, **127**, 130, 131
Physa acuta **34**: 37, **41**
Physa gyrina **35**: **184**
Physalia **21**: 104, 109, 110, 114, 115, 130, 132; **24**: 100, 108, 174–177, 183, 185, 186, 196, 200, 222; **25**: 5
  physalis **21**: 105, 108–110, 112, 114–116, 120, 130; **24**: 124, 126, 154, 155, 203, 205, 207–209, 215–218, 221–223, 225, 226, 229, 244; **25**: **31**, 33, **35**, 45
  utriculus **21**: 105
Physalia physalis **28**: **88**
Physeter catodon **39**: 280, 281; **20**: 271
Physeter macrocephalus **39**: 281
Physiculus hexacytus **32**: **170**, **194**, 206
Physiculus longicavis **32**: **170**, 188, **194**
Physiculus parini **32**: **170**
Physiculus sazonovi **32**: **170**
Physiculus sp. **32**: 167, **170**, 186, 191
Physobrachia douglasi **21**: 106, 132
Physophora **24**: 173, 184, 200
  hydrostatica **24**: 113, 115, 126, 178, 186, 198, 213, 217
Phytoplankton
  pigmentation in coral reef organisms **43**: 280, 308
  Southern Ocean pack ice **43**: 197, 198, 204, 218, 222, 238, 239
Pictodentalium **42**: *143*
Pictothiris picta **28**: *351*
Pigeon guillemot *see Cepphus columba*
Pigrogomitus **24**: 77
  timsanus **24**: 66, 69

Pilarella **33**: **51**, 73, 102
    longicornis **33**: 102, **498**
Pilayella littoralis **28**: *79*
Pimephales pomelas **30**: 261
Pimephales promelas **31**: 192
Pimpehales promelas **22**: 103, 175
Pinnipedia **40**: 26
Pinnipeds **43**: 47
Pinnotheres (crustacean)
    P. deccacensis **41**: *189*
    P. halingi **41**: *189*
    P. semperi **41**: *189*
Pipadentalium **42**: *143*
Piperaceae **40**: 128
Pisaster **23**: 77
    ochraceus **23**: 73
Pisaster ochraceus **43**: 136
Pisces **32**: **452, 453, 454**; **40**: 149–154
    Elasmobranchii **40**: 25
    Teleostei **40**: 25–27
Piscinoodinium **25**: 127, 134
Pisidium **35**: **169**
Pisonidae **29**: 195
Placida **42**: *116*, 133, 135
Placocalanus **33**: **51**, 66, 101, **498**, 501
    brevipes **33**: 101
    inermis **33**: 101
    insularis **33**: 101
    longicauda **33**: 101
    nannus **33**: 101
Placomium **30**: 113
Placopecten magellanicus **34**: 14; **35**: **168**,
    175, **184**; **43**: 101, 145, 149
Placozoa **43**: *21*
Plagiodinium belizeanum **40**: 124
Plagiogeneion **32**: 186, 218
Plagiogeneion geminatum **32**: **173**, 186,
    211, 218
Plagiogeneion unispina **32**: **173**, 208, 218
Plagiogenion rubiginosus **30**: 331
Plagioglypta **42**: *143*, 153
    P. iowensis **42**: 216, 217
Plaice *see Pleuronectes platessa*
Plakobranchoidea **42**: 80, 88, 112, 115
Plakobranchus **42**: *116*, 133, 135
Planaria gonocephala **21**: 133
    lugubris **21**: 133
Plankton
    see also Phytoplankton; Zooplankton
    pigmentation in coral reef organisms
        43: 280, 308

Southern Ocean pack ice **43**: 183, 197,
    198, 204, 218, 220–222, 231, 238,
    239, 245
Planktonetta atlantica **25**: 128
Planktoniella **32**: 560
Planktoniella sol **32**: *531*, 535, *551*, 554, 555
Planorbidella
    depressa **34**: **370**
    planispira **34**: **370**
Planorbis **21**: 158, 164
Planorbis contortus **34**: 18, 22, 37, **41**
Planulina ariminesis **30**: 53
Platenista gangetica **40**: 157
Platichthys flesus (flounder) 10, 18, 25, 54
Platichthys flesus **24**: 274; **25**: 3, 17, 19, *20*,
    21, 22, 29, 32, 47, 50; **26**: *87*, 139;
    28: *65*, *88*, 90; **30**: 220; **40**: 15;
    43: 68
    stellatus **24**: 274
Platichthys stellatus **31**: 8
Platichthys stellatus × *Kareius*
    bicoloratus **31**: 20
Platidia **28**: 229, 275
Platidia anomoides **32**: **165**
Platycephalidae **32**: **172**
Platychthys × *Pleuronectes* **31**: 27
Platycopia **33**: **51**, 76, 77, 99, **498**
    inornata **33**: 99
    orientalis **33**: 99
    perplexa **33**: 99
    pygmaea **33**: 99
    robusta **33**: 99
    sarsi **33**: 99
    tumida **33**: 99
Platycopiidae **33**: 49, **51**, 99, **489, 498**
Platycopioida **33**: 1, 2–3, *3*
    taxonomy **33**: **2**, **51**, *54*, *55*, 98, 99
Platycopioidea **33**: 99
Platygyra **22**: 40, 48
Platyhelminthes **43**: 7, 20, *22*
Platypodia granulosa **21**: 140, 141
Platytroctidae **32**: **169**
plebejus, Morphology **27**: 13, 4, *49*, 50, *51*,
    53
    Behaviour **27**: 336, 338, 340, 349, 352
    Food/feeding **27**: 322, 323, 326
    Life histories **27**: 297, 302, **305**, 309,
        310, 312, **313**
    Moulting/growth **27**: 214, **231**, **250**
    Parasites **27**: 380
    Physiology **27**: 191, **204**, 205, *205*,
        206, 208

Predation **27**: 373, 374, 375
Reproduction **27**: 260, 264, **271**, 273, **280**
Taxonomy **27**: 95, **112**, **116**
Zoogeography/Evolution **27**: 139, **144**, 147, **148**, 154, 155, 157
Plecoglossus altivelis **26**: **74**, **92**, 103
Plectorhynchus goldmani **44**: 231
Plectospira **39**: 184
Plectranthias **32**: 186
Plectranthias exsul **32**: **172**, 224
Plectranthias parini **32**: **172**
Plectrogenium barsukovi **32**: **172**
Plectropomus leopardus **44**: 216, 218, 223
Pleistophora **24**: 278, 291, 293; **40**: 4, 10, 30, 50
Pleistophora crangoni **38**: 180
Pleistophora gadi **40**: 4, 10, 12–13, 30
Pleistophora, Parasites **27**: 3*84*
Pleopis **31**: *84, 85, 86*, 102
Pleopis intermedius **31**: 130
Pleopis polyphemoides **31**: 84, 86, 90, 91, 92, 93, 101, 102, 103, 107, 111, 112, *113*, 115, 119, 122, 128, 130, 133, *136*, 138, 139, 140, 141, 142, *143*, 145–148
Pleopis schmackeri **31**: 84, 86, 133, 139
Pleoticus, Taxonomy **27**: 64
Plerocercoides **25**: 150
Plesiastrea **32**: 158
Plesionika aff. *williamsi* **32**: **163**
Plesionika edwardsii **32**: **162**
Plesionika ensis **32**: **162**
Plesionika martia **32**: **162**
Plesionika ocellus **32**: **162**
Plesionika sp. **32**: **129**, 166
Pleuragramma **24**: 334
 antarcticum **24**: 328, 333, 336
Pleuragramma antarcticum **43**: 220, 226, 229
Pleurehdera **42**: 114, 132
Pleurobema coccineum **22**: 103, 109, 148
Pleurobrachia **44**: 38; **25**: *6*, 149, 150; **20**: 53
 bachei **25**: 5, *28*
 pileus **25**: 30, 39, 43, 47, 149
Pleurobrachia bachei **44**: 129
Pleurobrachia pileus **32**: 40, *42*
Pleurobranchacae *see* Notaspidea
Pleurobranchaea **42**: 112, 113, *114*, 132
Pleurobranchaeidae **42**: 81, 88
Pleurobranchella **42**: 112, 113, *114, 116*, 132, 133, 135

Pleurobranchidae **42**: 81, 89, 96, 109, 113, 115, 129
Pleurobranchoidea **42**: 81, 112, 113
Pleurobranchomorpha **42**: 80
Pleurobranchus **42**: *114*, 115, *116*, 133,135
Pleurocapsales **42**: 212
Pleurocoela **42**: 110, 118
Pleurocoeles **42**: 118
Pleurocope **32**: **162**
Pleurogonium **32**: **162**
Pleuromamma **25**: 138; **33**: 203
 behaviour **33**: 407, 428, 434, 437, 438, 445, **446**
 ecology **33**: 363, 467, 486, 528
 morphology **33**: 17, 30, 36
 physiology **33**: 217
 reproduction **33**: 257
 taxonomy **33**: **51**, *56*, 56, 61, *62*, 62, 67, 109
Pleuromamma abdominalis **33**: **7**, **26**, **32**, 56, 62, 109
 borealis **33**: *56*, 109, **435**, 437, 438, **514**
 gracilis **33**: **7**, *56*, 109
 behaviour **33**: **412**, **426**, **435**, 438
 distribution **33**: **514**
 ecology **33**: **361**
 physiology **33**: 217
 weight **33**: **224**, **231**
 indica **33**: *56*, 109, **435**, **514**
 piseki **33**: *56*, 109, **231**, **435**, **514**
 quadrungulata **33**: 56, 109, **435**, **514**
 robusla **33**: **7**, *56*, 56, 109
 behaviour **33**: **417**, **435**, 444
 distribution **33**: **514**
 ecology **33**: 468, 522
 growth/development **33**: *315*, 315
 morphology **33**: 32, *33*, **34**
 size/weight **33**: **224**
 scutullata **33**: *56*, 56, 109, 407
 behaviour **33**: **412**, **417**, **426**, **435**
 distribution **33**: **514**
 ecology **33**: **485**
 size/weight **33**: **224**, **231**
 wolfendeni **33**: 109
 xiphias **33**: **7**, 37, 48, **56**, 109
 behaviour **33**: 409, **412**, 421, 422, **435**
 distribution **33**: **14**
 ecology **33**: 363, 364
 nutrition **33**: **194**
 physiology **33**: 217
 size/weight **33**: **231**
Pleuromamma spp. **32**: 29

Pleuroncodes planipes **29**: **133**, 153
Pleuronechthys decurrens **39**: 279
Pleuronectes
  americanus **30**: 229
  platessa **30**: 220
Pleuronectes americanus (winter
    flounder) **38**: 8, 54
Pleuronectes asper (yellowfin sole)
    39: 58–60
Pleuronectes bilineatus (rock sole)
    39: 58–60
Pleuronectes flesus **26**: **121**, 270
Pleuronectes hippoglossus **26**: 3
Pleuronectes platessa (plaice) **38**: 4, *16*, 54
  effect of fertilizer on **38**: 18, 19
  egg production **38**: 6
  North Sea and Baltic (1893–1990) 11–13
  rearing in UK (1957–1967) 14–16
  Scotland (1894–1920) 8, 9
Pleuronectes platessa **44**: 253, 254, 255;
      25: 3, *13*, 14, 17, 19, *20*, 21, 22, 29,
      32, *32*, 33, 47, 50, 59, 60, *60*, 64;
      26: **121**, 133, 138, 141, 152, 251,
      270; **28**: 19, 24, 64, 65, 67, *88*, **94**,
      108, **110**, 128; **29**: 216, 238, 252,
      254, 267, 272, 279, 282, 283, 295,
      296, 324, 334; **34**: 211; **40**: 57
Pleuronectidae **40**: 27
Pleuroprion **39**: 159
Pleurosapsaceae **42**: 212
Pleurosigma **40**: 124
Pleurotomaria diluculum **32**: **409**
Pleurotomariacea **32**: 409, **409**
Pleurotomariidae **32**: 409, **409**
Pleurotomarioidea **42**: 237–294
  see also Pleuromarioidean gastropods *in
    subject index*
Pleurotomella (Anomalotomella) minuta
    32: **164**
Plexechinus **32**: 366
Plinthaster sp. **32**: **165**, **166**, 178
Plistophora **25**: 144
Plocamium **21**: 93
  cartilagineum **21**: 93
Plumbaginaceae **40**: 93
Plutonaster bifrons **43**: 147
Pneumoderma **25**: 153
Pneumodermopsis **25**: 153
Pocillopora **22**: 10, 11, 14–16, 42; **24**: 60;
      31: 230, 246, 256; **32**: **159**
  damicornis **22**: 14, 15, 18, 23, 26, 28, 31,
      33, 41, 42, 47

  danae **22**: 11
  eydouxi **22**: 31, 54
  meandrina **22**: **10**, 11, 14, 15, 31, 33
Pocillopora damicornis **31**: 248, 252, 258,
      260, 271
Pocillopora Plate 4 **43**: 302, *305*
  P. damicornis **43**: 149, 150, 288
Pocillopora verrucosa **31**: 253, 254
Pocilloporidae **43**: 287, 288
Podarkeopsis brevipalpa **26**: **179**
Podocidaris sp. **32**: **165**
Podocoryne minima **25**: *128*, 131
Podocotyle **40**: 16
Podocotyle angulata **40**: 6, 11, 16, 27, 34
Podocotyle atomon **24**: 274, 288, 299; **40**: 6,
      16, 27, 31, 35, 36, 39, 43, 46
  reflexa **24**: 274, 288, 299
Podocotyle reflexa **40**: 6, 16, 27, 31, 35, 36
Podocotyle staffordi **40**: 11
Podolampas **32**: 560
Podon **31**: *84*, 86, 92, 102, 116, 125, 130
Podon camptonyx **31**: 107
Podon intermedius **31**: 85, 86, 92, 133, *136*,
      137, 139, 140, 148
Podon leuckarti **31**: 85, 86, 101, 111, 117,
      119, 120, 121, 130, *136*, 137, 139,
      145, 148, 152
Podonevadne **31**: 86, 106
Podonevadne angusta **31**: 120, 125
Podonevadne camptonyx **31**: 117
Podoplea **33**: **2**
Poecilasma kaempferi litum **32**: **161**
Poecilia reticulata **21**: 125; **34**: 253;
      40: 292–294
Poecilochaetus serpens **25**: 146
Poecilostomatoida **33**: **2**, 2, *3*
Poecilostomozoa **43**: 210
Pogonophora **32**: 329, 333, **335**, 408;
      43: *21*
Pojetaia **42**: 41
  P. runnegari **42**: 42
Polarella glacialis **43**: 201
Polinices **25**: 200
Polinices duplicatus **35**: **185**
Polinices josephinus **28**: **397**, 400, **405**,
      407–409, *417*, 419
Polinices strangei, see *Uber strangei*
Pollachius pollachius **40**: 14
Pollachius virens (pollock; saithe) **38**: 4, 54;
      24: 274; **25**: 10; **40**: 11, 31, 32, 49,
      50
Pollicipes **22**: 200

Polyacanthonotus cL *rissoanus* **34**: **380**
Polycelis cornuta **21**: 133
   nigra **21**: 133
Polychaeta **32**: **56**, *97*, **161**, 329, 330, 333, 335, 357, 362, 367, 429, **438**, **446**, **447**, **460**, 463, 467, 468, **483**, **486**, **492**, 496; **39**: 36; **40**: 26, 27
   fecundity
      environmental factors **43**: 143–145
      life-history strategies **43**: 115, 125, 130–132, 136, 137–139
      quantification methodologies **43**: 100, *102*, 103, 104, 106
   parasitism **43**: 69, 70, *94*, *95*
Polychaetes **37**: 130
Polycheira rufescens **21**: 146
Polycheria antarctica **39**: 204
Polycladida **43**: *22*
Polydactylus multiradiatus, Predation **27**: **363**
Polydactylus sexfilis (threadfin) **38**: 29, 30, 54
   survival and size at release **38**: **38**
   survival from hatchery releases **38**: **41**
Polydora **21**: 35; **31**: 92
Polydora ciliata **20**: 366
Polydora ligni **29**: 195; **43**: 144
Polydora spp. **44**: 22, 132
Polyipnus **32**: 198
Polyipnus inermis **32**: **169**, 189, **195**, 217
Polyipnus parini **32**: 217
Polykrikos schwartzi **21**: 73
Polymetme **32**: 187
Polymetme andriashevi **32**: **169**, 188, **195**
Polymixia **32**: 186
Polymixia japonica **30**: 334
Polymixia salagomezensis **32**: **171**
Polymixia yuri **32**: **171**
Polymixiidae **32**: **171**, 211
Polynoidae **32**: **161**, 455; **39**: 36
Polynoidae commensal **32**: **126**
Polyodon spathula **40**: 306, 309
Polyopisthocotylea **43**: *15*, *22*, 41, *42, 43*, 46
Polyorchis karafutoensis **24**: 43
Polypedilum **35**: **163**, **169**, *180*, **181**
Polypera simushirae **32**: 267
Polyphemoides pediculus **31**: 102
Polyphemus **31**: 95, 97, 122, 124, 125, 140; **30**: 198
Polyphemus exiguus **31**: 86, 117

Polyphemus pediculus **31**: 80, 86, 92, 94, 95, 96, 97, 98, 99, 100, 106, 116, 117, 118, 121, 122, 123, *123*, 130
Polyplacophora **31**: 39; **42**: 42
Polypocephalus sp., Parasites **27**: *386, 390*
Polypodiaceae **40**: 128
Polypodiophyta **40**: 93
Polyprion americanus **30**: 331; **39**: 280, 283
Polypterus senegalus **40**: 279, 302, 305, 319
Polyrachis **40**: 145
Polyrachis sokolova **40**: 146
Polyrhabdina **25**: 140
Polyschides **42**: *144, 177*
Polysiphonia **23**: 46
   havaniensis **23**: 108
   urceolata **23**: 64
Polysporoplasma sparis **43**: 65
Pomacentridae **43**: 16, 84
Pomacentrus amboinensis **44**: 248
Pomadasys argenteus, Predation **27**: **363**
   kaakan, Predation **27**: **363**
   olivaceum, Predation **27**: **353**
Pomadasys commersonni **25**: 195
Pomatocheles jeffreyseii **42**: 216
Pomatomus saltatrix **26**: 146; **39**: 280
Pomatoschistus minutus **26**: 138, 139
Pomolobus **20**: 34, 160
   aestivalis **20**: 160
Pomolobus pseudoharengus **29**: 279, 282, 317
Pomoxis annularis **25**: 15; **31**: 178, 186
Pomphorhynchus kostylewi **24**: 273
   laevis **24**: 273
Pomphorhynchus laevis **40**: 9, 19, 28, 34
Pomphorhynchus tereticollis **40**: 26
Pontella **33**: *16*
   ecology **33**: 486, **491**
   taxonomy **33**: **52**, *61*, 66, 95, 114, 115
Pontella agassizi **33**: 114, **514**
   alata **33**: 114
   andersoni **33**: 114
   asymmetrica **33**: 114
   atlantica **33**: **26**, 30, 114, *492*
   cerami **33**: 114
   chierchiae **33**: 114
   cristata **33**: 114
   danae **33**: 114, **514**
   denticauda **33**: 114, **514**
   diagonalis **33**: 114
   elegans **33**: 114
   fera **33**: 114, **514**
   forficula **33**: 114

gaboonensis **33**: 114
gracilis **33**: 114
hanloni **33**: 114
indica **33**: 114
inermis **33**: 114
investigatoris **33**: 114
karachiensis **33**: 114
kieferi **33**: 114
latifurca **33**: 114
lobiancoi **33**: 30, 114, *492*
marplatensis **33**: 114
meadii **33**: **26**, 114, 197, 217, **270**, **307**
mediterranea **33**: **26**, 30, **32**, 114, *492*
  behaviour **33**: 420
  reproduction **33**: 268, **270**, **288**
mimocerami **33**: 114
natalis **33**: 114
novae-zealandiae **33**: 114
patagoniensis **33**: 114
pennata **33**: 114
polydactyla **33**: 114
princeps **33**: **47**, 114, **430**, **514**
pulvinata **33**: 114
rostraticauda **33**: 114
securifer **33**: 114, **430**
sewelli **33**: 114
sinica **33**: 115
spinicauda **33**: 115
spinipedata **33**: 115
spinipes **33**: **32**, 115, **430**
surrecta **33**: 115
tenuiremis **33**: 115, **514**
tridactyla **33**: 115
valida **33**: 115, **514**
whiteleggei **33**: 115
Pontellidae **33**: 434
  ecology **33**: **489**, 490, **491**
  nutrition **33**: 141, **152**
  reproduction **33**: 44, 46, 257, 261, 268, **270**
  taxonomy **33**: **52**, 113–115
Pontellina **33**: 27, **52**, 95, 115
morii **33**: 115, **491**, **514**
platychela **33**: 115, **515**
plumata **33**: **7**, 30, 115
  behaviour **33**: **417**, **430**
  chromosomes **33**: **47**
  distribution **33**: **515**
  ecology **33**: 490, **491**
sobrina **33**: 115, **515**
Pontellopsis **33**: 17, **52**, 95, 115, 486, **491**
albatrossi **33**: 115

armata **33**: 115
bitumida **33**: 115
brevis **33**: **26**, 115
digitata **33**: 115
elongatus **33**: 115
globosa **33**: 115
herdmani **33**: **47**, 115
inflatodigitata **33**: 115
krameri **33**: 115
laminata **33**: 115
lubbockii **33**: 115
macronyx **33**: 115
occidentalis **33**: **26**, 115
pacifica **33**: 115
perspicax **33**: 115
pexa **33**: 115
regalis **33**: 30, **32**, 115, *492*
  behaviour **33**: **430**
  nutrition **33**: 166, 203
scotti **33**: 115
sinuata **33**: 115
strenua **33**: 115
tasmaniensis **33**: 115
tenuicauda **33**: 115, 506
villosa **33**: 30, 115, *492*, **515**
yamadai **33**: 115
Pontocaris rathbuni **32**: **163**
Pontocrates altamarinus **39**: 204
Pontocrates arcticus **39**: 204
Pontocrates arenarius **39**: 204
Pontogammarus crassus **39**: 204
Pontogammarus maeoticus **39**: 204
Pontogammarus robustoides **39**: 204
Pontogammarus subnudus **39**: 204
Pontogeloides latipes **39**: 212
Pontogeneia inermis **39**: 204
Pontogeniella brevicornis **39**: 204
Pontophilus gracilis juncaeus **32**: **163**
Pontophilus japonica **32**: 217
Pontophilus nikiforovi **32**: **163**, 217
Pontophilus prionolepis **32**: 217
Pontophilus spinosus **38**: 179
Pontoporeia **35**: 166, **169**, 174, **177**, *178*; **39**: 143; **40**: 26
Pontoporeia affinis **39**: 204
Pontoporeia femorata **39**: 204
Pontoporeia microphthalma **39**: 205
Pontoptilus **33**: **51**, 70, 74, 105
lacertosus **33**: 105
mucronatus **33**: 105
muticus **33**: 105
ovalis **33**: 105

pertenuis **33**: 105
robustus **33**: 105
Pontostratiotes **30**: 51
Porcellanasteridae **32**: 445, 464, 466, **483**,
    492; **43**: 107, *108*, 147
Porcellanopagurus foresti **32**: **163**
Porcellio **30**: 121
    scaber **30**: 119
Porcellio dilatatus dilatatus **39**: 110
Porcellio laevis **29**: 186
Porcellio scaber **39**: 127
Porichthys notatus **26**: 150; **39**: 279; **40**: 294,
    295, 324
Porifera **31**: 39; **32**: 329, 330, 333, 335, 344;
    40: 131–134; **43**: *21*, 150
Porites **22**: 42; **24**: 60; **31**: 275; **32**: **159**;
    43: 130, 292, 302, *305*, 307
    andrewsii **22**: 10, 14, 15, 23, 26, 41, 42
    astreoides **22**: 10, 15, 16, 26, 30, 33,
        38–40, 42, 43
    compressa **22**: 11, 28
    divaricata **22**: 38, 42
    furcata **22**: 38, 42
    lobata **22**: 10, 11, 33
    lutea (haddoni) **22**: 5, 11, 15, 16, 18, 21,
        22
    P. astreoides **43**: 134, 258
    porites **22**: 12, 37, 40
Porites astreoides **31**: 266, 269
Porites compessa **31**: 254
Porites porites **31**: 254
Poritidae **43**: 287, 288
Porocoma **30**: 50
Porolithon **23**: 113
    pachydermum **23**: 103
Porolithon Plate 4: **43**
Poromyidae **32**: 468
Porosira glacialis **32**: 535
Porosira pseudodenticulata **31**: 353
Porphyra **21**: 11; **23**: 27; **43**: 282
    yezoensis **23**: 33–35, 37
Porphyra tenera **20**: 321
Porpita **24**: 200
Porrocaecum **24**: 290
Portlandia aestuariorum **42**: 31
Portunus **26**: 27
    pelagicus **29**: **133**, 187
    sanguinolentus **29**: **133**, 134, 187
Poseidonamicus **34**: 376
Poseidonamicus sp. **32**: **127**
Posidonia **24**: 57
    oceanica **24**: 32, 57, 66, 67

Posidonia oceanica **34**: 222, 229
Postelsia palmaeformis **23**: 17, 18, 28, 77
Potamobius **29**: **132**, 160
Pourtalesia **32**: 466
Pourtalesia jeffreysi **32**: 358
Pourtalesiidae **32**: 466
Praunus flexuosus **29**: **131**, 185; **39**: 123,
    126, 127, 129, 131, 133, 148, 151,
    184, 190, 216
Praunus inermis **39**: 128–131, 134, 190,
    216
Praya **24**: 191, 199, 202
    dubia **24**: 191
    reticulata **24**: 224
Prayola tottoni **24**: 119
Prelametila **42**: 15
Preussia aurantiaca **40**: 120
Priacanthidae **32**: **173**
Priapulida **43**: *21*
Priapulidae **32**: 329, 366
Priapuloidea **32**: **335**
Priapulus caudatus **21**: 139
Priapulus tuberculatospinosus **32**: 367
Prickleback *see Stichaeus*
Primno **32**: 303
Primno abyssalis **39**: 207
Primno evansi **39**: 207
Primno johnsoni **39**: 207
Primno macropa **32**: 265, 266, 274, 306
Primnoa **30**: 338
Prionace glauca **39**: 279
Prionitis lanceolata **23**: 109
Prionospio **34**: **368**
    sandersi **34**: **368**
Prionotoleberis styx **34**: **376**
Prionotus **40**: 313
Pristigloma **42**: 24
Pristiglomidae **42**: *5, 6, 37*
Pristipomoides
    filamentosus **30**: 334
    sieboldii **30**: 334
    zonatus **30**: 334
Procellarea (petrels) **43**: 207
Procellaria aequinoctialis **39**: 281
Procellaria parkinsoni (Parkinson's
    petrel) **39**: 82
Procerodes lobata **21**: 133
Processa pygmaea **32**: **163**
Prochlorococcus (cyanobacteria) **41**: 58
Prochristianella hispida, Parasites **27**: 386,
    387
Prochromadorella **30**: **50**

Procladius **35**: 176, *180*, **181**
Prodajus **25**: 154
Prodajus bigelowiensis **39**: 184
Prodentaliidae **42**: *143*
Prodentalium **42**: *143*
Prodisco **33**: **51**, 76, 106
  princeps **33**: 106
  secundus **33**: 106
Profundiconus smirna **32**: **163**
Progadilina **42**: *144*
Progeryon mararae **32**: **163**
Progonus muelleri **40**: 6, 16, 27
Progymnoplea **33**: **2**, 99
Proisocrinus **32**: 410
Proisocrinus ruberrimus **32**: **409**
Projasus **32**: 216, 223
Projasus bahamondei **32**: 146, **163**, 174,
  175, 176, 178, *179*, 180, 181, *214*,
  216, 218, 220, 223, 227
Projasus parkeri **32**: 216
Proleptomysis rubra **39**: 216
Promethichthys prometheus **30**: 334;
  32: **174**
Prometor grandis **32**: **460**
Pronoprymna petrowi **24**: 277, 288
Pronucula insignis **42**: 41
Propallene **24**: 38
  ardua **24**: 76
  kempi **24**: 23, 34, 38, 43, 76
  longiceps **24**: 32, 34, 36–38, 40, 42–44
Prorocentrum **21**: 72; **31**: 309; **20**: 27, 28
  balticum **21**: 72
  cordatum **21**: 72
  linna **21**: 72
  minimum var *mariae-lebouriae* **21**: 72
Prorocentrum foraminosum **40**: 124
Prorocentrum formosum **40**: 124
Prorocentrum gracile **31**: 92
Prorocentrum lima **31**: 315
Prorocentrum maculosum **40**: 124
Prorocentrum minimum **37**: 67, 125
Prorodon **25**: 131
Proseriata **43**: *22*
Prosobranchia **42**: 99, 100, 102, 117
Prosorhynchoides basargini **24**: 275,
  284, 290
  graciliscens **24**: 275
Prosorhynchoides gracilescens **40**: 5,13–14,
  25, 33, 35, 36, 51
Prosorhynchus crucibulum **40**: 5, 16, 25,
  33, 46

Prosorhynchus squamatus **40**: 6, 13, 20, 22,
  23, 28, 43
Prostebbingia gracilis **39**: 205
Prostilifer (gastropod) **41**: *189*
Protanthea **22**: 73
Proteocephalus **24**: 293, 302
Protis hydrothermica **34**: **368**
Protobranchia **32**: 373, 463, **483**; **42**: *37*
  orders see Nuculoida; Solemyoida *see also*
    Protobranch bivalves in *subject*
    index
Protodrilidae **29**: 195
Protoeuglena noctilucae **25**: 124
Protohaustorius deichmannae **39**: 205
Protolamellodiscus **43**: *29*
Protolira sp. **32**: *127*, 130
Protolira valvatoides **32**: **127**; **34**: **371**
Protomystides **34**: **367**
  verenae **34**: 367
Protoodinium **25**: 126, 127
  chattoni **25**: *128*, 131
Protoperidinium **31**: 91; **40**: 124
Protophoxus australis **39**: 205
Protopterus annectens **40**: 305
Protorygaeus ammothelloides **24**: 14
Protothaca **21**: 178
Protothaca jedoensis **24**: 52
Protothaca staminea (littleneck clam)
  39: 15, 31, 32, 39; **37**: **10**, **110**
Protozoa **40**: 129; **43**: 17, 64, 68, 193, 208,
  209, 213, 214
  Apicomplexa **40**: 4
  Ciliophora **40**: 4
  Mastigophora **40**: 4
  Microsporidia **40**: 4
Protrachypene, Taxonomy **27**: 99, 106, 107,
  113
  Life-histories **27**: 285
  Physiology **27**: 203
  Reproduction **27**: 255
  Zoogeography/Evolution **27**: 150
  precipua, Taxonomy **27**: 99
  Zoogeography/Evolution **27**: **145**, 147
Provanna **32**: 130; **34**: **372**, 415
  buccinoides **34**: 372
  glabra **34**: 372
  goniata **34**: 372
  ios **34**: 372
  laevis **34**: 372
  muricata **34**: 372
  nassariaeformis **34**: 372
  segonzaci **34**: 372

variabilis **34**: 372
Provannidae **32**: 120, 124, 130
Prymnesiophyceae **31**: 309
Prymnesiophyceae **41**: 49
Prymnesiophyte **43**: 198, 237
Prymnesium **21**: 91; **31**: 309
 parvum **21**: 72, 85, 90, 91
Prymnesium parvum **29**: 79; **31**: 317
Prymnesium patelliferum **31**: 317
Psammacora **22**: 10
Psammechinus microtuberculatus **21**: 145
Psammokalliapseudes **39**: 159
Psammonyx nobilis **39**: 205
Psammonyx terranovae **39**: 205
Pselionema **30**: **50**, 50
Psenes sp. **26**: 145
Psenopsis humerosa **35**: **34**
Psetta maxima **40**: 17
Psettichthys melanostictus (sand
  sole) **38**: 178
Pseudactinia flagellifera **22**: 83
Pseudanabaena **29**: 80
Pseudanthocotyloides heterocotyle **24**: 264
Pseudarchaster **32**: 178
Pseudarchaster sp.n.aff. *pusillus* **32**: **166**
Pseudaugaptilus **33**: **51**, 56, 71, 74, 105
 longiremis **33**: 105
 orientalis **33**: 105
 polaris **33**: 105
Pseudeuchaeta **33**: 20, **53**, 78, 127, 128, 486
 arctica **33**: 127
 brevicauda **33**: *4*, 127
 flexuosa **33**: 127, **499**
 magna **33**: 128, **499**
 major **33**: 128
 spinata **33**: 128
Pseudevadne **31**: 84, 86
Pseudevadne tergestina **31**: 86, 87, 102,
  106, *108*, 111, 116, 122, 130, 132,
  133, *136*, 137, 139, 142, 145
Pseudhaloptilus **33**: **51**, 106
 longimanus **33**: 70, 106
Pseudidothea bonneri **39**: 158
Pseudidothea scultatus **39**: 158
Pseudinciola obliquua **39**: 205
Pseudione affinis **38**: 182
Pseudobathytanais **32**: 211
Pseudobathytanais schtokmani **32**: **161**
Pseudoboeckella **33**: **52**, 112, 113
Pseudocalanidae **33**: 3, **152**, 434
Pseudocalanus **25**: *28*, 133, 147, 151;
  26: 256, 258, 269; **28**: 100, 113;

 30: 249; **33**: 218, 236, 298, 346,
  516; **20**: 30
 behaviour **33**: 402, **408**, **413**
 biochemistry **33**: **239**, 247
 chromosomes **33**: 48
 ecology **33**: **361**, 366, 372, **378**, **397**, 469,
  480, 481, 482, 487, **493**, 495, 496,
  524, 527, 528
 growth/development **33**: 298, 299, **307**,
  338, **343**
 life history **33**: 347, 348, 389
 minutus **30**: 250
 morphology **33**: 8, 25
 nutrition **33**: 140, 165, 167, **187**, **191**, 207
 physiology **33**: 215
 reproduction **33**: 43, 46, 257, 259, 261,
  265, 267, 280, **284**, 289, 348
 size/weight **33**: **232**, *332*
 taxonomy **33**: **53**, 60, 64, 88, 89, 130
Pseudocalanus acuspes **33**: 43, 130
 life history **33**: **387**
 reproduction **33**: 254
 biochemistry **33**: 247
 ecology **33**: **361**, 364, **493**, 495
 growth/development **33**: **307**, **311**, **313**,
  338
Pseudocalanus elongatus **32**: 11, 22, 40
Pseudocalanus mimus **44**: 94–99
Pseudocalanus spp. **44**: 99, 100
Pseudocaranx dentex **30**: 334
Pseudoceratina crassa **21**: 98
Pseudocetonurus **32**: 186
Pseudocetonurus septifer **32**: **171**
Pseudochaenichthys georgianus **24**: 352
Pseudochirella **33**: 236
 ecology **33**: 486
 nutrition **33**: **152**
 reproduction **33**: 263, 266, 267
 taxonomy **33**: **53**, *61*, 87, 90, 92, 93,
  128
Pseudochirella accepta **33**: 128
 batillipa **33**: 128
 bilobata **33**: 128
 bowmani **33**: 128
 calcarata **33**: 128
 cryptospina **33**: 128
 dentata **33**: 128
 divaricata **33**: 128
 dubia **33**: 128
 elongata **33**: 128
 fallax **33**: 128
 formosa **33**: 128

gibbera **33**: 128
granulata **33**: 128
gurjanovae **33**: 128
hirsuta **33**: 128
limata **33**: 128
lobata **33**: 128
major **33**: 128
mariana **33**: 128
mawsoni **33**: 128
notacantha **33**: 128
obesa **33**: 128
obtusa **33**: 128
pacifica **33**: 128
palliata **33**: 128
polyspina **33**: 128
pustulifera **33**: 128
scopularis **33**: 128
semispina **33**: 128
spectabilis **33**: 128
spinosa **33**: 128
sqaulida **33**: 128, 262
tanakai **33**: 128
vervoorti **33**: 128
Pseudochromadora interdigitatum **40**: 135
Pseudocohnilembus **43**: 213
Pseudocuma longicorne **39**: 116
Pseudocuma longicornis **39**: 218
Pseudocyclopia **33**: **53**, 77, 133, **500**, 528
 caudata **33**: 133
 crassicornis **33**: 133
 giesbrechti **33**: 133
 insignis **33**: 133
 minor **33**: 133
 muranoi **33**: 133, 402
 stephoides **33**: 133
Pseudocyclopidae **33**: **51**, 100, 470, **489**, **498**
Pseudocyclopiidae **33**: 53, 133, 470, **489**, **500**
Pseudocyclopoidea **33**: **51**, *54*, *55*, 100, 101
Pseudocyclops **33**: 518
 behaviour **33**: 423
 ecology **33**: **489**, 497, **498**
 taxonomy **33**: **51**, 68, 100
Pseudocyclops arguinensis **33**: 100
 australis **33**: 100
 bahamensis **33**: 100
 bilobatus **33**: 100
 cokeri **33**: 100
 crassiremis **33**: 100
 gohari **33**: 100
 kulai **33**: 100
 lakshmi **33**: 100

latens **33**: 100
latisetosus **33**: 100
lepidotus **33**: 100
lerneri **33**: 100
magnus **33**: 100
mathewsoni **33**: 100
minya **33**: 100
mirus **33**: 100
obtusatus **33**: 100
oliveri **33**: 100
pacificus **33**: 100
paulus **33**: 100
pumilis **33**: 100
reductus **33**: 100
rostratus **33**: 100
rubrocinctus **33**: 100
simplex **33**: 100
spinulosus **33**: 100
steinitzi **33**: 100
umbraticus **33**: 100
xyphophorus **33**: 100
Pseudocyttus maculatus **35**: 37
Pseudodiaptomidae **33**: **52**, 115, 116, **498**
Pseudodiaptomus **33**: 23, 218, 298
Pseudodiaptomus **40**: 130
 behaviour **33**: 450
 distribution **33**: 506
 ecology **33**: 469, 497, **498**, 528
 growth/development **33**: 296, 321
 reproduction **33**: 257, 258, 259, 260, 262, 266, 267, 279
 taxonomy **33**: **52**, 67, 115, 116
Pseudodiaptomus acutus **33**: **26**, 115, **307**, **345**, 508
 americanus **33**: 115
 andamanensis **33**: 115
 annandalei **33**: 115
 ardjuna **33**: **26**, 115
 aurivilli **33**: **26**, **47**, 115
 australiensis **33**: 115
 batillipes **33**: 115
 bayli **33**: 115
 binghami **33**: **26**, 115
 bispinosus **33**: 115
 bowmani **33**: 115
 brehmi **33**: 115
 bulbiferus **33**: 115
 bulbosus **33**: 115
 burckhardti **33**: 115
 caritus **33**: 115
 charteri **33**: 115
 clevei **33**: 115

cokeri **33**: 115, **308**, **345**
colefaxi **33**: 115, 450, 497
compactus **33**: 115
cornutus **33**: 116, 497
coronatus **33**: **26**, 116, 297
  behaviour **33**: **417**, **433**, 450
  growth/development **33**: **308**, 321
  longevity **33**: **345**
  nutrition **33**: 166
  physiology **33**: 217
  reproduction **33**: **288**
cristobalensis **33**: 116
culebrensis **33**: 116
dauglishi **33**: 116
diadelus **33**: 116
dubius **33**: 116
euryhalinus **33**: **26**, 116, 260
forbesi **33**: 116, **507**
galapagensis **33**: 116
galleti **33**: 116
gracilis **33**: 116
griggae **33**: 116
hessei **33**: 116, **232**
  behaviour **33**: 451, 452
  ecology **33**: **379**
  growth **33**: **308**, **311**, **320**
heterothrix **33**: 116
hickmani **33**: 116
hypersalinus **33**: 116
incisus **33**: 116
inflatus **33**: 116
inflexus **33**: 116
inopinus **33**: 116, **507**
ishigakiensis **33**: 116
jonesi **33**: 116
lobipes **33**: 116
longispinosus **33**: 116
malayalus **33**: 116
marinus **33**: **26**, 116, 298
  biochemistry **33**: **239**
  distribution **33**: 506, **507**
  ecology **33**: **379**, 380
  growth/development **33**: **308**, **311**, **312**, **313**, **320**, **343**
  nutrition **33**: 157, **186**, **201**
  reproduction **33**: 280, 282, **284**
marshi **33**: 116
masoni **33**: 116
mertoni **33**: 116
mixtus **33**: 116
nankauriensis **33**: 116
nihonkaiensis **33**: 116, 429, **431**

nostradamus **33**: 116
occidentalis **33**: 116
ornatus **33**: 116
pacificus **33**: 116
panamensis **33**: 116
pankajus **33**: 116
pauliani **33**: 116
pelagicus **33**: 116
penicillus **33**: 116
philippinensis **33**: 116
poplesia **33**: 116
poppei **33**: 116
richardi **33**: 116
salinus **33**: 116
serricaudatus **33**: **47**, 116
sewelli **33**: 116
smithi **33**: 116
spatulus **33**: 116
stuhlmanni **33**: 116
tollingeri **33**: 116
trihamatus **33**: 116
trispinosus **33**: 116
wrighti **33**: 116
Pseudodiaptomus coronatus **29**: **131**, 181
Pseudodiaptomus hessei **44**: 100, 101
Pseudodiaptomus marinus **44**: 101–103
Pseudoeunotia doliolus **32**: 535, *542*, *550*, 554, 555, 558
Pseudoguinardia recta **31**: 352
Pseudojaera **39**: 118
Pseudojaera investigatoris **39**: 118
Pseudokuhnia **43**: 9
  P. minor **43**: 9, *12*, *13*, 24, *29*
Pseudolana cocinna **39**: 212
Pseudolana towrae **39**: 193, 212
Pseudolithophyllum cabiochae **24**: 57
Pseudolovenula **33**: **52**, 113
  magna **33**: 113
Pseudomacrochiron **25**: 153
Pseudomma **25**: 154
Pseudomma sp. **32**: **452**
Pseudomonas **29**: 194; **40**: 16, 116
Pseudomonas, Parasites **27**: *382*
Pseudomyona **42**: 41
Pseudo-nitzschia australis (= *Nitzschia pseudoseriata*) **31**: 311
Pseudo-nitzschia multiseries (formerly *Pseudonitzschia pungens* f. *multiseries*) **31**: 304, 311, 319, *320*
Pseudonybelinia **25**: 150, 151
Pseudopallene **24**: 8, 26

ambigua **24**: 16
circularis **24**: 11, 63
Pseudopecoelus **25**: 148
Pseudopentaceros wheeleri **30**: 329, *333*, 334, 335
Pseudopentaceros wheeleri **35**: 5, **45**, 50, 103
Pseudophaenna **33**: **53**, 85, 89, 135
typica **33**: 135
Pseudophyllidea **40**: 10; **32**: 211
Pseudoplectella dentatum **32**: **160**
Pseudopleuronectes americanus **25**: 9, 10; **26**: 97; 31: 171
Pseudoplexaura **22**: 49
Pseudorchestoidea brasiliensis **39**: 205
Pseudorhombus arsius, Predation **27**: **362**, 369
Pseudorimula **32**: 127, 130; **34**: 417
marianae **34**: **371**
midatlantica **34**: **371**
Pseudoscourfieldia marina **29**: 79
Pseudostichopus sp. **32**: **462**
Pseudostylochus oestreophagus **21**: 31
Pseudosuberites pseudos **21**: 97
Pseudotanais **39**: 160
Pseudotanais forcipes **39**: 175
Pseudoterranova **24**: 290
decipiens **24**: 274, 291
Pseudoterranova decipiens **40**: 8, 18, 22, 24, 26, 31, 36, 37, 40, 43–45, 57
Pseudotharybis **33**: **53**, 82, 128
brevispinus **33**: 128
dentatus **33**: 80, 128
magnus **33**: 128
robustus **33**: 80, 128
spinibasis **33**: 128
zetlandicus **33**: 128
Pseudotindaria erebus **42**: 24
Psilaster florae **35**: 31
Psilaster pectinatus **32**: **446**, **449**
Psolus **35**: 31
Psolus sp. **32**: **469**
Psychropotes **32**: **437**
Psychropotes raripes **32**: **461**
Psychropotidae **32**: **461**
Psychroptes **43**: 106
Psylaster charcoti **32**: 346
Pteraster jordani **35**: 31
Pteraster sp. **32**: **166**
Pterasteridae **32**: 445, 455
Pteria vulgaris **42**: 184
Pteriomorphia **42**: 4, *37*

Pterochirella **33**: **53**, 66, 128
tuerkayi **33**: *128*, **499**
Pterocladia capillacea **23**: 110
Pterocotyle morrhuae **40**: 11
Pteroma plagiophleps **40**: 178
Pteropoda **32**: **159**, **452**, **453**, **454**
Pteropodia **42**: 99
Pteropus alecto **40**: 157
Pteropus conspicillatus **40**: 157
Pterosagitta **25**: 151
Pterosagitta draco **44**: 119
Pterothrissa **20**: 160
Pterotrachea **25**: 142, 143
Pteroxena **25**: 153
Pterygioteuthis spp. **44**: 168
Pterygophora californica **23**: 39, 80
Pterygotrigla **32**: 186, 198
Pterygotrigla picta **32**: 146, **159**, **172**, **192**, 193, **194**, 218, 224, 227
Ptilanthura tenuis **39**: 114
Ptilota **23**: 46
Ptychodiscus breve **31**: 313, 314
Ptychodiscus brevis, see also, *Gymnodinium breve* **21**: 73
Ptychogonimus megastomus **42**: 214
Ptychoramphus aleuticus **20**: 245
Puchinia **33**: **53**, 82, 135
obtusa **33**: 135, *422*
Puerulus angulatus **29**: 165
Puffinus creatopus **39**: 280
Puffinus griseus **39**: 279; **20**: 61
Puffinus tenuirostris **32**: 270; **39**: 280
Pulmonata **42**: 68, 70
Pulsellidae **42**: *142, 144*
Pulsellum **42**: *144*, 191
feeding and digestion **42**: 163, *166*, 169
P. lofotensis **42**: 196
P. salishorum **42**: *154*, 164, 176, 192, 196, 215
Pumilus antiquatus **28**: 249, 250, 254, 265, 271, 329
Puncturella
n. sp. **34**: **371**
parvinobilis **34**: **371**
rimaiazenaensis **34**: **371**
solis **34**: **371**
Pungitius pungitius **40**: 15
Pupa **42**: 119
Pupa niveon **28**: *393*, **403**
Pycnogonida **32**: 404, 445
Pycnogonids **43**: 93
Pycnogonum (Nulloviger) africanum **24**: 11

anovigeum **24**: 11
elephas **24**: 11, 15
Pycnogonum **24**: 4, 11, 13, 22, 25, 38, 39, 71
  arbustum **24**: 11
  aurilineatum **24**: 45
  forte **24**: 71
  gaini **24**: 70
  littorale **24**: 15, 17, 21–25, 28, 32, 33, 34, 35, 37–39, 41, 42, 43, 44, 45, 49, 55, 63, 69
  nodulosum **24**: 69
  platylophum **24**: 60, 63
  plumipes **24**: 15, 57
  pusillum **24**: 15
  stearnsi **24**: 4, 10, 12, 18
  tuberculatum **24**: 13
Pycnopodia helianthoides (sunflower star) **39**: 38, 41; **21**: 144; **23**: 75–77, 84
Pycnosomia asterophila **24**: 52
  strongylocentroti **24**: 52
Pygoscelis
  P. adeliae **43**: 227, 243
  P. antarctica **43**: 227, 229
Pygoscelis adeliae **20**: 278
  antarctica **20**: 278
Pylochelidae **42**: 216
Pyramicocephalus phocarum **40**: 7, 17, 22, 23, 25, 30, 36, 37, 51
Pyramidellidae **42**: 111, 118
  see also Lower Heterobranchia
Pyramimonas virginica **29**: 79
Pyramodon **32**: 186
Pyramodon parini **32**: **171**, **194**
Pyrgocythara nodulosa **32**: **164**
Pyrgoma **22**: 204, 233
Pyrocystis **32**: 560
Pyrocystis noctiluca **32**: *532*, 535
Pyrodinium **31**: 356
Pyrodinium phoneus **21**: 73
  bahamense **21**: 73
Pyropelta **34**: 415
  bohlei **34**: **371**
  corymba **34**: **371**, 393
  musaica **34**: **371**
Pyropelta wakefieldi **35**: 24
Pyrosoma
  elegans **25**: 120, 139
  giganteum **25**: 137
Pyrostephos **24**: 200
  vanhoeffeni **24**: 117, 118, 141

Pyrrhophyta **43**: *283*
Python molorus **40**: 154

Quahog (hard clam) *see Mercenaria mercenaria*
Queubus **24**: 26
  jamesanus **24**: 71

Rabbit fish (*Siganus* spp.) **43**: 53, 66
Racovitzanus **33**: **53**, 78, 82, 84, 135
  antarcticus **33**: 135
  levis **33**: 135
  pacificus **33**: 135
  porrectus **33**: 135
Radianthus **21**: 132
  paumotensis **21**: 107, 132
Radiella sp. **32**: **160**
Raia radiata **40**: 22, 23
Rainbow trout *see Oncorhynchus mykiss*
Rainbow trout *see Salmo gairdneri*
Raja (skate) **38**: 178; **30**: 327; **34**: 248
  batis **34**: 237
  clavata **34**: 310
Raja radiata (starry ray) **38**: 175
Raja tengu, Predation **27**: 367
  texana, Predation **27**: 367
Ralfsia **23**: 22, 111; **34**: 48
  expansa **23**: 113
  spongiocarpa **23**: 111
  verrucosa **23**: 45, 113
Ramex californiensis **43**: 137
Ramnogaster arcuata **20**: 180
  melanostoma limnoica **20**: 180
Ramosia vitiazi **24**: 116
Rana **40**: 154
Rana hexadactyla **40**: 154
Rana pipiens **21**: 164
Randallia nana **32**: **163**
Rangia cuneata **37**: **10**, 68
Ranina ranina **29**: **133**, 134
Rapana thomasiana **21**: 31
Raphidascaris **24**: 278, 290
Raphidascaris acus **40**: 8, 10, 18, 34
Rastrelliger **43**: 56, 59
Rastrelliger kanagurta **25**: 9, 12
Rathkea octopunctata **25**: 31
Razor clam *see Siliqua patula*
Red algae (Rhodophyta) **43**: *20*, 281, *305*, 308
Red drum *see Sciaenops ocellatus*
Red sea bream *see Pagrus major*
Red-breasted merganser *see Mergus serrator*

Red-winged blackbird *see Agelaius phoeniceus*
Regadrella peru **32**: **160**
Reinhardtius hippoglossoides **26**: 43; **35**: **37**, 50, 83, 84, 103, 107; **38**: 175
Reinhardtius hippoglossoidesmatsuurae **32**: 267, **297**, 304
Renicola **24**: 276
Reophax **30**: 52
Reptilia **40**: 154
Retusa **42**: 83, 91, 92
  R. obtusa **42**: 94
Retusa obtusa **40**: 27
Retusa truncatula **40**: 27
Retusidae **42**: 87, 96
Rexea **32**: 186, 198
Rexea antefurcata **32**: **174**, 185, **192**, 193, **196**, 198, 218
Rexea brevilineata **32**: **174**, 189, **192**, **196**, 218
Rexea solandri **32**: 218
Rhabdidae **42**: *142, 143, 144*
Rhabditophora **43**: *21*
Rhabdocoma sp. **32**: **126**
Rhabdonella **25**: 129
Rhabdosargus globiceps, Predation **27**: **364**
Rhabdosargus holubi **25**: 195
Rhabdosoma **39**: 108, 157
Rhabdosoma brevicaudatum **39**: 207
Rhabdosoma whitei **39**: 183, 207
Rhabdus **42**: *143*
    predation **42**: 215
    sensation and innervation **42**: 191, 193, 194
    shell **42**: 149, 150, 151, 152
  R. dalli **42**: 191
  R. perceptum **42**: 191
  R. rectius Plate 1 **42**:
    associations **42**: *211*, 212, 214
    circulation **42**: 181, 185, 186, *187*
    excretion **42**: 188, 189
    feeding and digestion
      alimentary tract **42**: *174*, 175, 176, 178, 179
      behaviour **42**: 169, 172
      captacula **42**: *166*
      diet **42**: 163, 164
      gas exchange **42**: 159, 161
      locomotion **42**: 162
      mantle cavity **42**: 155, 156, *157, 158, 160*

Rhacophorus **40**: 154
Rhacophorus maculatus **40**: 154
Rhadinorhynchus trachuri **24**: 277, 290
Rhapidophorus **33**: **51**, 102
  wilsoni **33**: 68, 102, **498**
Rhepoxynius abronius **39**: 205
Rhigophila dearborni **24**: 328, 329, 336, 339, 340, 344, 345, 349, 350, 353
Rhincalanus
  ecology **33**: 486, 487, 528
  morphology **33**: 24
  nutrition **33**: 140, **152**
  taxonomy **33**: **52**, 76, 122
Rhincalanus cornutus **33**: *24, 26*, 122, **224**
  gigas **33**: 7, **26**, 122
    biochemistry **33**: 247
    ecology **33**: **361**, 399, 492, 494
    life history **33**: **387**, 390
    nutrition **33**: **191, 208**
    reproduction **33**: **255, 287**
    size/weight **33**: **224, 232, 234**
  nasutus **33**: *4, 7, 26*, 122, **232**, 297
    behaviour **33**: **413, 417, 418**, 422, 424, 455
    biochemistry **33**: **239**
    chromosomes **33**: 47
    distribution **33**: **515**
    ecology **33**: 365, 366, 371, 379, 481, **484, 485**, 503
    growth/development **33**: **308**, 321, **343**
    nutrition **33**: 159, 166, **191, 194**
    reproduction **33**: 265, **288**
  rostifrons **33**: 122, **515**
    chromosomes **33**: **47**
    distribution **33**: **515**
    nutrition **33**: **194**
Rhincalanus gigas **44**: 81; **32**: 26, 27, *28*; **43**: 217, 219
Rhincalanus nasutus **44**: 81; **32**: 36
Rhincalanus spp. **32**: 29
Rhinobatis **25**: 195
Rhinoceros sondaicus **40**: 157
Rhinoclama (Austroneaera) similis **32**: **164**
Rhinoclama brevirostris **32**: 216
Rhinoclama raoulensis **32**: 216
Rhinoclama similis **32**: 216
Rhinoclavis **28**: 402
Rhinoclavis aspera **28**: *392*, **403**
Rhinoclavis fasciata **28**: *392*
Rhinomaxillaris **33**: **53**, 123
  bathybia **33**: 123

Rhinoptera bobasus, Predation **27**: 36
Rhinosardinia **20**: 160
Rhipidocotyle **24**: 278, 284, 290
Rhithropanopeus harrisii **29**: **133**, 144
Rhizobium **40**: 115
Rhizocephala **22**: 200
Rhizocephala **43**: 45, 70
Rhizoctonia **40**: 121
Rhizophora **40**: 86, 88, 89, 91, 98,
    99,100–103, 107–110, 118, 124,
    125, 137, 152, 158, 177, 184
Rhizophora apiculata **40**: 95, 97,
    102,104–106, 114, 121, 122, 159,
    160, 167, 168, 170, 171, 173, 176
Rhizophora mangle **24**: 60; **40**: 95, 97, 101,
    103, 109–113, 120, 121, 129–136,
    150, 152, 157, 159–63,
    167–71,174–176, 178, 181
Rhizophora mucronata **40**: 95, 97, 105, 112,
    114, 138, 170, 176, 181
Rhizophora racemosa **40**: 95, 114, 165
Rhizophora samoensis **40**: 95
Rhizophora stylosa **40**: 96, 97, 99, 112–114,
    160, 165
Rhizophora × annamalayana **40**: 96, 97
Rhizophoraceae **40**: 95–97
Rhizophora × *harrissonii* **40**: 96, 97
Rhizophora × *lamarckii* **40**: 96, 97
Rhizophorales **40**: 93
Rhizophora × *selala* **40**: 96
Rhizophysa **24**: 108, 113, 116, 177, 185,
    191, 193
  eysenhardti **24**: 113, 174, 205, 208, 215,
    227, 229, 239
  filiformis **24**: 194, 195, 205, 212, 215,
    217, 229
Rhizophysa eysenhardti **21**: 105
  filiformia **21**: 105
Rhizopus nigricans **40**: 121
Rhizoselenia **29**: 44
  stolterfothii **29**: *8*
Rhizosolenia **40**: 124
Rhizosolenia acuminata **32**: 555
Rhizosolenia alata **32**: *530*, 535, 554
Rhizosolenia bergonii **32**: 555
Rhizosolenia castracanei **32**: 535, *538*, 554,
    555
Rhizosolenia curvata **32**: *532*, 542, 560
Rhizosolenia cylindrus **32**: 541
Rhizosolenia simplex **32**: *533*, 541, 555
Rhizosolenia sp. **31**: 92
Rhizosolenia stolterforthii **32**: 555

Rhizostoma **25**: 153, 156
  pulmo **25**: 156
Rhizostoma pulmo **21**: 106
Rhodactis howesi **21**: 106, 132
Rhodoglossum affine **23**: 109, 114
Rhodomonas salina **41**: 182
Rhodophyceae **32**: 176
Rhodophycota **32**: **160**
Rhodophyta **43**: *20*, 281, *283, 305*
Rhodopseudomonas **40**: 118
Rhodymenia pertusa **23**: 24
Rhogobius
  contractus **34**: **375**
  pressulus **34**: **376**
Rhomboxiphus **42**: *144*
Rhopalophthalmus dakini,
    Food/feeding **27**: 322
Rhopalophthalmus terranatalis **39**: 216
Rhopalorhynchus **24**: 9
  claudus **24**: 10, 12
Rhynchonerella
  fulgens **25**: 140
  petersii **25**: 140
Rhynchopelta **34**: 417
  concentrica **34**: 370
  nux **34**: **370**
Rhynchothorax **24**: 2, 8, 9, 14, 21, 22, 26,
    45, 54, 69
  alcicornis **24**: 8, 57
  anophthalmus **24**: 8, 32, 69
  australis **24**: 18, 19, 46
  mediterraneus **24**: 32, 57, 69
  monnioti **24**: 69
  philopsammum **24**: 8, 64
  unicornis **24**: 8
Rhyncopelta concentrica **35**: 50
Rhytiodentalium **42**: *143*
  R. kentuckyensis **42**: 216, 217
Rhyzophysa eysenhardti **25**: 30, **31**, 33,
    35, 45
Ribeirioida **42**: 145
Richtersia **30**: **50**
Rictaxis punctocaelatus **42**: 91
Ridgeia **23**: 324; **34**: 408, *409*; **35**: **41**, 50
  phaeophiale **35**: **41**
  piscesae **34**: **366**, 394, 398, 404, 407,
    408, *409*; **35**: **41**, **45**, 89
Ridgewayia **33**: **51**, 70, 73, 101, 497, **498**
  canalis **33**: 101
  flemingeri **33**: 101
  fosshageni **33**: 101, 503
  gracilis **33**: 101

klausruetzleri **33**: **26**, 101
krishnaswamyi **33**: 101
marki **33**: 101, **489**
marki minorcaensis **33**: 101
shoemakeri **33**: 101
typica **33**: 101
wilsoni **33**: 101
Ridgewayiidae **33**: 17, **51**, 100, 101, 470, **489**, **498**
Ridleia **21**: 98
Riftia **23**: 319, 320, 322, 325, 326, 329, 330, 338, 345; **35**: 22
  pachyptila **23**: 303, 311, 315, 321, 324, 333–335, 337, 343, 344; **35**: 27, 28, **41**, 90, 91
Riftia pachyptila **32**: 114; **34**: **366**, 398, 404, 407, 408, *409*; **43**: 105
Rimicaris
  aurantiaca **34**: **377**
  exoculata **34**: **377**, 398
Rimicaris exoculata **32**: 94, 99, *101*, 102, 105, 107, 108, 110, 110–114, 116, **117**, 118, 124, **129**, *134*; **35**: 19, *20*, **40**, **45**, 50, 63, 87, 89
Rimicaris sp. **32**: *104*, 107, 110, 114, 120, 130
Rimipontius **32**: 130
Rimipontius mediospinifer **32**: **128**; **34**: **376**
Ringed seal *see Phoca hispida*
Ringicula **28**: *393*; **42**: 97, 99, 112, *116*, 132, 135
Ringiculidae **42**: 75, 82, 87, 90, 109, 110, 129
Rissa tridactyla (black-legged kittiwake) **39**: 69–70, 280
Rissa tridactyla **34**: 268; **20**: 281
Rissoa **24**: 285
  membranacea **24**: 285
Rissoellidae *see* Lower Heterobranchia
River otter *see Lutra canadensis*
Rivularia **32**: **160**
Rivulus marmoratus **40**: 153
Roccus saxatalis **26**: **92**
Rocinela danmoniensis **40**: 9, 19
Rock sole *see Pleuronectes bilineatus*
Rockfish *see Sebastes*
Rockfish *see Sebastes schlegeli*
Rockling *see Ciliata mustela*
Roperia **32**: 560
Rosacea **24**: 151, 191, 199, 224

cymbiformis **24**: 174, 205–208, 212, 213, 216, 226, 227, 230, 232, 234, 239
flaccida **24**: 113, 115, 119, 215
plicata **24**: 145, 202
villafrancae **24**: 118
Rossella **32**: 346
Rostroconchia **42**: 145, 204
Rotifera **43**: *21*, 208, 214
Roughy, orange (*Hoplostethus atlanticus*) **43**: 62
Rouleina **32**: 186
Rouleina attrita **32**: **169**
Rouleina maderensis **32**: **169**
Rubiaceae **40**: 96
Ruddy turnstone *see Arenaria interpres*
Rudjakovia plicata **24**: 116
Rugogaster hydrolagi **43**: 41
Runcina **42**: *89*, 115
Runcinacea **42**: 74, *89*, 111
Runcinidea **42**: 109, 110, 112, 129
Runcinoidea **42**: 98, 112
Ruppia maritima, Food/feeding **27**: 321
Rutilus rutilus **26**: 83, 97, **120**, **121**, 126; **40**: 313
Ruvettus pretiosus **32**: **174**, 190, 191, 192, **196**
Rye grass *see Elymus*
Ryocalanidae **33**: **52**, 122, 123, **499**
Ryocalanoidea **33**: **52**, *54*, *55*, 122, 123
Ryocalanus **33**: **52**, 78, 80, 122, 123, **499**
  admirabilis **33**: 122
  asymmetricus **33**: 122
  bicornis **33**: 123
  bowmani **33**: 123
  infelix **33**: 123
Rythabis **33**: **53**, 84, 138
  atlantica **33**: 138, **501**

Sabatieria **30**: 48, **50**, 50
Sabellaria alveolata **24**: 45
Sabellaria spinulosa **34**: 219
Sabellidae **39**: 36
Sabinea septemcarinata **32**: **448**
Sablefish *see Anoplopoma fibriata*
Saccardoella **40**: 119
Saccharomyces cerevisiae **44**: 302
Saccocirrida **29**: 195
Saccorhiza polyschides **23**: 11, 137
Saccostrea commercialis **37**: **10**, 16, 95
  bioassays **37**: 43, **44**, 72
Saccostrea sakhalinensis **37**: **9**
Sacculina **20**: 366

Sacculina carcini **43**: 70
Sacoglossa **42**: 71, 133
  see also Oxynoe
  characters and evolutionary
    pathways **42**: 83, 87, 88, 90–100
    passim **42**: 102, 103
  morphology **42**: 78–80, *79*
  taxonomic history **42**: 73, *89*, 107, 108,
    110, 111, 112, 115, *116*
Saduria **30**: 91
Saduria entomon **39**: 119, 120, 121, 126,
    127, 133, 136, 137, 139, 158, 212
Sagamicadulus **42**: *144*
Sagamichthys abei **32**: **169**
Sagartia **22**: 73
  troglodytes **22**: 72, 73, 74
Sagartia elegans **24**: 45; **21**: 106
  longa **21**: 106
Sagina elegans **32**: 22, 55
Sagitta **25**: *6*, **31**, 122, 125, 141, 147, 148,
    149, *149*, 150, 151; **28**: *87*; **29**: 196;
    20: 175
  elegans **23**: 234; **25**: 7, *28*, **34**, 138, 148,
    151, 157, 158; **20**: 52; **33**: 372
  friderici **23**: 235
  enflata **33**: 160
  hispida **29**: 196
  hispidia **25**: 7
  lyra **25**: 141
  minima **25**: 138
  punctate **25**: 141, *142*
  setosa **23**: 235; **25**: 7, **34**, 150; **20**: 52
Sagitta crassa **44**: 119
Sagitta elegans **44**: 119, 120; **26**: *131*
Sagitta hispida **44**: 22, 120, 121
Sagitta setosa **32**: 40
Sagitta sp. **32**: 289, 302
Sagmalia hinomaru **24**: 118
Saithe *see Pollachius pollachius*
Saithe *see Pollachius virens*
Salenia scrippsae **32**: **159, 165**
Salenidae **32**: 466
Salenocidaris nudispina **32**: **165**
Salmo gairdneri **26**: **74**, 79, **92**; **32**: 269,
    283; **40**: 286, 291
Salmo salar **24**: 273; **25**: 36, 63, 64; **26**: **74**,
    79, 130; **40**: 18, 23
  trutta **24**: 273
Salmo trutta **26**: **92**; **40**: 32, 41, 260, 263
Salmon
  Atlantic (*Salmo salar*) **43**: 65
  Chinook *see Oncorhynchus tshawytscha*

chum *see Oncorhynchus keta*
  coho *see Oncorhynchus kisutch*
  pink *see Oncorhynchus gorbuscha*
  sockeye (*Oncorhynchus nerka*) **43**: 60
Salpa **25**: 156
  democratica **25**: 137
  fusiformis **25**: *155*
Salpa thompsoni **43**: 219, 220, 223, 224
Salpingoeca pelagica **25**: 120
Salvelinus alpinus **26**: **92**; **30**: 263; **40**: 56,
    58
Salvelinus malma (Dolly Varden char)
    39: 17, 58, 59
Samaridae **32**: **174**
Samariscus nielseni **32**: **174**
Samariscus sp. **32**: 206
Sambucus nigra **40**: 259
Sand hopper *see Talitrus saltator*
Sand lance *see Ammodytes hexapterus*
Sand sole *see Psettichthys melanostictus*
Sanderia malayensis **21**: 106
Sapphirina **25**: 142
Sarcophyton glaucum **21**: 106, 126
Sarda chiliensis **20**: 171
  sarda **20**: 171
Sarda sarda **40**: 20
Sardina **34**: 257; **20**: 17, 22, 177
  pilchardus **20**: 6, 11, 18, 39, 86, 102, 155,
    171
Sardina pilchardus **24**: 278, 305; **23**: 235;
    25: 134; **28**: 25, 60, 61, **64**; **29**: 280;
    31: 196
Sardinella **20**: 21, 160, 177
  anchovia **20**: 7
  aurita **20**: 7, 102, 160, 180
  cameronensis **20**: 160
  eba **20**: 180
  longiceps **20**: 7, 11
  maderensis **20**: 102
  marguesensis **20**: 143
Sardinella aurita **31**: 196
Sardinella brasiliensis **28**: 49
Sardinella maderensis **31**: 196
Sardinia pilchardus **26**: 260
Sardinops **32**: 206; **20**: 14, 17, 20, 21, 38,
    177
  caerulea **30**: 229
  caerulea **20**: 5, 6, 11, 37, 102, 131, 246
  melanosticta **30**: 334; **20**: 5, 258
  neopilchardus **20**: 173
  ocellata **20**: 5, 6, 11, 18, 37, 171, 247
  sagax **20**: 29, 66, 167, 171

Sardinops caerulea **26**: 259; **28**: 25, 43, 54, **56**, 59, *89*, 128
Sardinops caeruleus (California sardine) **38**: 3, 53
Sardinops melanosticta **26**: 137; **28**: 55; **38**: 3, 54
Sardinops melanostitictus **39**: 276
Sardinops ocellata **28**: 15
Sardinops sagax **25**: *3*, 14, 21, 48; **28**: *89*, 90; **31**: 196; **34**: 276;
Sardinops sagax caeruleus **39**: 277
Sarepta hadalis **42**: 39
Sareptidae and Sareptinae **42**: 5, *6*, 37–39, 40
Sargassum **23**: 65; **24**: 60, 66; **43**: 308;
  muticum **23**: 28, 29
  sagamianum **23**: 63
  sinclairii **23**: 10, 11
  vulgare **23**: 45
Sarortia elegans **21**: 132
Sarotherodon mossambicus **34**: 83–85
Sarsarietellus **33**: **51**, 68, 74, 103
  abyssalis **33**: 103, **498**
  natalis **33**: 103
Sarsia tubulosa **44**: 122; **25**: **31**; **28**: *88*
Saurida spp., Predation **27**: 369
Saurida undosquamis **30**: 334
Saury *see Cololabis saira*
Saxidomus **21**: 178
Saxidomus giganteus (butter clam) **39**: 15, 31, 32, 39
Saxipendium coronatum **34**: 379; **23**: 331
Scaeurgus **32**: 206
Scaeurgus pacifica **32**: 220
Scaeurgus unicirrhus patagiatus **32**: **164**, 205, 206, 220
Scallop **43**: 132, 140, 149
  giant **43**: 101, 145
Scallop *see Chlamys rubida*
Scalpellida **32**: **409**
Scalpellidae **32**: **409**, 468
Scalpellum abyssicola **32**: 506
Scalpellum regium **32**: 503
Scalpellum vitreum **32**: 503
Scaphander **28**: *393*, 412, 421; **42**: *89*, 95, 97, 98
  S. lignarius **42**: *72*, 73
Scaphander punctostriatus **24**: 53
Scaphandridae **42**: 76, 109, 129
Scapharca inaequivalis **35**: **185**
Scaphocalanus
  ecology **33**: 486, **500**

  taxonomy **33**: **53**, 57, *61*, 80, 82, 84, 85, 134, 135, 136
Scaphocalanus acuminatus **33**: 135
  acutocornis **33**: 135
  affinis **33**: 135
  amplius **33**: 135, 136
  angulifrons **33**: 136
  antarcticus **33**: 135
  bogorovi **33**: 136
  brevicornis **33**: 135, 479
  brevirostris **33**: 135
  californicus **33**: 136
  curtus **33**: 135, 136
  difficilis **33**: 135
  echinatus **33**: 135
  elongatus **33**: 136
  farrani **33**: 135
  insignis **33**: 136
  insolitus **33**: 136
  invalidus **33**: 135
  longifurca **33**: 135
  magnus **33**: 135, *441*
  major **33**: 135
  parantarcticus **33**: 135
  paraustralis **33**: 135
  pseudobrevirostris **33**: 135
  similis **33**: 135, 136
  subbrevicornis **33**: 135
  subcurtus **33**: 136
  subelongatus **33**: 136
Scaphopoda **31**: 39; **32**: **335**, **504**; **42**: 42, 137–236
  see also Scaphopoda *in subject index*
Scardinius erythrophthalmus **26**: **120, 121**, 126
Scarus spp. **44**: 218
Scenedesmus obliquus **21**: 71
Schedophilus
  huttoni **35**: **34**
  labyrinthicus **35**: **42**, 43
Schindleria **32**: 186
Schindleria praematurus **32**: **174**, 198, **218**
Schistocephalus **40**: 7, 17, 25, 29, 34
Schistomysis kervillei **39**: 187, 216
Schistomysis ornata **39**: 216
Schistomysis spiritus **39**: 216
Schizochytrium (alga) **41**: 170
Schizochytrium mangrovei **40**: 119
Schizodentalium **42**: *143*
Schizoscelus ornatus **24**: 205
Schizothrix calcicola **21**: 71, 89, 90
schmitti, Taxonomy **27**: 98, **112**, 123

Behaviour **27**: 336, 355
Food/feeding **27**: **318**
Life histories **27**: 284, 293
Moulting/growth **27**: 239
Reproduction **27**: **280**
Zoogeography/Evolution **27**: **146**
Sciaenops ocellata, Predation **27**: 358, 361, **364**
Sciaenops ocellatus (red drum) **38**: 54
  in Gulf of Mexico **38**: 27, 28
  survival and size at release **38**: **38**
  survival from hatchery releases **38**: 41
Sciaenops ocellatus **28**: 121; **30**: 239; **31**: 186
Scillaelepas **32**: **409**
Scillaelepas n. **34**: **374**
Scissuradentalium **42**: *143*
Scissurellidae **32**: **127**
Scleractinia **43**: 290
  S. australasicus **43**: 6, *9, 10, 12, 15*, 24, 62
  S. brasiliensis **43**: *19*
  S. cavalla **43**: *19*
  S. commerson **43**: *19*
  S. concolor **43**: *19*
  S. guttatus **43**: *19*
  S. japonicus **43**: 6, 9, *10, 12*, 24
  S. koreanus **43**: *19*
  S. lineolatus **43**: *19*
  S. maculatus **43**: *19*
  S. multiradiatus **43**: *19*
  S. munroi **43**: *19*
  S. niphonius **43**: *19*
  S. plurilineatus **43**: *19*
  S. queenslandicus **43**: *19*
  S. regalis **43**: *19*, 59, *60*
  S. scombrus **43**: 9, *10*, 23, *24*
  S. semifasciatus **43**: *19*
  S. sierra **43**: *19*
  S. sinensis **43**: *19*
  S. tritor **43**: *19*
  Scomber **43**: 29, 56, 59
Sclerasterias sp. **32**: **166**, 178
Sclerocrandon abyssorum **32**: **446**, **448**
Sclerocrandon communis **32**: **446**, **448**
Sclerocrandon derjugini **32**: **446**, **448**
Sclerocrandon salebrosa **32**: **446**, **448**
Sclerocrandon zenkevitschi **32**: **446**, **448**
Scolecithircella **25**: 138
Scolecithricella
  ecology **33**: 486, **500**
  taxonomy **33**: **53**, 57, 61, 80, 81, 84, 136

Scolecithricella abyssalis **33**: 136
  aspinosa **33**: 136
  avia **33**: 136
  canariensis **33**: 136
  cenotelis **33**: 136
  curticauda **33**: 134
  dentata **33**: 136
  denticulata **33**: 134
  glacialis **33**: 494
  globulosa **33**: 136
  incisa **33**: 134
  lanceolata **33**: 134
  lobata **33**: 134, 136
  lobophora **33**: 136, 503
  marquesae **33**: 134, 136
  minor **33**: **7**, 136
  modica **33**: 136
  neptuni **33**: 136
  obscura **33**: 136
  orientalis **33**: 136
  ovata **33**: 136
  pacifica **33**: 136
  paramarginata **33**: 136
  pearsoni **33**: 136
  polaris **33**: 136
  profunda **33**: 136
  propinqua **33**: 134
  pseudoculata **33**: 136, **500**
  schizosoma **33**: 136
  spinacantha **33**: 136
  spinata **33**: 134
  timida **33**: 134
  tropica **33**: 136
  unispinosa **33**: 136
  vespertina **33**: 136
  vittata **33**: 136
Scolecithrix
  behaviour **33**: 407, **417**
  ecology **33**: 486
  taxonomy **33**: **53**, 57, 78, 80, 84, 136
Scolecithrix aculeata **33**: 134
  birshteini **33**: 136
  bradyi **33**: 136, **485**
  ctenopus **33**: 136
  danae **33**: **7**, 136, *422*
    behaviour **33**: **413**
    ecology **33**: **485**, 503
    nutrition **33**: 154, **194**
    reproduction **33**: 41
    size/weight **33**: 222, **224**
  elaphas **33**: 134
  fowleri **33**: 136

grata **33**: 136
laminata **33**: 136
longipes **33**: 136
longispinosa **33**: 136
magnus **33**: 134
marginata **33**: 136
maritima **33**: 136
medius **33**: 134
mollis **33**: 134
nicobarica **33**: 136
subdentata **33**: 136
subvittata **33**: 136
tenuipes **33**: 136
tenuiserrata **33**: 136
valens **33**: 134
valida **33**: 134
Scolecitrichidae **33**: 29, 340, 434
  ecology **33**: **500**
  nutrition **33**: **152**
  taxonomy **33**: **53**, 134–137
Scolecocalanus **33**: **53**, 80, 81, 137
  galeatus **33**: 137
  lobatus **33**: 137
  spinifer **33**: 137
Scolelepis squamata **29**: 195
Scolex pleuronectis **24**: 276, 277, 284, 289,
        295, 296, 304; **25**: 151; **26**: 38;
        28: 82; **40**: 7, 17, 25, 33, 36
  polymorphus **24**: 277
  socialis **24**: 284
Scolex polymorphus **40**: 17
Scoloplos ehlersi **34**: **366**
Scolopostoma **39**: 114
Scomber **25**: *3*, 12, 14, **52**; **32**: 206; **40**: 20
  japonicus **25**: 9, *13*, 16, 21, *22*; **30**: 334
  scombrus **25**: 37, 149; **30**: 249
Scomber japonicus **26**: 260; **28**: 55; **20**: 171,
        249
  scombrus **20**: 279
Scomber scombrus **24**: 277, 299; **23**: 197,
        234; **26**: 149; **28**: 15, 97, *89*, 127;
        29: 280, 286, 317, 334; **34**: 264;
        40: 16, 18
Scomberesocidae **32**: 62
Scomberesox saurus **25**: 3
Scomberjaponicus **32**: 200, 209, 220
Scomberoides lysan, Predation **27**: 366
Scomberomorus
  sierra **30**: 228, 279
Scomberomorus commersoni **31**: 5, 19, 24
Scomberomorus guttatus **31**: 19, 24

Scomberomorus lineolatus (= *S.*
        commersoni × *S. guttatus*) **31**: 5,
        8, 20
Scombridae **32**: 62; **43**: 9, 17, 59
  see also Grammatorcynus; *Scomber*;
        Scomberomorus
Scombrolabracidae **32**: **174**
Scombrolabrax heterolepis **32**: **174**
Scopalatum **33**: **53**, 81, 82, 137, **500**
  dubia **33**: 137
  farrani **33**: 137
  gibbera **33**: 137
  smithae **33**: 137
  vorax **33**: 137, **152**, 160, 503
Scopelocheiros shellenbergi **32**: 73, 369
Scophthalmus **31**: 27
Scophthalmus maximus (turbot) **38**: 23, 36,
        54
  stock enhancement studies **38**: **31**
  survival from hatchery releases **38**: **41**
Scophthalmus maximus **24**: 265
  rhombus **24**: 265
Scophthalmus maximus **44**: 297; **25**: 17, 19;
        26: 73, **74**, **92**; **29**: 217, 323,
        325, 333; **30**: 224
Scophthalmus rhombus **40**: 17
Scorpaena **32**: 186
Scorpaena uncinata **32**: **172**, 224
Scorpaenidae **32**: 167, **172**
Scorpin *see Myxocephalus scorpius*
Scoter *see Melanitta*
Scotoecetes introrsus **34**: **376**
Scotoplanes hanseni **32**: **461**
Scotoplanes kurilensis **32**: **461**
Scotothuria herringi **32**: **77**
Scottocalanus **25**: 138, 139; **33**: **53**, 82, 84,
        137, 486
  corystes **33**: 137
  dauglishi **33**: 137
  farrani **33**: 137
  helenae **33**: 137
  infrequens **33**: 137
  investigatoris **33**: 137
  longispinus **33**: 137
  persecans **33**: 137
  rotundatus **33**: 137
  securifrons **33**: 137, 444
  sedatus **33**: 137
  setosus **33**: 137
  terranovae **33**: 137
  thomasi **33**: 137
  thorii **33**: 137

Scottolana **33**: **184**
Scottolana canadensis **43**: 149
Scottula
  abyssalis **33**: 103
  inaequicornis **33**: 102
Scrippsechinus **32**: 211
Scrippsechinus fisheri **32**: **159**, **165**, 174,
    175, 180, 223
Scrobicularia plana **22**: 103–105, 109, 110,
    113–121, 123, 125, 127, 129,
    132–147, 150–161, 163–167,
    171, 175
Scrobicularia plana **37**: **10**
  bioassay procedures **37**: 70, 72, 73
  toxicity testing **37**: **91**, **99**
Scrobiculariidae **32**: 463
Scutigerella
  immaculata **29**: 148
  silvestrii **29**: 148
Scutogerulus **33**: **51**, 68, 103
  pelophilus **33**: 101, 103, **498**
Scyliorhinus caniculus **26**: 141
Scyliorhinus stellaris **40**: 278, 279
Scylla serrata **44**: 230; **29**: **133**, 134–137,
    138, **139**, 142–145, 153, **188**, 189,
    194; **40**: 140, 141
Scylla serrata, Predation **27**: 360
Scyllarisdes squammosus **34**: 224
Scyphipora **40**: 108
Scyphipora hydrophyllacea **40**: 96
Scyphozoa **32**: *333*
Scytonema **32**: **160**; **40**: 116
Scytosiphon **23**: 25
Sea bass *see Dicentrarchus labrax*
Sea bream *see Sparus aurata*
Sea bream, gilthead (*Sparus auratus*)
    43: 65
Sea otter *see Enhydra lutris*
Sea robin *see Prionotus*
Sea urchins **37**: 7, 131
Seabirds **43**: 34, 44, 207, 226
  see also Penguin
Seagrass **43**: 306
  Enhalus **41**: *150*, 175, 199
  Syringodium **41**: 154
  Thalassia hemprichii **41**: 149,174, 175
  Zostera **41**: 149, 150
Seal
  Antarctic fur (*Arctocephalus gazella*)
    43: 227
  crabeater (*Lobodon carcinophagus*)
    43: 227, 242, 243

leopard (*Hydrurgu leptonyx*) **43**: 228,
    242, 243
Weddell (*Leptonychotes weddellii*)
    43: 228, 229, 242–243
Seal, harbour *see Phoca vitulina*
Seaside sparrow *see Ammodramus maritimus*
Seastar *see Dermasterias*; *Evasterias*
seaweed (*Eucheuma cottonii* and
    Gracilaria) **41**: 200
Seba profundus **34**: **377**
Sebastes (rock fish) **39**: 17, 279; **25**: **52**;
    28: *46*; **30**: 313, 338; **35**: 69, **70**
  aleutianus **30**: 334
  aleutianus **35**: 30, **38**, 103
  alutus **30**: 335; **35**: **38**, 103
  elongatus **35**: **38**
  entomelas **30**: 335
  fasciatus **35**: **38**
  helvomaculatus **30**: 334; **35**: 30, **38**, 103
  jordani **30**: 335
  marinus **35**: **38**, 43, 66, 67, 103
  "*marinus* type" **35**: **38**
  melanops **30**: 334; **35**: 30, **38**, 103
  mentella **35**: **38**, 43, 67, 103
  "*mentella* type" **35**: **38**
  norvegicus **35**: 43
  proriger **30**: 334; **35**: 30, **38**, 103
  rosaceous **30**: 335
  variegatus **30**: 335
  viviparus **35**: **38**, 43, 103
Sebastes alutus **34**: 256
Sebastes diploproa **26**: 147
Sebastes flavidus **26**: 132
Sebastes marinus **26**: 33, *34*; **40**: 23
Sebastes schlegeli (rockfish) **38**: 31, 44, **54**
Sebastes sp. **26**: **120**, 123, 132
Sebastolobus **35**: 26
  alascanus **35**: 26, 30, **38**, 103
  altilevis **35**: 26, **38**, 50
  macrochir **35**: **38**
Sebatolobus alascanus **30**: 334
Segonzacia mesatlantica **32**: *100*, *101*, **129**;
    34: **378**
Segonzacia sp. **32**: 99, 102, 103, *104*, 107,
    109, **129**, 130
Selea **40**: 178
Selenoxiphus **42**: *144*
Semibalanus
  balanoides **34**: 50
  cariosus **34**: 48, 49
Semibalanus balanoides **39**: 26, 29

semisulcatus, Taxonomy **27**: 96, 98, **112**,
    123, 124, 125, 126
  Behaviour **27**: 333, 335, 336, 337, 340,
    341, 344, *344*, 345, 346, 348, 354,
    356
  Food/feeding **27**: **318**, 320, 322, 324,
    326, 328, 330, 331
  Life histories **27**: 286, 287, 290, 293,
    294, **305**, 310, **313**, 314
  Moulting/growth **27**: 226, **228**, **231**,
    235, **240**, 243, **250**
  Parasites **27**: 380, 388, 389
  Physiology **27**: 206
  Predation **27**: 375
  Reproduction **27**: 257, **259**, 260, 262,
    263, **271**, 277, **280**
  Zoogeography/Evolution **27**: **144**, **147**
Senecella **33**: 3, **53**, 128
  calanoides **33**: 128, **413**
  siberica **33**: 128
Sentusidium pelionense **31**: 306
Sepia **25**: 87, *95*, 103, 105, 108
  officinalis **25**: 99, 101, 103
Sepia aculeata, Predation **27**: 359, 360
  officinalis, Predation **27**: 360, 374
Sepia officinalis **44**: 161, 166, 177; **21**: 183,
    191; **30**: 235
Sepialla **25**: 87
  japonica **25**: 101, 104, 105
Sepiella inermis, Predation **27**: 359
Sepietta oweniana (sepiolid squid) **38**: 178
Sepiidae **32**: 289
Sepiola **44**: 171
Sepiola affinis **44**: 165
Sepiola robusta **44**: 172
Sepiola spp. **44**: 158
Sepiolid squid *see Sepietta oweniana*
Sepioloidea **32**: 206, 211, 220
Sepioloidea pacifica **32**: 211
Sepioloidea sp.cf. pacifica **32**: **164**
Sepioteuthis lessoniana **44**: 173, 174, 176;
    25: 96
Sepioteuthis sepioidea **30**: 163
Septibranchia **32**: **210**, 470
Septicollarina hemiechinata **32**: 217, **409**,
    410, *411*
Septicollarina oceanica **32**: **165**, 217, **409**,
    410
Septioteuthis lessoniana **39**: 274
Sergestes **25**: 119, 136, 141, 142, 144
  lucens **25**: 151

Sergestes, Zoogeography/Evolution
    27: 193
Sergestidae **32**: 62, 193
Sergia lucens **30**: **164**
Seriatopora **22**: 11
  hystrix **22**: 11, 48
Sericosura
  cochleifovea **34**: **374**
  cyrtoma **34**: **374**
  heteroscela **34**: **374**
  mitrata **34**: **374**
  venticola **34**: **374**
Sericosura heteroscela **32**: **127**
Sericosura mitrata **32**: **127**, 130
Seriola **21**: 62; **32**: 186
  lalandii **30**: 331
  quinqueradiata **30**: 223
Seriola lalandi **32**: **173**, 192, **196**
Seriolella
  brama **35**: **34**, 103, 109
  caerulea **35**: **34**, 103, 109
  punctata **35**: **34**, 103, 109
Seriolella labyrinthica **32**: **174**, 218, 224
Serolidae **39**: 158
Serolis cornuta **39**: 212
Serolis polita **39**: 212
Serolis tropica **39**: 212
Serpulidae **32**: *97*, **126**, **492**
Serpulorbis squamigerus **34**: 31
Serranidae **32**: 167, **172**, 211
Serripes groenlandicus (clam) **39**: 39
Sesarma angustipes **44**: 210
Sesarma guttatum **40**: 141
Sesarma leptosoma **40**: 142, 143
Sesarma meinertii **40**: 141, 144
Sesarma messa **40**: 142
Sesarma smithii **40**: 142
Sesbania **40**: 115
Sesuvium portulacastrum **40**: 99, 116
Setarches guentheri **32**: **172**
setiferus **27**: 1
    Behaviour **27**: 333, 336, 338, 340,
      343, 346, 347, 349, 350, 352,
      354
    Food/feeding **27**: **318**, 322, 323, 325,
      327, 330
    Life histories **27**: 285, 296, 302, **305**,
      306, 308, 309, 314
    Morphology **27**: 21, 33, 44
    Moulting/growth **27**: 215, 225, **232**,
      233, 239, 243, 245, **250**
    Parasites **27**: 383, 384, 385, 386, 387

Physiology **27**: **161**, 162, 163, 164, **168**, 187, 190, 193, 197, 200, **204**, 206, 209, 210
Predation **27**: 374, 375
Reproduction **27**: *254*, 257, **259**, 264, 265, **271**
Taxonomy **27**: 98, 107, **112**, 123, 124, 125
Zoogeography/Evolution **27**: 139, **146**
Sewellodhiron **25**: 153
Sexanymphon **24**: 26
mirabilis **24**: 69, 70
Shad *see Alosa sapidissima*
Shanny *see Blennius pholis*
Sheepshead *see Archosargus probatocephalus*
Shinkai
longipedata **34**: **367**
semilonga **34**: **367**
Shinkailepas **34**: 416
kaikatensis **34**: **370**
tufari **34**: **370**
Shrimp
caridean **43**: 98, 135, 142
Chorismus antarcticus **43**: 143, 148
Notocrangon antarcticus **43**: 143, 148
pandalid **43**: 148
Parapasiphae sulcatifrons **43**: *97*, 98
snapping (*Betaeus emarginatus*) **43**: 148, 149
Shrimp *see Pandalus*
Shulzeviella gigas spinosum **32**: **160**
Siboglinum **29**: 196
Siboglinum atlanticum **24**: 434; **35**: 30, **33**
Sicyonia **29**: 172
ingentis **29**: **132**, 171, 172, **188**, 189
Sicyonia nasica **32**: **163**
Sicyonia, Taxonomy **27**: 115, 123
brevirostris, Physiology **27**: 201
carinata, Reproduction **27**: 276
ingentis, Morphology **27**: 31, 32, 33
Reproduction **27**: 252, *252*, 263, 264, 265, *265*
Life histories **27**: 302
Predation **27**: 370
Sida **31**: 93
Siderastrea siderea **22**: 11, 12, 14, 17, 26, 30
Sigalionidae **39**: 36
Siganus
S. *doliatus* **43**: 53
S. *rivulatus* **43**: 66
S. *sutor* **43**: 53

Siganus canaliculatus **44**: 231
Siganus lineatus **44**: 231
Siganus punctatus **44**: 231
Sigmadocia **21**: 98
silasi, Taxonomy **27**: 96
Zoogeography **27**: **144**
Silicula **42**: 18, 20, 25
Siliculidae **42**: *5, 6*, 37
Siliqua patula (razor clam) **39**: 15
Sillaginidae **43**: 55, 56, *57, 58*
S. *aeolus* **43**: *58*
S. *burrus* **43**: *58*
S. *chrondropus* **43**: *58*
S. *ciliata* **43**: *58*
S. *lutea* **43**: *58*
S. *macrolepis* **43**: *58*
S. *shama* **43**: *58*
Sillago
Sillago analis, Predation **27**: 366
sihama, Predation **27**: 361, **364**, 366, 368
Sillago spp. **44**: 222
Silurus glanis **40**: 309, 317, 318
Silver gull (*Larus argentatus*) **43**: 44
Simenchelys parasitica **34**: **380**
Simenchelys parasiticus **32**: **168**
Simrothiella **23**: 324
Simulium **35**: *180*, **181**
Sinocalanus **33**: **52**, 67, 113
doerrii **33**: 113, **507**
sinensis **33**: 113, **507**
solstitialis **33**: 113
tenellus **33**: 113, **232**, 330
biochemistry **33**: **239**, **241**
ecology **33**: 365
growth/development **33**: **308**, **311**, **343**
nutrition **33**: 159, **201**
reproduction **33**: **255**, 260, 266, **270**, **287**, **288**
Sinophysis microcephalus **40**: 124
Sinularia abrupta **21**: 106, 113
Siphonaria **23**: 97; **34**: 45
atra **34**: 21, 22, 40
japonica **34**: 21, 22, 40
sirius **34**: 21, 22
Siphonobrachia lauensis **34**: **365**, 404
Siphonodentaliidae **42**: *177*; **42**: *144*
Siphonodentalium **42**: *144, 177*
S. *dalli antarcticus* **42**: 163
S. *galatheae* **42**: 205
Siphonostomatoida **32**: **128**; **33**: **2**, 2, *3*
Sipuncula **32**: 329, 366, 429, **438**, **504**
Sipunculoidea **32**: 333, **335**, **460**, **483**

Sipunculus **28**: 420
Sipunucula **43**: *21*
Siriella **25**: 154
Siriella armatai **39**: 142
Siriella chierchiae **39**: 217
Siriella clausii **39**: 217
Sitka black-tailed deer *see Odocoileus hemionus sitkensis*
Situla **32**: 373
Skate *see Raja*
Skeletonema **31**: 148, 363
Skeletonema costatum **33**: 290; **28**: 79; **29**: 36, 38; **31**: 92, 93; **20**: 40;
Skeletonema, Food/feeding **27**: 319
Skeneidae **32**: **127**
Sluiterina sp. **32**: 506
Smaragdinella **42**: 105
Smaragdinellidae **42**: 75
Smenospongia **21**: 98
Snapper *see Lutjanus griseus*
Snapping shrimp **43**: 148, 149
Snelliaetideus **33**: 125
Sognocalanus **33**: **53**, 87, 90, 123
  confertus **33**: 123
Solaster papposus **21**: 145, 168
Solasteridae **32**: 445, 455
Sole *see Hippoglossoides*; *Pleuronectes*
Sole *see Solea solea*
Solea
  aegyptiaca **29**: 221
  impar **29**: 221
  kleini **29**: 221, **222**
  lascaris **29**: 221, **222**, 229
  nasuta **29**: 221
  senegalensis **29**: 221, **222**
  solea, see Bristol Channel, sole
Solea solea (sole) **38**: 4, 34, 35, 54
  rearing in UK (1957–1967) 14–16
Solea solea **44**: 215, 226; **26**: **74**, 76, 141; **28**: 64, 67; **34**: 211
Solea vulgaris **26**: 127
Solemya **23**: 339; **42**: 31, 39, 42
  anatomy and morphology **42**: 8, *10, 12*, 14, 16
  development **42**: *27*, 28, 29
  ecology **42**: 34
  feeding **42**: *18*, 21, 22, 23
  reproduction **42**: 25, 26
  S. bartschii **42**: 29
  S. occidentalis **42**: 8, 22, 23
  S. reidi **42**: *10*, 22, 23, 25, 26, 28, 42
  S. togata **42**: *12*

S. velum **42**: 21, 22, 26, *27*, 28, 31, 39, 42
Solemya johnsoni **34**: **374**
Solemyoida **42**: 2, 4, *5*, 7, 9, 11, 24, 42, 43
  Nucinellina **42**: *5*
  Solemyina
    see also Solemya
    Solemyacea and Solemyidae **42**: 4, *5*, 6, 7, 8, 10–14, *10, 12*, 32, 37, 40
Solenocera, Taxonomy **27**: 115
  Predation **27**: 370
Solenodinium **25**: 127
  fallax **25**: *128*, 135
Solenopsis **40**: 145
Solmissus sp. **32**: 72
Solutigyra **34**: 415, 416
  reticulata **34**: **371**
Somateria mollissima **34**: 271
Somniosus **32**: 198
Somniosus microcephalus **26**: 41
Somniosus rostratus **32**: **168**
Sonneratia **40**: 99–100, 102, 107, 111, 178, 184
Sonneratia acida **40**: 106
Sonneratia alba **40**: 96, 104, 114, 163, 170
Sonneratia apetala **40**: 96, 99, 101, 111, 112
Sonneratia caseolaris **40**: 96, 112, 175
Sonneratia griffithii **40**: 96
Sonneratia lancelata **40**: 96, 170
Sonneratia ovata **40**: 96
Sonneratiaceae **40**: 96
Sonneratia × *gulngai* **40**: 96
Spadella **25**: 122, 151; **29**: 196
Spadentalina **42**: *144*
Sparidae **43**: 25, 65; **24**: 436
Spartina **31**: 49, 54; **39**: 43; **40**: 128
Spartina alterniflora **31**: 21, 22
Spartina anglica **31**: 23
Spartina maritima **31**: 21, 22
Spartina, Food/feeding **27**: 317, 319
  Predation **27**: 373, 374
Spartina ×. townsendii **31**: 23
Sparus aurata (sea bream) **38**: 33, 54
Sparus auratus **43**: 65
Spatangus purpureus **21**: 145
Spathebothrium simplex **40**: 7, 13, 21
Spelaemysis longipes **39**: 127
Spelaeogriphacea **39**: 107, 171
Speleonectes
  bevyjamini **29**: **131**, 186
  lucayesis **29**: **131**, 186
Sphacelaria **23**: 107
Sphaeragalma rotunda **24**: 116

Sphaerechinus granularis **21**: 146, 154
Sphaeripara **25**: 127, 128, 135
  catenata **25**: 134
Sphaerolaimus **30**: **50**
Sphaeroma **39**: 119, 128, 178
Sphaeroma hookera **39**: 212
Sphaeroma peruvianum **40**: 139
Sphaeroma quadridentatum **39**: 213
Sphaeroma rugicauda **39**: 152, 177, 212
Sphaeroma serratum **39**: 155, 178, 213
Sphaeroma terebrans **40**: 139
Sphaeromatidae **39**: 158
Sphaeronectes **24**: 107, 108, 154, 225;
  25: 124
  bougisi **24**: 121
  fragilis **24**: 121
  gamulini **24**: 121
  gracilis **24**: 121, 230, 234, 239, 244
  irregularis **24**: 121, 169
Sphaeronella **39**: 184
Sphaerosyllis ridgensis **34**: **368**
Sphaerothuria bitentaculata **32**: **461**
Sphaerotrichia divaricata **21**: 92, 93
Sphaerozoum **25**: 125, 127
Spheciospongia vesparium **21**: 97
Sphenaria ezogremena **32**: 408, **409**
Spheniscus demersus **20**: 247
Spheniscus magellanicus **39**: 281
Sphincterochila boisseri **34**: 32
Sphraenops bairdianus **32**: **172**
Sphyrapus **39**: 159
Sphyrion lumi **40**: 9, 21
Sphyrna lewini **44**: 223; **39**: 280
Spicipes **33**: **53**, 66, 130
  nanseni **33**: 130
Spiliphera **30**: **50**
Spilocuma salomani **39**: 218
Spilocuma watlingi **39**: 218
Spinocalanidae **33**: **52**, 123, 124, **152**, **499**
Spinocalanoidea **33**: **52**, **53**, *54*, *55*, 123, 124
Spinocalanus **25**: 138
  ecology **33**: 486, **499**, 528
  taxonomy **33**: **53**, 78, 80, 124
Spinocalanus abruptus **33**: 124
  abyssalis **33**: 124
  angusticeps **33**: 124
  antarcticus **33**: 124
  aspinosus **33**: 124
  brevicaudatus **33**: 124, 479, 502
  brevicornis **33**: 502
  dispar **33**: 124
  hirtus **33**: 124

  hoplites **33**: 124
  horridus **33**: 124
  longicornis **33**: 124, 502
  macrocephalon **33**: 124
  magnus **33**: 124
  oligospinosus **33**: 124
  polaris **33**: 124
  profundalis **33**: 124
  similis **33**: 124
  spinosus **33**: 124
  terranovae **33**: 124
  usitatus **33**: 124
  validus **33**: 124
Spinosella vaginalis **21**: 97
Spinula **32**: 349; **42**: 14
Spinula calcar **32**: 503
Spinula oceanica **32**: 503
Spinula vityazi **32**: 370
Spinulidae **42**: *6*; **42**: *5*
Spiny dogfish *see Squalus acanthias*
Spio spp. **44**: 132
Spionidae **32**: **126**; **39**: 36; **43**: 143
Spirastrella inconstans **21**: 97
Spirillum, Parasites **27**: 382
Spirobidae **39**: 36
Spirobrachia **32**: 373
Spirocodon saltatrix **43**: 102
Spiromonas **25**: 124
Spirontocaris lilljeborgii **38**: 179
Spirontocaris sp. **38**: 105
Spirontocaris spinus **38**: 179
Spironucleus barkhanus **40**: 56
Spironucleus torosa **40**: 4, 11, 13–14, 30,
  49, 55, 56
Spirula **25**: 87, 101, 102, 109
Spirulina subsala **21**: 87
Spirurida **24**: 278
Spiruroideorum **40**: 8
Spisula **21**: 178; **35**: **163**, **169**
  sachalinesis **35**: **185**
Spisula sachaliensis and *S. solidissima*
  43: 104
Spisula solidissima **22**: 103, 129, 130, 134;
  37: **10**, 43
  bioassay procedures **37**: 52, 56, 67, 68
  toxicity testing **37**: **91**, 95
Splendrillia basilirata **32**: **164**
Splendrillia obscura **32**: **164**
Sponges **43**: 150
Spongia **32**: **335**, 357, 467, 469, **483**, **486**,
  492, 495, **504**
Spongia officinalis **21**: 99, 103

Spongicola parvispina **32**: **162**
Spongosorites suberitoides **42**: 216
Spotted seatrout *see Cynoscion nebulosus*
Sprattus **20**: 21
  fuegensis **20**: 180
  sprattus **20**: 6, 11, 18, 39, 86, 280
Sprattus sprattus **24**: 264; **25**: **35**; **28**: 27,
  118
Squalidae **32**: 167, **168**, 211
Squalogadus **32**: 186
Squalogadus modificatus **32**: **171**
Squalus **32**: 186
Squalus acanthias (spiny dogfish) **38**: 178;
  26: 144; **34**: 280
Squalus mitsukurii **32**: **168**, **192**, 193, **194**,
  223
Squamariaceae **32**: 175
Squid *see Loligo*; *Nototodarus*;
  Ommastrephes
Squilla holochista **29**: **131**, 184, *185*
Staphylococcus **40**: 115, 116
Starry ray *see Raja radiata*
Staurophora mertensi **25**: 23
Stechodactyla haddoni **21**: 107
Steganoderma formosum **40**: 6, 13, 21
Stegastes planifrons **44**: 228
Stegocephalidae **32**: **128**
Stegocephaloides christianiensis **39**: 205
Stegocephalus inflatus **39**: 114, 205
Steleuthera ecoprophycea **32**: **128**; **34**: **377**
Stellarima microtrias **31**: 353
Stellaroidea **31**: 39
Steller sea lion *see Eumetopias jubatus*
Stellifer lanceolatus, Predation **27**: **364**
Stenakron vetustum **40**: 6, 13, 21
Stenetrium **39**: 118
Stenobrachius leucopsarus **32**: 266, 305,
  306
Stenobrachius nannochir **32**: 306
Stenobrachius spp. **32**: 303
Stenocollum fragile **40**: 6, 11
Stenogammarus compressus **39**: 205
Stenogammarus kereuschui **39**: 205
Stenogammarus macrurus **39**: 205
Stenogammarus similis **39**: 205
Stenopodidea **32**: **162**
Stepanjantsia polymorpha **24**: 116
Stephalia corona **24**: 118, 200
  (Stephonalia) bathyphysa **24**: 118
Stephanophyes **24**: 200
  superba **24**: 119, 200, 225, 232, 234, 239
Stephanopyxis palmeriana **31**: 352

Stephanoscyphus simplex **32**: 333
Stephanostomum baccatum **40**: 6, 16, 27
Stephanostomum caducum **40**: 6, 13–14
Stephanostomum pristis **40**: 6, 13, 27, 33, 36,
  37
Stephidae **33**: **53**, 99, 137, 138, 470, **489**,
  500
Stephos
  behaviour **33**: 402
  ecology **33**: 469, 497, **500**, 501, 528
  taxonomy **33**: **53**, 85, 88, 89, 137, 138
Stephos antarcticus **33**: 137
  arcticus **33**: 137
  balearensis **33**: 137, **489**
  canariensis **33**: 137, **500**
  deichmannae **33**: 137
  exumensis **33**: 137
  fultoni **33**: 137
  gyrans **33**: 137
  kurilensis **33**: 137
  lamellatus **33**: 137
  longipes **33**: 137, **361**, **387**, **493**, 494
  lucayensis **33**: 137
  maculosus **33**: 137
  margalefi **33**: 137, **489**
  minor **33**: 137
  morii **33**: 137
  pacificus **33**: 137
  pentacanthos **33**: 137
  robustus **33**: 137
  rustadi **33**: 137
  scotti **33**: 137
  seclusum **33**: 138
  tropicus **33**: 138
  tsuyazakiensis **33**: 138
Stephos longipes **43**: 210, 211
Stercorarius
  parasiticus **34**: 268
  skua **34**: 268
Stercorarius parasiticus **20**: 282
Sterculiaceae **40**: 96
Stereocidaris **32**: *214*
Stereocidaris creptiferoides **32**: 217
Stereocidaris japonica **32**: 217
Stereocidaris nascaensis **32**: **165**, 174, 175,
  178, 180, 182, 217
Stereolepis gigas **34**: 237
Steringophorus furciger **40**: 6, 13, 21
Sterna hirundo **40**: 155
Sterna paradisaea **34**: 268
Sternoptychidae **32**: 62, **169**, 211, 266
Sternopygus **40**: 308

Stethopristes eos **32**: *172*, 188, **218**
Sthenoteuthis oualaniensis **32**: 259;
    39: 274
Stichaeus (prickleback) **39**: 48, 49
Stichaeus punctatus (Arctic shanny)
    39: 49–51
Stichaster australis **23**: 73
Sticholonche zanclea **25**: 128
Stichopathes **30**: 319
Stichopus chloronotus **21**: 146
    japonicus **21**: 146, 159
    variegatus **21**: 146, 148, 156, 169
Stichopus hermani **41**: 176
Stichopus japonicus **32**: 457
Stiliola subula **32**: **453**
Stizostedian vitreum **25**: *13*, 64, *65*
Stoichactis **21**: 125; **22**: 92
    helianthus **21**: 124, 125; **22**: 93
    kenti **21**: 124
Stoichactis helianthus **24**: 222; **33**: 503
Stolephorus **20**: 14, 15, 21, 22
    purpureus **20**: 6, 11, 39, 107
Stolephorus purpureus **25**: 41, 42; **28**: 10
Stolidobranchiata **32**: **469**
Stolonifera **43**: *291*
Stoloteuthis sp. **32**: 206, 220
Stoloteuthis sp.n.aff. *leucoptera* **32**: **164**,
    204
Stolothrissa **20**: 160
Stomacontion **39**: 114
Stomias boa ferox **26**: **120**
Stomiatidae **32**: 62
Stony corals **43**: 290
Storthyngura **32**: 359, **446**, 463
Storthyngura bicornis **32**: **446**
Storthyngura birsteini **39**: 213
Streblospio benedicti **33**: 293; **43**: 125, 131,
    132, 136, 137, 143
Streptodajus **25**: **154**
Streptotheca **32**: 560
Strictosiphonia kelanensis **40**: 125
Striocadulus **42**: *144*
Striodentalium **42**: *143*
Striopulsellum **42**: *144*
Striped bass *see Morone saxatilis*
Striped mullet *see Mugil cephalus*
Strombidium **25**: 131; **29**: 105, *106*
Strombidium sulcatum **33**: 164
Strombilidium **25**: 131
Strombus gigas **34**: 23, 33
Stronglyocentrotus franciscanus **31**: 25

Strongylocentrotus **24**: 52; **35**: **163**; **37**: 131;
    43: 127, 128
    droebachiensis **23**: 86; **35**: 92
    purpuratus **23**: 75
    S. *droebachiensis* **43**: 146, 147, 149
    S. *franciscanus* **43**: 146
Strongylocentrotus drobachiensis **21**: 146
    purpuratus **21**: 146, 150, 156
Strongylocentrotus droebachiensis **31**: 16;
    34: 233; **39**: 41, 73, 77
Strongylocentrotus pallidus **31**: 16
Struthiolaria **28**: *392*
Sturgeon *see Acipenser*
Stygiomysidae **39**: 160
Stygiopontius **34**: 402, 419
    appositus **34**: **376**
    brevispina **34**: **376**
    bulbisetiger **34**: **376**
    cinctiger **34**: **376**
    cladarus **34**: **376**
    flexus **34**: **376**
    hispidulus **34**: **376**
    latulus **34**: **376**
    lauensis **34**: **376**
    lumiger **34**: **376**
    mirus **34**: **376**
    mucroniferus **34**: **376**
    paxillifer **34**: **376**
    pectinatus **34**: **376**, 419
    quadrispinosus **34**: **376**
    regius **34**: **376**
    rimivagus **34**: **376**
    sentifer **34**: **376**
    serratus **34**: **376**
    stabilitus **34**: **376**
    teres **34**: **376**
    verruculatus **34**: **376**
Stygiopontius bulbisetiger **32**: **128**
Stygiopontius cladarus **32**: **128**
Stygiopontius latulus **32**: **128**
Stygiopontius mirus **32**: **128**
Stygiopontius pectinatus **32**: **128**, 131
Stygiopontius regius **32**: **128**
Stygiopontius serratus **32**: **128**
Stygiopontius teres **32**: **128**
Stygocyclopia **33**: **53**, 78, 134
    balearica **33**: 134, **489**
Stylaster **43**: 292
    S. *elegans* **43**: *291*
    S. *roseus* **43**: *291*
    S. *sanguineus* **43**: *291*

Stylasterina **43**: *291*
stylirostris, Taxonomy **27**: 97, **112**, 123
  Food/feeding **27**: 321, 326, 330
  Life histories **27**: 307
  Moulting/growth **27**: 215, **228**, **232**, 235, 239, **250**
  Parasites **27**: 380, 381, 382
  Physiology **27**: **161**, 163, **168**, **204**, 205, 206, 210
  Predation **27**: 359
  Reproduction **27**: **280**
  Zoogeography/Evolution **27**: **145**
subtilis, Taxonomy **27**: 97
  Food/feeding **27**: 319
  Zoogeography/Evolution **27**: **146**
teraoi (*see Penaeus marginatus*)
trisulcatus (*see Penaeus kerathurus*)
vannamei, Taxonomy **27**: 97, **112**, 123
  Behaviour **27**: 337
  Food/feeding **27**: **318**, 327
  Life histories **27**: **305**
  Moulting/growth **27**: **228**, **232**, 235, 237, 239, 241, **250**
  Parasites **27**: 380, 384, 385
  Physiology **27**: 160, **161**, 162, **168**, **204**, 205, 208
  Predation **27**: 358, 359
  Reproduction **27**: 260, 262, **280**
  Zoogeography/Evolution **27**: **145**
Stylocheilus longicauda **21**: 88, 181
Stylocheiron **25**: 141; **32**: 193;
  carinatum **25**: 154
  longicorne **25**: 154; **29**: **131**, 183
Stylochus **21**: 31
  iijimai **21**: 31
  neapolitanus **21**: 133
  ostreophagus **21**: 31
Stylodactylus libratus **32**: 217
Stylodactylus multidentatus **32**: 217
Stylodactylus pubescens **32**: **163**, 216
Stylodinium **25**: 126
  gastrophilum **25**: 131
Stylomesus inermis **32**: 366
Stylomysis **25**: 137
Stylopallene longicauda **24**: 31
Stylophora **22**: 11, 17; **31**: 230; **32**: **159**
  pistillata **22**: 11, 16, 19, 46
Stylophora pistillata **31**: 247, 260, 264; **32**: 158, **159**
Stylosatomum ellipse **21**: 133
Stypocaulon scoparium **24**: 57
Styracaster chuni and *S. horridus* **43**: 147

Styracaster sp. **32**: **437**, **462**
Suaeda **40**: 163
Subcancilla filaris **28**: *393*
Suberites domunculus **21**: 97, 99
  inconstans **21**: 96, 100
Subeucalanus **33**: 122
Suevidontus **42**: *144*
Sula bassana **34**: 268; **20**: 281
  capensis **20**: 247
Sulcanidae **33**: **52**, 116
Sulcanus **33**: **52**, 66, 116
  conflictus **33**: 116, **308**, **311**, 469
Sulcogadila **42**: *144*
Sulculeolaria **24**: 139, 179, 188, 199, 205, 232; **25**: 131
  angusta **24**: 120
  bigelowi **24**: 120
  biloba **24**: 120
  brintoni **24**: 119, 120
  chuni **24**: 120
  monoica **24**: 120
  pacifica **24**: 119, 120
  quadrivalvis **24**: 120, 224, 230, 234
  tropica **24**: 119, 120
  turgida **24**: 120
Sunflower star *see Pycnopodia helianthoides*
Surf bird *see Aphriza virgata*
Surf scoter *see Melanitta perspicillata*
Sursamucro **33**: **53**, 80, 88, 128
  spinatus **33**: 128
Sutilizona sp. **32**: **127**
Sutilizona theca **34**: **371**
Swampomyces armeniacus **40**: 120
Sycia **25**: 141
Sygnathus fuscus **44**: 229
Syllidae **39**: 36
Sylon hippolytes **38**: 182
Syltodinium listii **33**: 524
Sylvilagus palustris hefneri **40**: 158
Symbiodinium **25**: 125; **43**: 285, 292
Symbolophorus **32**: 63
Symmetromphalus **34**: 416
  hageni **34**: **369**
  regularis **34**: **369**
Sympagurus **32**: 166, 216
Sympagurus affinis **32**: **163**, 216
Sympagurus africanus **32**: **163**, 216
Sympagurus boletifer **32**: **163**, 216
Sympagurus rectichela **32**: **163**
Sympagurus ruticheles **32**: **163**, 216
Sympagurus wallisi **32**: **163**, 216
Symphorus spilurus **44**: 231

Symphysanodon **32**: 192
Symphysanodon maunaloae **32**: **173**, **195**, **218**
Sympronoe parva **24**: 205
Synactinomyxon **25**: 146
Synallactidae **32**: 466
Synaphobranchidae **32**: **129**, **168**
Synaphobranchus **32**: 109, 198
Synaphobranchus affinis **32**: **168**, 192, **194**, 206
Synaphobranchus kaupi **34**: **380**
Synaphobranchus sp. **32**: **129**, 186
Synapseudes **39**: 159
Synapta maculata **21**: 148
Synaptidae **32**: **462**, 466
Synaptobothrium caudiporum **40**: 6, 13, 20
Synaptura lusitanica **29**: 221
Syndermata **43**: *21*
    see also Acanthocephala; Rotifera
Syndinium **25**: 127, 133; **33**: 524; **38**: 180
    oikopleurae **25**: 134
Synechococcus (cyanobacterium) **41**: 47, 58, 64, 65
    S. bacillaris **41**: *36*
Synechococcus **29**: 80, 91
Syngnathus fuscus **44**: 210
Synhemidinium **25**: 127
Synodontidae **32**: **169**
Synodus doaki **32**: **169**, **218**
Synodus foetens, Predation **27**: **365**
Syringodium (seagrass) **41**: 154; **24**: 60, 76
Syringolaimus **30**: 48, **50**
Systellapsis braueri **34**: **378**; **23**: 330
Systellaspis debilis **43**: *97*

Taaningichthys **32**: 63
Tachypleus gigas **21**: 140
Tacita **32**: 366
Takifuga pardalis **40**: 282
Talacalanus **33**: **53**, 81, 82, 132
    greeni **33**: 132
    maximus **33**: 132
Talismania **32**: 206
Talismania bussingi **32**: **169**, 220
Talitrus **30**: 108
Talitrus saltator (sand hopper) **39**: 109, 125, 126, 145, 156, 185, 205
Talorchestia
    capensis **30**: 114, 119
    martensii **30**: 112
Talorchestia margaritae **39**: 205
Talorchestia martensii **39**: 205

Tamaria sp. **32**: **166**, 176
Tamoya gargantua **21**: 105
    haplonema **21**: 105
Tanaidacea **32**: **161**, **216**, 329, **335**, 347, 360, 362, 407; **39**: 107, 159, 186, 187, 189, 193, 219
Tanais **39**: 151, 160
Tanais dulongi **39**: 176, 180, 181
Tanypenaeus, Taxonomy **27**: 99–100, 107, **113**
    caribeus Taxonomy **27**: 100
    Zoogeography/Evolution **27**: **146**
    Physiology **27**: 203
    Zoogeography/Evolution **27**: **146**
Tanystylum **24**: 4, 8, 11, 26, 69, 70
    brevipes **24**: 71
    cavidorsum **24**: 55, 60
    conirostre **24**: 57, 66
    oedinotum **24**: 60
    orbiculare **24**: 39, 44, 66
    styligerum **24**: 60
Tanytarsus **35**: **169**, 176, *180*, **181**
Tapes (Ruditapes) philippinarum **24**: 52
Tapes **21**: 179
    japonica **21**: 179, 189
    semidecussata **21**: 170
Tapes philippinarum **34**: 211; **37**: **10**, 46
Tautogolabrus adspersus **44**: 217, 218
Tealia **22**: 74
Tectibranchia **42**: 67, 70, 75
    see also Opisthobranchia
    characters and evolutionary
        pathways **42**: 99, 100, 102, 103, 104, 105
    taxonomic history **42**: 73, 107, 108, 111, 112
Tecticeps japonicus **39**: 110
Tectovalopsis
    diabolus **34**: **377**
    wegeneri **34**: **377**
Tectura (Collisella) scutum **34**: 49
Tectura persona (limpet) **39**: 24, 26, 72
Tedania nigrescens **21**: 96, 97
    toxitalis **21**: 95, 97
Teleost fish **43**: 25, *26, 32*, 47, 51
    see also Acanthopagrus
Teleostei **43**: *29*
Tellina **35**: **169**
    fabula **35**: **168**, **173**, *174*
    tenuis **35**: **173**, *174*
Tellina fabula **22**: 103, 109
Tellina perna **24**: 53

Tellinacea **32**: 463
Tellinidae **39**: 36
Telmapsylla **40**: 145
Telmatactis vermiformis **21**: 106
Telmessus **39**: 32, 41, 42, 49, 52
Telmessus cheiragonus (helmet crab) **39**: 37,
    38, 41
Temnocephalida **43**: *22*
Temnocinclis euripes **34**: **371**
Temnozaga parilis **34**: **371**
Temora **25**: 147, 151
    behaviour **33**: 402, 407, 423
    ecology **33**: **397**, 480, 481, 486, 490, 497,
        524, 528
    life history **33**: 347, 389
    nutrition **33**: 152
    reproduction **33**: 257
    size **33**: *332*
    taxonomy **33**: **52**, 76, 93, 117
Temora discaudata **33**: **47**, 117, **515**
    kerguelensis **33**: 117
    longicornis **33**: 6, **7**, **26**, 298
        behaviour **33**: 403, **408**, **413**, 414, **417**,
            418, 419, 420, 434, 444, 454
        biochemistry **33**: 247, 251
        distribution **33**: **515**
        ecology **33**: **362**, 366, 369, 372, 373,
            379, 380, **397**, 469, 472, 475, 481
        growth/development **33**: **309**, **311**, **312**,
            313, **320**, **343**
        life history **33**: 340, 388
        nutrition **33**: 143, 145, 150, 156, 159,
            161, 164, 166, 170, **184**, **186**,
            187, **191**, 199, **201**, 204, 205,
            206, 207
        reproduction **33**: 257, **271**, 282, **287**,
            291
        size/weight **33**: **232**, **331**, 332, 333,
            335, **337**
        taxonomy **33**: 49, 63, 117
    stylifera **33**: **7**, **26**, 117, 236
        behaviour **33**: **408**, **413**, **446**
        chromosomes **33**: **47**
        distribution **33**: **515**
        ecology **33**: **362**, 364, 370, **379**, **397**,
            480, **484**
        growth/development **33**: **309**, **311**, **312**,
            315, 316, 338
        nutrition **33**: 181, **184**, **191**, **201**, **206**,
            208
        physiology **33**: 215

    reproduction **33**: 43, **255**, 260, 265,
        283, **287**, 290
    size/weight **33**: **225**, 335
    turbinata **33**: **7**, **26**, 117
        behaviour **33**: 403, **446**, 447, 451
        biochemistry **33**: 251
        chromosomes **33**: **47**
        distribution **33**: **515**
        growth/development **33**: **309**, **320**, **343**
        life history **33**: 383
        nutrition **33**: **198**, **201**
Temora longicornis **44**: 104–108; **29**: *48,
    55*; **30**: 250
Temora stylifera **44**: 108
Temora turbinata **44**: 108, 109
Temoridae **33**: 3, **52**, 116, 117, 268, 434,
    473
Temorites **33**: **52**, 76, 122, **498**
    brevis **33**: 122
    discoveryae **33**: 66, 122
Temoropia **33**: **52**, 77, 117
    mayumbaensis **33**: 117
    minor **33**: 117
    setosa **33**: 117
Tenagomysis macropsis **39**: 217
Tenagomysis tasmaniae **39**: 217
Tenellia aspera **43**: 146
Teneriforma **33**: **53**, 87, 89, 124, **499**
    meteorae **33**: 124
    naso **33**: 87, 124
    pentatrichodes **33**: 124
Tentacularia coryphaenae **40**: 7, 17, 25, 33
Terapon jarbua **44**: 222
Terapon jarbua, Predation **27**: 365, 368
Terebellidae **32**: 463
Terebra **25**: 186; **28**: *393*
Terebra affinis **28**: *393*
Terebra babylonia **28**: **403**, 410
Terebra cinerea **28**: *393*, **403**, **405**, 408, 410,
    413–420
Terebra cingulifera **28**: 410
Terebra glauca **28**: *393*
Terebra gouldi **28**: 410, 411
Terebra maculata **28**: *393*, **403**, 410
Terebralaia palustris **40**: 148
Terebratalia **28**: *353*
Terebratalia coreanica **28**: *213*, 275, *353*
Terebratalia transversa see subject index
Terebratella **28**: **202**, **206**, 231, 234
Terebratella inconspicua **28**: *319*
Terebratella sanguinea **28**: 178, **204**, 221,
    245, 249, 265, 271, 276, 319

Terebratulina **28**: 229, 250, 275
Terebratulina cailleti **28**: 211
Terebratulina retusa
  nomenclature **28**: 178
Terebratulina septentrionalis **28**: 211, *213*,
    265, 275, 306, 309, 326, 327, 330,
    341, 348, 349
Terebratulina unguicula **28**: 223, 224, 237,
    265, 275, 283, 285, 309, 312, 316,
    319, 330
Teredo furcifera **40**: 149
Tergestia **25**: 148
Tesseracme **42**: *143*
Testicardines **32**: **409**
Tethya actinia **21**: 99
  aurantia **21**: 99
  crypta **21**: 102
Tethys leporina **24**: 52
Tetraclita **22**: 229, 233, 234
  purpurascens **22**: 204
  stalactifera **22**: 228
Tetragantha nitens **40**: 177
Tetragonurus cuvieri **35**: **39**
Tetrahymena pyriformis **21**: 164
Tetraphyllidea **32**: 306
Tetraselmis chui **31**: 92
Tetraselmis chuii (alga) **41**: 182, 199
Tetraselmis suecica **25**: 227; **34**: 28; **37**: 26,
  39
Tetrochetus **25**: 148
Tevnia **23**: 325, 335; **34**: 408
  jerichonana **34**: **366**, *409*
Tevnia jerichonana **35**: 50
Thais **21**: 170
  carnifera **25**: 225
  clavigera **21**: 31
  floridana **21**: 177
  haemastoma **21**: 177
  lapillus **25**: 225
Thais chocolata **35**: **185**
Thais emarginata **24**: 285
  lamellosa **24**: 285
Thais emarginata and *T. lamellosa* **43**: 146
Thais kiosquiformis **40**: 148
Thais lapillus **23**: 87
Thalamita crenata **40**: 141
Thalassarche chrysostoma **39**: 281, 286
Thalassarche melanophrys **39**: 281
Thalassia **24**: 60; **38**: 115
Thalassia hemprichii (seagrass) **41**: 149,
    174, 175; **40**: 127
Thalassia testudinum **23**: 100, 102; **40**: 128

Thalassicolla **25**: 125, 127, 135
  nucleate **25**: *128*
  spumida **25**: *128*
Thalassionema nitzschioides **32**: *552*;
    40: 124
Thalassiophyllum clathrus **23**: 82
Thalassiosira (diatom) **41**: *34*, 36, 62
  T. oceanica **41**: *31*, *34*, 35, *36*, 65, *66*, 69,
    70
  T. pseudonana **41**: 30, *31*, *34*, 35, *36*, 37,
    38, 65, *66*, 69, *70*
  T. weissflogii **41**: 26, 27, *26*, *31*, 32, *34*,
    65, *66*, 69, *70*, 71, 76
Thalassiosira **31**: 92; **33**: 332; **20**: 27, 40
  partheneia **33**: 155
  rotula **20**: 40
  wessflogii **33**: 164, 181
Thalassiosira antarctica **31**: 353; **32**: 535,
    551
Thalassiosira gracilis **32**: *552*
Thalassiosira gravida **32**: 529, 535, *543*,
    551, 555
Thalassiosira hyalina **32**: *549*
Thalassiosira nordenskioldii **32**: *531*, 536,
    549
Thalassiosira oestrupii **32**: *550*
Thalassiosira pacifica **32**: 529, 544, 549
Thalassiosira pseudonana **29**: 78; **31**: 93, 149
Thalassiosira weissflogii **31**: 93, 149
Thalassiothrix frauenfeldii **40**: 124
Thalassiothrix karstenii **32**: *550*
Thalassiothrix lanceolata **32**: *550*
Thalassoalaimus **30**: **50**
Thalassodendron ciliatum **40**: 128
Thalassolampe margarodes **25**: 124
Thalassoma bifasciatum **44**: 230, 248;
    34: 255
Thalassomyces **25**: 136
  albatrossi **25**: 137
  fagei **25**: 137
  fasciatus **25**: 137
  marsupii **25**: 137
Thalassophyes ferrarii **24**: 121
Thalassophysa **25**: 125
Thalia **25**: 156
  democratica **25**: 139
Thalia democratica **44**: 9, 22, 31, 134, 135
Thalicola **25**: 143
  ensiformis **25**: 143
  filiformis **25**: 143
Thallasiosira weisflogi **29**: 12
Thalyseurypon **21**: 99

Tharybidae **33**: **53**, 99, 138, **152, 500**, 502
Tharybis **33**: **53**, 80, 84, 87, 90, 138
  altera **33**: 138
  angularis **33**: 138, **501**
  asymmetrica **33**: 138
  compacta **33**: 138
  crenata **33**: 138, **501**
  fultoni **33**: 138
  macrophthalma **33**: 138
  magna **33**: 138
  megalodactyla **33**: 138
  minor **33**: 138
  neptuni **33**: 138
  sagamiensis **33**: 138
Thecacineta **25**: 138
Thecidellina **28**: **183**, 265, 275, 340
Thecosomata **42**: 73, 99, 111
Thelenota ananas **21**: 147, 148, 169; **41**: 190
Thelohamia hereditaria **39**: 113
Thelohania butleri **38**: 180
Thelohania sp. (*see Agmasoma*)
  duorara (*see Agmasoma*)
Themisto **32**: 303
Themisto japonica **44**: 126, 127; **32**: 17, 265, 266, 274, 303, 306
Themisto libellula **32**: 19, 265
Themisto pacifica **32**: 265, 266, 269, 274, 303
Themisto sp. **26**: 131
Thenus orientalis **29**: **132, 163**, 164, 194
Theora lata **24**: 52
Theragra chalcogramma (Alaska pollock) **38**: 3 53, 178
Theragra chalcogramma (walleye pollock) **39**: 60
Theragra chalcogramma **28**: 51–53, **53**, 106; **44**: 250; **25**: 40, 59; **26**: 129; **30**: 235; **32**: 266, 267, 281, 283, 303, 304; **35**: **35, 45**, 50, 84
Theragra chalcogramma see Walleye pollock *in subject index*
Theragra finmarchicus **37**: 183
Theristus **30**: **50**, 50
Thermaces **23**: 337
Thermaloniscus cotylophorus **34**: **377**
Thermanermertes valens **34**: **365**
Thermarces andersoni **35**: 28
Thermarces cerberus **34**: **381**
Thermiphione
  fijiensis **34**: **368**
  tufari **34**: **368**
Thermobiotes mytilogeiton **34**: **380**

Thermochiton undocostatus **34**: **369**
Thermopalia **24**: 198
  taraxaca **24**: 113, 114, 118, 154
Thermopalia taraxaca **23**: 319; **34**: **365**
Thermopolynoë branchiata **34**: **368**
Thermosbaena **39**: 172
Thermosbaena mirabilis **39**: 118, 120, 122, 156
Thermosbaenacea **39**: 107, 161
thermydron **23**: 303, 330, 331, 344
Thimopides **34**: **378**
Thimopides sp. **32**: **129**
Thiomicrospira **23**: 316; **35**: 49
Thioploca **35**: 22
Thiothrix **23**: 316
Tholozodium ocellatum **39**: 213
Thoracaster magnus **32**: **462**
Thoracaster sp. **32**: **462**
Thoracica **22**: 200
Thoracotylidae **43**: 9
Thraustochytridae **43**: 207
Thraustochytrium striatum **40**: 119
Threadfin *see Polydactylus sexfilis*
Thryssa baelama, Predation **27**: 358, **363**
  hamiltoni, Predation **27**: 368
Thunnus
  alalunga **30**: 334
  albacares **30**: 334
  thynnus **30**: 334
Thunnus alalunga **28**: 99; **32**: 270; **39**: 281
Thunnus albacares **39**: 280, 281; **44**: 252
Thunnus maccoyii **28**: **89**, 99, 122
Thunnus obesus **39**: 280, 281, 283
Thunnus thynnus **44**: 254; **28**: 56, 57, 101; **34**: 259
Thyasiridae **32**: 125; **42**: 23, 24
Thynnascaris **25**: 151; **32**: 308
Thyropus edwardsii **24**: 205
Thyrsites atun **20**: 252; **35**: 109
Thysanoessa **25**: 141, 151, 154; **32**: 303
  inermis **25**: 137, 144, 151
  inermis **30**: 176, 178
  longicauda **25**: 152
  longicaudata **30**: 176
  longipes **25**: 152
  raschi **25**: 23, **34**, 137, 152
  raschii **30**: 178
  spinifera **30**: 176, 187
Thysanoessa inermis **44**: 128; **32**: 265, 266, 274, 303
Thysanoessa inspinata **32**: 266, 274

Thysanoessa longipes **32**: 265, 266, 274, 303, 306
Thysanoessa macrura **43**: 219
Thysanoessa raschi **44**: 128; **32**: 265, 266, 274, 303
Thysanoessa spinifera **44**: 128
Thysanopoda **25**: 150
   aequalis **29**: **131**, 183
   orientalis **29**: **131**, 183
   tricuspidata **29**: **131**, 183
Thysanopoda sp. **32**: 116, 193
Thysanoteuthis rhombus **44**: 168
Thysanozoon brocchi **21**: 133
Tiaropsis multicirrata **25**: **31**
Ticosina **28**: *353*
Tigriopus **20**: 183
   californicus **33**: 155
   fulvus **33**: 218
   japonicus **33**: 364
Tigriopus fulvus **44**: 302
Tigriopus japonicus **44**: 300, 308
Tigriopus spp. **44**: 300, 311
Tilapia
   mossambica **34**: 83, 84, 85
   rendalli **34**: 84
Tilapia **43**: 65
Tilapia mossambica **26**: 152
Tilapia *see Oreochromis niloticus*
Timea bairdii **25**: 156
Tinca tinca **26**: 84; **40**: 260, 261, 268
Tindaria **32**: **437**
   T. callistiformis **42**: 24, 29
   T. cervola **42**: 26
Tindariidae **42**: *5*, 43
Tinogullmia riemanni **30**: 53
Tintinnopsis **25**: 129; **40**: 129
Tintinnopsis lobiancoi **29**: 105, *106*
Tintinnus **25**: 129
   fraknoii **25**: *130*
Tironucula jugata **42**: 9
Tisbe **44**: 301, 304, 305, 310, 311; **20**: 183
Tisbe carolinensis **44**: 300, 302
Tisbe clodiensis **44**: 300
Tisbe furcata **44**: 300, 307
Tisbe holothuriae **29**: **131**, 136, 179, 183; **44**: 298, 299, 300, 302, 303, 304, 306, 307, 308, 309, 312
Tisbe sp. **26**: *131*
Tisbe spp. **44**: 305, 307, 311
Tisbidae **43**: 216
Tjaernoeidae *see* Lower Heterobranchia
Tmetonyx **39**: 205

Tmetonyx cicada **40**: 11
Todarodes pacificus **32**: 259, 270, 275; **39**: 264, 267, 273, 276, 279, 283, 288, 289; **44**: 151, 154, 156, 160, 162, 180
Todarodes sagittatus **39**: 274
Tokophrya **25**: 138
Tomoarus **29**: 148
Tomopteris **32**: 302
Tomthompsonia **42**: 81, 88, *114*, 130, 132
   T. spiroconchalis **42**: 86, 113
Tonatus oregonensis **32**: 249, 291
Tornatinidae **42**: 109, 110, 129
Torpedinidae **32**: **168**, 211
Torpedo **32**: 186; **40**: 309, 310
Torpedo microdiscus **32**: **168**, 188
Torpedo semipelagica **32**: **168**, 188, 190, 192, **194**
Tortanidae **33**: **52**, 117, 118, 148, **152**, **271**
Tortanus **33**: **52**, 95, 117, 118, 166, 480
   barbatus **33**: **47**, 117, 447
   bonjol **33**: 117
   bowmani **33**: 117
   brevipes **33**: 118
   capensis **33**: 118
   compernis **33**: 118
   denticulatus **33**: 118
   derjugini **33**: 118, **271**
   dextrilobatus **33**: 118
   digitalis **33**: 118
   discaudatus **33**: 7, **26**, 118
     distribution **33**: **515**
     growth/development **33**: **311**
     nutrition **33**: 159, 163, **188**
     reproduction **33**: **271**, 279, 294
   erabuensis **33**: 118
   forcipatus **33**: 7, 22, 118, **232**
     behaviour **33**: **433**
     biochemistry **33**: **239**, **241**
     chromosomes **33**: **47**
     nutrition **33**: 162, 163
     reproduction **33**: 268, **271**
   giesbrechti **33**: 118
   gracilis **33**: **26**, **47**, 118, 163, **433**
   longipes **33**: 118, **431**
   lophus **33**: 118
   murrayi **33**: 118
   recticauda **33**: 118
   rubidus **33**: 118, **431**
   ryukyuensis **33**: 118
   scaphus **33**: 118
   setacaudatus **33**: 118

sheni **33**: 118
sinensis **33**: 118
spinicaudatus **33**: 118
tropicus **33**: 118
vermiculus **33**: 118
Tortanus discaudatus **25**: 48
Torticaecum **25**: 148
Tortinata **42**: 83
Tottonia contorta **24**: 116
Toxopneustes elegans **21**: 146
  pileolus **21**: 146, 154, 156, 168
Trabala krishna **40**: 178
Trachinotis falcatus, Predation **27**: 368
  hamiltoni, Predation **27**: **363**
Trachinotus rhodopus **20**: 129
Trachinus **21**: 61
Trachonurus **32**: 186
Trachonurus villosus **32**: **171**, **218**
Trachurus **32**: 193, 206; **34**: 265
Trachurus mediterraneus **26**: **120**
Trachurus mediterraneus ponticus **26**: 127
Trachurus novaezalandiae **24**: 366–368
Trachurus sp. **26**: 145
Trachurus symmetricus **25**: 42, 49, 50,
    49, 58; **26**: **75**, 145; **28**: 87; **30**: **221**,
    334; **32**: **196**, 209; **20**: 171
  trachurus **20**: 171, 247
Trachurus symmetricus murphyi **32**: 146,
    190, 200, 220, 227
Trachypenaeopsis sp., Taxonomy **27**: **112**
  minicoyensis, Taxonomy **27**: 101–2
  mobilispinis, Taxonomy **27**: 101
    Zoogeography/Evolution **27**: **146**
  richtersii, Taxonomy **27**: 101
    Zoogeography/Evolution **27**: **144**
Trachypenaeopsis, Taxonomy **27**: 100–102,
    112, **113**
  Life histories **27**: 285
  Physiology **27**: 203
  Zoogeography/Evolution **27**: 141, 150,
    154
Trachypenaeus spp., Predation **27**: 370
  albicomus, Taxonomy **27**: 104, 105
    Zoogeography/Evolution **27**: **145**
  anchoralis, Taxonomy **27**: 104
    Reproduction **27**: 264
    Zoogeography/Evolution **27**: **145**
  asper (*see Trachypenaeus curvirostris*)
  brevisuturae, Taxonomy **27**: 104
    Zoogeography/Evolution **27**: **145**
  byrdi, Taxonomy **27**: 105
    Zoogeography/Evolution **27**: **145**

  constrictus, Taxonomy **27**: 105, 107, **112**,
    124
    Reproduction **27**: 268, **271**
    Zoogeography/Evolution **27**: **146**
  curvirostris, Taxonomy **27**: 103, 105, **112**
    Food/feeding **27**: **318**
    Moulting/growth **27**: **232**
    Predation **27**: 360, 367
    Zoogeography/Evolution **27**: 141, **145**,
      147
  faoe, Taxonomy **27**: 105
    Zoogeography/Evolution **27**: **145**
  fulvus, Taxonomy **27**: *103*, 104, 105,
    112
    Food/feeding **27**: **318**
    Moulting/growth **27**: **232**, 245, **250**
    Zoogeography/Evolution **27**: **145**
  fuscina, Taxonomy **27**: 105
    Zoogeography/Evolution **27**: **145**
  gonospinifer, Taxonomy **27**: 104
    Zoogeography/Evolution **27**: **145**
  granulosus, Taxonomy **27**: 104, **123**
    Reproduction **27**: 264
    Zoogeography/Evolution **27**: **144**
  longipes, Taxonomy **27**: 104
    Zoogeography/Evolution **27**: **145**
  malaianus, Taxonomy **27**: 105 (*see also*
    Trachypenaeus curvirostris)
  pacificus, Taxonomy **27**: 105
    Zoogeography/Evolution **27**: **145**, 147,
      154
  pescadoreensis, Taxonomy **27**: 104
    Moulting/growth **27**: **232**
    Zoogeography/Evolution **27**: **145**
  sedili, Taxonomy **27**: 104
    Zoogeography/Evolution **27**: **145**
  similis, Taxonomy **27**: 105, 124
    Life histories **27**: 286
    Physiology **27**: 203, **204**
    Zoogeography/Evolution **27**: **146**, 147,
      154
  villaluzi, Taxonomy **27**: 103
    Zoogeography/Evolution **27**: **145**
Trachypenaeus, Morphology **27**: 42
  Food/feeding **27**: **318**, 321
  Life histories **27**: 285, 286, 287, 302
  Moulting/growth **27**: **232**, *247*
  Physiology **27**: 203, **204**
  Predation **27**: 370
  Reproduction **27**: 255, 260, 267, 268,
    271

Taxonomy **27**: 85, 99, 100, 102–105, 106, 107, **112**, **113**, 118, 119, *120*, 121, 123
Zoogeography/Evolution **27**: 141, **145**, 150, 154
Trachyscorpia cristulata **34**: **381**; **32**: **172**
Transtectonia torrentis **34**: **377**
Traumatiophora **25**: 137
Travisia forbesii **32**: **460**
Travisia profundi **32**: **460**
Trematodes **43**: 11, *14*, 16, 23, 32, 45, 59, 67
  see also Curcuteria; Gynaecotyla; Microphallus; Rugogaster
Trematomus **24**: 328, 329, 332, 346, 359
  bernacchii **24**: 326–329, 333, 340, 344, 347, 348, 355, 361, 369, 374
  bernachii **35**: 37
  centronotus **24**: 328, 333, 355, 356
  cf. *bernachii* **35**: 37
  hansoni **24**: 327, 328, 340, 341, 347, 348, 355, 361, 364, 371; **35**: **37**
  lepidorhinus **24**: 328
  loennbergi **24**: 328, 334; **35**: **37**, 52
  newnesi **24**: 328, 348
  nicolai **24**: 328
  pennelli **35**: **37**
Trematospaeria lineolatispora **40**: 119
Tremoctopus **25**: 90
Tremoctopus violaceus **24**: 207
Tresus capax and *T. nuttali* **37**: **10**, **110**
Tribolium **30**: 189
Trichinella **43**: 68, 69
  T. nativa **43**: 69
Trichiuridae **32**: **174**
Trichiurus lepturus **39**: 280
Trichiurus lepturus, Predation **27**: 358, **365**
Trichodesmium (cyanobacterium) **41**: 40, 49, 58
Trichodesmium **21**: 90; **24**: 416, 432, 433
  erythreum **21**: 71
Trichodesmium thiebautii **31**: 316, 352
Trichodina **25**: 139; **40**: 4, 48, 55; **42**: *157, 159, 211*; **43**: 68
Trichodina claviformis **40**: 4, 10, 15, 30
Trichodina cooperi **40**: 4, 12–13, 30, 55
Trichodina cottidarum **40**: 4, 13, 19, 30
Trichodina domerguei saintjohnsi **40**: 4, 15, 30
Trichodina elegini **40**: 4, 10, 13–14, 30
Trichodina jadranica **40**: 4, 10, 15, 33

Trichodina murmanica **40**: 4, 15, 30, 35–37, 55
Trichodina ploveri **24**: 291
Trichodina raabei **40**: 4, 10, 15, 33
Trichophoxus epistomus **39**: 205
Trichophrya salparum **25**: 139
Tricladida **43**: *22*
Tricolia pulloides **34**: 24
Tricoma **30**: **50**
Tricystis planctonis **25**: 141
Tridacna maxima **21**: 170
Trigonocidaris albida **32**: **165**
Trimusculus reticulatus **34**: 31, 33
Trinectes maculatus **26**: 138
Tringa totanus **20**: 229
Tripedalia cystophora **40**: 130
Triphoturus mexicanus **35**: **36**, 86, 92
Triploblasts **43**: 22
Tripneustes gratilla **21**: 146, 155, 156, 168, 169
  ventricosus **21**: 146, 169
Triposolenia **32**: 545, 560
Tripterophycis **32**: 218
Tripterophycis svetovidovi **32**: **170**, 218
Trisopterus esmarki **34**: 257; **26**: **121**; **20**: 291
Trisopterus luscus **26**: 270; **40**: 14
Tritaxis conica **30**: 53
Tritonia **42**: *116*, 133, 135
Trizocheles sakii **42**: 216
Trochactaeonidae **42**: 87
Trochidae **39**: 36
Trochostoma sp. **32**: **461**
Tromikosoma hispidum **32**: **165**, 213, 216
Tromina abyssalis **32**: 371
Trygaeus communis **24**: 8, 44
Trypanodinium **25**: 127
Trypanophis **25**: 125
  grobbeni **25**: 124
  major **25**: 124
Trypanophis grobbeni **24**: 207
Trypanorhynchidea **32**: 306
Trypanosoma murmanensis **40**: 4, 15, 24, 30, 36, 43, 44, 48, 50, 53, 55
Tryphosella kergueleni **39**: 205
Tryphosella nanoides **40**: 9, 10, 32
Tuarangia **42**: 41
Tubbia tasmanica **35**: **34**
Tubifex **25**: 146; **21**: 158, 164; **34**: *409*
Tubipora **43**: 292
  T. musica **43**: *291*
Tubularia **21**: 33; **24**: 46, 175

Tuna (*Euthynnus affinis*) **43**: 84
Tunicata **32**: 467; **43**: 110
Tunicophrya sessilis **25**: 120, 139
Turbellaria **43**: 7, 8, 20, 208, 209, 215
  see also Fecampiida; Ichthyophaga;
    Urastoma
Turbinaria **43**: 308
Turbinaria ornata **21**: 71
Turbinaria turbinata **23**: 103
Turbo sarmaticus **35**: **184**
Turbonilla (Strioturbonilla) **32**: **159**
Turbot *see Psetta maxima*
Turbot *see Scophthalmus maximus*
Turf algae **43**: 281
Turridae **32**: **163**, **210**
Turritella **28**: *392*, 411, 423
Turritopsis nutriculus **32**: 72
Tursiops spp., Predation **27**: 372
turtlegrass *(Halodule* and *Halophila)*
  41: *151*
Tylodina **42**: *80*, 132, 133, 135
  characters and evolutionary pathways
    42: *84, 85*, 99, 105
  taxonomic history **42**: 112, *114, 116*
Tylodinoidea **42**: 81
Tylos **30**: 89–153, *96*
  australis **30**: 92
  capensis **30**: 95, 97, 107, 110, 117, 119,
    122, 122, *123*, 123, 124, 127, 128,
    129, 130, 137, 141
  europaeus **30**: 93, 94, 97, 98, 107, 108,
    110, 111, 124, 125, 131
  granulatus **30**: 91, *92*, 93–95, 97–99,
    100, 100, 101, *102*, 103, *104, 105*,
    105–107, 110–114, *115*,
    115–117, *118*, 118, 119, *120*, 121,
    123, 125, *126*, 126, 130–132, *133*,
    134–138, 141
  latreillei **30**: 93, 94, 103, 110, 113, 124
  latreillei europaeus **30**: 93
  latreillei sardous **30**: 93
  punctatus **30**: 91, 92, 95, 97, 98,
    106–108, *109*, 109, 113–116, 118,
    119, 123, 124, 127, *128*, 128, 131,
    132, 134, 135
  sardous **30**: 93, 94, 98, 107, 108, 111,
    124, 125, 131
Typhloplanida **43**: *22*
Typhlotanais **39**: 160, 182
Typhlotanais brevicornis **39**: 219
Typhlotanais inermis **32**: **161**
Typhlotanais magnifica **39**: 160, 182, 219

Typhlotanais sp. **32**: **128**
Typhlotanais spinicauda **32**: **161**
Typlotanais sp. **34**: **377**

Uber strangei **28**: 407
Uca
  lactea **29**: **133**, 135
  lactenus **29**: 135
Uca pugilator **21**: 114
Uca urvillei **40**: 140, 141
Ucides cordatus **40**: 140
Udonella **43**: 31
Uhlorchestia spartinophila **39**: 205
Ulkenia **40**: 119
Ulosa rutzleri **40**: 134
Ulothrix flacca **23**: 17
Ulva **44**: 304; **23**: 13, 113–115, 135, 141;
    30: 113
  lactuca **23**: 13–16, 98
  lobata **23**: 134
  pertusa **23**: 64
Ulva pertosa **21**: 92
Umbellula **32**: 328, 359
Umbellula durissima **32**: 359
Umbellula magniflora **32**: 359
Umbellula thompsoni **32**: 333
Umbonium **28**: *39234*: **31**
  vestiarium **34**: 31, 32
Umbraculidae **42**: 90, 109, 129
Umbraculoidea **42**: 81, 96, 100, 103, 117
Umbraculomorpha **42**: 80, 81
Umbraculum **42**: 130, 131, 132, 133, 135
  characters and evolutionary
    pathways **42**: 81, 83, 86, 90
  taxonomic history **42**: *89*, 112, *114*, 115,
    116
  U. umbraculum **42**: *72*, 73
Undaria pinnatifida **20**: 320
Undeuchaeta **33**: 236
  behaviour **33**: 407
  ecology **33**: 486, 528
  taxonomy **33**: **53**, 87, 93, 129
Undeuchaeta bispinosa **33**: 129
  incisa **33**: 129
  intermedia **33**: 129
  magna **33**: 129
  major **33**: 129, **515**
  plumosa **33**: 6, **7**, 129, **417**, 444, **515**
Undinella **33**: **53**, 87, 90, 93, 138, **501**
  acuta **33**: 138
  altera **33**: 138, **501**
  brevipes **33**: 138

compacta **33**: 138, **501**
frontalis **33**: 138
gricei **33**: 138
hampsoni **33**: 138, **501**
oblonga **33**: 138
spinifer **33**: 138
stirni **33**: 138
Undinothrix **33**: **53**, 84, 137
spinosa **33**: 137
Undinula **25**: **133**; **33**: **52**, 68, 119, 486
vulgaris **33**: **7**, **26**, 119
  behaviour **33**: **417**, **446**
  biochemistry **33**: 251
  distribution **33**: **515**
  ecology **33**: **379**, **484**
  growth/development **33**: **309**, **343**
  nutrition **33**: **186**, **201**, **206**, **208**
  reproduction **33**: 263, **287**
  size/weight **33**: **245**
Undinula vulgaris **44**: 71, 72
Unicapsula **40**: 5, 10
Unio tumidus **22**: 103, 156, 157
Urastoma **43**: 20, 22
Urechinidae **32**: 466
Urechinus **32**: 366
Urechinus loveni **32**: **462**
Uria **34**: 268; **39**: 69–70
Uria aalge (murre) **39**: 76, 280
Uria aalge **20**: 228
  lomvia **20**: 228
Uria lomvia (Brünnich's guillemot) **38**: 175
Uria lomvia **32**: 267, 270
Urohaustorius metungi **39**: 205
Urolophus **40**: 309
Uronema rabaudi **25**: 138
Urophycis chuss **26**: 147
Urophycis floridae, Predation **27**: **363**
Uroptychus
  bicavatus **34**: **378**
  n. sp. **34**: **378**
  thermalis **34**: **378**
Urothoe brevicornis **39**: 205
Ursus americanus (black bear) **39**: 15, 17,
  43, 44
Ursus arctus (brown bear) **39**: 15, 17, 43, 44
Urticina **24**: 49
  felina **24**: 45
Utriculastra **42**: 87

Vaccinia **40**: 106
Valdiviathyridae **32**: **409**
Valdiviathyris quenstedti **32**: 409, **409**

Valdiviella **33**: 236
  ecology **33**: 486
  reproduction **33**: 267
  taxonomy **33**: **53**, 63, 93, 129
Valdiviella brevicornis **33**: 129
  brodskyi **33**: 129
  ignota **33**: 129
  imperfecta **33**: 129
  insignis **33**: *22*, 22, 129
    ecology **33**: 529
    life history **33**: 392
    reproduction **33**: 274, *277*, *279*, *281*,
      283
  minor **33**: 129
  oligarthra **33**: 129
Vallentinia gabriellae **40**: 134
Valonia **43**: 282
Valvatidae *see* Lower Heterobranchia
Valvifera **39**: 158
Vampyrophrya pelagica **25**: 137
Vampyroteuthidae **32**: 63
Vampyroteuthis **25**: 100, 109
  infernalis **25**: 88
Vampyroteuthis infernalis **44**: 169
Vanadis crystallina **25**: 140
Varanus bengalensis **40**: 154
Varanus flavescens **40**: 154
Varanus salvator **40**: 154
Vaucheria **43**: 146
Velella **24**: 200
Velella velella **25**: 125
Vema (Laevipilina) hyalina **32**: 408
Venerupis decussata **22**: 103, 129
Venerupis decussata **35**: **185**
Venerupis pullastra **34**: 13
Ventiella sulfuris **32**: 113; **34**: **377**, 398;
  35: **40**, 88, 89, 91
Ventnfossa teres **32**: **171**, 189
Ventrifossa **32**: 167, 191
Ventrifossa johnboborum **32**: **171**, 189, **218**
Ventrifossa macrodon **32**: **171**, 189, **195**
Ventsia tricarinata **34**: **371**
Ventulonia phalcata **34**: **371**
Venus **35**: **163**
Venus striatula **22**: 103, 109
Vermetidae **43**: 93
Verongia **21**: 104
Verruca (Verruca) scrippsae **32**: **159**, **161**
Verruca **22**: 232
  n. sp.1 **34**: **375**
  n. sp.2 **34**: **375**
  stroemia **22**: 214, 234

Verruculina enalia **40**: 120, 121
Verticordiidae **32**: 468
Vesicomya **35**: 50
 cordata **35**: **40**
 gigas **35**: 23, 24, **40**
 kilmeri **35**: 24
Vesicomya cordata **23**: 339
Vesicomya gigas **32**: 131
(Vesicomya) gigas **34**: **373**
(Vesicomya) lepta **34**: **373**
Vesicomyidae **32**: **127**, 130; **42**: 24
Vestimentifera **32**: 130; **43**: *21*, 105
Vetigastropda **42**: 181
Vexillum exasperatum **28**: *392*
Vexillum michaudi **28**: *392*
Vibalia armata **39**: 207
Vibalia propinqua **39**: 207
Vibilia **25**: 141, 142, *142*, 154, 156
Vibrio (bacterium) **41**: *189*
 V. harveyi **41**: 188, 198
Vibrio **25**: 119, 120; **29**: 194; **40**: 116
 parachaemolyticus **25**: 119, 120
Vibrio fischeri **44**: 171
Vibrio logei **44**: 171
Vibrio spp. **44**: 170
Vibrio, Parasites **27**: 382
 alginolyticus, Parasites **27**: 382
 anguillarum, Parasites **27**: 382
 parahaemolyticus, Parasites **27**: 382
Vinciguerria **30**: 279
Vipera trimeresurus **40**: 154
Vireo pallens **40**: 156
Virgularia mirabilis **34**: 223
Viridiplantae **43**: *20*
viruses **43**: 207
Vitjazaster djakonovi **32**: **462**
Vitjaziana gurjanovae **32**: 73
Vogtia **24**: 139, 202, 203
 glabra **24**: 160, 161, 202
 kurae **24**: 119
 serratia **24**: 145
 spinosa **24**: 202
Volvaria canaliculata **42**: 87
Volvatella **42**: 71, 92
Volvatellidae **42**: 80
Volvulella vigourouxi kawamurai **42**: *72*, 73
Vorticella **29**: 105, *106*

Waldheimia see Macandrevia; *Magellania*
Waldoia antillaria **23**: 49
Walleye pollock *see Theragra*
 chalcogramma

Waltonia inconspicua **28**: 178, **194**, **219**,
 221, 245, 249, 250, 261, 265, 276,
 281, 307, 329, 336, 342, 345,
 351–353
Watasenia scintillans **32**: 63, 266, 303
Wellington flying squid *see Notodarus*
 sloanii
Wemersoniella **42**: *144*
Wemersoniellidae **42**: *142, 144*
West coast mussel *see Mytilus trossulus*
Westwoodilla caecula **39**: 205
Whale **43**: 242
 blue (*Balaenoptera musculus*) **43**: 229,
 230
 fin (*Balaenoptera physalus*) **43**: 229, 230
 humpback (*Megaptera novaengliae*)
 43: 229, 230
 killer (*Orcinus orca*) **43**: 228, 229
 killer whale *see Orcinus*
 minke (*Balaenoptera acutorostrata*)
 43: 227–229
 sperm whale *see Physeter catodon*
Whelk *see Nucella lamellosa*
Whelk, common (*Buccinum undatum*)
 43: 67
White sea bass *see Atractoscion nobilis*
White whale *see Delphinapterus leucas*
Whitefish *see Coregonus lavaretus*
Whiting **43**: 55, 56
Whiting *see Merlangius merlangus*
Wilsonidius **33**: **53**, 92, 129
 alaskaensis **33**: 129
Winter flounder *see Pleuronectes*
 americanus
Wrasse *see Xyrichthys novacula*
Wrasse, cleaner (*Labroides miniatus*) **43**: 35

Xandaros acanthodes **34**: **368**
Xantharus **33**: **53**, 81, 84, 132
 formosus **33**: 132
Xantho **31**: 26
Xanthocalanus
 ecology **33**: 486, **500**, 522, 528
 taxonomy **33**: **53**, 61, 80, 81, 82, 84, 85,
 133
Xanthocalanus agilis **33**: 133
 alvinae **33**: 133, **500**
 amabilis **33**: 133
 antarcticus **33**: 133
 borealis **33**: 133
 claviger **33**: 133
 cornifer **33**: 133

crassirostris **33**: 133
difficilis **33**: 133
dilatus **33**: 133
distinctus **33**: 133, **500**
echinatus **33**: 133
elongatus **33**: 133, **500**
fallax **33**: 26, 133, **500**
giesbrechti **33**: 133
gracilis **33**: 133
greeni **33**: 132
groenlandicus **33**: 133, 138
harpagatus **33**: 133
incertus **33**: 133
irritans **33**: 133
kurilensis **33**: 133
legatus **33**: 133
macilental **33**: 133
macrocephalon **33**: 133, 138, **500**
marlyae **33**: 133, **500**
medius **33**: 133
minor **33**: 133, **500**
mixtus **33**: 133
multispinus **33**: 133
muticus **33**: 133
obtusus **33**: 133
oculata **33**: 133
paraincertus **33**: 133, 138
pavlovskii **33**: 133
pectinatus **33**: 133
penicillatus **33**: 133
pinguis **33**: 133
polaris **33**: 133
profundus **33**: 133, 522
propinquus **33**: 133
pulcher **33**: 133
rotunda **33**: 133
serrata **33**: 133
simplex **33**: 133
soaresmoreirai **33**: 133
squamatus **33**: 133
subagilis **33**: 133
tenuiremis **33**: 133
tenuiserratus **33**: 133
typicus **33**: 133
Xanthodaphne tropica **32**: **164**
Xantus murelets **20**: 61
Xenarcturus **39**: 159
Xenodermichthys copei **26**: **120**
Xenomystus nigri **40**: 304, 305
Xenophora **32**: **159**
Xenophyophoria **32**: 465
Xenopus laevis **30**: 200

Xenoturbella **42**: 35, 44
Xestospongia muta **21**: 99
  subtriangularis **21**: 98
Xiphias gladius **44**: 254; **26**: 124, *125, 135*;
      39: 281, 283
Xiphocephalus whitei **39**: 183, 207
Xiphopenaeus, Taxonomy **27**: 99, 106, **112**,
      113, 118
  kroyeri, Taxonomy **27**: 106, *106*, **112**
    Reproduction **27**: 268, **271**
    Zoogeography/Evolution **27**: 147, **146**
    Life histories **27**: 284
    Physiology **27**: 203
    Reproduction **27**: 255, 268, **271**
    Zoogeography/Evolution **27**: 150, 154
  riveti, Taxonomy **27**: 106
    Zoogeography/Evolution **27**: **145, 147**
Xylocarpus **40**: 99, 108
Xylocarpus granatum **40**: 95, 99, 111
Xylocarpus mekongensis **40**: 95, 158
Xylocarpus moluccensis **40**: 95
Xylocythere **34**: 384
  n. sp. **34**: **376**
Xylodiscula major **34**: **372**
Xylodisculidae *see* Lower Heterobranchia
Xylophaga **32**: 328; **34**: 384
Xyrichthys novacula **40**: 287
Xyrichtys splendens **44**: 216

Yellowfin sole *see Pleuronectes asper*
Yoldia **39**: 39
  anatomy and morphology **42**: 11, 16
  development **42**: 26, *27*, 29
  ecology **42**: 31, 32, 33, 34, 35
  feeding **42**: 19, 20
  Y. eightsi **42**: 19, 32, 34, 35
  Y. ensifera **42**: 19
  Y. limatula **42**: 16, 25, 30
    development **42**: 26, *27*, 29
    ecology **42**: 31, 32, 33
    feeding **42**: 19, 20
    zoogeography **42**: 39, 40, 42
  Y. notabilis **42**: 29, 34, 35
  Y. scissurata **42**: 20
  zoogeography **42**: 39, 40, 42
Yoldia notabilis **43**: 105, 132, 133, 149
Yoldiella **42**: 14
  Y. curta **42**: 36
  Y. ella **42**: 36
  *Y. inconspicua inconspicua***42**: 36
  Y. jeffreysi **42**: 26, 36
  Y. lata **42**: 36

Yoldiellinae **42**: *5*, 43
Yoldiidae **42**: 7; **42**: *5*
Ypsilothuria talismani **43**: 106
Yungia aurantiaca **21**: 133
Zalophus californianus **39**: 280
Zalopus californianus **20**: 246
Zanclea costata **25**: 157
Zeidae **32**: **171**
Zenion **30**: 334; **32**: 186, 198
Zenion hololepsis **32**: **172**, **195**
Zeniotidae **32**: **172**
Zenkevitchiella **33**: **52**, 77, 122, **498**
  abyssalis **33**: 122
  atlantica **33**: 122
  crassa **33**: 122
  tridentae **33**: 76, 122
Zenopsis **32**: 186, 187
Zenopsis nebulosa **30**: 334
Zenopsis oblongus **32**: 146, **172**, 189, 190,
        191, **192**, 193, **196**, 206, 227
Zernovia volgensis **39**: 205
Zeus faber **29**: 323
Zeuxapta seriolae **43**: *43*
Ziba fulgetrum **28**: *393*

Zoanthus sociatus **24**: 63
Zoarces viviparus **40**: 21, 24
Zoarcidae **32**: **129**, 304
Zooplankton **43**: 183, 216, 220, 221, 226
Zoothamnium **25**: 138
  alternans **25**: 120
  pelagicum **25**: 120
Zoothamnium niveum **40**: 116
Zoothamnium sp., Parasites **27**: 385
Zooxanthella **22**: 98, 90
Zoroaster fulgens **35**: **31**
Zoroaster sp. **32**: **166**
Zoroasteridae **32**: 445, 455
Zosimus aeneus **21**: 76, 140, 141
Zostera (seagrass) **41**: 149, 150
Zostera **21**: 19; **38**: 115
Zostera cruris **24**: 432
  maina **24**: 432
Zostera marina **30**: 113, 134; **39**: 34, 35
Zostera, Food/feeding **27**: 316
  capricorni, Food/feeding **27**: 320
    Predation **27**: 374
Zschokkella hildae **40**: 5, 30, 36, 37, 43
Zschokkella russelli **40**: 5, 15, 30

CONTRIBUTORS TO VOLUMES 20–44

L. AINIS, *Department of Animal Biology and Marine Ecology, University of Messina, Faculty of Science, I-98166 Messina, Italy* **40**: 253

H. B. AKBERALI *Departments of Zoology and Botany, University of Manchester, Manchester M13 9PL, England* **22**: 101

A. D. ANSELL *Dunstaffnage Marine Laboratory, PO Box 3, Oban PA34 4AD, Scotland* **28**: 175

F. ARNAUD *Station Marine d'Endoume, F-13007 Marseille, France* **24**: 1

K. M. BAILEY *Resource Assessment and Conservation Engineering Division, Alaska Fisheries Science Center, 7600 Sand Point Way NE, Seattle WA 98115, USA* **25**: 1; **37**: 179

R. N. BAMBER *Marine Biology Unit, CEGB, Fawley, Southampton, Hants SO4 1TW, UK* **24**: 1

R. BEIRAS *Area de Ecoloxia, Universidade de Vigo, 36200 Galicia, Spain* **37**: 1

P. BENTZEN *Marine Molecular Biology Laboratory, School of Fisheries, University of Washington, Seattle WA 98195, USA* **37**: 179

B. I. BERGSTRÖM *The Royal Swedish Academy of Sciences, Kristineberg Marine Research Station, Kristineberg 2130, S 450 34 Fiskebäckskil, Sweden* **38**: 57

B. J. BETT *Institute of Oceanographic Sciences Deacon Laboratory, Brook Road, Wormley, Godalming, Surrey GU8 5UB, UK* **30**: 1

B. L. BINGHAM, *Huxley College of Environmental Studies, Western Washington University, Bellingham, WA 98225, USA* **40**: 81

J. H. S. BLAXTER *Scottish Association for Marine Science, Dunstaffnage Marine Laboratory, PO Box 3, Oban, Argyll PA34 4AD, UK* **20**: 1; **38**: 1

S. V. BOLETZKY *C.N.R.S., Observatoire Océanologique de Banyuls, Laboratoire Arago, F-66651 Banyuls-sur-Mer, France* **25**: 86; **44**: 143

E. BOURGET *Département de Biologie, Université Laval, Québec G1K 7P4, Canada* **22**: 200

T. BREY *Alfred Wegener Institute for Polar and Marine Research, PO 120161, D-27515 Bremerhaven, Germany* **35**: 153

A. S. BRIERLEY *Gatty Marine Laboratory, School of Biology, University of St'Andrews, St Andrews, Fife, KY16 8LB, UK* **43**: 173

R. O. BRINKHURST *Ocean Ecology Laboratory, Institute of Ocean Sciences, Sidney, British Columbia, Canada* **26**: 170

A. C. BROWN *Department of Zoology, University of Cape Town, Rondebosch 7700, South Africa* **25**: 180; **28**: 389; **30**: 89

B. E. BROWN *Department of Marine Sciences and Coastal Management, University of Newcastle-upon-Tyne, Newcastle-upon-Tyne, NE1 7RU, UK* **22**: 1; **31**: 221

B. J. BURD *Galatea Research Inc., Brentwood Bay* **26**: 169

A. R. O. CHAPMAN *Department of Biology, Dalhousie University, Halifax, Nova Scotia, Canada, B3H 4JI* **23**: 1

M. J. COLLINS *Department of Geology, University of Bristol, Bristol BS8 1RJ, UK* **28**: 175

C. CONAND *Université de La Réunion, Laboratoire d'Écologie Marine, 15 Avenue René Cassin, Saint-Denis, Cedex 9, La Réunion 97715, France* **41**: 131

S. S. CREASEY *Institute of Biological Sciences, University of Wales, Aberystwyth, Ceredigion, SY23 3DA, and Marine Biological Association, Citadel Hill, Plymouth PL1 2PB, UK, and Oceanography Department, University of Southampton, Southampton Oceanography Centre, Empress Dock, Southampton SO17 1BJ, UK* **35**: 1

D. J. CRISP *Natural Environment Research Council, Unit of Marine Invertebrate Biology, Marine Science Laboratories, Menai Bridge, Gwynedd LL59 5EH, UK* **22**: 199

G. B. CURRY *Department of Geology and Applied Geology, University of Glasgow, Glasgow GT2 8QQ, Scotland* **28**: 175

D. H. CUSHING *198 Yarmouth Road, Lowestoft, Suffolk, U.K. NR32 4AB* **26**: 250

C. J. CUTTS *Seafish Marine Farming Unit, Ardtoe, Acharacle, Argyll PH36 4LD, United Kingdom* **44**: 295

W. DALL *CSIRO Marine Laboratories, PO Box 120, Cleveland, Queensland 4163, Australia* **27**: 1

M. S. DAVIES *Ecology Centre, University of Sunderland, Sunderland SR1 3SD, UK* **34**: 1

A. DINET *Laboratoire Arago, F 66650 Banyuls sur Mer, France* **30**: 1

J. F. DOWER *Department of Biology, Queen's University, Kingston, Ontario, Canada, K7L 3N6* **31**: 169

D. A. EGLOFF *Department of Biology, Oberlin College, Oberlin, OH 44074-1082, USA* **31**: 79

C. D. ELVIDGE, *Office of the Director, NOAA National Geophysical Data Center, 325 Broadway, Boulder, CO 80303, USA* **39**: 261

S. FASULO, *Department of Animal Biology and Marine Ecology, University of Messina, Faculty of Science, I-98166 Messina, Italy* **40**: 253

T. FERRERO *Department of Zoology, The Natural History Museum, Cromwell Road, London SW7 SBD, UK* **30**: 1

A. FERRON *Department of Biology, McGill University, Montréal, Quebec, Canada, H3A 1B1* **30**: 217

J. LE FÈVRE *Laboratoire d'Océanographie Biologique, Université de Bretagne Occidentale, F-29287 Brest Cedex, France* **23**: 163

P. W. FOFONOFF *Smithsonian Environmental Research Center, PO Box 28, Edgewater, Maryland 21037-0028, USA* **31**: 79

R. W. FURNESS *Zoology Department, Glasgow University, Glasgow G12 8QQ, Scotland* **20**: 225

S. V. GALKIN *P.P. Shirshov Institute of Oceanology, Russian Academy of Sciences, Nakhimovsky Prospekt 36, Moscow 117851, Russia* **32**: 93

J. P. A. GARDNER *Island Bay Marine Laboratory, School of Biological Sciences, Victoria University of Wellington, PO Box 600, Wellington, New Zealand* **31**: 1

A. V. GEBRUK *P.P. Shirshov Institute of Oceanology, Russian Academy of Sciences, Nakhimovsky Prospekt 36, Moscow 117851, Russia* **32**: 93

A. J. GOODAY *Institute of Oceanographic Sciences Deacon Laboratory, Brook Road, Wormley, Godalming, Surrey GU8 5UB, UK* **30**: 1

W. S. GRANT *Conservation Biology Division, Northwest Fisheries Science Center, 2725 Montlake Blvd., Seattle WA 98112, USA* **37**: 179

J. F. GRASSLE *Woods Hole Oceanographic Institution, Woods Hole, Massachusetts 02543, USA* **23**: 301

J.-F. HAMEL *Society for the Exploration and Valuing of the Environment (SEVE), 655 rue de la Riviére, Katevale (Québec), Canada J0B 1W0* **41**: 131

M. G. HARASEWYCH *Department of Systematic Biology, National Museum of Natural History, Smithsonian Institution, Washington, DC 20560-0118, USA* **42**: 237

T. HAUG *Department of Marine Biology, Tromsø Museum, University of Tromsø, Norway* **26**: 1

S. J. HAWKINS *School of Biological Sciences, University of Southampton, Southampton SO17 1BJ, UK* **34**: 1

M. R. HEATH *SOAFD Marine Laboratory, Victoria Road, Aberdeen, Scotland* **28**: 1

J. D. HEDLEY *Tropical Coastal Management Studies, Department of Marine Sciences and Coastal Management, Ridley Building, University of Newcastle, NE1 7RU, UK* **43**: 279

W. HEMMINGSEN, *Institute of Biology, University of Tromsø, N-9037 Tromsø, Norway* **40**: 1

B. J. HILL *CSIRO Marine Laboratories, PO Box 120, Clevedon, Queensland 4163, Australia* **27**: 1

A. G. HIRST *British Antarctic Survey, High Cross, Madingley Road, Cambridge, CB3 0ET, United Kingdom* **44**: 1

E. HIS *IFREMER, Quai Silhouette, 33120 Arcachon, France* **37**: 1

I. HOLMEFJORD *The Agricultural Research Council of Norway, Institute of Aquaculture Research, Sunndalsøra, Norway* **26**: 71

J. HORWOOD *Ministry of Agriculture, Fisheries and Food, Directorate of Fisheries Research, Fisheries Laboratory, Lowestoft, Suffolk NR33 0HT, UK* **29**: 215

E. D. HOUDE *Chesapeake Biological Laboratory, University of Maryland, Solomons, Maryland 20688, USA* **25**: 1

L. S. HOWARD *Department of Zoology, University of Newcastle upon Tyne, Newcastle upon Tyne NE1 7RU, England* **22**: 1

J. R. HUNTER *Southwest Fisheries Center, P.O. Box 271, La Jolla, California 92038, U.S.A.* **20**: 255

M. A. JAMES *Portobello Marine Laboratory and Department of Zoology, University of Otago, Dunedin, New Zealand* **28**: 175

S. JENNINGS *School of Biological Sciences, University of East Anglia, Norwich NR4 7TJ, UK* **34**: 203

W. S. JOHNSON *Department of Biological Sciences, Goucher College, Towson, MD 21204, USA* **39**: 107

J. C. JOYEUX *North Carolina State University, Department of Zoology, Box 7617, Raleigh, NC 27695, USA (Present address: Université Montpellier II,*

Laboratoire d'Hydrobiologie Marine et Continentale, CNRS UMR 5556, case 093, Place E. Bataillon, 34095 Montpellier Cedex 5, France) **34**: 73

M. J. KAISER *School of Ocean Sciences, University of Wales, Bangor, Menai Bridge, Anglesey, LL59 5EY, UK* **34**: 203

B. G. KAPOOR, *UGC Project, School of Studies in Zoology, Jiwaji University, Gwalior (M.P.), India* **40**: 253

K. KATHIRESAN, *Center of Advanced Study in Marine Biology, Annamalai University, Parangipettai 608 502, India* **40**: 81

T. KIØRBOE *Danish Institute for Fisheries and Marine Research, Charlottenlund Castle, DK-2920 Charlottenlund, Denmark* **29**: 1

E. KJØRSVIK *Norwegian College of Fishery Science, University of Tromsø, Tromsø* **26**: 71

H. KUOSA *Finnish Institute of Marine Research, PO Box 33, SF-00931 Helsinki, Finland* **29**: 73

J. KUPARINEN *Finnish Institute of Marine Research, PO Box 33, SF-00931 Helsinki, Finland* **29**: 73

P. J. D. LAMBSHEAD *Department of Zoology, The Natural History Museum, Cromwell Road, London SW7 5BD, UK* **30**: 1

R. S. LAMPITT *George Deacon Division for Ocean Processes, Southampton Oceanography Centre, Empress Dock, Southampton, SO14 3ZH, United Kingdom* **44**: 1

W. C. LEGGETT *Department of Biology, McGill University, Montréal, Quebec, Canada, H3A 1B1* **30**: 217

W. C. LEGGETT *Department of Biology, Queen's University, Kingston, Ontario, Canada, K7L 3N6* **31**: 169

R. MALCOLM LOVE *Formerly Torry Research Station, Abbey Road, Aberdeen, Scotland* **36**: 1

J. A. MACDONALD *Department of Zoology, University of Auckland, Private Bag, Auckland, New Zealand* **24**: 321

K. MACKENZIE *DAFS Marine Laboratory, Victoria Road, Aberdeen, UK* **24**: 263

K. MACKENZIE, *Department of Zoology, The University of Aberdeen, Tillydrone Avenue, Aberdeen, AB24 2TZ, UK* **40**: 1

G. O. MACKIE *Department of Biology, University of Victoria, PO Box 1700, Victoria, British Columbia, Canada V8W 2Y2* **24**: 97

A. MANGOR-JENSEN *Institute of Marine Research, Austevoll Aquaculture Research Station, Storebø* **26**: 71

J. MAUCHLINE *Dunstaffnage Marine Research Laboratory, Oban, Scotland* **33**: 1

C. A. McALPINE *Landscape Ecology Group, Department of Geographical Sciences and Planning and The Ecology Centre, The University of Queensland, Brisbane 4072, Australia* **44**: 205

A. G. McARTHUR *Josephine Bay Paul Centre for Comparative Molecular Biology and Evolution, Marine Biological Laboratory, Woods Hole, MA 02543-1015, USA* **34**: 355

D. MᴄHᴜɢʜ *Department of Organismic and Evolutionary Biology, Museum of Comparative Zoology, Harvard University, Cambridge, MA 02138, USA* **34**: 355

A. Mᴇʀᴄɪᴇʀ *Society for the Exploration and Valuing of the Environment (SEVE), 655 rue de la Riviére, Katevale (Québec), Canada J0B 1W0; International Center for Living Aquatic Resources Management (ICLARM), Coastal Aquaculture Centre, PO Box 438, Honiara, Solomon Islands* and *Institut des Sciences de La Mer de Rimouski (ISMER), 310 allée des Ursulines, Rimouski (Québec), Canada G5L 3A1* **41**: 131

P. M. Mɪᴋᴋᴇʟsᴇɴ *Division of Invertebrate Zoology, American Museum of Natural History, Central Park West at 79th Street, New York, NY 10024-5192, USA* **42**: 69

T. J. Mɪʟʟᴇʀ *Chesapeake Biological Laboratory, Center for Environmental and Estuarine Studies, University of Maryland, Solomons, Maryland 20688-0038, USA* **31**: 169

A. N. Mɪʀᴏɴᴏᴠ *P.P. Shirshov Institute of Oceanology, Russian Academy of Sciences, Nakhimovsky Prospekt 36, Moscow 117851, Russia* **32**: 147

J. C. Mᴏɴᴛɢᴏᴍᴇʀʏ *Department of Zoology, University of Auckland, Private Bag, Auckland, New Zealand* **24**: 321

L. I. Mᴏsᴋᴀʟᴇᴠ *P.P. Shirshov Institute of Oceanology, Russian Academy of Sciences, Nakhimovsky Prospekt 36, Moscow 117851, Russia* **32**: 93

P. J. Mᴜᴍʙʏ *Tropical Coastal Management Studies, Department of Marine Sciences and Coastal Management, Ridley Building, University of Newcastle, NE1 7RU, UK* **43**: 279

J. D. Nᴇɪʟsᴏɴ *Marine Fish Division, Canada Department of Fisheries and Oceans, Biological Station, St. Andrews, New Brunswick, Canada* **26**: 115

A. Nᴇᴍᴇᴄ *International Statistics and Research Corp., Brentwood Bay* **26**: 169

K. N. Nᴇsɪs *P.P. Shirshov Institute of Oceanology, Russian Academy of Sciences, Nakhimovsky Prospekt 36, Moscow 117851, Russia* **32**: 147

F. J. Oᴅᴇɴᴅᴀᴀʟ *Department of Zoology, University of Cape Town, South Africa 7700* **30**: 89

T. Oɴʙᴇ́ *Faculty of Applied Biological Science, Hiroshima University, Higashi-Hiroshima 739, Japan* **31**: 79

N. J. P. Oᴡᴇɴs *Institute for Marine Environmental Research, Prospect Place, The Hoe, Plymouth PL1 3DH, UK* **24**: 389

N. V. Pᴀʀɪɴ *P.P. Shirshov Institute of Oceanology, Russian Academy of Sciences, Nakhimovsky Prospekt 36, Moscow 117851, Russia* **32**: 147

D. L. Pᴀᴡsᴏɴ *National Museum of Natural History, Smithsonian Institution, Mail Stop 163, Washington DC, 20560-0163, USA* **41**: 131

L. S. Pᴇᴄᴋ *British Antarctic Survey, High Cross, Madingley Road, Cambridge CB3 0ET, UK* **28**: 175

R. I. Pᴇʀʀʏ *Marine Fish Division, Canada Department of Fisheries and Oceans, Biological Station, St. Andrews, New Brunswick, Canada* **26**: 115

C. H. Peterson *University of North Carolina at Chapel Hill, Institute of Marine Sciences, Morehead City, North Carolina 28557, USA* **39**: 1

O. Pfannkuche *Forschungzentrum für Marine Geowissenschaften, GEO MAR Abt. Marine Umweltgeologie, Universität Kiel, Wischhofstr, 1–3, Kiel, Germany* **30**: 1

S. J. Pittman *Landscape Ecology Group, Department of Geographical Sciences and Planning and The Ecology Centre, The University of Queensland, Brisbane 4072, Australia* **44**: 205

P. R. Pugh *Institute of Oceanographic Sciences, Wormley, Godalming, Surrey GU8 5UB, UK* **24**: 97

J. E. Purcell *Horn Point Environmental Laboratories, University of Maryland, PO'Box 775, Cambridge, Maryland 21613, USA* **24**: 97

T. J. Quinn II *Juneau Center, School of Fisheries and Ocean Sciences, University of Alaska Fairbanks, 1120 Glacier Highway, Juneau AK 99801-8677, USA* **37**: 179

E. Ramirez Llodra *School of Ocean and Earth Science, University of Southampton, Southampton Oceanography Centre, European Way, Southampton, SO14 3ZH, UK* **43**: 87

P. D. Reynolds *Biology Department, Hamilton College, 198 College Hill Road, Clinton, NY 13323, USA* **42**: 139

M. C. Rhodes *Academy of Natural Sciences, Nineteenth and the Parkway, Philadelphia, Pennsylvania 19103, USA* **28**: 175

K. Richardson *Danish Institute for Fisheries Research, Charlottenlund Castle, DK-2920, Charlottenlund, Denmark* **31**: 301

D. A. Ritz *Zoology Department, University of Tasmania, Box 252C, GPO, Hobart, Tasmania 7001, Australia* **30**: 155

P. O. Rodhouse *British Antarctic Survey, Natural Environment Research Council, High Cross, Madingley Road, Cambridge CB3 0ET, UK* **39**: 261

J. C. Roff *Department of Zoology, University of Guelph, Ontario N1G 2W1, Canada* **44**: 1

A. D. Rogers *School of Earth and Ocean Sciences, University of Southampton, Southampton Oceanography Centre, Empress Dock, Southampton SO17 1BJ, UK (formerly Marine Biological Association, Citadel Hill, Plymouth PL1 2PB, UK)* **30**: 305; **35**: 1

K. Rohde *Division of Zoology, School of Biological Sciences, University of New England, Armidale, NSW 2351, Australia* **43**: 1

P. C. Rothlisberg *CSIRO Marine Laboratories, PO Box 120, Cleveland, Queensland 4163, Australia* **27**: 1

F. E. Russell *College of Pharmacy, University of Arizona, Tucson, Arizona 85721, U.S.A.* **21**: 60

M. N. L. Seaman *Institute of Marine Research, 24105 Kiel, Germany* **37**: 1

H. J. Semina *P.P. Shirshov Institute of Oceanology, Russian Academy of Sciences, Nakhimovsky Prospekt 36, Moscow 117851, Russia* **32**: 527

D. J. Sharples *CSIRO Marine Laboratories, PO Box 120, Cleveland, Queensland 4163, Australia* **27**: 1

G. E. SHULMAN *Institute of Biology of the Southern Seas, Sevastopol, Republic of Ukraine* **36**: 1

M. N. SOKOLOVA *P.P. Shirshov Institute of Oceanology, Russian Academy of Sciences, Nakhimovsky Prospekt 36, Moscow 117851, Russia* **32**: 429

T. SOLTWEDEL *Institut für Hydrobiologie und Fischereiwissenschaft, Universität Hamburg, Zeiseweg 9, 22765 Hamburg, Germany* **30**: 1

A. J. SOUTHWARD *Marine Biological Association of the UK, Citadel Hill, Plymouth, PL1 2PB, UK* **32**: 93

J. M. E. STENTON-DOZEY *Department of Zoology, University of Cape Town, Rondesbosch, 7700, South Africa* **25**: 179

M. STEVENS *Department of Biology, Ripon College, 300 Seward Street, Ripon, WI 54971, USA* **39**: 107

T. SUBRAMONIAM *Department of Zoology, University of Madras, Guindy Campus, Madras 600 025, India* **29**: 129

J. THÉODORIDÈS *Laboratoire d'Évolution des Êtres Organisés, Université P & M Curie, Paris, France* **25**: 117

D. N. THOMAS *School of Ocean Sciences, University of Wales-Bangor, Menai Bridge, Anglesey, LL59 5EY, UK* **43**: 173

P. N. TRATHAN *British Antarctic Survey, Natural Environment Research Council, High Cross, Madingley Road, Cambridge CB3 0ET UK* **39**: 261

E. R. TRUEMAN *Department of Zoology, University of Manchester, Manchester M13 9PL, England* **22**: 101

E. R. TRUEMAN *Department of Zoology, University of Cape Town, Rondebosch 7700, South Africa* **25**: 180; **28**: 389

V. TUNNICLIFFE *School of Earth and Ocean Sciences, University of Victoria, Victoria, B.C., Canada V8W 3N5* **34**: 355

M. VAN-PRAËT *Laboratoire de Biologie des Invertébrés Marins et Malacologie, Muséum National d'Histoire Naturelle, 75005 Paris, France* **22**: 65

A. VANREUSEL *University of Gent, Zoology Institute, Marine Biology Section, K. L. Ledeganckstraat 35, B 9000 Gent, Belgium* **30**: 1

R. F. VENTILLA *Sea Fish Industry Authority, Marine Farming Unit, Ardtoe, Acharacle, Argyll, PH36 4LD, Scotland* **20**: 310; **21**: 1

A. L. VERESHCHAKA *P.P. Shirshov Institute of Oceanology, Russian Academy of Sciences, Nakhimovsky Prospekt 36, Moscow 117851, Russia* **32**: 93

M. VINCX *University of Gent, Zoology Institute, Marine Biology Section, K. L. Ledeganckstraat 35, B 9000 Gent, Belgium* **30**: 1

M. E. VINOGRADOV *P.P. Shirshov Institute of Oceanology, Russian Academy of Sciences, Nakhimovsky Prospekt 36, Moscow 117851, Russia* **32**: 1

N. G. VINOGRADOVA *P.P. Shirshov Institute of Oceanology, Russian Academy of Sciences, Nakhimovsky Prospekt 36, Moscow 117851, Russia* **32**: 325

A. B. WARD *North Carolina State University, D.H. Hill Library, Box 7111, Raleigh, NC 27695, USA* **34**: 73

L. WATLING *School of Marine Science, Darling Marine Center, University of Maine, Walpole, ME 04573, USA* **39**: 261

R. M. G. WELLS *Department of Zoology, University of Auckland, Private Bag, Auckland, New Zealand* **24**: 321

M. WHITFIELD *Marine Biological Association of the United Kingdom, The Laboratory, Citadel Hill, Plymouth PL1 2PB, UK* **41**: 1

G. ZACCONE, *Department of Animal Biology and Marine Ecology, University of Messina, Faculty of Science, I-98166 Messina, Italy* **40**: 253

J. D. ZARDUS *Pacific Biomedical Research Center, Kewalo Marine Laboratory, University of Hawaii, 41 Ahui St, Honolulu, HI 96813, USA* **42**: 1

O. N. ZEZINA *P.P. Shirshov Institute of Oceanology, Russian Academy of Sciences, Nakhimovsky Prospekt 36, Moscow 117851, Russia* **32**: 389